Italian Society of Chemistry
Division of Organic Chemistry
Division of Medicinal Chemistry
Division of Mass Spectrometry

TARGETS IN HETEROCYCLIC SYSTEMS

Chemistry and Properties

Volume 13 (2009)

Reviews and Accounts on Heterocyclic Chemistry
http://www.soc.chim.it/it/libriecollane/target_hs

Editors

Prof. Orazio A. Attanasi
Institute of Organic Chemistry, University of Urbino
Urbino, Italy

and

Prof. Domenico Spinelli
Department of Organic Chemistry, University of Bologna
Bologna, Italy

Published by:

Società Chimica Italiana

Viale Liegi, 48

00198 Roma

Italy

Volume 1 (1997)	First edition 1997	ISBN 88-86208-24-3	
	Second edition 1999	ISBN 88-86208-24-3	
Volume 2 (1998)	First edition 1999	ISBN 88-86208-11-1	
Volume 3 (1999)	First edition 2000	ISBN 88-86208-13-8	
Volume 4 (2000)	First edition 2001	ISBN 88-86208-16-2	
Volume 5 (2001)	First edition 2002	ISBN 88-86208-19-7	
Volume 6 (2002)	First edition 2003	ISBN 88-86208-23-5	
Volume 7 (2003)	First edition 2004	ISBN 88-86208-28-6	ISSN 1724-9449
Volume 8 (2004)	First edition 2005	ISBN 88 86208-29-4	ISSN 1724-9449
Volume 9 (2005)	First edition 2006	ISBN 88 86208-31-6	ISSN 1724-9449
Volume 10 (2006)	First edition 2007	ISBN 978-88-86208-51-2	ISSN 1724-9449
Volume 11 (2007)	First edition 2008	ISBN 978-88-86208-52-9	ISSN 1724-9449
Volume 12 (2008)	First edition 2009	ISBN 978-88-86208-56-7	ISSN 1724-9449
Volume 13 (2009)	First edition 2010	ISBN 978-88-86208-62-8	ISSN 1724-9449

Printed and bound in Italy by:

Arti Grafiche Editoriali s.r.l.

Via S. Donato, 148/C

61029 Urbino (Pesaro-Urbino)

Italy

April 2010

Indexed/Abstracted in: Chemical Abstracts; Current Contents; ISI/ISTP&B online database; Index to Scientific Book Contents; Chemistry Citation Index; Current Book Contents; Physical, Chemical & Earth Sciences; Methods in Organic Synthesis; The Journal of Organic Chemistry; La Chimica e l'Industria; Synthesis.

Preface

Heterocyclic derivatives are important in organic chemistry as products (including natural) and/or useful tools in the construction of more complicated molecular entities. Their utilization in polymeric, medicinal and agricultural chemistry is widely documented. Both dyestuff structures and life molecules frequently involve heterocyclic rings that play an important role in several biochemical processes.

Volume 13 (2009) keeps the international standard of THS series and contains twelve chapters, covering the synthesis, reactivity, and activity (including medicinal) of different heterorings. Authors from Czech Republic, France, Germany, Italy, Russia, Spain and Switzerland are present in this book.

Comprehensive Reviews reporting the overall state of the art on wide fields as well as personal Accounts highlighting significant advances by research groups dealing with their specific themes have been solicited from leading Authors. The submission of articles having the above-mentioned aims and concerning highly specialistic topics is strongly urged. The publication of Chapters in THS is free of charge. Firstly a brief layout of the contribution proposed, and then the subsequent manuscript, may be forwarded either to a Member of the Editorial Board or to one of the Editors.

The Authors, who contributed most competently to the realization of this Volume, and the Referees, who cooperated unselfishly (often with great patience) spending valuable attention and time in the review of the manuscripts, are gratefully acknowledged.

The Editors thank very much Dr. Lucia De Crescentini for her precious help in the editorial revision of the book.

Orazio A. Attanasi and Domenico Spinelli

Editors

Table of Contents

C2-Functionalized furans as dienes in [4+3] cycloaddition reactions 231

Ángel M. Montaña, Pedro M. Grima and Consuelo Batalla

V

Stereoselective synthesis of optically active pyridyl alcohols. Part I: pyridyl *sec*-alcohols　　　　303

Giorgio Chelucci

THE N→C=O INTERACTION: A WEAK BOND WITH ATTRACTIVE PROPERTIES FOR THE DELIBERATE EXPLOITATION IN MEDICINAL AND SUPRAMOLECULAR CHEMISTRY

Jens Hasserodt,* Arnaud Gautier, Romain Barbe and Michael Waibel

Laboratoire de Chimie, Université de Lyon - Ecole Normale Supérieure de Lyon
46 Allée d'Italie, F-69364 Lyon, France (e-mail: jens.hasserodt@ens-lyon.fr)

Abstract. *The literature on the weak bond resulting from the interaction between a tertiary amine and a carbonyl group is reviewed from its first observation in 1920 to modeling attempts by modern methods. The article then reviews the own recent investigations by the authors into the deliberate incorporation of the N→C=O interaction into monocyclic and polycyclic systems in order to exploit its unique stereoelectronic properties in the domains of medicinal and supramolecular chemistry.*

Contents

1. The N→C=O interaction in natural products

1.1. Discovery of a class of medium-sized cyclic alkaloids

The existence of a weak interaction between a tertiary amine and a carbonyl group has been first proposed in order to explain the unusual properties of some natural alkaloids with medium-sized aminoketone rings (Scheme 1). In 1920, Gadamer suggested that the conjugated acids of *protopine* and *clivorine*, ten- and eight-membered rings, respectively, were in fact bicyclic systems with a quaternary nitrogen bridge.[1] In theory, one might expect that the N→C=O interaction would lead to a decreased electrophilic character of the carbonyl carbon, as well as a less nucleophilic and less basic nitrogen, and indeed this was confirmed experimentally by Kermack and Robinson in 1922. They studied the alkaloid *protopine* containing a cyclic 10-membered aminoketone (Scheme 1) and *cryptopine*, an 8-membered, unsaturated aminoketone (Scheme 3).[2] In both cases, they found an abnormally low reactivity of the carbonyl group and weak basicity for the NCH₃ group which prompted them to propose the existence of an electronic interaction across the ring. In the same line, Huisgen *et al.* advanced in 1949 the explanation that an electronic interaction within the 9-membered cyclic alkaloid *vomicine* is responsible for the absence of ketone reactivity and for the weak nucleophilicity of the amine towards methyl iodide.[3] Similar absence of ketone reactivity towards lithium aluminum hydride was found for example for *N-methyl-sec-isopseudostrychnine*, an alkaloid with a structure close to that of *vomicine*.[4] Since these pioneering works, crystal structures of many natural medium-sized aminoketone alkaloids showing a shorter distance between the tertiary amine and the carbonyl group than the sum of the Van der Waals radius (320 pm) have been reported, including *retusamine*,[5] *protopine*,[6] *cryptopine*,[7] *clivorine*[8] and *senkirine*[9] thereby confirming the existence of an interaction between the tertiary amine and the carbonyl group.

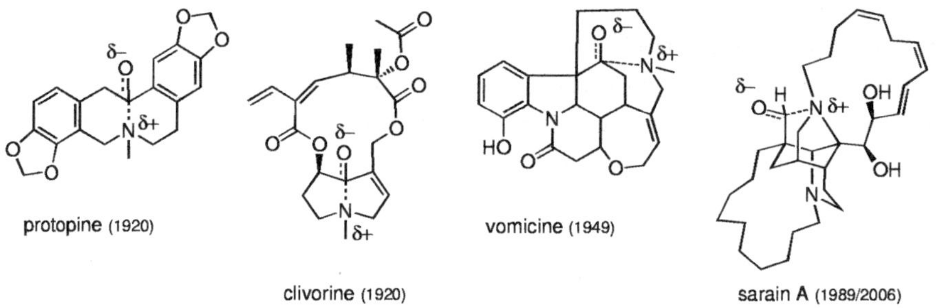

Scheme 1. Structures of some natural alkaloids with medium-sized aminoketone rings displaying the N→C=O interaction.

Natural products displaying the N→C=O interaction continue to be discovered sporadically. *Sarain A* is a prominent example, discovered in a marine sponge from the bay of Naples by Cimino *et al.* in 1989.[10] Its total synthesis by Overman *et al.* reported in 2006 caught much recent attention.[11]

1.2. Crystal structure characterisation: the Bürgi-Dunitz angle

In the early seventies, Bürgi and Dunitz compiled structural data by X-ray diffraction analysis primarily from the alkaloid examples mentioned above. These examples displayed the intramolecular interaction between a tertiary amine and a carbonyl group at N-C distances ranging from 150 to 300 pm. They took these data as an experimental basis to map the reaction coordinate (corresponding to the minimal energy pathway) of the bimolecular attack of a nucleophile on a carbonyl centre.[12] They considered each example as a frozen state on the reaction coordinate, which allowed them to correlate the lengths of the N-C and the C-O bond with the pyramidalization of the carbonyl group (Scheme 2). This led to the establishment of a constant angle of attack of a nucleophile on a carbonyl group, *i.e.* 107°, now universally recognized as the "Bürgi-Dunitz angle" (α, Scheme 2 A).[13] The latest addition (2002) to the small collection of known X-ray crystal structure determinations of molecules displaying an N→C=O interaction is that of a (non-natural) aza-adamantane (**1**) synthesized by Kirby *et al.* (Scheme 2 B and C).[14]

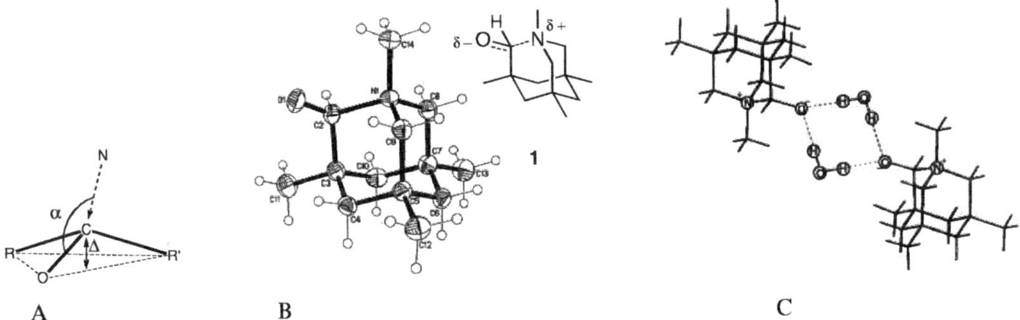

Scheme 2. (A) the Bürgi-Dunitz angle; (B) the most recent crystal structure (2002) of a molecule displaying the N→C=O interaction: aza-adamantane **1** (reproduced with kind permission from the RSC publishers); (C) dimeric structure of **1** in the crystal lattice, showing the central H-bonding motif including two distinct water molecules.

2. Physicochemical studies: dipole moment, pKa and spectroscopy

Beginning with the fifties, Leonard *et al.* were the first to explore the phenomenon of the N→C=O interaction in detail, by looking at so-called "trans-annular interactions".[15,16] In studying medium-sized cyclic aminoketones, they accumulated evidence for a weak N→C=O interaction by observing unexpected bands in the corresponding infrared spectra.[17] In CCl_4 solution, the conjugate bases of cryptopin, protopin and *N*-methyl-pseudostrychnidine showed IR stretching frequencies for the C=O bond about 20–30 cm^{-1} lower than those expected for a regular ketone. Moreover, the spectra of the perchlorate salts (conjugate acids) of cryptopine (Scheme 3 A) and protopine (Scheme 1), as well as that of the iodohydrate of N-methyl-pseudostrychnidine did not show any C=O stretching band. Rather, they exhibited a signal corresponding to the stretching vibration of an O-H bond, suggesting that these compounds displayed a N^+-C-OH moiety. Taking the C=O stretching frequencies as an indicator of the presence or not of a transannular interaction, they then evaluated the factors influencing the transannular N→C=O interaction within a family of 8-, 9-, 10- and 11-membered cyclic aminoketones.

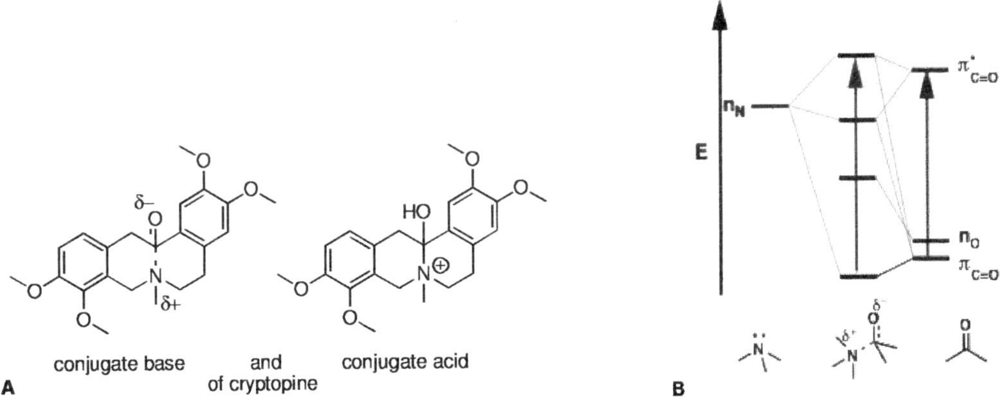

Scheme 3. (A) Passage from a weak interaction to a fully covalent bond by protonation; (B) energy levels of the N→C=O molecular orbitals constructed from the nitrogen *n* orbital and the carbonyl π orbitals and observed transitions.

They demonstrated that the presence of the interaction depended on ring size and steric constraints. Indeed, the N→C=O interaction occurred only for the 8-, 9- and 10-membered rings that could form thermodynamically favoured 5- or 6-membered rings through trans-annular interaction between diametrically opposed carbonyl and tertiary amine functions.[17,18] Electronic factors could also be identified, in particular those associated with the nucleophilicity of the nitrogen: 8- and 9-membered rings with tertiary amines displaying low basicity (*e.g.* aromatic amines) did not give any spectroscopic evidence for the transannular interaction even though the previously established rules in size and geometry were in favour of such an interaction.[19] Leonard *et al.* also determined the electronic spectra (UV) of their series of compounds.[15,18] Those that displayed the transannular interaction according to the IR data were analyzed in their basic forms and gave an absorption band that was more intense and shifted towards shorter wavelengths than that for a regular carbonyl group; no absorption band was observed for the conjugate acid. The behavior for the basic form was interpreted with the presence of a highly polar N→C=O moiety within the medium-sized rings. This intense band can be explained in terms of molecular orbitals with the homoconjugation of the non-bonding n orbital of the nitrogen atom and the anti-bonding π orbitals of the carbonyl group (Scheme 3 B). The corresponding $\pi \rightarrow \pi^*$ transition of the carbonyl group thereby experiences a larger energy gap leading to the observed hypsochromic shift. By contrast, an absorption corresponding either to the N→C=O interaction or the carbonyl function could not be observed in case of the conjugate acids. Interestingly, Kirby *et al.* invoke a generalised anomeric effect to explain structural features of the N→C=O system: interaction between the n_O- orbital and the π^*_{C-N} orbital causes a lengthened N-C bond and a shortened C-O bond.[14]

In the 1960s, Archer *et al.* investigated 3α-phenyl-3β-tropanylphenylketone (**2**) for its pharmaceutical properties (Scheme 4).[20] IR and UV data of **2** in methanol showed abnormal values which prompted them to suspect an N→C=O interaction. However, this implied that they were the first to observe the N→C=O interaction outside the context of a transannular interaction, but rather one, where an exocyclic carbonyl group was involved. In 2006, we have observed that varying proportions of the peptidomimetic γ-amino-aldehydes **3a–c** adopt a cyclized, N→C=O interacted form in methanol (Scheme 4).[21,22] Their UV spectra showed two bands, one at 245 nm that was attributed to the aldehyde absorption, and a very intense band at 220 nm corresponding to the N→C=O interaction. Very recently, we introduced the class of ureas **4a–e** (Scheme 4) and found them to exist solely in their cyclized constitution in methanol. Their UV spectra thus show an intense single band at 208 nm, while no aldehyde absorption can be observed (Scheme 5 A).[23,24]

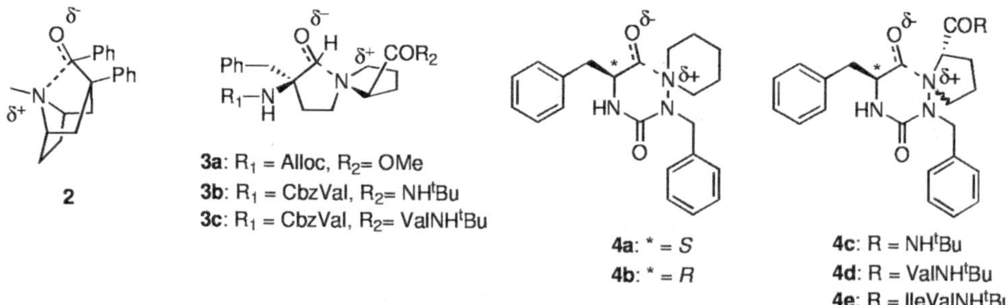

3a: R₁ = Alloc, R₂= OMe
3b: R₁ = CbzVal, R₂= NH^{t}Bu
3c: R₁ = CbzVal, R₂= ValNH^{t}Bu

4a: * = *S*
4b: * = *R*

4c: R = NH^{t}Bu
4d: R = ValNH^{t}Bu
4e: R = IleValNH^{t}Bu

Scheme 4. Molecules whose UV absorptions are discussed.

Leonard *et al.* also determined the dipole moment of their molecules under study in order to characterize the degree of charge separation.[25] The bicylic azanonane **5** (11-methyl-11-azabicyclo[5.3.1]-hendecan-4-one, Scheme 5 B) needs to adopt a quasi-boat conformation in order to have the two functional groups interact with one another. Bicyclic azanonane **5** thus shows a dipole moment of 4.87 D, although calculations on the non-interacting version furnished a value of only 3.6 D. This increase nicely illustrates the consequences of such a weak interaction and hints at other conditions that might favour it (*vide infra*).

In 1983, the first NMR studies of the N→C=O interaction were undertaken by McCrindle and McAlees at the example of simple aminoaldehyde **6**, soluble in water (Scheme 5 C).[26] In place of a signal in the typical chemical shift range for aldehyde functions, they observed one at 5.4 ppm and ascribed it to the methine proton of the N→C=O interaction. Another adventitious observation of the N→C=O interaction was made by Carroll *et al.*.[27] They investigated compound **7** (Scheme 5 C) in water by NMR spectroscopy only to discover that in place of a signal corresponding to the aldehyde function, the ^{13}C-NMR spectrum showed two signals at 100.49 ppm and 100.05 ppm, *i.e.* right in the middle between the olefinic/aromatic and the aliphatic region. It may be concluded that the carbon corresponding to this signal adopts a hybridization state between those of sp^2 and sp^3. The split signals may be explained with the fact that a new stereogenic centre has been formed and that two diasteromers were generated. Accordingly, the ^{1}H-NMR spectrum in water lacked a signal in the habitual region for aldehydic protons while displaying one at 6.45 ppm.

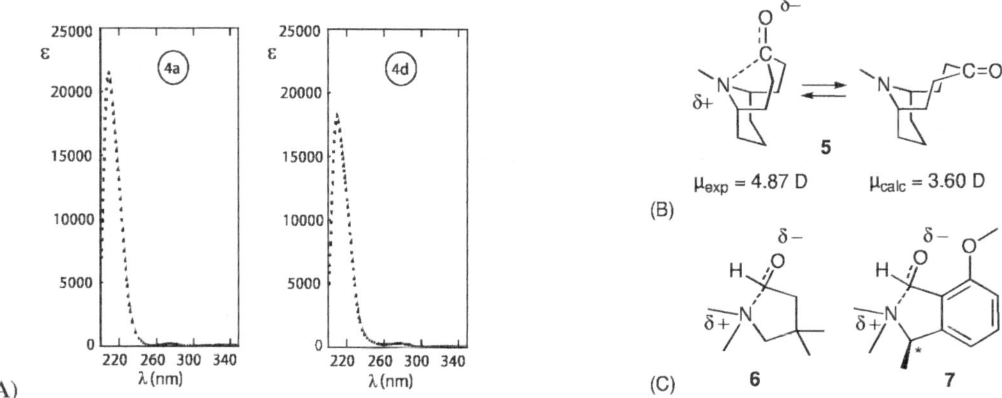

Scheme 5. (A) UV spectra of compounds **4a** and **4d** in MeOH; (B) dipole-moment evolution upon interaction; (C) structures whose NMR investigations are discussed.

McCrindle and McAlees also measured a pKa of 8.7 for **6** by titrating an aqueous solution with hydrochloric acid.[26] Kirby *et al.* determined the pKa value of the HBF$_4$ salt of their aza-adamantane **1** (Scheme 2) by potentiometric titration with base. They obtained a value of 10.1 at 25 °C.[14] This value falls in the region expected for such highly polar species and can be compared to the value of 9.33 determined by Hine and Kokesh for the parent compound Me$_3$N$^+$CH$_2$OH.[28] Finally, Leonard *et al.* found a pKa for the conjugate acid of compound **9** (Scheme 7) of 9.2 in water.[29]

When we recently investigated our enzyme inhibitor candidates for their propensity to undergo an N→C=O interaction, both ^{1}H- and ^{13}C-NMR spectra of peptidomimetics **3a–c** and hydrazino ureas **4a–e** in methanol showed an upfield shift of the methine proton and carbon (compared to the aldehyde signals) and thus in perfect congruence with the earlier reports discussed above.[22–24] As it was observed for compound **7** (Scheme 5 C), the signal corresponding to the methine carbons of the hydrazino ureas **4a–e** are split, indicating that two diasteromers were generated.[24] Indeed, all signals were split, as illustrated at the example of **4c** in Scheme 6. Futher evidence for the N→C=O interaction in methanol could be obtained by 2D NMR spectroscopy. In fact, the ^{1}H- ^{13}C-HMBC spectrum of compound **4a** showed [3]J coupling between the methine proton and the carbon atoms adjacent to the piperidine nitrogen. A similar cross-peak between the N-methyl protons and the N→C=O carbon was also observed by Kirby *et al.* during HMBC analysis of **1**.[14]

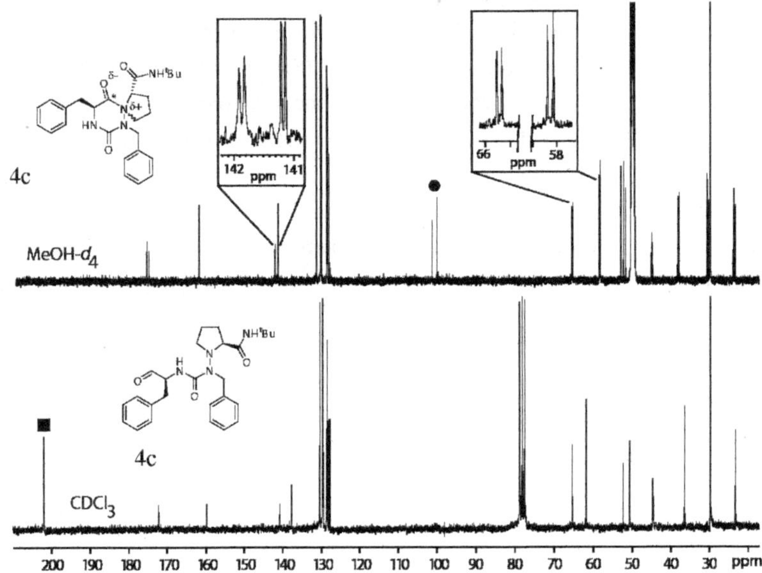

Scheme 6. ^{13}C-NMR spectra (127 MHz) of **4c** in MeOH-d_4 and CDCl₃; ● signals corresponding to two N→C=O carbons belonging to two diastereomers of comparable stability; ■ signal stemming from the free aldehyde group.

Finally, in the 1990s Rademacher *et al.* chose to further explore the character of the N→C=O interaction with photoelectron spectroscopy in the gas phase, a method that gives insight into the orbital distribution of the entire molecule, but it requires the study of rather simple compounds for ease of interpretation.[30] By combining their results with those obtained from theoretical studies, they were able to propose molecular orbital diagrams for medium-sized cyclic aminoketones and correlate their evolution with the N-C distance.[31] This may eventually serve to model the transition states of the diverse class of amine-carbonyl reactions.

3. Factors favouring the observation of the N→C=O interaction

3.1. Preorganization

The N→C=O interaction was first observed in medium-sized cyclic alkaloids where thermo-dynamically favoured 5- or 6-membered rings are formed by a transannular interaction. Leonard *et al.* extended their chemistry to simpler, artifical systems that feature 8-,9-,10- and 11-membered rings and may or may not display the interaction (**8–13**, Scheme 7).[15,29] They made some crucial observations on the influence by ring size and sterics (CCl$_4$ solution). By using the wave number shift as an indicator, they concluded that the interaction is only present in 8-, 9- and 10-membered rings. The resulting bicyclic systems display only 5- or 6-membered rings, which requires that the amino- and ketone functions are located at exactly opposite positions on the medium-sized ring, interacting in a truly trans-annular fashion.[17,32] Electronic factors, especially the nucleophilicity of the tertiary nitrogen, also have an influence on the interaction. For example, cyclopropylamine **13**, being less basic than methylamine **9**, does not show any transannular interaction.[19] Attention should also be drawn to the presence of Z-configured double bonds in the alkaloids discussed above (Scheme 1) that further enhance preorganization and thus favour the interaction.

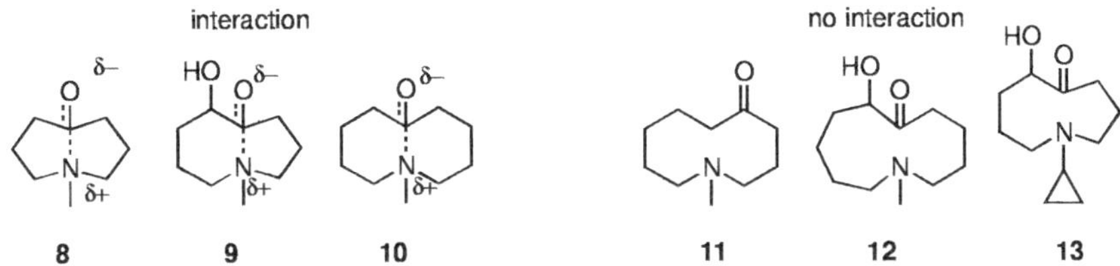

Scheme 7. Systematic study to determine the prerequisites for the observation of the interaction in a trans-annular context.

From these studies, it was not obvious that the interaction would eventually be observed also in **acyclic** systems (that would of course become monocyclic in the process). Leonard *et al.* used optical rotatory dispersion as a means to demonstrate that cyclization of a simple δ-aminoketone does not occur in methanol.[33] This example ranks with a plethora of other ones contained in the general literature that demonstrate the subtle interplay of enthalpic and entropic contributions to decide whether a given chemical process is thermodynamically favoured or not. The first instance where a system, still macrocyclic, exhibits an N→C=O moiety incorporating an exocyclic carbonyl group was already treated above (**2**, Scheme 4).[20]

The example of 3,3-dimethyl-4-dimethylaminobutanal (**6**, Scheme 5 C), as reported by McCrindle and McAlees,[26] constitutes a comparatively simple acyclic system. Their NMR investigations proved the existence of the folded form in water and its absence in chloroform. By contrast, the conjugate acid is cyclized in both media. Importantly, this system benefits from the so-called Thorpe-Ingold effect[34] by sporting a quaternary carbon with two methyl substituents. An even more pronounced Thorpe-Ingold effect can be identified in the structure of the well-known artificial opioid *methadone* (6-dimethylamino-4,4-diphenyl-3-heptone, discovered in 1936) that displays a quaternary carbon with two phenyl substituents (Scheme 8). Bürgi and Dunitz explained the fact that methadone efficiently crosses the blood brain barrier (BBB) with the folded status in aqueous physiological media. An N-C distance (291 pm) shorter than the van der Waals contact (320 pm) was also observed in the crystal structure. They proposed that the cyclic conformation of methadone exhibits an increased lipophilicity thus facilitating its diffusion through biological membranes.[35]

The display of the two interacting groups on a rigid scaffold should promote the interaction even further. The example of the *o*-aminomethyl-phenylcarbaldehyde reported by Carroll *et al.* has already been introduced above (**7**, Scheme 5 C).[27] Its constitution allows formation of a five-membered ring, like that reported by McCrindle and McAlees and that of the methadone base. But **7** also greatly benefits from the presence of one Z-configured rigid bond being part of the phenyl ring. It does not come as a surprise that the system adopts a cyclic conformation in water which causes the carbonyl carbon to experience a remarkable upfield shift of 70 ppm in the ^{13}C-NMR spectrum when adopting the new pseudo-tetrahedral configuration.

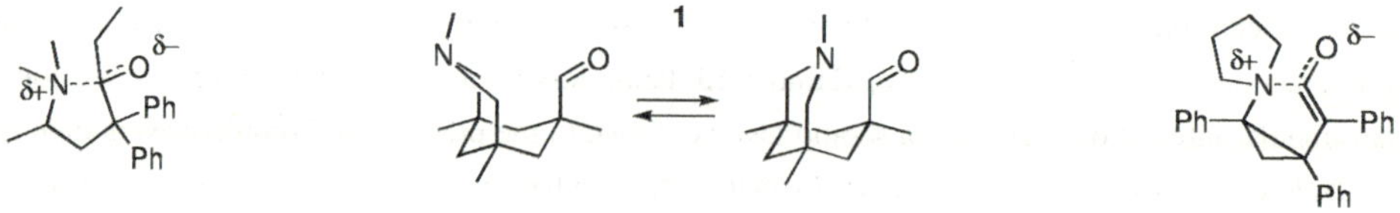

Scheme 8

Contrary to linear systems, preorganization is brought to the extreme with the previously introduced example of sarain A (Scheme 1) and especially that of the aza-adamantane system (**1**, Schemes 2 and 8) reported by Kirby *et al.* in 2002.[14,36] In this latter case, the display of the two interacting groups on a scaffold as rigid as that of Kemp's acid[37] assures maximal interaction of the two groups, especially in view of the only other choice of the system, that of the piperidine ring contained in the bicyclic system adopting a boat conformation (Scheme 8).

3.2. Influence by the nature of the amine and carbonyl substituents

Rather few examples exist for systems with carbonyl groups other than ketone or aldehyde that still show a strong propensity to interact. While Leonard and Dunitz did look at carboxylic acids, esters and carboxamides, it can be concluded that they are not suitable for the reliable observation of the interaction, even in protic solvents. They also extended their studies to the variation of the nucleophilic part, replacing the amine with ether or thioether units,[16] concluding however that these measures did not favour an interaction either. Accidentally, in 1974 Eicher and Bohm happened upon a unique system consisting of a cyclopropane as the platform on which a ketene substituent and a tertiary amine are displayed (**14**, Scheme 8).[38] This highly preorganized system showed a strong N→C=O interaction. An example where a regular tertiary amine has been replaced by a hydrazine structure was only reported by us as a result of our recent search for novel protease inhibitors (**4a–e**, Scheme 4).[23]

3.3. Solvent influence and constitutional equilibria

Systems principally capable of showing the N→C=O interaction may show various ratios of the bound (cyclic) and non-bound (linear) forms, even in media favouring the interaction. As we shall see, even in unfavourable media, where cyclization may be enforced by turning the system into its conjugate acid, varying proportions of cyclized and non-cyclized form, it may be observed.

The first hint at a solvent influence was recognized by McCrindle and McAlees who did not observe the N→C=O interaction of compound **6** (Schemes 5 C and 9) in chloroform as evidenced by UV spectral analysis. However, in water the same molecule was converted to more than 95% to the cyclized form. They proposed that polar protic solvents favour the N→C=O interaction by stabilizing the charge separation through an appropriate solvation mechanisms. A more pronounced, amounting to decisive, effect was observed for **7** (Scheme 9) when compared in chloroform and water. The proportion of 81% found for extremely preorganized **1** (Scheme 9) in dichloromethane containing only six equivalents of water shows how few protic molecules need to be present to favour the interaction (see Section 4 for comparative insights gained by molecular modeling).

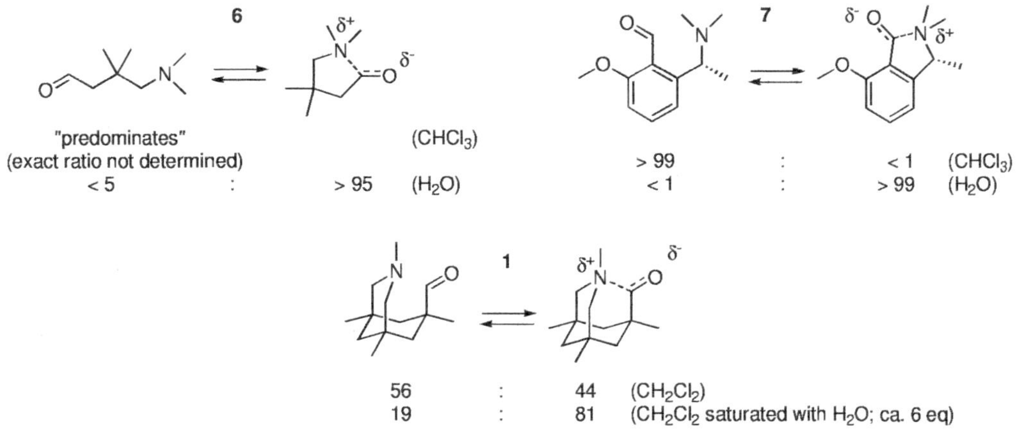

Scheme 9. Interaction (constitutional) equilibria in aprotic solvents compared to protic ones, as determined by UV or NMR analysis.

Finally, the equilibria of compound **3a** and **4a** are completely shifted to the open chain version in chloroform (Scheme 10). In methanol, **3a** adopts to 70 % the cyclized form and compound **4a** exists solely in the cyclized form.[21,22] The quaternary carbon bearing the aldehyde, amine and benzyl substituent of course exercises the Thorpe-Ingold effect as much as the examples described above and it should be largely at the origin of the observed folding tendency for the otherwise insignificantly preorganized **3a**.

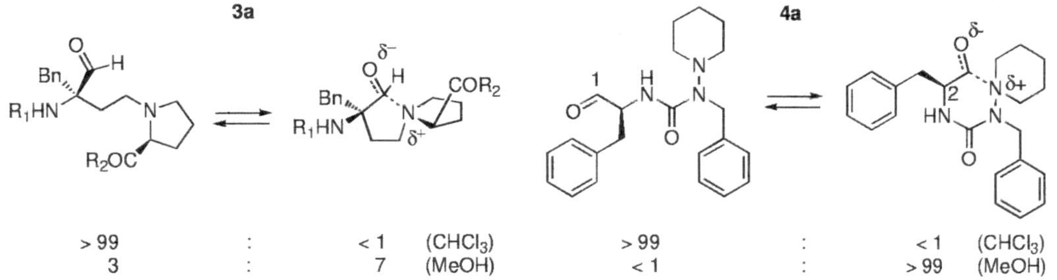

Scheme 10. Interaction (constitutional) equilibria for compounds designed for aspartic protease inhibition.

The folding of linear **3a** is associated with small rotational barriers compared to urea **4a**. Indeed, we were able to monitor by NMR the folding to the new N→C=O bound cyclic isomer of **4a** (Scheme 11) when we dissolved the linear form in MeOH-d_4. Scheme 11 illustrates the unmistakable progression of the cyclization at room temperature. Forty percent of the system is cyclized after 10 minutes, 70% conversion is reached after 30 minutes and cyclization is complete in 3 hours. These kinetics may be explained with the

9

existence of a configurational equilibrium present in **4a** (Scheme 10). In the absence of a polar protic medium, the thermodynamic form is *E*-**4a**, where the planar secondary amide of the urea moiety is *E*-configured. In MeOH-*d₄*, this configurational equilibrium is shifted completely to cyclized *Z*-**4a**. The activation barrier of the *E/Z* equilibrium of regular ureas has been determined to be approximately 13 kcal/mol[39] and thus to be significantly lower than that of the carboxamide bond (around 20 kcal/mol).[40,41] Still, a barrier of 13 kcal/mol explains the relative slowness of the process observed here at room temperature.

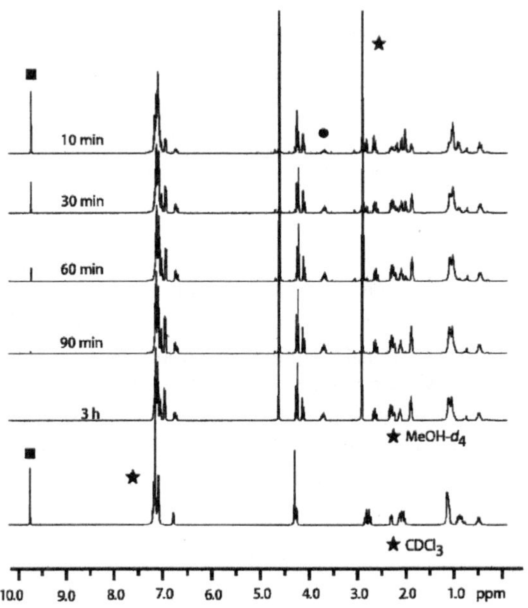

Scheme 11. ¹H-spectra (500 MHz) of **4a** in CDCl₃ and at certain time intervals beginning with initial dissolution in MeOH-d₄. ■ *E*-**4a** (position 1 proton, see Scheme 10), ● cyclized *Z*-**4a** (position 2 proton).

In the special case of a highly preorganized 1,8-disubstituted naphthalene (**15**, Scheme 12), but with apparently limited capacity for approach of the two functions, Dunitz *et al.* observed a significantly shortened N-C distance (2.56 Å) in the solid state, compared to a non-bound contact distance of 3.2 Å.[42] This picture is somehow confirmed in CDCl₃ solution, where the authors detected a broad peak for the two methyl groups at 37 °C that can be completely resolved by cooling the sample to −36 °C, indicating at least restricted rotation of both substituents (⇒ "peri-strain"). When Kirby *et al.* looked at the aldehyde derivative of this system (**16**, Scheme 12) ten years later, the NMR spectrum of the **conjugate acid** in dichloromethane containing 10% TFA proved that only two thirds of the compound existed in the N-C bound form, the other being the protonated amine.[43] This illustrates the requirement for unrestricted approach of both functional groups in order to observe a strong preference for the bound form over that of the open-chain form. In this regard, it would be interesting to analyze the **conjugate base** of **15** in protic organic solvents or outright water in order to verify if the most favourable solvent known for the N→C=O interaction may energetically overcome the strain and favour only one isomer, *i.e.* the folded one; but this has to await future studies.

15

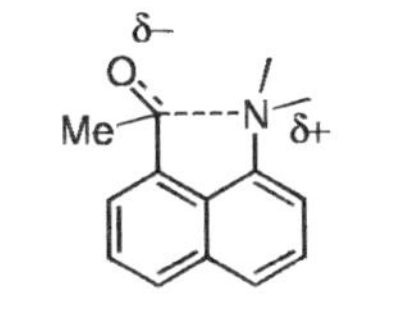

in the solid state,
possibly also in CDCl$_3$

16

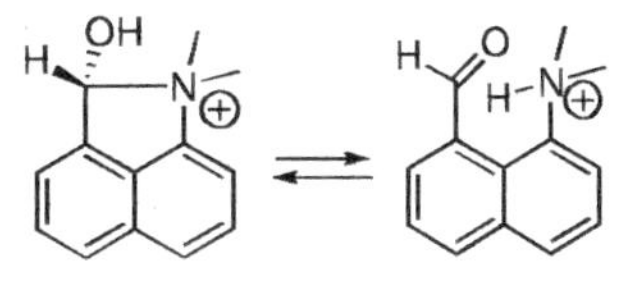

2 : 1
(in TFA/CH$_2$Cl$_2$ 1:10)

Scheme 12. Strain in a 1,8-disubstituted naphthalene ketone and aldehyde
and its impact on the N→C=O interaction.

Finally, the observation by Kirby *et al.* that cpd **1** (Scheme 9) exists to 81% in the N-C bound form in an aprotic solvent just "doped" with 6 equivalents of water is particularly instructive in grasping the fundamental nature of the N→C=O interaction (*vide infra*).

4. Molecular modeling: MM/QM

In view of the relative rarity of the N→C=O interaction, very few modeling studies have been reported to date. As briefly mentioned above, Rademacher *et al.* have initiated such studies at the end of the 80s.[30,31,44] They applied semi-empirical methods (MNDO et AM1) to the interaction of model compounds trimethylamine and acetone in order to calculate the energy evolution of the non-bonding oxygen and nitrogen orbitals with diminishing N-C distance. They also measured the ionization potentials of the respective orbitals for published macrocyclic aminoketones by photo-electron spectroscopy and discovered that they fit the theoretical values fairly well (Scheme 13 A).

Even though it should *a priori* be possible to employ molecular mechanics (MM2) to obtain low-energy conformations, Rademacher *et al.* did not report the determination of any force field and no parameter set was furnished to model the interaction. A more recent study has since compared numerous force fields and the progressive adaptation of their parameters has allowed to accurately reproduce experimental results of certain alkaloids, such as clivorine or cryptopine.[45] However, this approach is of limited interest because molecular mechanics take into account neither the large variations in N-C bond length known (from 160 to 290 pm) nor environmental influences (*e.g.* solvent effects).[46]

Even more recently, a renewed interest in the N→C=O interaction has prompted theoretical chemists to carry out more advanced modeling experiments. Thus, *ab initio* calculations were conducted on the interaction between ammonia and formaldehyde, as well as on the influence of cinchonidine on the asymmetric hydrogenation during heterogeneous catalysis (Scheme 13 C).[47,48]

The object of these studies was to determine the fundamental nature of the interaction, *i.e.* its molecular orbitals as well as its electrostatic character (Scheme 13 B). Depending on the distance between the functional groups and the angle of attack, either the dipole-dipole or the orbital interaction is favoured. The orbital overlap is in fact maximized if the non-bonding lone pair of the nitrogen attacks the carbonyl double bond at a perfectly right angle. At this orientation, the interaction of the highest occupied molecular orbital of the nitrogen and the lowest unoccupied molecular orbital of the carbonyl group is predominant, while that of the dipole moments is zero thanks to the perfectly orthogonal position. At the other extreme, with an NCO angle at 180°, the orbital interaction becomes zero while that of the dipoles is maximal.

The most frequently encountered understanding of the N→C=O interaction is that of a true bond caused by orbital interaction (n_N→π^*_{CO}), which corresponds to the observed geometries as well as to the

11

negligible reactivity of the carbonyl function.[47] In 2007, a study of the model system γ-(dimethylamino)-butyraldehyde that combined DFT calculations with the electron localization function (ELF) and with an explicitly correlated wavefunction (MRCI) analysis has cast doubt on this explanation and proposed a more electrostatic character for the N-C bond. According to these calculations, N→C=O bond formation is driven by the enhancement of the ionic contribution C^+-O^- by the strong polarization effect of the neighboring N lone pair, thereby creating an electrostatic bond.[49] Whatever the respective conclusions, all of these works have in common the all-important influence by the solvent. For the above study,[49] the quality of the modeling experiments is significantly improved by switching from gas-phase conditions, over the simulation of water as a continuum, to the distinct situation *where three to four water molecules* interact directly with the N→C=O interaction (see Schemes 2 C and 9 for comparison).

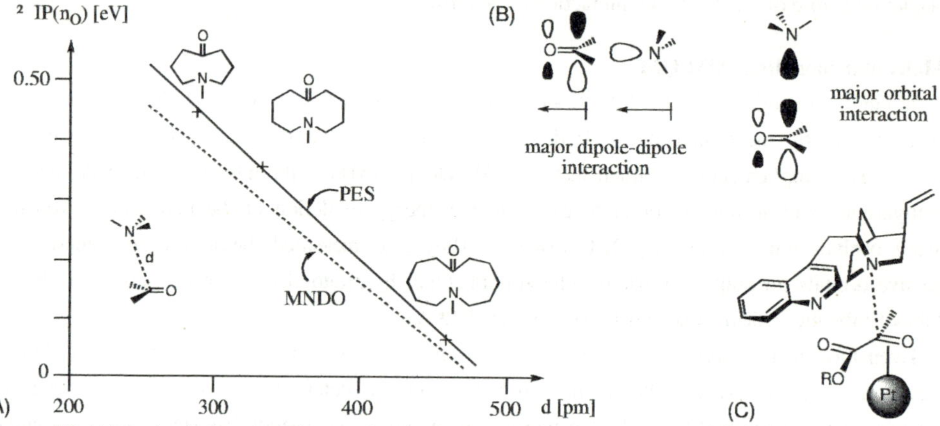

Scheme 13. (A) Ionisation potentials correlated with N-C distance calculated for a model interaction and measured by photoelectron spectroscopy (PES) for puplished cyclic aminoketones (adapted from ref. 30); (B) two extreme representations of the N→C=O interaction; (C) one of the proposed conformers that could explain the observed stereoselectivity.

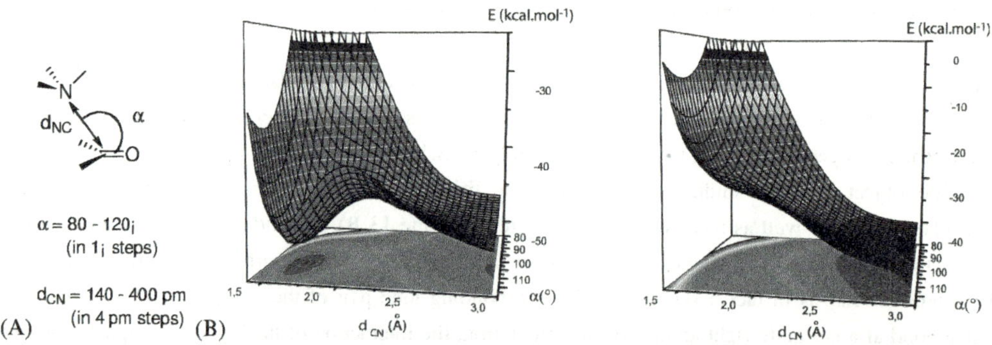

Scheme 14. (A) Semi-empirical exploration of the N→C=O interaction (reminiscent of the Bürgi-Dunitz trajectory); (B) potential energy surfaces calculated with the semi-empirical method "PM5-water" (left, with local minimum representing the N→C=O interaction) and "PM5" (right, no local minimum).

Finally, numerous semi-empirical methods have been tested to model molecules whose size precluded any *ab initio* calculations using DFT. Thus, MNDO,[44] and particularly AM1,[30,47,50] has been applied with more or less convincing results to larger molecules. It has been found that the Hamiltonian according to AM1 reproduces the expected properties more faithfully than PM3 or MNDO. We obtained the best results by use of the PM5 method in conjunction with the simulation of an aqueous environment.[46] In fact, the potential energy surface of the trimethylamine-acetone system as a function of the NCO angle and the C-N bond distance (Scheme 14) presents a minimum whose structural parameters (α=106°; d_{CN}=164 pm) reproduce those obtained from much more powerful methods remarkably well. The combination of molecular mechanics with a semi-empirical method such as "PM5-water" (software package "SciGress Explorer") thus allows for the modeling of a fairly complex molecule in acceptable computation times.

5. Deliberate incorporation of the N→C=O motif by molecular design

5.1. Enzyme inhibitor design

5.1.1. General notions about aspartic peptidases

Inhibitors of aspartic peptidases have great importance in the treatment of a variety of diseases, such as malaria, cancer, hypertension, AIDS and Alzheimer's disease.[51,52] Aspartic peptidases are endopeptidases that are capable of complexing six to ten amino acid residues of the peptide substrate. They generally do so by interacting in a beta sheet-like fashion with a linear conformation of the substrate.[53] The active site of these enzymes contains two aspartic acid residues that are responsible for catalytic activity, hence the name of this enzyme class. Next to these two residues, an enzyme bound (or "catalytic") water molecule is present, which is the second substrate in the bimolecular hydrolysis reaction. A particular feature of HIV-1 protease is the presence of two enzymatic loops, so-called "flaps". During catalysis the flaps indirectly bind the peptide substrate through hydrogen bonding of two lysine residues *via* a localized water molecule (Scheme 15). This molecule is also called the "flap water" and it is the second localizable enzyme bound water molecule found in HIV-1 protease-ligand complexes besides the catalytic water that was mentioned above.

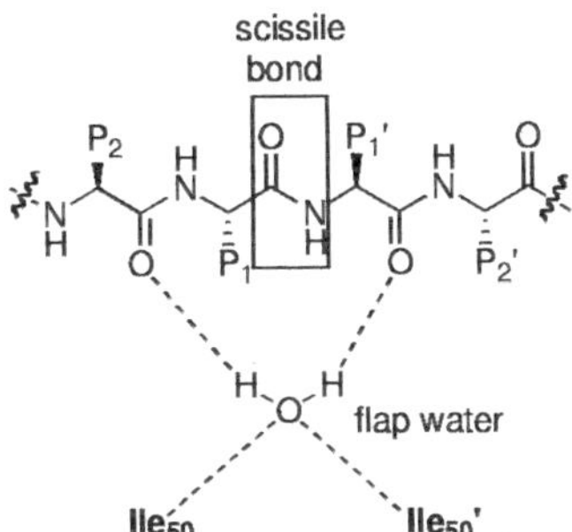

Scheme 15. Peptide substrate bound by the "flaps" of HIV-1 protease through the flap water molecule; side chain assignment (P1, P1', etc.) according to the nomenclature of Schechter and Berger.[54]

Aspartic peptidases exert their catalytic activity through a number of interactions with the peptide substrate and the catalytic water molecule. The generally accepted hydrolysis mechanism is shown in Scheme 16:[55–58] (1) addition of the water molecule to the carbonyl group of the peptide bond leads past transition state 1 (TS1) to a tetrahedral intermediate (INT); (2) protonation of the nitrogen and cleavage of the former amide bond produces the reaction products through TS2. The two transition states have pseudo-tetrahedral geometry as a result of a hybridization state of the former carbonyl carbon located between sp^2

and sp^3. They are characterized by a partial negative charge on the former carbonyl oxygen and a particularly long C=O bond (bond order between 1 and 2). The nitrogen atom of the second transition state is partially positively charged and the C-N bond elongated (bond order between 0 and 1).

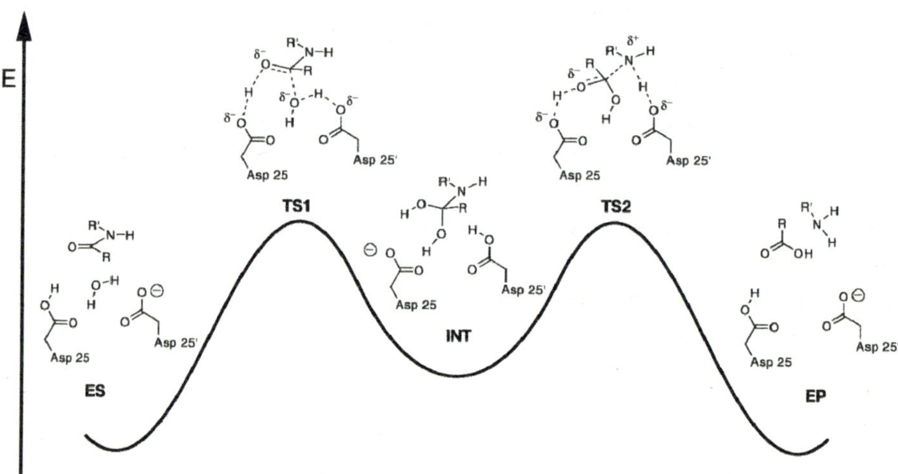

Scheme 16. Mechanism of the peptidolysis reaction catalyzed by aspartic peptidases.

5.1.2. Design of inhibitors for HIV-1 peptidase

The observation that Nature relies heavily on host (receptor)-guest (ligand) interactions has undoubtedly led to the foundation of the field of supramolecular chemistry. The same principles of weak interactions and complementarity are employed for the design of competitive, reversible inhibitors as for that of entirely artificial host-guest systems. Devising an analog for the transition state (TS) of enzymatic peptide-bond hydrolysis has long been used in the development of potent inhibitors of aspartic peptidases as part of a fundamental teaching of enzymology.[59–61] A transition state analog is defined as a stable compound with a structure similar to the transition state of the corresponding enzyme-catalyzed reaction. Its affinity to the enzyme active site arises from non-covalent interactions, such as hydrogen bonds and coulombic or Van der Waals interactions. In order to be a promising inhibitor candidate, a molecule should resemble the transition state in its steric and electrostatic features. The common approach in designing aspartic peptidase inhibitors has been the incorporation of a TS analog into a peptidomimetic that reflects the linear shape of the biologically occurring peptide substrate. This strategy evolved from the discovery that pepstatine, a natural peptide containing a non-proteinogenic amino acid called statine, inhibits most aspartic peptidases (Scheme 17 A).[62] Crystal structures of the enzyme-inhibitor complexes revealed that the secondary hydroxyl of statine interacts with the two aspartic acid residues (Scheme 18).

It is also generally maintained that the tetrahedral nature of the secondary alcohol function in the statine unit reproduces the geometry of the (pseudo)tetrahedral transition states. The hydroxyethyl unit is thus generally referred to as a transition state mimic of peptide hydrolysis. The interaction of the hydroxyl function in statine with the aspartic acid residues results in the extrusion of the catalytic water molecule mentioned above, resulting in a considerable entropic advantage for increased affinity. Many other such transition state-mimicking motifs have been explored, for example the hydroxyethylamine and the hydroxyethylene units that contain a secondary alcohol, as statine does (Scheme 17 B). The latter two are

14

among the most frequently employed in inhibitors that are currently used in clinical treatment of AIDS (Scheme 17 C).[51,52,63] The crucial role of HIV-1 peptidase in the replication of Human Immunodeficiency Virus I (HIV-I) has made this aspartic peptidase an important target for the design of anti-HIV agents. Notably, the replacement of the preferred cleavage site of HIV-1 peptidase, namely phenylalanine-proline (Phe-Pro), by a TS mimic has led to the most successful inhibitors.[64]

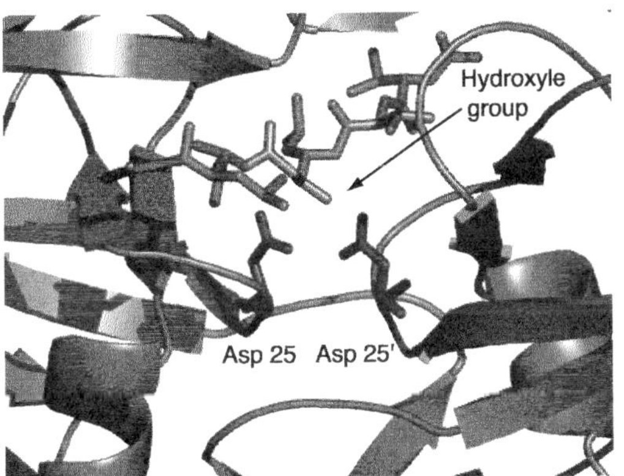

Scheme 17. (A) Aspartic peptidase substrate and pepstatine; (B) common transition state-mimicking motifs; (C) two clinically used HIV-1 peptidase inhibitors containing the hydroxyethylamine and hydroxyethylene motif.

Scheme 18. Pepstatine complexed to HIV-1 peptidase (PDB entry: 5HVP).

5.1.3. The N→C=O motif as a transition state mimic in the design of HIV-1 peptidase inhibitors

The appearance of peptidase mutants, being resistant to existing drugs, has become a challenge in inhibitor design and necessitates the development of novel concepts.[65] In the context of enzyme inhibition, it has been found regrettable that current transition-state isosteres are often far from faithfully reproducing steric and electrostatic elements of the respective transition state(s).[22] Therefore, we have been interested in exploring the N→C=O moiety for its capacity to reproduce steric and electronic properties of the transition states of peptide hydrolysis. Its attractive features for this project are as stated earlier : (a) an N-C bond order between 0 and 1; (b) a carbon hybridization between sp^2 and sp^3; (c) a C-O bond order between 1 and 2; (d) partial charges on the oxygen and the nitrogen, resulting in an enhanced dipole moment. This strongly polarized unit may thus form particularly strong hydrogen bonds and electrostatic interactions with the catalytic machinery of aspartic peptidases. Scheme 19 illustrates the formal resemblance between the cyclized form of our first-generation inhibitors **3a–c**[21,22] and the TS2 of peptide hydrolysis by aspartic peptidases, without taking into account the unknown conformation of the TS and the presence of the ethylene bridge that is essential to the observation of the N→C=O interaction. Inhibitors **3a–c** are peptidomimetics incorporating a mimic of the Phe-Pro dipeptide sequence. The valine residues at the P2 and P2' positions and the Cbz and the NH-*t*-Bu units were chosen for their presence in several potent aspartic peptidase inhibitors.[64] Valine is also found as the P_2 or $P_2^{'}$ residue in some of the preferred cleavage sites of HIV-1 peptidase.

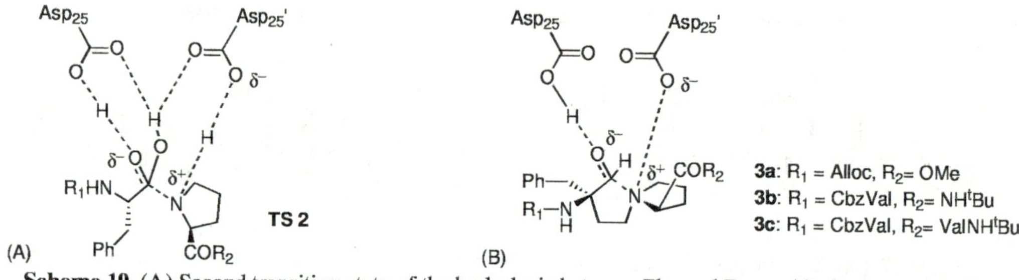

Scheme 19. (A) Second transition state of the hydrolysis between Phe and Pro and its hydrogen bonding with the Asp-Asp catalytic machinery; (B) hypothetical hydrogen-bonding of compounds **3a–c** with the enzyme catalytic site.

In the synthesis of **3a–c**, compound **21** is the common synthetic intermediate (Scheme 20). The quaternary carbon centre in **18** was obtained using the method of 'self-regeneration of stereocentres' developed by Seebach.[66] *cis*-Oxazolidinone **17** was obtained from *L*-phenylalanine following literature protocols.[67] The corresponding enolate was generated by treatment with potassium hexamethyldisilazide in THF at −78 °C. The diastereoselective alkylation of the resulting chiral anion with 2-bromoethyl triflate afforded **18** in 85% yield as a single diastereomer with retention of configuration at the starting stereocentre. The stereochemistry of **18** was assigned by its NOESY spectrum, which revealed NOE contacts between the *t*-Bu group and the benzyl methylene group, and between the methine and the methylene group of the introduced moiety. Bromide **18** was then reacted with commercially available H-Pro-O-*t*-Bu in the presence of Et₃N to give **19** in 82% yield. Reduction of **19** with lithium aluminum hydride at −78 °C afforded oxazolidinol **20**, which was converted into dithioketal **21** by a BF₃·OEt₂-catalyzed exchange reaction with 1,2-ethanedithiol. Methylation of **21** with trimethylsilyldiazomethane in benzene/methanol produced ester **22**. Final deprotection of **22** with HgO/BF₃·OEt₂ gave aldehyde **3a**. Target compound **3a** was thus prepared *via* an efficient six-step synthesis with 8.7% overall yield.

The synthesis of compounds **3b** and **3c** started with the coupling of **21** with *t*-butylamine and H-Val-NH-*t*-Bu, respectively, using the peptide-coupling cocktail DCC/HOBt to produce **23** and **25**

(Scheme 21). The *alloc* protecting group was then selectively removed *via* Pd(Ph$_3$)$_4$-catalyzed allyl transfer in the presence of dimedone.[68] Peptide coupling under classic coupling conditions with the liberated amine function adjacent to a quaternary centre failed. We successfully coupled the sterically hindered free amine with N-Cbz-*L*-valine using the reagent cocktail CIP/HOAt to afford **24** and **26** in 65% and 64% yield, respectively.[69] The synthesis was finished with their deprotection by use of HgO/BF$_3$·OEt$_2$ and aldehydes **3b** and **3c** were obtained in 61% and 47% yield, respectively.

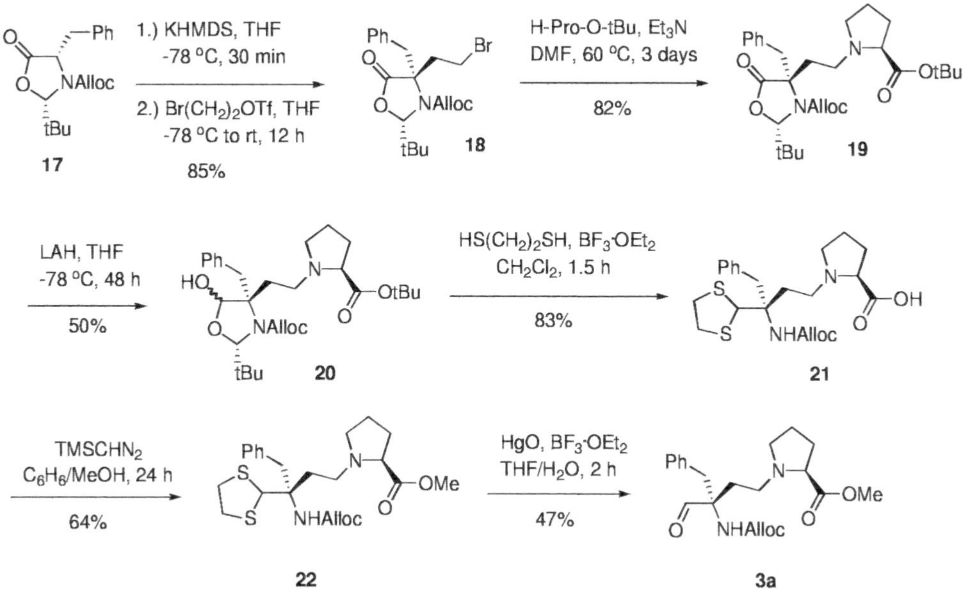

Scheme 20. Synthesis of first-generation inhibitor candidate **3a**.

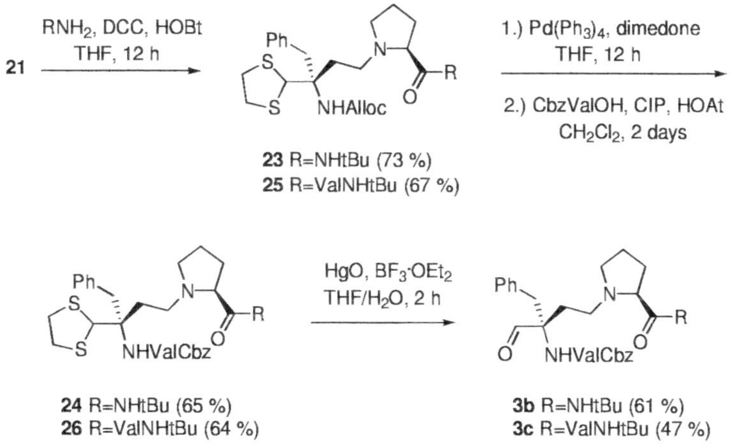

Scheme 21. Synthesis of first-generation inhibitor candidates **3b** and **3c**.

The inhibitory potency of compounds **3a–c** was tested with HIV-1 peptidase using a well-established continuous assay based on FRET (fluorescence resonance energy transfer) technology.[70,71] Compound **3a** showed an IC$_{50}$ value greater than 1 mM, thus representing negligible inhibitory activity (Table 1).[22] The K_i values of compounds **3b** and **3c** were determined to be 574 µM and 99 µM, respectively. These values are at least five orders of magnitude higher than those found for classic inhibitors (see pepstatine as reference compound in Table 1). Moreover, the inhibitory potency in the series **3a–c** increases with the number of amino acid residues (valine) attached to the core. This suggests that affinity with the enzyme is not primarily achieved by the core, but by the side chains satisfying the recognition subpockets. Structural incongruency of the core with the active site may have been caused by a counterproductive conformation of the newly formed heterocycle. More importantly, molecules **3a–c** exist as an equilibrium of the open form and a cyclized form that displays the N→C=O interaction (see above in Sections 2 and 3). The presence of an open form rules out any exploration as a drug candidate since aldehydes are justly regarded as too reactive towards nucleophiles in live organisms.

An alternative approach in HIV-1 peptidase inhibitor design has been the use of heterocyclic non-peptidic templates, such as a cyclic urea (**14**)[72] or a 5,6-dihydro-4-hydroxy-2-pyrone (**15**)[73,74] (Scheme 22). The two main characteristics of these inhibitor core units are their elevated rigidity and the carbonyl function that mimics the hydrogen-bonding capacities of the flap water molecule. This water molecule is found in most crystal structures of HIV-1 peptidase complexed with different linear peptidomimetic inhibitors. Incorporation of these types of heterocycles into an inhibitor can contribute to better binding energy due to a favourable entropic effect in that (a) the structural water molecule is replaced by the inhibitor[72,75] and (b) less degrees of freedom have to be frozen during complexation.[72,76] Beyond these features, tipranavir (**15**) is likely the only commercialized HIV PR inhibitor whose hydroxyl group should have a significantly lower pKa than its competitors that exhibit regular secondary alcohol groups (pKa≈16). This is due to its being part of a vinylogous carbonic acid monoester. This higher acidity may reflect that of the oxy-anion of the transition states of peptide hydrolysis more faithfully than that of secondary alcohol groups and may thus be suspected to be a non-negligible element in the high affinity of **15** to its target.

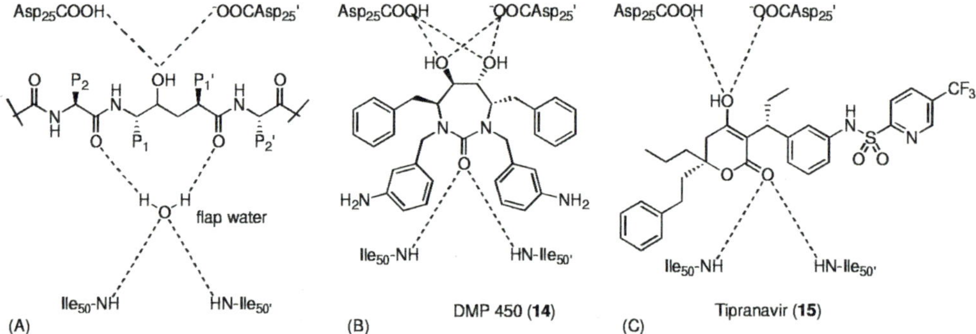

Scheme 22. Hydrogen bonding between the catalytic diad (aspartic acid residues) as well as the enzyme flaps (isoleucines) with (A) a peptidomimetic inhibitor; (B) cyclic urea inhibitor DMP 450 (14); (C) heterocyclic inhibitor Tipranavir (15).

We wished to have our second-generation inhibitor candidate to benefit from these same advantages while displaying the N→C=O interaction to 100% in a protic environment. We thus opted for the incorporation of a urea motif that promised a strong preorganizing effect for the formation of the target heterocycle while providing with the already proven capacity to bind to the flaps (⇒ DMP 450). This forced us to use a hydrazine moiety in order to provide the system with an amine donor at least as strongly nucleophilic as a regular amine. As described in Sections 2 and 3, the resulting range of compounds **4a–e** was found to exist to 100% in their N→C=O interacted form in polar protic media.[23,24,77]

In the synthesis of **4a**, aldehyde **27**[78] was protected by acid-catalyzed acetalization furnishing compound **28** in a convenient overnight reaction at room temperature (Scheme 23). Removal of the Cbz group using straightforward catalytic hydrogenation quantitatively yielded amine **29**. Reaction with carbonyl-diimidazole (CDI), activation of the resulting **30**[79] with methyl iodide as an imidazolium ion and subsequent treatment of this intermediate with hydrazine **32** yielded hydrazino urea **33**. Benzyl hydrazine **32** was obtained by reacting commercially available hydrazine **31** with benzaldehyde, followed by *in situ* reduction of the intermediate hydrazone. Removal of the dimethyl acetal protecting group was achieved by exposure to *in situ* generated TMSI,[80] furnishing desired target compound **4a** in excellent yield.

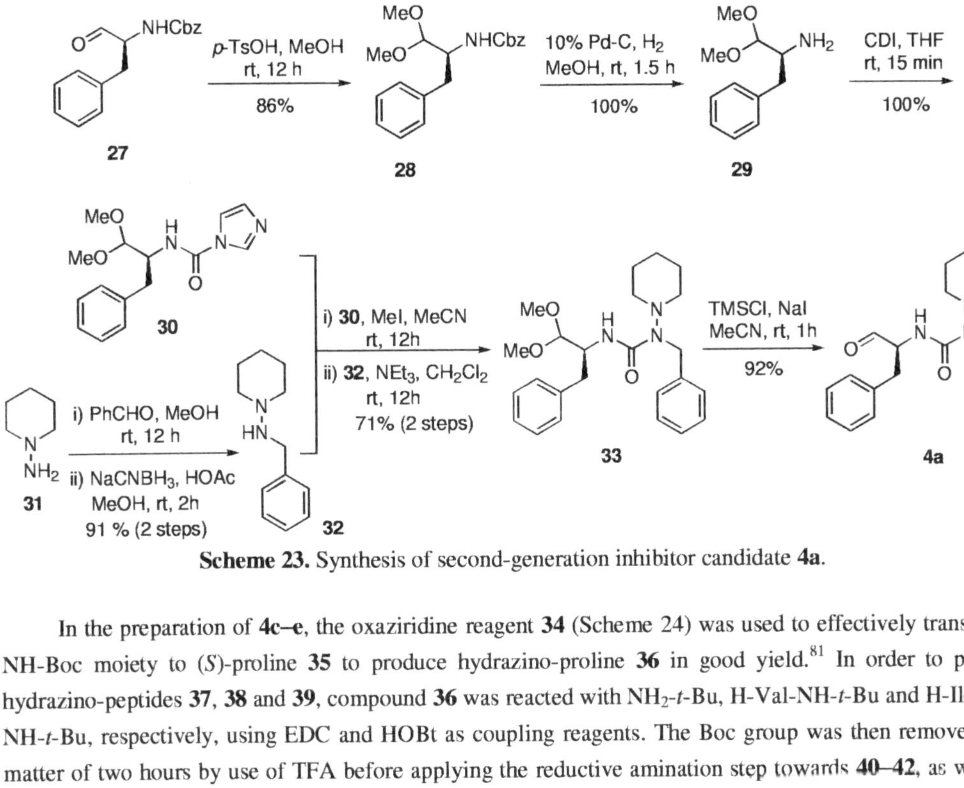

Scheme 23. Synthesis of second-generation inhibitor candidate **4a**.

In the preparation of **4c–e**, the oxaziridine reagent **34** (Scheme 24) was used to effectively transfer an NH-Boc moiety to (*S*)-proline **35** to produce hydrazino-proline **36** in good yield.[81] In order to prepare hydrazino-peptides **37**, **38** and **39**, compound **36** was reacted with NH₂-*t*-Bu, H-Val-NH-*t*-Bu and H-Ile-Val-NH-*t*-Bu, respectively, using EDC and HOBt as coupling reagents. The Boc group was then removed in a matter of two hours by use of TFA before applying the reductive amination step towards **40–42**, as worked out in the preparation of **4a**. The preparation of unsymmetrical ureas **43–45** was carried out in an identical manner as used for **33**, by reacting hydrazines **40–42** with methylated **30**. Deprotection of the dimethyl acetal using NaI and TMSCl gave target compounds **4c–e** in excellent yield.

In general, the compounds of the hydrazino urea series have better inhibitory potency than the first generation inhibitors **3a–c**, which could be due to a favourable urea carbonyl-flap interaction, but remains to be confirmed in future studies (Table 1). Structure activity relationships in the hydrazine urea series are quite instructive: the (*R*)-enantiomer **4b** had the lowest inhibitory effect of the new compounds with a K_i about

20-fold higher than that of **4a**. Clearly an (*S*)-configured phenylalanine-derived motif is much more tolerated by HIV PR. In this context, it is interesting to note that subtle configurational changes such as a single stereocentre inversion can have a huge impact on affinity. In an extreme case a K_i difference of approximately seven orders of magnitude was reported.[82–84] The smallest compounds **4a** and **4c** without long peripheral side chains were the best of the hydrazino urea series, indicating that the core adopts a conformation with promising affinity to the catalytic centre. On the other hand, attaching side chains of one or two amino acid residues results in decreased potency, by one order of magnitude (**4d, 4e**). This indicates that an unsuitable geometry of the side chain–core connection impairs the rather favourable interactions of the core.

40 R= NHtBu (83 %)
41 R= ValNHtBu (93 %)
42 R= IleValNHtBu (80 %)

43 R= NHtBu (66 %)
44 R= ValNHtBu (70 %)
45 R= IleValNHtBu (72 %)

4c R= NHtBu (83 %)
4d R= ValNHtBu (91 %)
4e R= IleValNHtBu (78 %)

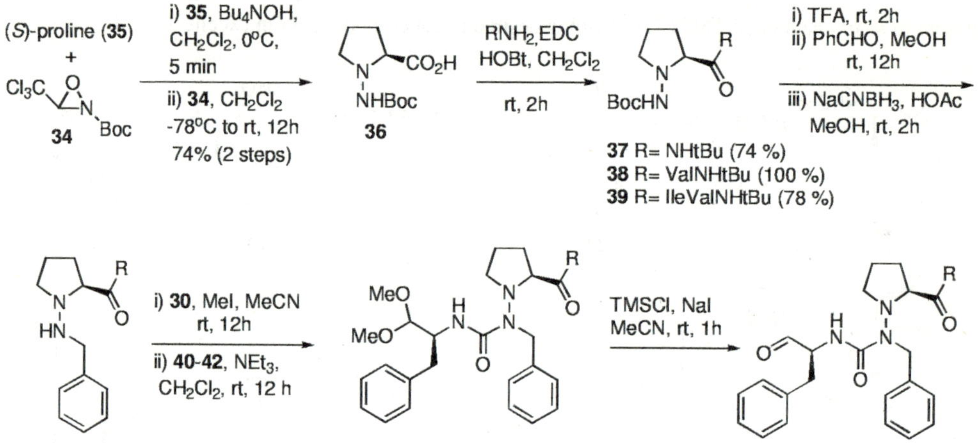

Scheme 24. Synthesis of second-generation inhibitor candidates **4c–e**.

Table 1. Parameters for candidates **3** and **4** characterizing their inhibition of HIV-1 PR proteolytic activity.

Cpd.	IC_{50} (μM)	K_i (μM)	MW (g/mol)
4a	37.2	26.2	365
4b	543	461.0	365
4c	44.5	34.3	450
4d	341	292.0	549
4e	169	148.6	662
3a	> 1000	-	388
3b	545	574	578
3c	204	99	677
pepstatine	0.036	0.036	686

This study broadens our exploration of the N→C=O motif as a potential isostere that reproduces key stereoelectronic properties of the transition states of peptide hydrolysis. In contrast to our first-generation inhibitors, only one constitutional component is present in all molecule candidates in a protic medium, such as provided by methanol. In no instance we have noticed an impairment to the mutual approach of the two interacting functional groups for steric reasons. It should be stressed that the demonstrated absence of any aldehydic component constitutes a major stride towards the consideration of the N→C=O motif as a future pharmacophore. The most potent inhibitor in this study is so small (**4a**: 365 g/mol) that great improvements in affinity and ADME properties may be obtained by introduction of additional molecular bulk without violating Lipinski's rule of remaining within the limit of 500 g/mol.[85] Computational and synthetic studies with a number of diverse substitution patterns of this hydrazino urea motif are currently underway.

5.2. Foldamer design

The phenomenon of secondary structure adoption by biopolymers has inspired innovation in the field of supramolecular chemistry probably as much as ligand recognition by protein receptors. For example, multicellular organisms use short peptides to defend themselves against microbes.[86] These oligomers tend to fold into amphiphilic helices in order to assure a binding mode where the cationic N terminus binds to the anionic surface of the bacterial cell membrane and the hydrophobic side chains interact with the lipid bilayer. This compromises the impermeability of the cell wall. Such peptides are of course highly interesting as a point of departure for the design of synthetic analogs that will increase our power to fight resistance to common antibiotics.

In order for this class of compounds to be useful as drugs, two major hurdles have to be overcome: (a) loss of secondary structure in aqueous media (and concomitant loss of amphiphilic lipophilicity) and (b) loss of primary structure by proteolytic degradation through endogenous peptidases. The class of beta-peptides has been designed to respond to the latter point because they show a much higher resistance to hydrolytic degradation than do alpha-peptides.[87] Unsurprisingly, they have thus been promoted as a class of new drugs. However, even when they are water soluble, and this is often not easy to achieve, they reliably suffer loss of structural integrity in this medium.[88] The intramolecular hydrogen bonds that guarantee defined folding of the oligomer are in competition with intermolecular bonds between the respective sites and water molecules. Since these bonds are of comparable energy, the gain in enthalpy upon folding is too low to compensate the loss in entropy associated with the adoption of a rigid secondary structure. Beta-peptide design thus attempts to favour hydrosolubility by introducing hydrophilic side chains while at the same time ruling out the total solvation of the backbone. Three strategies have been explored to this end: (1) the first has been inspired by work on alpha-peptides[89,90] and consists in locking the helix by the construction of an additional covalent bond (Scheme 25 A). This strategy has been called "molecular stapling" and is currently very much in vogue for the discovery of new drug principles.[91,92] (2) The second strategy has been developed simultaneously by the groups of Seebach[93] and De Grado.[94] The latter in particular coined the term "foldamer" for their synthetic targets in view of the principal aim of their research. It consists in the introduction of judiciously placed ammonium and carboxylate functions that stabilize the secondary structure of the foldamer *via* the formation of salt bridges as well as the compensation of the macro-dipole of the helix by inversely orientated dipoles that these salt bridges constitute.[95] An additional advantage is seen in the improved hydrosolubility that comes with such a modification (Scheme 25 B).

A more drastic solution to circumvent loss of structural integrity when passing from organic to aqueous media has been proposed by the group of Gellman. They synthesized oligomers with significantly more rigid backbones by employing units of *trans*-aminocyclohexane carboxylic acid (ACHC) or *trans*-aminocyclopentane carboxylic acid (ACPC) (*e.g.* Scheme 25 C).[96–99] Of course, all of the above strategies limit the choice of backbone sequences and possible monomers and often constitute a real challenge in synthetic chemistry. In any case, while alpha helices have been commonly observed in the beta-peptide realm, to our knowledge, no folding into beta sheets has ever been reported for this class. For these reasons, a new type of foldamer may be proposed that takes advantage of the N→C=O interaction. As it has been stated on numerous occasions in this manuscript, the interaction is particularly favoured in water and should be sufficiently strong (on the order of 40 kJ/mol, to be compared with ca. 12–32 kJ/mol for the H bond) to allow the oligomer to adopt a defined conformation in physiological media *via* the iterative energy-minimizing process that characterizes the cooperative effect. No competition with solvent molecules needs to be feared.

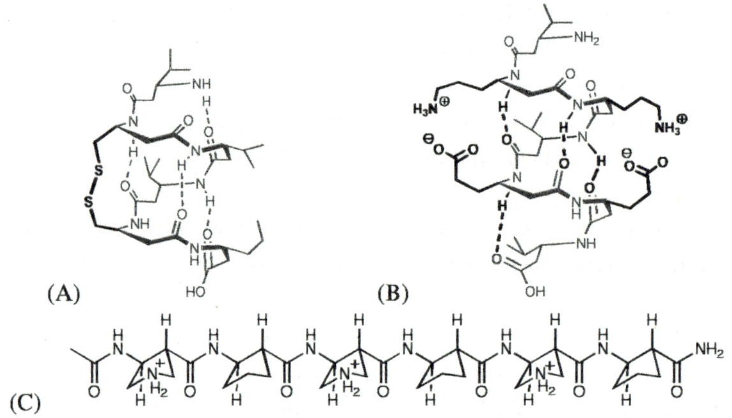

(A)   (B)   (C)

Scheme 25. Stabilisation of a beta-peptide by (A) introduction of a macrocylic disulfide bridge and (B) by salt bridges; (C) highly preorganised backbone leads to reliable folding in aqueous media while significantly reducing hydrosolubility.

We have designed a first backbone that incorporates alternating tertiary amine and ketone sites and that should adopt a beta sheet-type secondary structure in aqueous media forming enchained units of six-membered rings. Such a folded oligomer should be in equilibrium with its open-chain form at pH 9 in view of the estimated pKa of 9 for the interaction (*vide supra*). But even at neutral/physiological pH (7.4) where the interaction will become covalent and the resulting motif a cation, a state of equilibrium with open-chain forms may be envisaged to allow the system to find the lowest energy constitution and conformation. The pKa values of neighboring moieties should of course suffer more or less perturbation by the presence of a cation in their immediate vicinity.

The modeled target compounds (Scheme 26)[100] constitute a significant synthetic challenge in view of the functional group density. However, their design includes several features that make them attractive objects of an initial study: (a) a high frequency of tertiary amine and ketone sites on the backbone for a significant enthalpic advantage to folding; (b) formation of energetically favourable six-membered rings; (c) alternating donor and acceptor sites to mimic the situation found in peptidic beta sheets and to compensate dipoles on site; and lastly (d) a beta-turn like unit that ensures folding of the molecule on itself. While the modeling experiments gave conclusive results, the synthetic development of the molecules has not yet advanced to the point of a first object of study.[100]

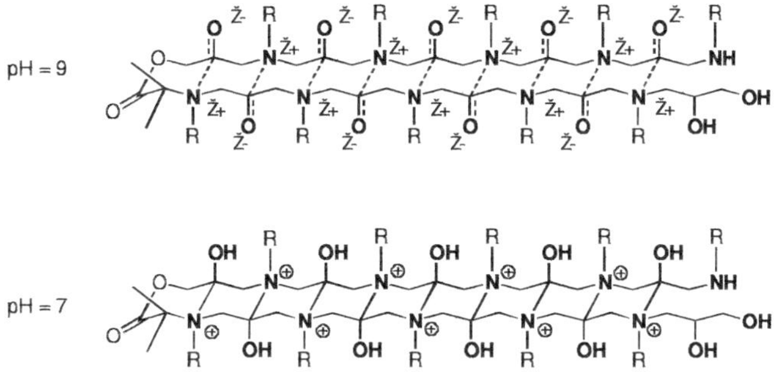

Scheme 26. Target molecules not yet realized that should fold in aqueous media as depicted.[46,100]

6. Conclusions

The N→C=O interaction can be counted among the weak interactions but distinguishes itself from its better known counterparts by the capacity to turn into a covalent one when pH is lowered. At this point, it has to be regarded as an addition product of a protonated ammonium ion on a carbonyl compound and can be compared to hemiaminals that are known to be in rapid equilibrium with their elimination products. Sufficiently preorganized systems capable of showing the N→C=O motif will exist solely in this interacted form if dissolved in protic polar media and our results have demonstrated that this is possible also for monocyclic *six-membered* rings. In aprotic media, these systems prefer the non-interacted, often "linear" form, and can then exert their full aldehyde reactivity. The elaboration by experts of force fields for the interaction is underway and molecular mechanics packages will eventually be equipped for convenient modeling of molecules cabable of showing the interaction. We have equally demonstrated the feasibility of exploring the binding properties of the N→C=O motif in enzyme inhibition and we have embarked on the ambitious synthetic project of observing the interaction for the first time in multiple copies on one molecule. The stage is now set to deliberately incorporate the N→C=O motif into scaffolds of one's choice and explore the unique and hopefully beneficial properties thereby conferred on the molecule in its entirety.

Acknowledgments

A. Gautier, R. Barbe and M. Waibel thank the Ministere de la Recherche for a PhD fellowship, respectively. Financial support from the French Ministry, the C.N.R.S. and the Region Rhone-Alpes is gratefully acknowledged. This report is dedicated to the memory of Nelson Leonard who kindly communicated with us in the early stages of our projects described above.

References

1. Gadamer, J. *Arch. Pharm.* **1920**, *258*, 148–167.
2. Kermack, W.; Robinson, R. *J. Chem. Soc.* **1922**, *121*, 427–440.
3. Huisgen, R.; Wieland, H.; Eder, H. *Ann.* **1949**, *561*, 193–215.
4. Boit, H. G.; Paul, L. *Chem. Ber.-Recl.* **1954**, *87*, 1859–1864.
5. Wunderlich, J. *Acta Crystallogr.* **1967**, *23*, 846–855.
6. Hall, S. R. *Acta Crystallogr., Sect. B* **1968**, *24*, 337–346.

7.   Hall, S. R.; Ahmed, F. R. *Acta Crystallogr., Sect. B* **1968**, *24*, 346–355.

8.   Birnbaum, K. B.; Klasek, A.; Sedmera, P.; Snatzke, G.; Johnson, L. F.; Santavy, F. *Tetrahedron Lett.* **1971**, 3421–3424.

9.   Birnbaum, G. I. *J. Am. Chem. Soc.* **1974**, *96*, 6165–6168.

10.   Cimino, G.; Destefano, S.; Scognamiglio, G.; Sodano, G.; Trivellone, E. *Bull. Soc. Chim. Belg.* **1986**, *95*, 783–800.

11.   Becker, M. H.; Chua, P.; Downham, R.; Douglas, C. J.; Garg, N. K.; Hiebert, S.; Jaroch, S.; Matsuoka, R. T.; Middleton, J. A.; Ng, F. W.; Overman, L. E. *J. Am. Chem. Soc.* **2007**, *129*, 11987–12002.

12.   Burgi, H. B.; Dunitz, J. D.; Shefter, E. *J. Am. Chem. Soc.* **1973**, *95*, 5065–5067.

13.   Burgi, H. B.; Dunitz, J. D. *Acc. Chem. Res.* **1983**, *16*, 153–161.

14.   Kirby, A. J.; Komarov, I. V.; Bilenko, V. A.; Davies, J. E.; Rawson, J. M. *Chem. Commun.* **2002**, 2106–2107.

15.   Leonard, N. *Record Chem. Progr.* **1956**, *17*, 243–257.

16.   Leonard, N. J. *Acc. Chem. Res.* **1979**, *12*, 423–429.

17.   Leonard, N. J.; Oki, M.; Brader, J.; Boaz, H. *J. Am. Chem. Soc.* **1955**, *77*, 6237–6239.

18.   Leonard, N. J.; Oki, M. *J. Am. Chem. Soc.* **1955**, *77*, 6239–6240.

19.   Leonard, N. J.; Oki, M. *J. Am. Chem. Soc.* **1955**, *77*, 6245–6246.

20.   Bell, M. R.; Archer, S. *J. Am. Chem. Soc.* **1960**, *82*, 151–155.

21.   Gautier, A. PhD thesis, Ecole Normale Superieure de Lyon, 2005.

22.   Gautier, A.; Pitrat, D.; Hasserodt, J. *Bioorg. Med. Chem.* **2006**, *14*, 3835–3847.

23.   Waibel, M.; Hasserodt, J. *J. Org. Chem.* **2008**, *73*, 6119–6126.

24.   Waibel, M.; Pitrat, D.; Hasserodt, J. *Bioorg. Med. Chem.* **2009**, *17*, 3671–3679.

25.   Leonard, N. J.; Morrow, D. F.; Rogers, M. T. *J. Am. Chem. Soc.* **1957**, *79*, 5476–5479.

26.   McCrindle, R.; McAlees, A. J. *J. Chem. Soc., Chem. Commun.* **1983**, 61–62.

27.   Carroll, J. D.; Jones, P. R.; Ball, R. G. *J. Org. Chem.* **1991**, *56*, 4208–4213.

28.   Hine, J.; Kokesh, F. C. *J. Am. Chem. Soc.* **1970**, *92*, 4383–4388.

29.   Leonard, N. J.; Fox, R. C.; Oki, M. *J. Am. Chem. Soc.* **1954**, *76*, 5708–5714.

30.   Rademacher, P. *Chem. Soc. Rev.* **1995**, *24*, 143–150.

31.   Spanka, G.; Rademacher, P. *J. Org. Chem.* **1986**, *51*, 592–596.

32.   Leonard, N. J.; Oki, M.; Chiavarelli, S. *J. Am. Chem. Soc.* **1955**, *77*, 6234–6337.

33.   Leonard, N. J.; Adamcik, J. A.; Djerassi, C.; Halpern, O. *J. Am. Chem. Soc.* **1958**, *80*, 4858–4862.

34.   Beesley, R. M.; Ingold, C. K.; Thorpe, J. F. *J. Chem. Soc., Trans.* **1915**, *107*, 1080–1106.

35.   Burgi, H. B.; Dunitz, J. D.; Shefter, E. *Nature-New Biology* **1973**, *244*, 186–188.

36.   Davies, J. E.; Kirby, A. J.; Komarov, I. V. *Helv. Chim. Acta* **2003**, *86*, 1222–1233.

37.   Kemp, D. S.; Petrakis, K. S. *J. Org. Chem.* **1981**, *46*, 5140–5143.

38.   Eicher, T.; Bohm, S. *Chem. Ber.-Recl.* **1974**, *107*, 2186–2214.

39.   Zhao, Y. M.; Raymond, M. K.; Tsai, H.; Roberts, J. D. *J. Phys. Chem.* **1993**, *97*, 2910–2913.

40.   Drakenberg, T.; Dahlqvist, K.-I.; Forsen, S. *J. Phys. Chem.* **1972**, *76*, 2178–2183.

41.   Scherer, G.; Kramer, M. L.; Schutkowski, M.; Reimer, U.; Fischer, G. *J. Am. Chem. Soc.* **1998**, *120*, 5568–5574.

42.   Schweizer, W. B.; Procter, G.; Kaftory, M.; Dunitz, J. D. *Helv. Chim. Acta* **1978**, *61*, 2783–2808.

43.   Kirby, A. J.; Percy, J. M. *Tetrahedron* **1988**, *44*, 6903–6910.

44.   Spanka, G.; Boese, R.; Rademacher, P. *J. Org. Chem.* **1987**, *52*, 3362–3367.

45.   Griffith, R.; Yates, B. F.; Bremner, J. B.; Titmuss, S. J. *J. Mol. Graph. Model.* **1997**, *15*, 91–99.

46.   Barbe, R. PhD thesis, Ecole Normale Superieure de Lyon, 2006.

47.   Carneiro, J. W. d. M.; de Oliveira, C. d. S. B.; Passos, F. B.; Aranda, D. A. G.; de Souza, P. R. N.; Antunes, O. A. C. *J. Mol. Cat. A Chem.* **2001**, *170*, 235–243.

48.   Studer, M.; Blaser, H.-U.; Exner, C. *Adv. Synth. Catal.* **2003**, *345*, 45–65.

49.   Pilme, J.; Berthoumieux, H.; Robert, V.; Fleurat-Lessard, P. *Chem. Eur. J.* **2007**, *13*, 5388–5393.

50.   Rashidi-Ranjbar, P.; Mohajeri, A.; Ghiaci, M. *Iranian J. Chem. Chem., Engl. Int. Ed.* **2001**, *20*, 102–105.

51. Bursavich, M. G.; Rich, D. H. *J. Med. Chem.* **2002**, *45*, 541–558.
52. Brik, A.; Wong, C. H. *Org. Biomol. Chem.* **2003**, *1*, 5–14.
53. Tyndall, J. D. A.; Nall, T.; Fairlie, D. P. *Chem. Rev.* **2005**, *105*, 973–999.
54. Schechter, I.; Berger, A. *Biochem. Biophys. Res. Comm.* **1967**, *27*, 157–162.
55. Allen, F. H.; Kennard, O.; Watson, D. G.; Brammer, L.; Orpen, A. G.; Taylor, R. *J. Chem. Soc., Perkin Trans. 2* **1987**, S1–S19.
56. Venturini, A.; Lopez-Ortiz, F.; Alvarez, J. M.; Gonzalez, J. *J. Am. Chem. Soc.* **1998**, *120*, 1110–1111.
57. Okimoto, N.; Tsukui, T.; Hata, M.; Hoshino, T.; Tsuda, M. *J. Am. Chem. Soc.* **1999**, *121*, 7349–7354.
58. Piana, S.; Bucher, D.; Carloni, P.; Rothlisberger, U. *J. Phys. Chem. B* **2004**, *108*, 11139–11149.
59. Pauling, L. *Chem. Eng. News* **1946**, *24*, 1375–1377.
60. Jencks, W. P. In *Current Aspects of Biochemical Energetics*; Kaplan, N. O., Kennedy, E. P., Eds.; Academic Press: New York, 1966; pp. 273–298.
61. Lienhard, G. E. *Science* **1973**, *180*, 149–154.
62. Babine, R. E.; Bender, S. L. *Chem. Rev.* **1997**, *97*, 1359–1472.
63. De Clercq, E. *Nature Rev. Drug Disc.* **2007**, *6*, 1001–1018.
64. Erickson, J. W.; Eissenstat, M. A. In *Proteinases of Infectious Agents*; Dunn, B. M., Ed.; Academic Press: San Diego, CA, 1999; pp. 1–60.
65. Boden, D.; Markowitz, M. *Antimicr. Ag. Chemother.* **1998**, *42*, 2775–2783.
66. Seebach, D.; Sting, A. R.; Hoffmann, M. *Angew. Chem. Int. Ed.* **1996**, *35*, 2708–2748.
67. Smith, A. B.; Guzman, M. C.; Sprengeler, P. A.; Keenan, T. P.; Holcomb, R. C.; Wood, J. L.; Carroll, P. J.; Hirschmann, R. *J. Am. Chem. Soc.* **1994**, *116*, 9947–9962.
68. Kunz, H.; Unverzagt, C. *Angew. Chem. Int. Ed.* **1984**, *23*, 436–437.
69. Akaji, K.; Kuriyama, N.; Kiso, Y. *J. Org. Chem.* **1996**, *61*, 3350–3357.
70. Matayoshi, E. D.; Wang, G. T.; Krafft, G. A.; Erickson, J. *Science* **1990**, *247*, 954–958.
71. Gershkovich, A.; Kholodovych, V. *J. Biochem. Biophys. Meth.* **1996**, *33*, 135–162.
72. DeLucca, G. V.; EricksonViitanen, S.; Lam, P. Y. S. *Drug Disc. Today* **1997**, *2*, 6–18.
73. Thaisrivongs, S.; Strohbach, J. W. *Biopolymers* **1999**, *51*, 51–58.
74. Hagen, S.; Prasad, J. V. N. V.; Tait, B. D. In *Advances in Medicinal Chemistry*; 2000; Vol. 5, pp. 159–195.
75. Dunitz, J. D. *Science* **1994**, *264*, 670–670.
76. Dunn, B. *Chem. Rev.* **2002**, *102*, 4431–4458.
77. Waibel, M. PhD thesis, Ecole Normale Superieure de Lyon, 2008.
78. Lam, P. Y. S.; Ru, Y.; Jadhav, P. K.; Aldrich, P. E.; DeLucca, G. V.; Eyermann, C. J.; Chang, C. H.; Emmett, G.; Holler, E. R.; Daneker, W. F.; Li, L. Z.; Confalone, P. N.; McHugh, R. J.; Han, Q.; Li, R. H.; Markwalder, J. A.; Seitz, S. P.; Sharpe, T. R.; Bacheler, L. T.; Rayner, M. M.; Klabe, R. M.; Shum, L. Y.; Winslow, D. L.; Kornhauser, D. M.; Jackson, D. A.; EricksonViitanen, S.; Hodge, C. N. *J. Med. Chem.* **1996**, *39*, 3514–3525.
79. Grzyb, J. A.; Shen, M.; Yoshina-Ishii, C.; Chi, W.; Brown, R. S.; Batey, R. A. *Tetrahedron* **2005**, *61*, 7153–7175.
80. Brown, H. C.; Krishnamurthy, S. *Tetrahedron* **1979**, *35*, 567–607.
81. Vidal, J.; Hannachi, J. C.; Hourdin, G.; Mulatier, J. C.; Collet, A. *Tetrahedron Lett.* **1998**, *39*, 8845–8848.
82. Schramm, V. L.; Baker, D. C. *Biochem.* **1985**, *24*, 641–646.
83. Rich, D. H.; Sun, C. Q.; Prasad, J.; Pathiasseril, A.; Toth, M. V.; Marshall, G. R.; Clare, M.; Mueller, R. A.; Houseman, K. *J. Med. Chem.* **1991**, *34*, 1222–1225.
84. Wolfenden, R. *Bioorg. Med. Chem.* **1999**, *7*, 647–652.
85. Bleicher, K. H.; Bohm, H. J.; Muller, K.; Alanine, A. I. *Nature Rev. Drug Disc.* **2003**, *2*, 369–378.
86. Williams, R. W.; Starman, R.; Taylor, K. M.; Gable, K.; Beeler, T.; Zasloff, M.; Covell, D. *Biochem.* **1990**, *29*, 4490–4496.
87. Seebach, D.; Overhand, M.; Kuehnle, F. N. M.; Martinoni, B. *Helv. Chim. Acta* **1996**, *79*, 913–941.
88. Gung, B. W.; Zou, D.; Stalcup, A. M.; Cottrell, C. E. *J. Org. Chem.* **1999**, *64*, 2176–2177.

89. Jackson, D. Y.; King, D. S.; Chmielewski, J.; Singh, S.; Schultz, P. G. *J. Am. Chem. Soc.* **1991**, *113*, 9391–9392.
90. Andrews, M. J. I.; Tabor, A. B. *Tetrahedron* **1999**, *55*, 11711–11743.
91. Walensky, L. D.; Kung, A. L.; Escher, I.; Malia, T. J.; Barbuto, S.; Wright, R. D.; Wagner, G.; Verdine, G. L.; Korsmeyer, S. J. *Science* **2004**, *305*, 1466–1470.
92. Kutchukian, P. S.; Yang, J. S.; Verdine, G. L.; Shakhnovich, E. I. *J. Am. Chem. Soc.* **2009**, *131*, 4622–4627.
93. Arvidsson, P. I.; Rueping, M.; Seebach, D. *Chem. Commun.* **2001**, 649–650.
94. Cheng, R. P. *Curr. Op. Struct. Biol.* **2004**, *14*, 512–520.
95. Hart, S. A.; Bahadoor, A. B. F.; Matthews, E. E.; Qiu, X. J.; Schepartz, A. *J. Am. Chem. Soc.* **2003**, *125*, 4022–4023.
96. Appella, D. H.; Barchi, J. J., Jr.; Durell, S. R.; Gellman, S. H. *J. Am. Chem. Soc.* **1999**, *121*, 2309–2310.
97. Appella, D. H.; LePlae, P. R.; Raguse, T. L.; Gellman, S. H. *J. Org. Chem.* **2000**, *65*, 4766–4769.
98. Wang, X.; Espinosa, J. F.; Gellman, S. H. *J. Am. Chem. Soc.* **2000**, *122*, 4821–4822.
99. Lee, H.-S.; Syud, F. A.; Wang, X.; Gellman, S. H. *J. Am. Chem. Soc.* **2001**, *123*, 7721–7722.
100. Barbe, R.; Hasserodt, J. *Tetrahedron* **2007**, 2199–2207.

MULTICOMPONENT REACTIONS INVOLVING HETEROCYCLIC SURROGATES OF OXOCARBENIUM AND IMINIUM IONS

Federica Catti[a] and Rodolfo Lavilla*[a,b]

[a]*Institute for Research in Biomedicine, Barcelona Science Park, Baldiri Reixac 10–12, E-08028 Barcelona, Spain*

[b]*Laboratory of Organic Chemistry, Faculty of Pharmacy, University of Barcelona, Avda Joan XXIII sn, E-08028 Barcelona, Spain (e-mail: rlavilla@pcb.ub.es)*

Abstract. *The participation of heterocyclic surrogates of activated carbonyls and imines in Povarov and isocyanide multicomponent reactions yields a wide range of products in a straightforward manner. The rich reactivity of heterocyclic systems in this context often gives rise to unexpected domino processes, which are studied to improve the knowledge about this chemistry and eventually to contribute to the design of new productive reactions. The mechanistic rationale for these processes is analyzed together with their synthetic applications.*

Contents

1. Introduction

Achieving the ideal synthesis of organic compounds is a highly challenging goal and requires substantial improvements in efficiency, selectivity and sustainability, among others. Apart from these concepts, the newly introduced step economy is gaining acceptance as a measure of the synthetic feasibility, especially in the discovery and preparation of drugs. In this context, multicomponent reactions (MCRs) play key roles as they feature the formation of many bonds in a single transformation and simultaneously allow the incorporation of meaningful parts of the starting materials into the final adduct.[1–6] Also, the modular character of these processes allows a high degree of structural variability which is critical in the exploration of biological and chemical space, an important issue in drug discovery process.[7–11] In fact, the increasing demand for new and safer drugs, together with the myriad of small molecules needed to decipher complex biological processes through chemical genetics, makes the use of fast and straightforward synthetic methodology a critical requirement.[12–14]

Furthermore, heterocycles constitute the most common motifs found in drugs and bioactive compounds. Incidentally, the pioneers in this field (Hantzsch, Biginelli, etc.) provided fundamental contributions for the development of powerful general and selective synthetic methodology that leads to these heterocyclic targets through MCRs.[15,16] An attractive and complementary approach consists of using heterocycles as the reactants for these processes,[17] a fruitful strategy which takes advantage of the rich and sometimes unique reactivity of these structures (Scheme 1).

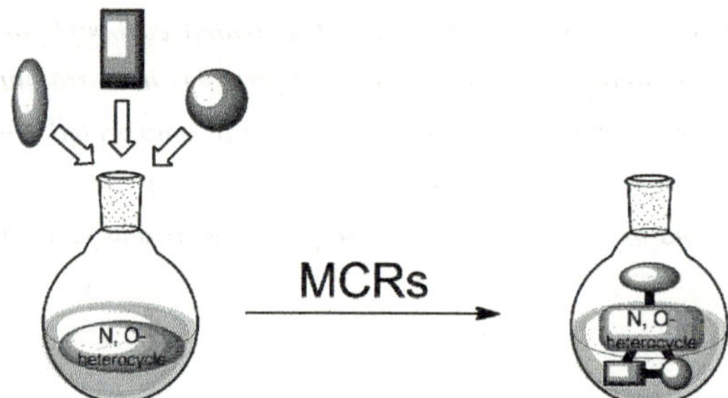

Scheme 1. MCRs with heterocyclic reactants.

Several known MCRs are based on the reactivity of activated carbonyls and iminium ions and mainly involve nucleophilic additions.[18] This chapter is devoted to our results in the development of new synthetic methodology based on the generation and reactivity of heterocyclic surrogates of such species (oxocarbenium and iminium ions) from common heterocycles (Scheme 2).

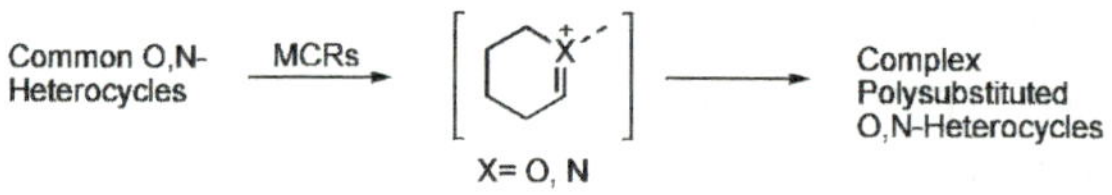

Scheme 2. MCRs based on heterocyclic surrogates of oxocarbenium and iminium ions.

2. Povarov-type reactions

The Povarov reaction consists of the interaction between imines and activated alkenes. Povarov's first approach reported the reactions of aldimines and electron-rich alkenes in the presence of a Lewis acid to yield a tetrahydroquinoline adduct.[19–21] Its multicomponent version was elaborated thereafter and the imine prepared *in situ* reacts with the electron-rich olefin in a formal imino Diels-Alder cycloaddition, as it was initially termed.[22,23] The development of new and powerful catalysts, especially lanthanide triflates,[24–27] together with improvements in the substrate range and in reaction conditions (including green chemistry approaches),[28–32] have allowed the general use of this process. Recently, catalytic enantioselective versions have been described.[33,34]

The reaction mechanism[35,36] starts with the initial condensation of aniline (**1**) and aldehyde (**2**) to form an imine (**A**), which is activated by a Lewis acid to increase its electrophilic character and favour the nucleophilic addition of an activated alkene (**3**) (Scheme 3). The resulting cationic species (**B**) undergoes an intramolecular aromatic electrophilic substitution on the aniline ring to yield the tetrahydroquinoline **4**, in a MCR sequence connecting Mannich and Pictet-Spengler-type processes. Several examples of Povarov MCRs involving heterocyclic precursors and mechanistic modifications are addressed in this section.

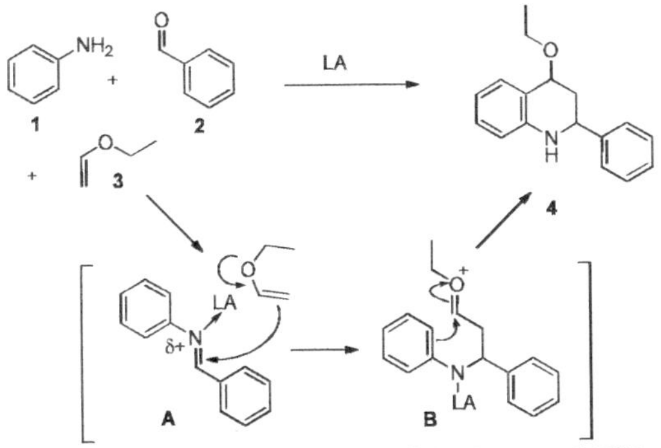

Scheme 3. Formal mechanism of the Povarov MCR.

2.1. Dihydropyridine-based Povarov reactions

The diversity of the olefin component in Povarov MCRs has been explored extensively. Initial results from cycloalkenes and enol ethers paved the way for the use of diverse structural patterns regarding this input. In this respect, dihydropyridines (DHPs) are particularly attractive because of their high degree of substitution, their availability (DHPs are easily prepared from commercial pyridines and several other precursors) and their reactivity which allows a wide range of synthetic transformations.[37] The main synthetic interest on the use of DHPs deals with their capacity to serve as piperidine precursors, particularly in the case of polysubstituted-polyfunctionalized targets, which are difficult to access from other substrates. This property has been exploited in the preparation of complex natural products and bioactive compounds (Scheme 4).

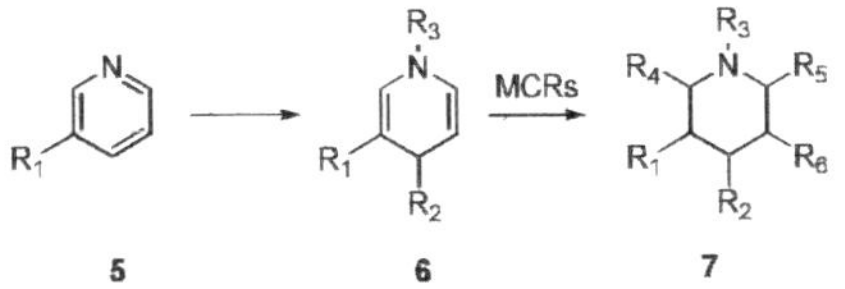

Scheme 4. DHPs as reactants in MCRs.

However, the major drawback of DHPs in organic synthesis is their easy oxidation to the corresponding pyridinium salts. In fact, this is their natural role, as NAD(P)H is a 1,4-DHP derivative and is present in many reductases as the stoichiometric reducing cofactor for the transformation of carbonyl and imine substrates into the corresponding alcohols and amines, normally with high enantioselectivity (Scheme 5). It should be mentioned that, under the appropriate conditions in the presence of Mg^{2+} or other Lewis acids, NAD(P)H alone (or synthetic analogs displaying a 1,4-DHP unit) promotes these reductions. This observation is relevant in this context, since we planned to perform MCRs with 1,4-DHPs as inputs together with carbonyls and amines, in the presence of Lewis acids, therefore facing potentially unproductive redox interferences.

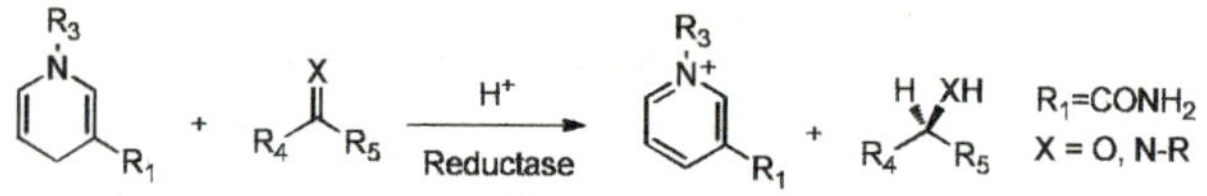

Scheme 5. Biomimetic oxidation of DHPs.

With these concepts in mind, we examined the reactivity of DHPs acting as the electron-rich olefin component in Povarov MCRs. The interaction between a *N*-alkyl-1,4-DHP (**6b**), an aldehyde (**2a**) and an aniline (**1a**) under Lewis acid catalysis would eventually provide straightforward access to benzonaphthyridine adducts[38] (Table 1). 1,4-DHPs are routinely prepared by $Na_2S_2O_4$ reduction of the corresponding *N*-alkylpyridinium salts, whereas their γ-substituted derivatives can also be accessed by nucleophilic addition to the same starting materials.[37,39] Initial experiments were carried out using different *standard* acid catalysts such as TFA, $BF_3 \cdot Et_2O$ and $Mg(ClO_4)_2$. These treatments lead to the polymerization of the somewhat acid-labile DHP, or to the biomimetic reduction of the intermediate imine with concomitant isolation of the corresponding pyridinium salt. Several Lewis acids were tested for the capacity to activate the imine and let it interact in the process, without damaging the DHP. Lanthanide cations and related species display the most suitable properties for these requirements,[40,41] and were therefore tested. Among these, $InCl_3$, $Sc(OTf)_3$, $Y(OTf)_3$ and $Yb(OTf)_3$ successfully promoted Povarov MCRs, the two latter being the most efficient. Thus, the process with DHP **6b**, *p*-methylaniline **1a** and ethyl glyoxalate **2a** led to a 2:1 diastereomeric mixture of the corresponding adduct **11a** when the reaction was carried out in dry CH_3CN using $InCl_3$ or $Sc(OTf)_3$ in the presence of molecular sieves in respectively 65% and 87% overall yield (Table 1, entries 1 and 2). As expected, the ring fusion was *cis* (arising from the most favourable cyclization route) and the major isomer was the one bearing the ethoxycarbonyl group in the β face.

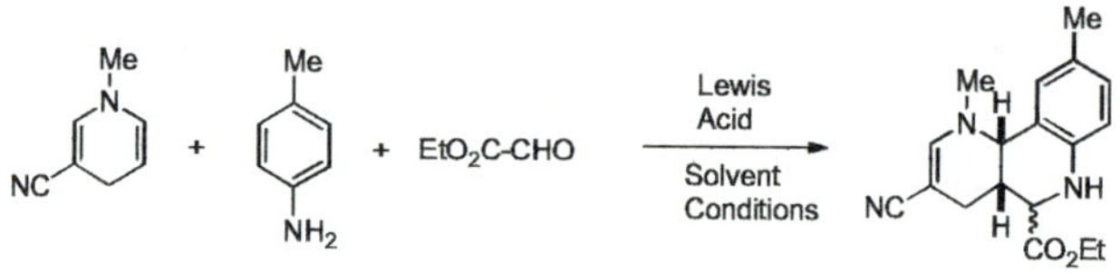

Table 1. Catalyst and reaction conditions screened for Povarov-DHP MCR.

Entry	Solvent	Conditions	Lewis Acid	Yield %	Selectivity (β/α)
1	CH_3CN	r.t., 12 h	$InCl_3$	65	2:1
2	CH_3CN	r.t., 12 h	$Sc(OTf)_3$	87	2:1
3	H_2O/SDS	r.t., 12 h	$InCl_3$	56	1:1
4	H_2O/SDS	r.t., 12 h	$Sc(OTf)_3$	67	1:1
5	CH_3CN	HP, 13 kbar	$Sc(OTf)_3$	97	1:2
6	CH_3CN	MW, 10 W, 5 min	$Sc(OTf)_3$	98	2:1
7	CH_3CN/THF	r.t., 12 h	Yb(III)P.S.	96	2:1

After studying several reaction conditions, it was determined that DCM and THF were not suitable solvents, whereas water in an anionic micellar system (Table 1, entries 3 and 4) yielded the desired compound albeit in a lower yield. High pressure conditions (13 kbar, entry 5) cleanly led to an almost quantitative yield of the adduct with an inversion in the stereoselectivity. Microwave irradiation was also efficient in promoting the MCR (entry 6, Table 1) notably reducing the reaction time to 5 minutes, in this case without detectable changes in stereoselectivity. Another practical improvement involved the use of a solid-supported catalyst (Yb^{3+} associated with a sulfonate polystyrene resin),[42] which allowed efficient recovery and reuse of the catalyst, without significant loss in activity for several cycles.

The scope of the reaction was established by systematically studying modifications affecting the four diversity points (R$_1$–R$_4$)[43] (Scheme 6). Regarding the DHP component, the reaction worked well with different R$_1$ substituents attached to the nitrogen (Me and Bn) and with electron-withdrawing groups linked to the β-position of the DHP ring (R$_2$=CN, CO$_2$Me, COMe, CONH$_2$) which are required for the stability of this input, otherwise the heterocyclic substrate is prone to suffer polymerization or oxidation even under mild conditions.

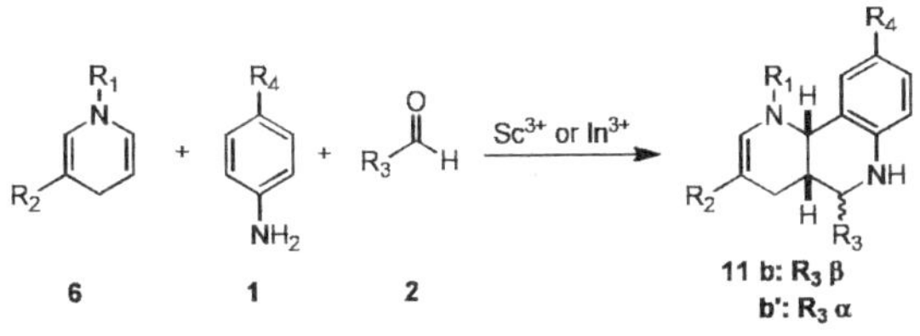

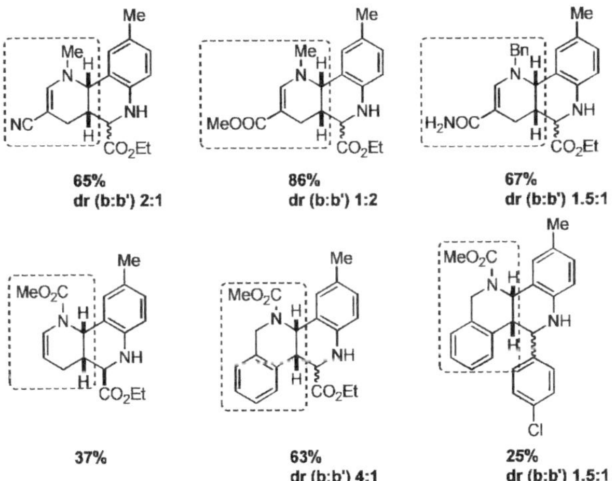

Scheme 6. Povarov-DHP MCR.

However, when the heterocyclic N atom was acylated, no substituents were required and H or fused benzene rings were tolerated (Scheme 7).

Scheme 7. Range of DHPs in Povarov MCRs.

A range of aldehydes (R$_3$) were tested as inputs in the MCR (Scheme 8). Apart from the highly activated ethyl glyoxalate (**2a**), aromatic aldehydes substituted with electron-withdrawing and electron-

31

donating groups yielded the desired products, although the yields tended to be higher with the former. The reactions generally afforded the usual 2:1 diastereomeric mixtures. Heteroaromatic aldehydes also reacted well. Remarkably, isatin afforded the expected spiroadducts, although this particular reaction required prolonged heating and afforded a low overall yield. Formaldehyde and *trans*-cinnamaldehyde failed to react under the usual conditions. Aldehydes displaying α-hydrogens were excluded from these experiments, as it is known that this set of reactants presents problems (due to the enamine-imine isomerization) and requires special conditions to be engaged in Povarov processes.[44]

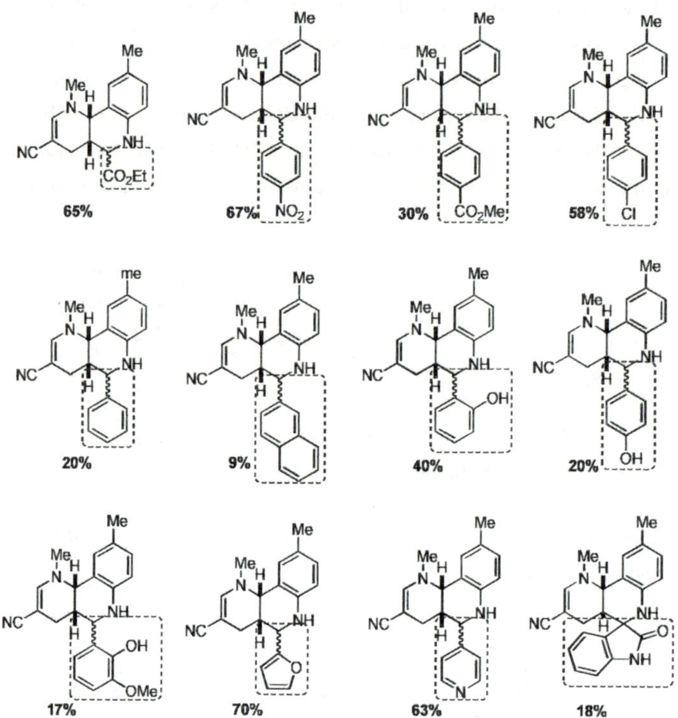

Scheme 8. Variation of the aldehyde component in the Povarov-DHP MCR.

Furthermore, the range of anilines showed a clear indication that the activation level of the substituents was critical for reactivity: mild activating and deactivating groups allowed the reaction, whereas strongly activated or deactivated aromatic rings were not productive. These experimental observations are in good agreement with the expected requirements for the final electrophilic cyclization of the iminium ion on the aniline ring (Scheme 3), where poorly nucleophilic aromatic rings would fail to trap the electrophilic moiety, whereas extremely activated systems would react preferentially with the activated imine, thereby kinetically bypassing the DHP[45] (Scheme 9).

Thus, alkyl groups, halides and activating substituents, such as acetoxy-, hydroxy- and acetamido-groups afforded the desired compounds (Scheme 10). For unknown reasons, aniline itself was not reactive. Deactivating groups such as carboxylate, carboxamide and methoxycarbonyl were tolerated and the higher yields were found in the mid activation range. Nitroanilines and *m*-methoxyaniline failed to yield the Povarov adduct for the above mentioned reasons.

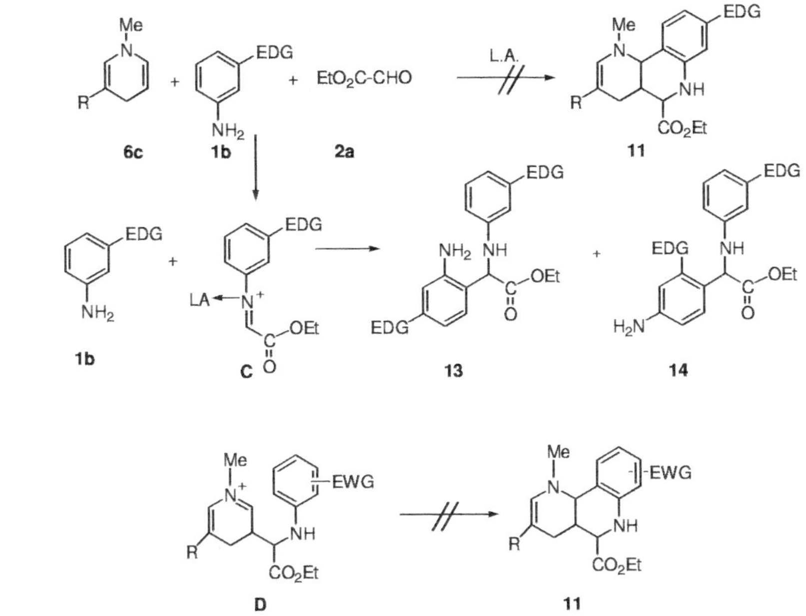

Scheme 9. Side-reactions with activated and deactivated anilines.

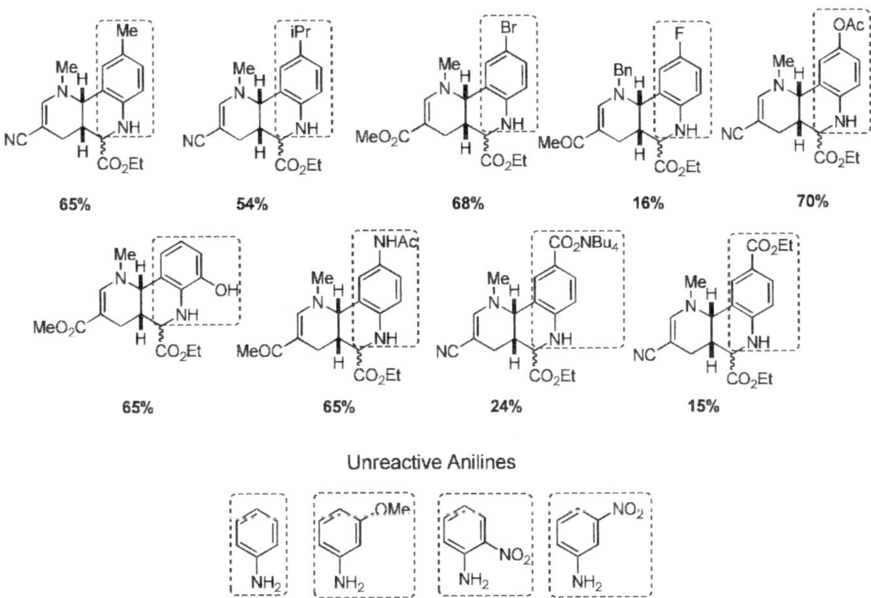

Scheme 10. Variation of the aniline component in the Povarov-DHP MCR.

Solid-phase methods were designed and implemented to facilitate the elaboration of the final adducts (Scheme 11). Thus, following known protocols,[46] the immobilization of the glyoxalate on a polystyrene resin afforded the functionalized polymer **15**. Thus, when resin **15** reacted with the DHP (**6**) and the aniline (**1**), the Povarov adduct attached to the solid support (**16**) was generated. An aqueous basic cleavage cleanly yielded the corresponding carboxylates **17** with high purity in the usual 2:1 diastereomeric ratio.[43] A new diversity point was introduced at this stage by promoting alternative cleavages from distinct nucleophilic species and therefore ester and amide derivatives **11** were also accessible. In a complementary manner, the aniline component was also linked to the solid support, thereby expanding the application range of this methodology to the Povarov-DHP process.[47]

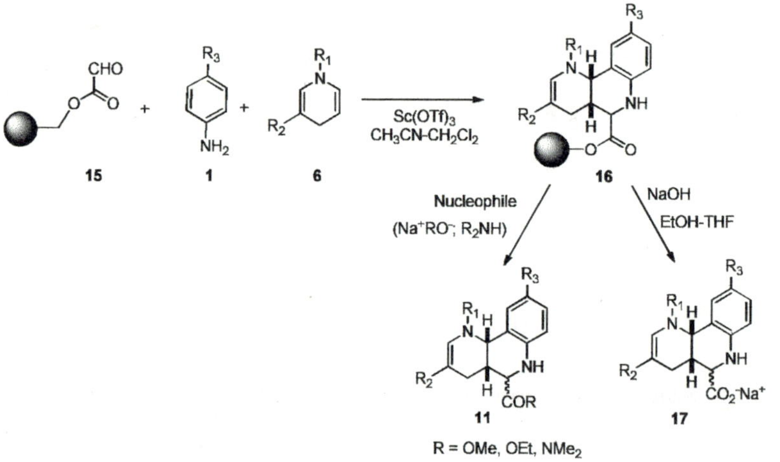

Scheme 11. Solid-phase approach.

Tandem processes were explored to allow the introduction of additional substituents at previously untouched positions of the naphthyridine scaffold. The goal was to link the *in situ* generation of the reactive DHP **6** (accessed from pyridines **5** by nucleophilic addition upon pyridinium salts) with the Povarov MCRs and finish the process with an acylation of the aniline nitrogen of the Povarov adducts.[48,49] By performing all steps in a sequential manner, without isolation of the intermediates, it should be feasible to develop 6-component one-pot protocols displaying high degrees of structural diversity (Scheme 12).

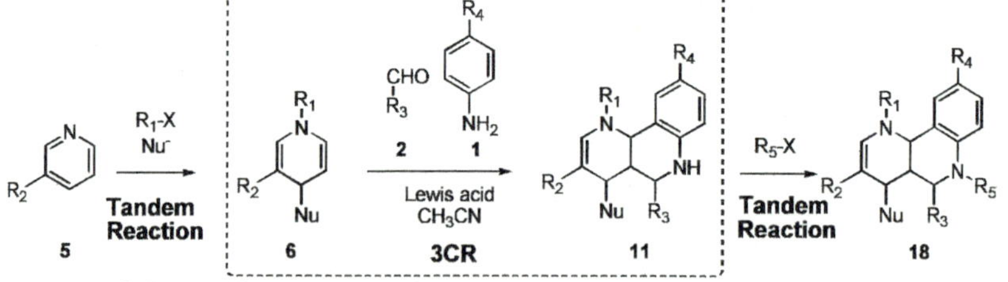

Scheme 12. Tandem processes leading to 6-component protocols.

The key point in this context is to find compatible conditions (*i.e.*, a common solvent) for all the steps in a one-pot sequence. As the Povarov process is, in practice, restricted to the use of CH_3CN, the flanking reactions should ideally be carried out in this solvent. Nucleophilic attack upon the *N*-alkylpyridinium salts in CH_3CN allowed the regioselective formation of the desired γ-substituted DHPs **6**, which were subjected *in* *situ* to the Povarov-MCR, simply by adding the Sc^{3+} catalyst, the aniline and the aldehyde (Scheme 12). Thus, several nucleophiles (cyanide, trimethyl phosphite and aminouracil) were successfully incorporated into the final MCR adduct to yield compounds **18a–c** (Figure 1).

Additional diversity was introduced through acylation-type reactions at the secondary amine site. Directly after the MCR step, two different electrophilic reagents (AcCl and TosCl) were added to the reaction vessel (together with the stoichiometric amounts of a tertiary amine) to yield the desired MCR-adducts **18d** and **18e** in a one-pot procedure[48](Figure 1).

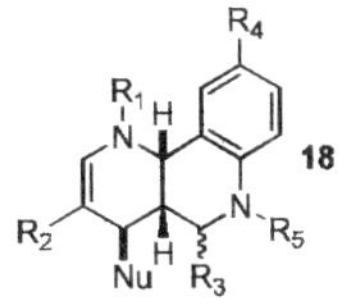

Figure 1. Tandem protocol adducts.

An extension of this methodology involved the generation of the reactive DHP directly from pyridines by the Reissert reaction.[50,51] Thus, treating pyridine or isoquinoline (**19**) with a chloroformate **20** and TMSCN **21** (as a soluble form of cyanide) in CH_3CN, the *N*-acyl DHP **E** was engaged *in situ* with the Povarov reaction, which was followed by the acylation step to afford the 6-component adduct in decent yield, directly from commercially available starting materials (Scheme 13). This methodology therefore allows the rapid assembly of a wide range of azine derivatives, anilines, nucleophiles, aldehydes and acylating agents, which generate up to 6 bonds in a sequential manner, in acceptable yields and with a broad scope of structural and functional diversity.

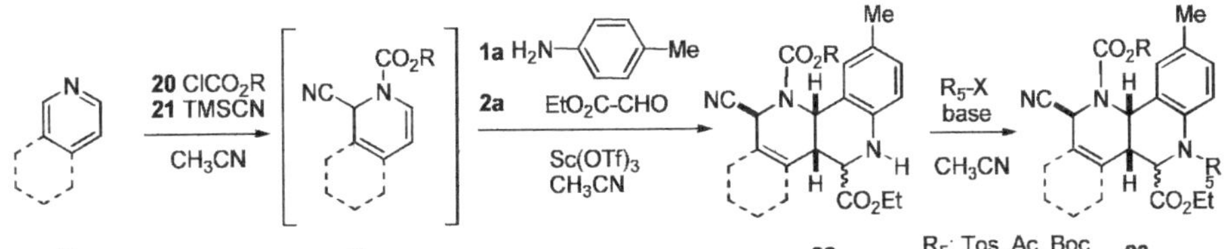

Scheme 13. Reissert-Povarov-acylation sequence.

Taking into account the mechanistic rationale for the Povarov process and the reactivity range of the aniline component, we explored a modified version of the parent reaction. Thus, the initial Mannich step was followed by an electrophilic cyclization from an aromatic ring attached to the heterocyclic nitrogen, which would be in competition with the aniline to trap of the iminium intermediate **F**[38] (Scheme 14).

When the tryptophyl-DHP (**24**), chosen for the presence of a suitable indole nucleophile, was reacted with ethyl glyoxalate (**2a**) and *p*-toluidine (**1a**) under the usual conditions, two products were isolated in equimolecular amounts: the standard Povarov adduct **26a** and the indoloquinolizidine **25a** (Scheme 14). This observation indicates a similar rate for both cyclization pathways. However, when a deactivated aniline was used (methyl *p*-aminobenzoate), the indoloquinolizidine **26b** was selectively formed.[38]

Next, we considered the use of aliphatic amines, naively assuming that these species would act as the imine precursors as well as external trapping agents for the iminium intermediate. Interestingly, when *n*-butylamine **27a** was reacted with DHP **6b** and ethyl glyoxalate **2a**, the bicyclic adduct **29a** was isolated together with the rearranged tetrahydropyridine **28a**, both in low amounts[52] (Scheme 15).

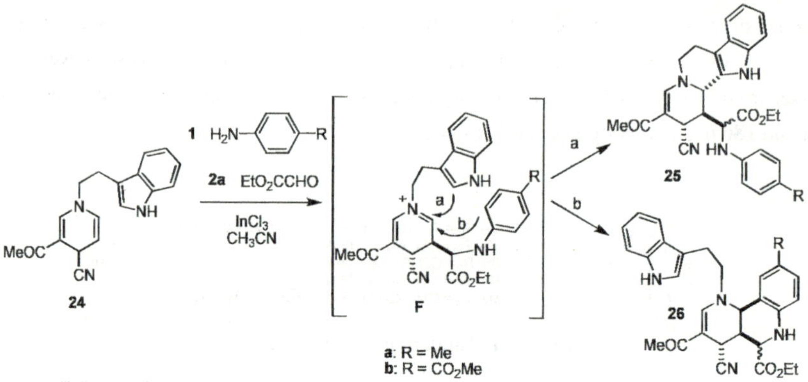

Scheme 14. Alternative cyclization pathways in Povarov-DHP processes.

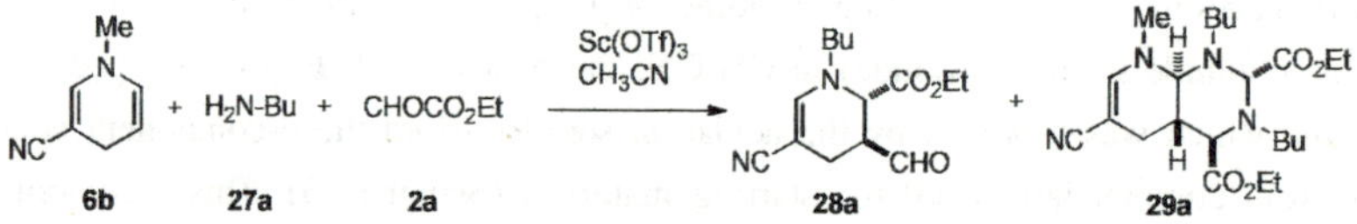

Scheme 15. Aliphatic amines in Povarov-DHP reactions.

After tuning the reaction conditions, it was found that the use of 2.2 equivalents of amine and 1.2 equivalents of ethyl glyoxalate (**2a**), with prolonged heating, resulted in the formation of the aldehyde **28a** (40% yield), whereas bicyclic derivative **29a** (40%) was better obtained in the presence of 4 Å molecular sieves at room temperature, using 2 equivalents of amine and 2 equivalents of ethyl glyoxalate, in shorter reaction times.

A likely mechanistic hypothesis is based on the initial condensation of the amine and the aldehyde moieties to generate the corresponding imines (Scheme 16).

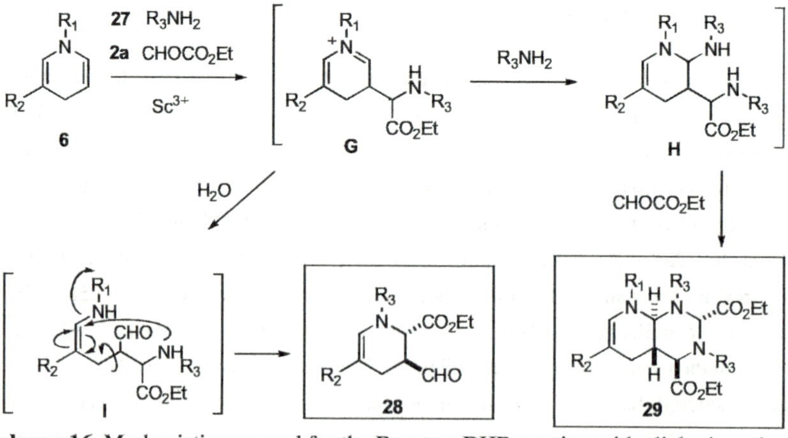

Scheme 16. Mechanistic proposal for the Povarov-DHP reaction with aliphatic amines.

36

Once activated by the Lewis acid, these imines would be attacked by the DHP, to form the iminium ion **G**, which, through a nucleophilic attack by a second equivalent of the amine, would generate a diamino intermediate **H**, which may interact with a second unit of glyoxalate to form the bicyclic compound **29**.[53] Alternatively, in the presence of water, intermediate **A** could suffer hydrolysis leading to aldehyde **I**, which may rearrange through an intramolecular addition-elimination sequence to close the tetrahydropyridine ring **28**. Interestingly, these MCRs display high stereoselectivity, which favours the *trans* isomers (probably because of the intermolecular nature of the iminium ion trappings and a thermodynamic equilibrium leading to the more stable stereochemistries).

2.2. Four-component reactions with enol ethers

The precedent results indicated the feasibility of trapping the iminium ion intermediate through an external nucleophilic *terminator*.[54,55] We decided to explore new aspects of this idea by using cyclic enol ethers. These substrates have been particularly productive in Povarov MCRs. In this respect, special mention is given to the pioneering work of Povarov, Lucchini[56] and Li[57,58] (Scheme 17). In this context, it is necessary to consider that enol ethers, in the presence of appropriate acid catalysts, may also act as surrogates of the aldehyde component as demonstrated by Li.[57] Interestingly, Batey observed this behaviour also with cyclic enamines, which led to the tricyclic core of martinelline and eventually to the total synthesis of this alkaloid.[59–61]

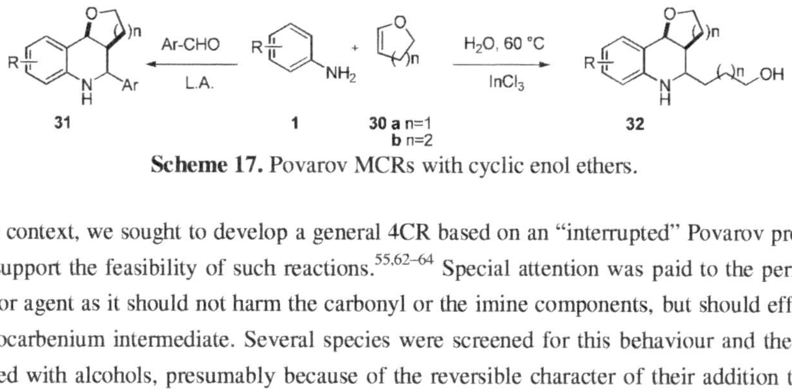

Scheme 17. Povarov MCRs with cyclic enol ethers.

In this context, we sought to develop a general 4CR based on an "interrupted" Povarov process. Some precedents support the feasibility of such reactions.[55,62–64] Special attention was paid to the performance of the terminator agent as it should not harm the carbonyl or the imine components, but should efficiently trap the final oxocarbenium intermediate. Several species were screened for this behaviour and the best results were obtained with alcohols, presumably because of the reversible character of their addition to aldehydes and imines. Thus, preliminary experiments were carried out using equimolar amounts of 3,4-dihydro-2*H*-pyran (**30a**), *m*-nitroaniline (**1c**), ethyl glyoxalate (**2a**) and ethanol as the terminator (**33a**)[65] (Scheme 18).

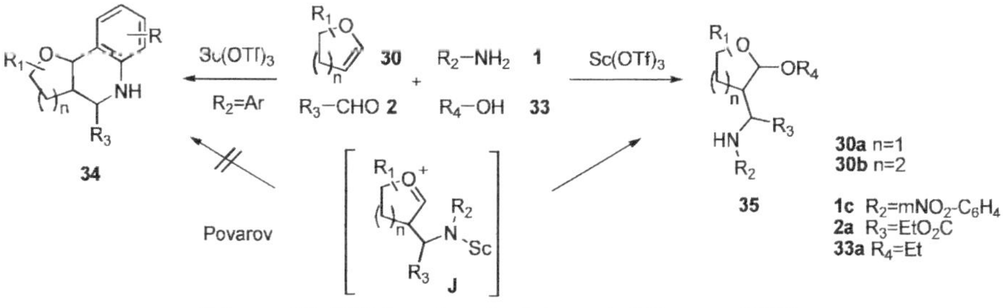

Scheme 18. 4CR based on the interrupted Povarov MCR.

37

The reaction afforded the desired 4CR-adducts as a mixture of diastereoisomers in high yield. In a related process, when glyoxylic acid was used (to act as a carbonyl and terminator agent), the stereochemistry of the bicyclic adduct was conveniently assigned and it was also chemically correlated with the previously obtained 4CR products. Thus, we determined that the stereochemical trends for this process were similar to the normal Povarov reactions and led to the usual mixtures of diastereoisomers (Scheme 19).

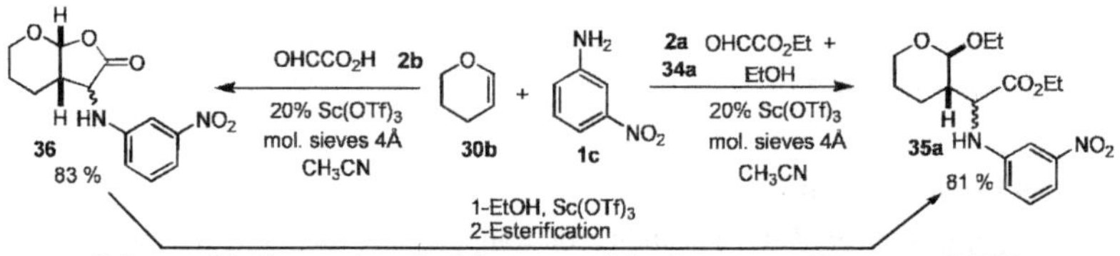

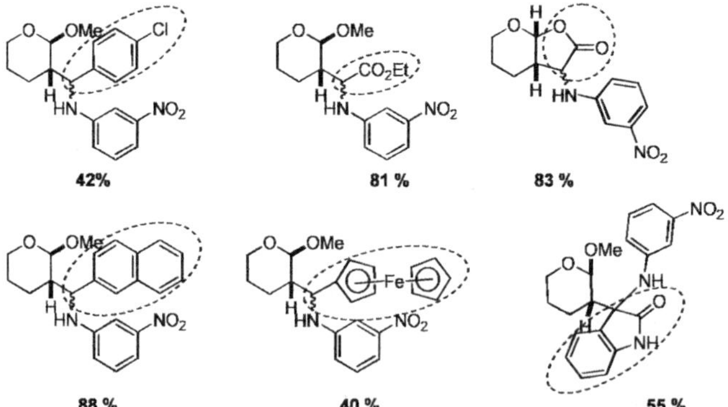

Scheme 19. Stereochemical features of the interrupted Povarov MCR.

Remarkably the reaction has a general character and it is applicable to a wide range of substrates regarding all components. The scope of amines tested (Figure 2) included deactivated and activated anilines as well as aliphatic derivatives.

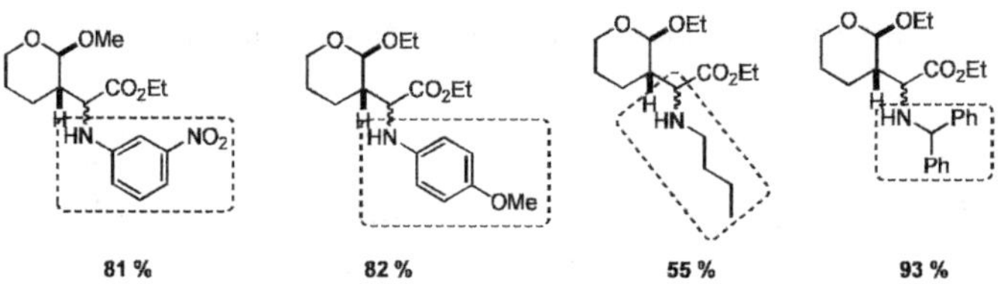

Figure 2. Range of amines in the interrupted Povarov MCR.

A range of carbonyl derivatives was also screened. These compounds followed similar trends to those found in standard Povarov reactions: aromatic aldehydes and activated carbonyls being the most reactive inputs (Figure 3).

Figure 3. Set of carbonyl derivatives in the interrupted Povarov MCR.

In a similar manner, the terminator range was established (Figure 4). Primary alcohols were efficient, whereas secondary derivatives reacted albeit in lower yields. Water afforded a somewhat labile hemiacetal derivative which was stabilized by further interaction with formaldehyde. Other oxygen-based species with reduced nucleophilicity (carboxylic acids, phenols, trifluoroethanol) did not afford the expected 4CR-adducts. In these cases, the cationic intermediate evolved towards the formation of the standard Povarov products. Interestingly ethanethiol performed well as a terminator.

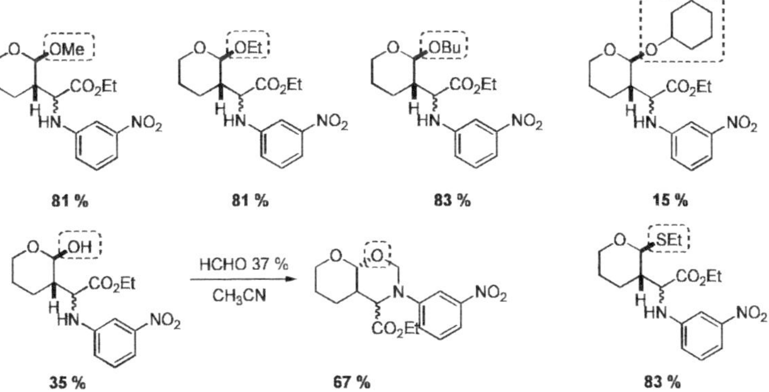

Figure 4. Set of terminators in the interrupted Povarov MCR.

This synthetic transformation accepts a wide range of cyclic enol ethers, ranging from the classic substrates (dihydrofurans and dihydro-2*H*-pyrans) to substituted derivatives and glycals (Figure 5). In the latter case, extremely stereoselective reactions were found, leading to the enantiopure adducts, although under standard conditions the reactions were very slow and low yielding, presumably because of the higher steric hindrance of the substrates. However, microwave activation allowed rapid and efficient transformations (Figure 5).

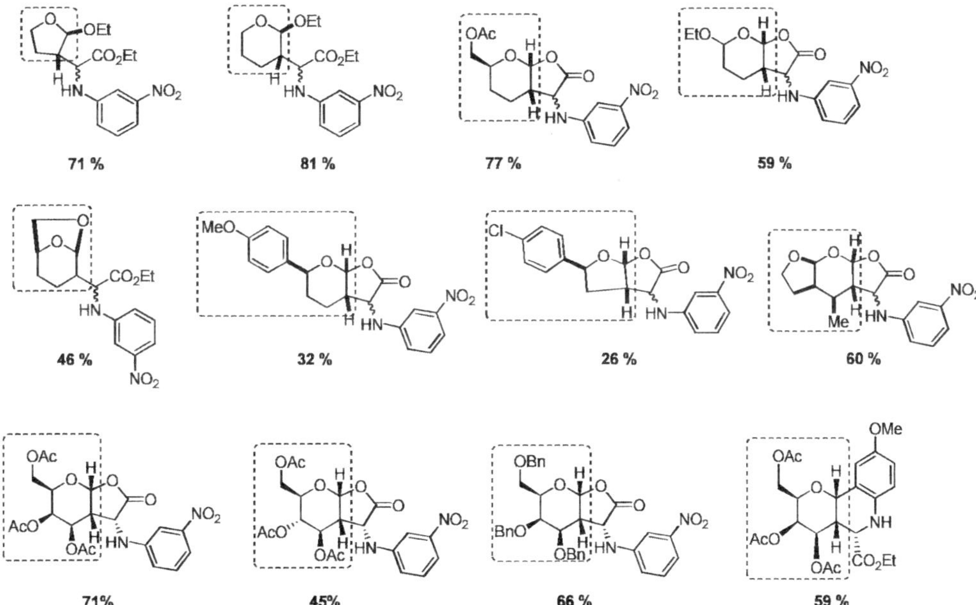

Figure 5. Enol ether range in the interrupted Povarov MCR.

It is expected that the scope of this multicomponent transformation will be widened by recent advances in the field, in particular the incorporation of aliphatic aldehydes containing α-hydrogen atoms.[44,66,67] The development of protocols with high stereoselectivity, like the enantioselective catalytic methodology developed by Zhu[33] and Akiyama,[34] are already milestones in this field. The elegant access to a wide set of enantiopure diamines designed by Zhu using an interrupted Povarov MCR is of note.[68]

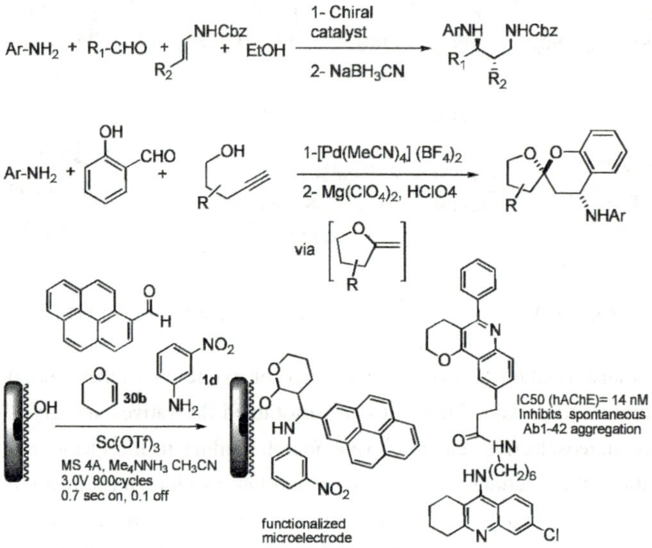

Scheme 20. Recent applications of Povarov-type MCRs.

Recently, Moeller described an impressive application of this MCR to the functionalization of microelectrode arrays. This approach features an innovative and precise timing of the reaction based on the oxidation state of the Sc species, which is controlled electrically.[69] Moreover, the implementation of this chemistry allowed the rapid preparation of a potent and selective acetylcholinesterase inhibitor which also inhibits the β-amyloid aggregation and therefore it has potential therapeutic use as anti-Alzheimer agent.[70] Recently, Barluenga and co-workers reported remarkable examples of Povarov and interrupted Povarov MCRs, where an enol ether with an *exo* double bond is generated *in situ* from a metal-catalyzed intramolecular hydroalkoxylation of alkynols. This process affords complex spiroadducts (acetals and quinolines) in a stereoselective manner[71,72] (Scheme 20).

2.3. Enol esters in Mannich-type multicomponent reactions

The structural similarity of enol esters with the usual substrates of the Povarov MCRs led us to consider their suitability as inputs for these processes. When 3,4-dihydro-6-methyl-2*H*-pyran-2-one (**37**) was treated with *p*-toluidine (**1a**) and ethyl glyoxalate (**2a**) under Sc(OTf)$_3$ catalysis, the major product was not the expected Povarov compound **40**, but the *cis-trans* isomers of the *N*-aryl lactam **38** (Scheme 21).[73] The X-ray analysis of a single crystal of one adduct allowed the assignment of their relative stereochemistry, the *cis* isomer being the major compound. Moreover, it was found that an acid-catalyzed epimerization cleanly led to *trans* stereochemistry (**39**) (Scheme 21).

The introduction of the carbonyl moiety in the electron-rich olefin triggered a dramatic change in the reaction mechanism. The usual Mannich process generated the highly reactive intermediate **K**, ready to undergo a facilitated lactamization, rather than the usual aromatic electrophilic substitution, which would lead to the fused tetrahydroquinoline adduct **40**, not detected in the reaction (Scheme 21).

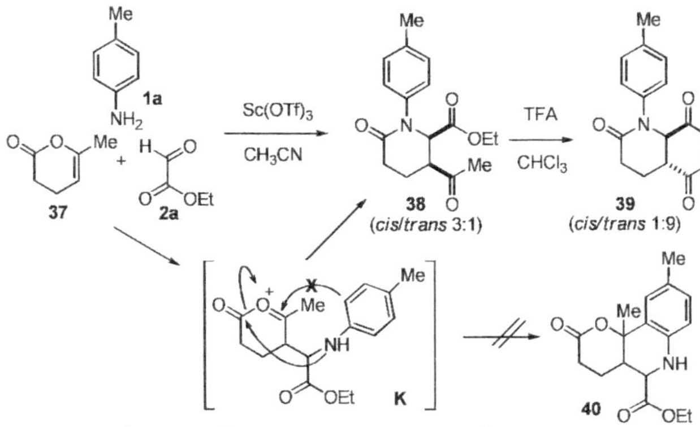

Scheme 21. Povarov reactions with enol esters.

The scope of the reaction was then studied and it was found that a wide range of anilines allowed the process; however the carbonyl component was restricted to ethyl glyoxalate, although aromatic aldehydes were tolerated in some entries. The range of enol esters was studied and 5- and 6-membered unsaturated lactones gave the corresponding *N*-arylated lactams, whereas isopropenyl acetate (**43**) gave rise to the corresponding open chain-type adduct **44** (Scheme 22). Vinyl acetate (**41**), however, afforded quinolines **42**, probably as a result of spontaneous AcOH elimination and oxidation to the fully aromatic nucleus from Povarov adduct (**L**)[73] (Scheme 22).

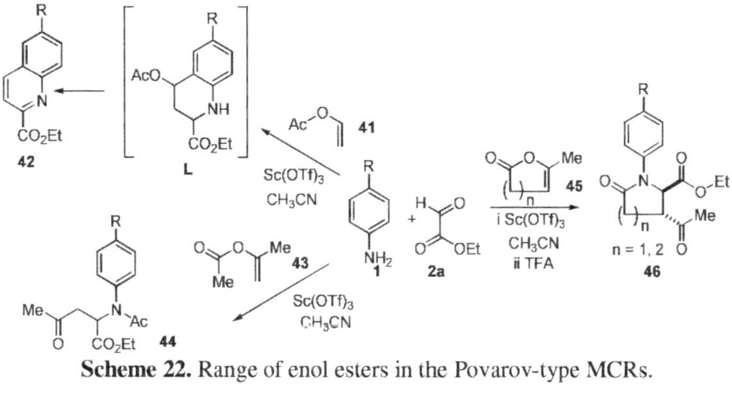

Scheme 22. Range of enol esters in the Povarov-type MCRs.

The use of cyclic enamides as electron-rich olefins has been fruitfully applied in synthesis.[61,74] However, in all previous examples, the carbonyl moiety was exocyclic and therefore the logical step was to determine whether an endocyclic carbonyl influences the reaction pathway, as seen in the oxygen analogs. Thus, several 5- and 6-membered ring enamides displaying *endo* or *exo* carbon-carbon double bonds with distinct substitution patterns were evaluated under standard Povarov reaction conditions. All these enamides were satisfactorily engaged in the MCRs and gave rise to the corresponding adducts, which displayed a remarkable degree of structural diversity[75] (Scheme 23).

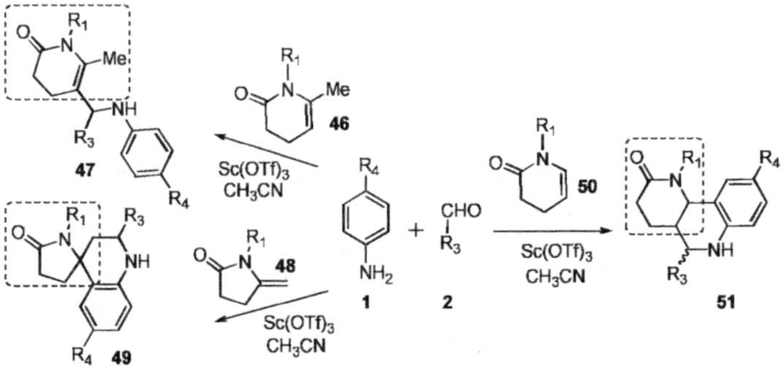

Scheme 23. Povarov MCRs with cyclic enamides.

3. Isocyanide-based multicomponent reactions

Isocyanides are the most frequently used functional group in MCRs because of their synthetic versatility, their bond-forming capacity and their availability. The Passerini and Ugi reactions are the most emblematic MCRs and hold a privileged position in medicinal and combinatorial chemistry with regards to their usefulness in the preparation of bioactive compounds and drugs.[3,9,76] These processes share common mechanistic profiles, where the key step is the isocyanide attack on an oxocarbenium (Passerini) or iminium (Ugi) ion intermediate (Scheme 24). Following our research program, we devised heterocyclic versions of these reactions, where the activated cationic species were prepared from fundamental heterocycles.

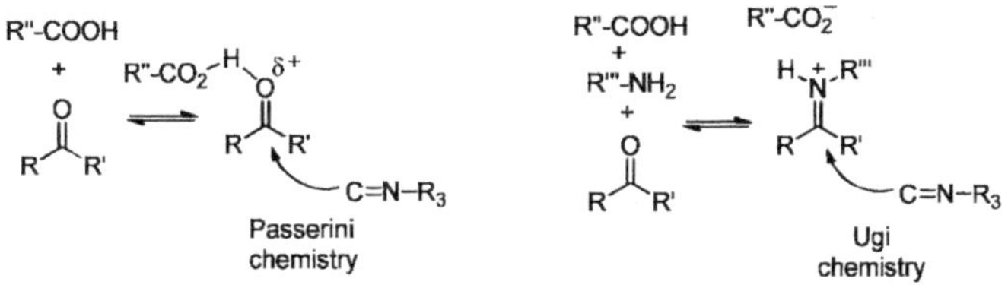

Scheme 24. Cationic intermediates in Passerini and Ugi MCRs.

3.1. Hydro-, halo- and seleno-carbamoylation of cyclic enol ethers and dihydropyridines

Cyclic enol ethers and DHPs were identified as suitable precursors of the cationic intermediates, as the latter could be generated by interaction of electrophilic species and, in this manner, achieve the heterocyclic Ugi-Passerini MCRs (Scheme 25).

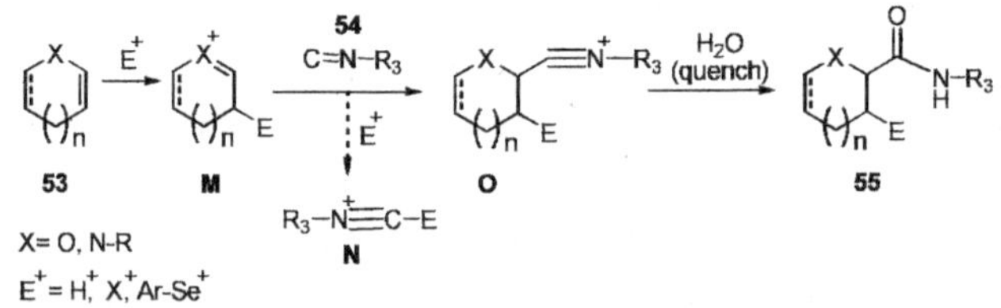

Scheme 25. Heterocyclic versions of Passerini and Ugi MCRs.

A crucial point in this approach is the competition of the substrate with the isocyanide for the capture of the initial electrophile, thereby leading to species **N** in an unproductive manner.

Initial studies on these processes started with a proton as the simplest electrophile. The use of aqueous acids was not efficient and the formamides arising from the acid-catalyzed hydrolysis of the isocyanides were obtained as the major compounds. However, the hydrocarbamoylation of cyclic enol ethers and DHPs can be conveniently carried out using an anhydrous acid source (*p*-toluenesulphonic acid) at low temperature.[77] The heterocyclic substrates reacted with the isocyanides by addition of a stoichiometric amount of acid and α-carboxamido derivatives were obtained in good yields after the final aqueous quench (Scheme 26). The scope of the process tolerated a reasonable range of cyclic enol ethers, DHPs and isocyanides. Interestingly, the intermediate nitrilium ion **O** seems to be stabilized by the sulfonate anion in the reaction medium, to be subsequently hydrolyzed during the work-up stage (Scheme 25).

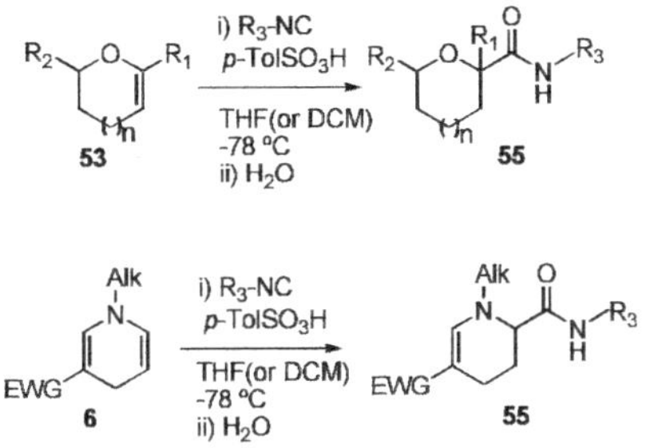

55a n=1, R_1/R_2=H/H, R_3=*tert*-Bu 71%
55b n=1, R_1/R_2=H/H, R_3=C_6H_{11} 65%
55c n=1, R_1/R_2=H/CH_2OAc, R_3=*tert*-Bu 56%
55d n=0, R_1/R_2=H/H, R_3=*tert*-Bu 65%
55e n=0, R_1/R_2=H/H, R_3=C_6H_{11} 60%
55f n=0, R_1/R_2=Me/H, R_3=*tert*-Bu 70%

55g Alk=Me, EWG=CN, R_3=*tert*-Bu 97%
55h Alk=Me, EWG=CO_2Me, R_3=C_6H_{11} 89%
55i Alk=Me, EWG=CO_2Me, R_3=Bn 75%
55j Alk=Bn, EWG=COMe, R_3=*tert*-Bu 70%
55k Alk=Bn, EWG=COMe, R_3=C_6H_{11} 85%
55l Alk=Bn, EWG=$CONH_2$, R_3=C_6H_{11} 64%

Scheme 26. Hydro-carbamoylation of cyclic enol ethers and DHPs.

Next we studied the participation of electrophilic selenium species to perform the analogous seleno-carbamoylations.[78] When the substituted cyclic enol ethers and DHPs were subjected to reaction with isocyanide and phenylselenium chloride, the corresponding seleno-carbamoyl adducts **55m–o** and **55p–q** were formed chemoselectively, the *trans* stereoisomers being the exclusive or major compounds (Scheme 27).

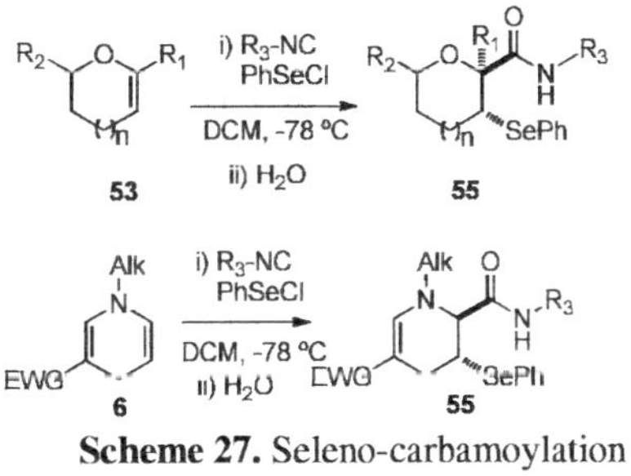

55m n=1, R_1/R_2=H/H, R_3=C_6H_{11} 25%
55n n=0, R_1/R_2=H/H, R_3=C_6H_{11} 37%
55o n=0, R_1/R_2=Me/H, R_3=C_6H_{11} 11%

55p Alk=Me, EWG=CN, R_3=C_6H_{11} 44%
55q Alk=Me, EWG=CO_2Me, R_3=C_6H_{11} 41%

Scheme 27. Seleno-carbamoylation of cyclic enol ethers and DHPs.

Similarly, the halo-carbamoylation of these heterocyclic substrates was explored. The reactions were carried out with the isocyanide using Br_2 or ICl as the halonium source, followed by a final water quench. The cyclic enol ethers yielded the desired β-halo-α-carbamoyl derivatives **55r** and **55s** (Scheme 28). However, these adducts proved somewhat unstable at the purification stage, thus leading to low isolated yields, although the overall transformation was quite satisfactory. As expected, the process was stereoselective, affording the *trans* stereoisomers. With respect to the DHPs, the reaction was also successful and a wide range of α-carbamoylated-β-halosubstituted tetrahydropyridines was generated in good to excellent yields after the final treatment with H_2O. Three observations deserve special mention. These MCRs efficiently by-passed dihaloimine formation, what would be the result of the unwanted interaction of the halogen with the isocyanide (a well documented process which often takes place violently).[79,80] Also, the nitrilium intermediates are temporarily trapped by the halide anion present in the reaction medium and afterwards the imidoyl halide is hydrolyzed to the corresponding amide. Finally, although halogens rapidly oxidize DHPs to the corresponding pyridinium salts, the electrophilic interaction to yield the β-haloiminium ion (**M** in Scheme 25) and its subsequent trapping by the isocyanide must take place at faster rates and the non-biomimetic oxidation of DHPs can proceed.[81]

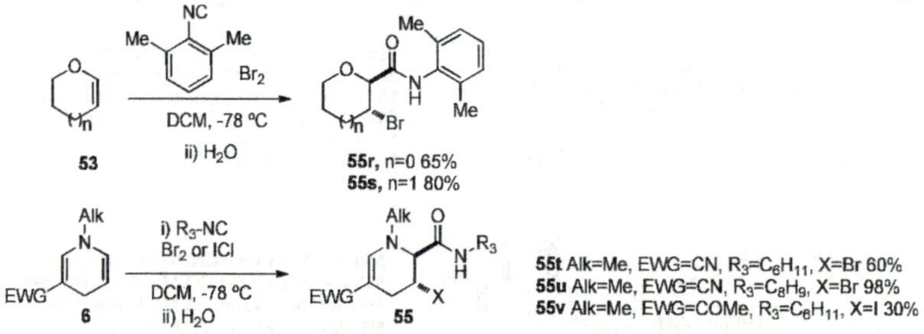

Scheme 28. Halo-carbamoylation of cyclic enol ethers and DHPs.

Some post-condensation transformations on the halo- and seleno-MCR adducts were therefore investigated (Scheme 29). The tetrahydropyridine-bromoamide **55t** was treated with NaH to afford the corresponding elimination product, the α-carbamoylated 1,4-DHP **56**, while the cyclic ether analog **55r** selectively yielded the β-lactam **57** under the same reaction conditions. Moreover, the oxidative elimination of the seleno-derivatives **55m** and **55o** was tested and, under standard conditions (*in situ* conversion to the selenoxides with H_2O_2), the corresponding conjugated and non-conjugated unsaturated derivatives **58** and **59** were produced in good yields. These results showed the feasibility of engaging the newly described MCRs in short synthetic sequences with preparative purposes.

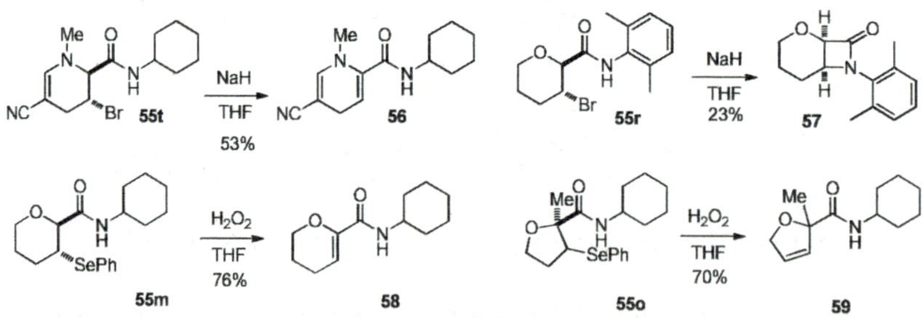

Scheme 29. Post-condensation transformations on the halo- and seleno-MCR adducts.

3.2. Benzimidazolium salts

As previously described, the interaction between DHPs **6** and cyclohexylisocyanide (**54a**) in the presence of bromine afforded the β-bromo-α-carbamoylated tetrahydropyridines **55** (Scheme 30). However, when the same reaction was carried out using iodine, the major isolated compounds were, unexpectedly, the benzimidazolium salts **60**.[82]

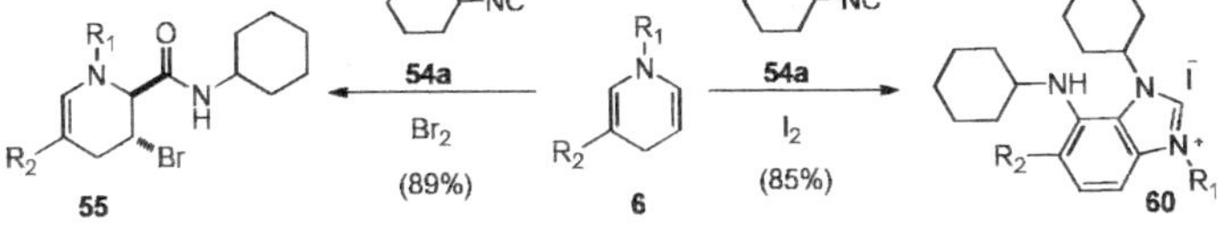

Scheme 30. Halogen interaction with isocyanides and DHPs.

The scope of this unprecedented transformation was studied through a systematic variation of the substitutents at the DHP and the isocyanide components (Scheme 31).

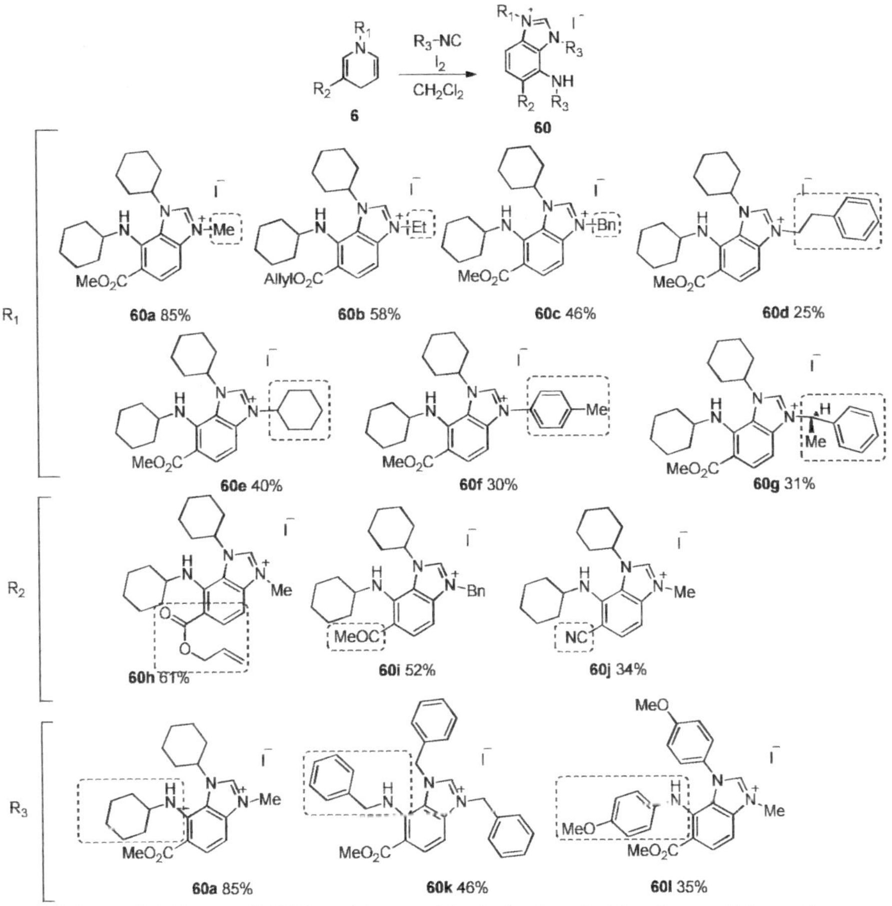

Scheme 31. Range of DHPs and isocyanides in the benzimidazolium salt formation.

A broad range of substitutions at the nitrogen of the DHP was well tolerated, including linear alkyl-, cycloalkyl, benzyl, aromatic and homochiral arylalkyl groups. Again β-substitution at the heterocycle was restricted to some electron-withdrawing groups (to guarantee the stability of the DHP), such as alkoxycarbonyl, cyano and acetyl groups; whereas DHPs bearing formyl and carboxamido groups did not undergo the transformation to the benzimidazolium salts. The first probably suffered oxidation to the carboxylic acid stage under the reaction conditions and the latter facilitated the biomimetic oxidation of the DHP ring to the pyridinium salt. Cyclohexyl-, benzyl- and aromatic isocyanides afforded the desired products, whereas *tert*-butylisocyanide, TOSMIC and isocyano acetates were not productive. Overall, the yields obtained ranged from practically quantitative to moderate or low in some reactant combinations.

Structural assignment was performed with the help of mono- and bi-dimensional NMR experiments and diagnostic information came from the use of labeled precursors (deuterated DHPs and ^{13}C-labeled isocyanides), which also afforded valuable data for the mechanistic proposal (Scheme 32). Several additional experiments also shed some light in this respect. For instance, the corresponding pyridinium salts (arising from the biomimetic oxidation of the DHPs), although isolated in small amounts in these reactions, are not intermediates *en route* to the benzimidazolium adducts, as determined in independent reactions. Second, the analogous reactions using ICl, as a alternative iodonium source, cleanly yielded the β-iodo-α-carbamoylated tetrahydropyridine. However, using bromine, although the halo-carbamoylation was the predominant process, traces of the corresponding benzimidazolium bromide were detected. Thus, the working hypothesis starts with the interaction of iodine with the reactive double bond of DHP (**6**) to generate the iodoiminium ion **P**, as previously described in the literature.[53,81] An isocyanide nucleophilic attack upon this intermediate would generate a nitrilium ion (**Q**), which then may undergo a new isocyanide addition to give raise to cation **R**,[83–86] which can be intramolecularly trapped by the enamine system to form a bridged intermediate (**S**). This species may experience iodide-promoted fragmentation (**T**) followed by ring closure and aromatization to access the final product **60**. Iodide anions should play a critical role in the reaction. They may stabilize the iminium and nitrilium intermediates in a different way than other halides do, thereby allowing a convenient rate for their formation without blocking their reactivity, thus promoting a double isocyanide incorporation followed by the complex rearrangement. At the same time, they may also participate in the ring opening of the byciclic intermediate, which leads to an imidoyl iodide ready to close the imidazolium ring. Although hypothetical, this proposal describes a unique domino sequence that justifies the chemical reactivity and the connectivity pattern observed in the process.

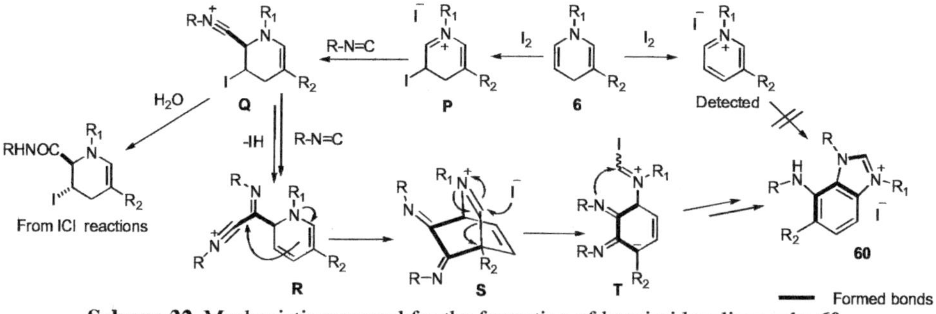

Scheme 32. Mechanistic proposal for the formation of benzimidazolium salts **60**.

The mechanistic complexity does not preclude the general usefulness of this reaction in the synthesis of a wide range of benzimidazolium derivatives that have this substitution pattern, which would otherwise be difficult to access by classical routes from diaminobenzene derivatives. Incidentally, during the course of a screening campaign for small molecule, non-peptidic inhibitors of the human Prolyl Oligopeptidase (an enzyme expressed in the central nervous system, with important effects on behaviour and mood), it was found that one of the previously prepared salts (**60i**) displayed selective and potent activities at the micromolar level. In addition, this compound is blood-brain barrier permeable, a requisite for bioactivity in the central nervous system[87] (Scheme 33).

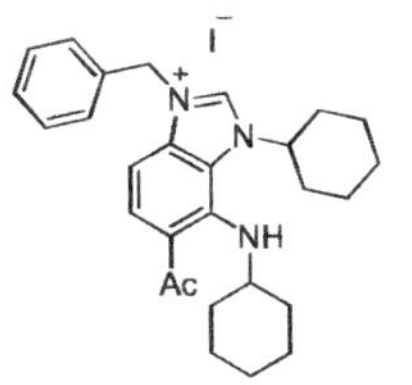

Scheme 33. Prolyl Oligopeptidase inhibition by benzimidazolium salt **60i**.

3.3. Isocyanide addition to pyridinium salts

Notably, the simple addition of isocyanides to *N*-alkylpyridinium salts is not described in the literature. In order to expand our knowledge on the use of isocyanides in heterocyclic MCRs and taking into consideration that pyridinium salts contain a reactive iminium-type moiety, we addressed these potentially useful reactions with the aim to achieve the direct carbamoylation of the heterocyclic systems[88] (Scheme 34).

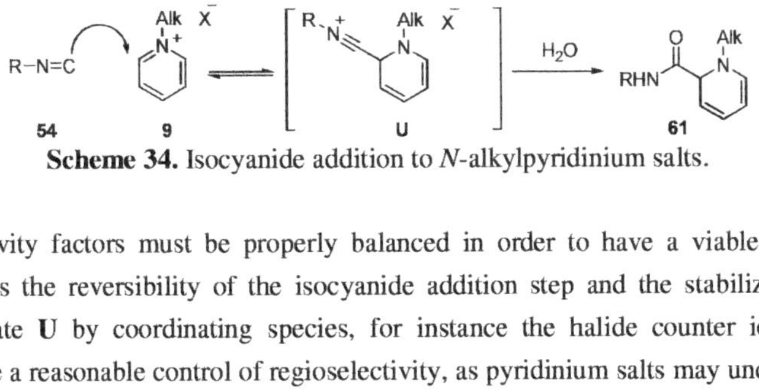

Scheme 34. Isocyanide addition to *N*-alkylpyridinium salts.

Several reactivity factors must be properly balanced in order to have a viable process. The most important is perhaps the reversibility of the isocyanide addition step and the stabilization of this DHP-nitrilium intermediate **U** by coordinating species, for instance the halide counter ion. Moreover, it is important to achieve a reasonable control of regioselectivity, as pyridinium salts may undergo addition at the α and γ positions. Initial results showed a lack of reactivity under standard conditions and several β-substituted pyridinium salts did not yield the desired adduct, as shown by the recovery of unchanged starting materials. This result may indicate that the reversibility of the addition step favours the stable reactants because of a deficient stabilization of the nitrilium ion. In our search for a practical solution to this problem, we were inspired by the work of Zhu, who has developed a general method based on the intramolecular participation of a carboxamido group (linked to the isocyanide moiety) to trap the nitrilium intermediate. This step leads to the formation of an oxazole ring, thereby resulting in the implementation of a family of reactions with extremely high synthetic versatility.[89–92] Thus, nicotinamide salts **9a**, bearing a carboxamido group suitably located near the cationic centre, were chosen as substrates for these transformations (Scheme 35). According to our expectations, the *N*-benzyl salt **9a** was treated with isocyanides in MeOH with 2 equivalents of AcONa to yield the γ-carbamoylated-β-cyano-1,4-dihydro-pyridine **62** through the initial formation of a bis-iminofuran-type adduct (**V**), which underwent an acid-base catalyzed isomerization to yield the β-cyano-γ-carboxamido-DHP[88,93] (Scheme 35). The process took place with high regioselectivity, favouring the generation of the more stable 1,4-DHPs. The isomeric 1,2-derivatives were not detected in these cases. Several nicotinamide salts and isocyanides were tested and provided successful results. Of note, the *umpolung* reactivity of the isocyanide which acts as the nucleophile in this process, ends up as a carboxamido group directly linked to the heterocycle.

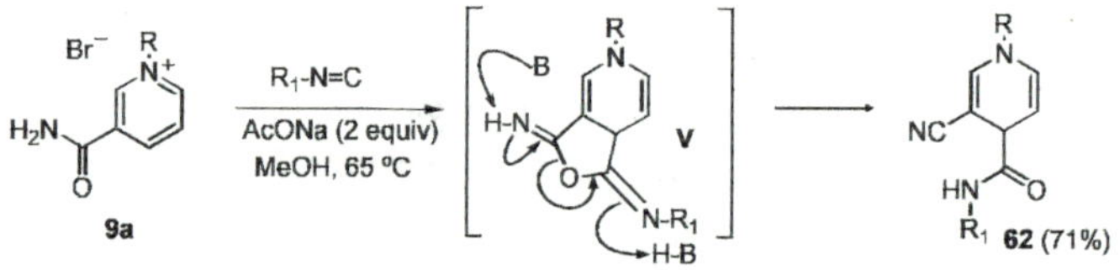

Scheme 35. Isocyanide additions to nicotinamide-derived pyridinium salts.

3.4. Reissert-Ugi reactions

Along these lines, we explored the reactivity of *in situ* generated *N*-acylazinium ions towards the isocyanide component. Thus, we promoted the interaction of an azine (such as isoquinoline) and an acylating agent to form the *N*-acylazinium salt[94] and evaluated its reactivity as a new source of iminium ions in what could be considered a Reissert-Ugi process (Scheme 36).

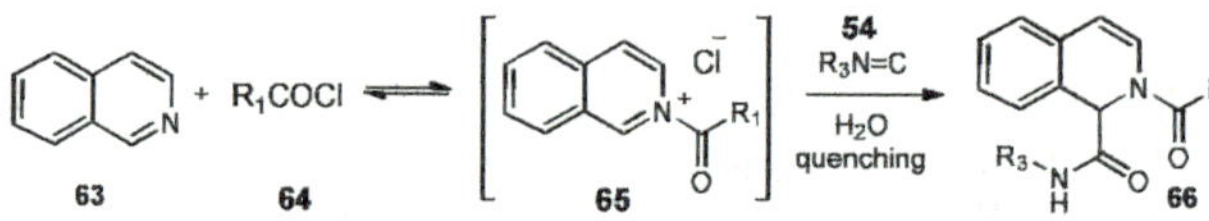

Scheme 36. Reissert-Ugi MCR.

The reaction of isoquinoline (**63**), methyl chloroformate **64a** and *tert*-butylisocyanide successfully yielded the α-carbamoylated dihydroisoquinoline **66a** (Scheme 37). Optimization of the process, using a stoichiometric amount of isocyanide and running the reaction in dichloromethane, increased the yield up to 90%.[95] The scope of the reaction was therefore explored and the three components were screened to establish the practical applicability of the process. Several isocyanides gave successful results: *tert*-butyl, cyclohexyl-, benzyl- and tosylmethyl-isocyanide (Scheme 37). Particular attention was given to the reactivity of the heteroaromatic substrate. Substitution of isoquinolines was allowed and, while quinolines and phenanthridines were also reactive; acridine and 2-methylquinoline, lacking reactive sites at the α-position, were recovered unchanged. Pyridine, together with diversely substituted derivatives, was inert under the usual conditions, in analogy with the classical Reissert process, where its reactivity requires special activation.[96] With respect to the acylating agent, a variety of chloroformates were found to be productive (benzyl-, phenyl- and allyl-chloroformate); Boc$_2$O reacted with isoquinoline and *tert*-butyl-isocyanide in decent yield, although requiring longer reaction time. Also, benzoylchloride and tosylchloride were successfully engaged in these MCRs, albeit affording lower yields of the corresponding adducts.

Successful applications of solid-phase Reissert processes have been reported.[97,98] In an attempt to improve the practical performance of our transformation, we performed the Reissert-Ugi MCR on a solid support (Scheme 38). Thus, the *N*-acylazinium salt was generated by adding the isoquinoline to a chloroformate resin (**64b**) and the process took place in the presence of the isocyanide, followed by the standard water termination. The MCR adduct (**66n**) remained attached to the polymer and oxidative cleavage yielded the carbamoylated quinoline **67** in high yield and purity. Also the union-concept (linking two or more MCRs in order to gain structural diversity)[3,16] was applied to the Reissert-Ugi adducts and they were used as substrates in the Povarov-DHP reaction, as they display an enamine moiety. Thus, the complex polyheterocyclic system **68** was conveniently prepared as the usual mixture of diastereomers.

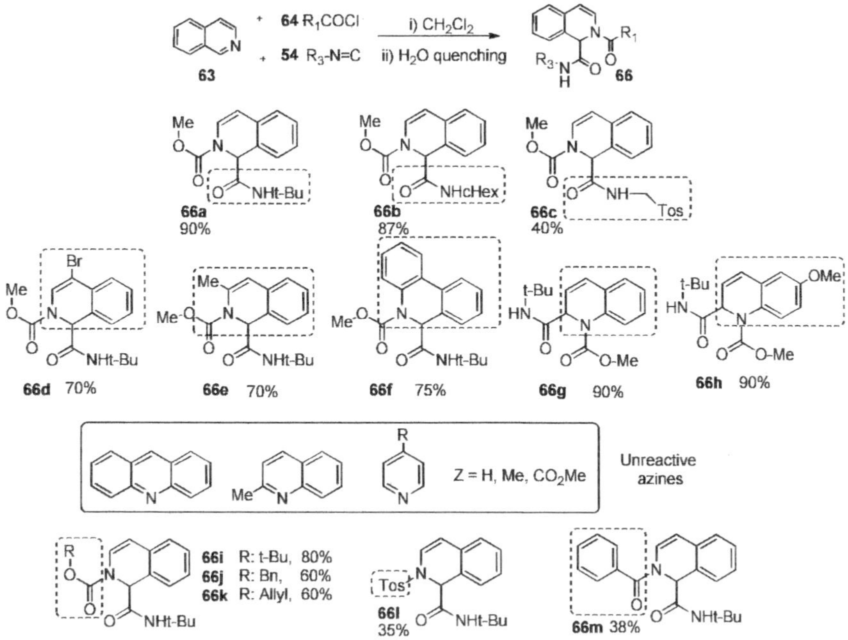

Scheme 37. Range of isocyanides, azines and acylating agents in the Reissert-Ugi MCR.

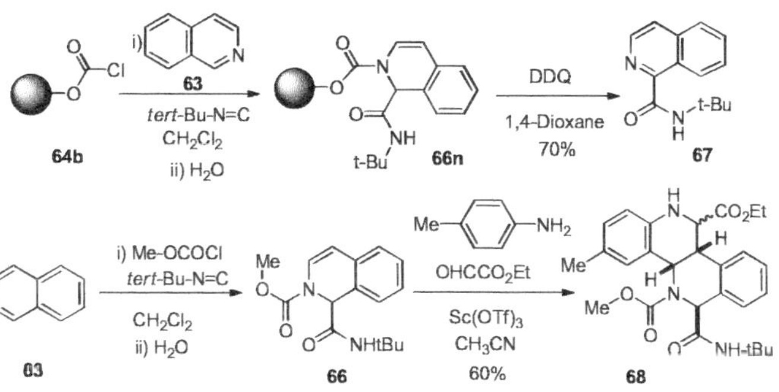

Scheme 38. Solid-phase Reissert-Ugi MCR and union with the Povarov MCR.

Looking for reactivity in the pyridine series, we sought the beneficial effect of the carboxamido group in the stabilization of the nitrilium intermediate. When nicotinamide (**5a**) was reacted with methyl chloroformate and cyclohexylisocyanide, good yields of the γ-addition adduct (**6d**) were obtained. The process took place through the previously described mechanism leading to the cyano derivatives. The reaction upon isoquinolines occurred with an inverted regioselectivity with respect to the Reissert-Ugi MCR

49

on pyridines, leading to α-substituted adducts. The pyridine adducts **6f** were transformed in subsequent reactions into *N*-H DHPs and pyridines[88] (Scheme 39).

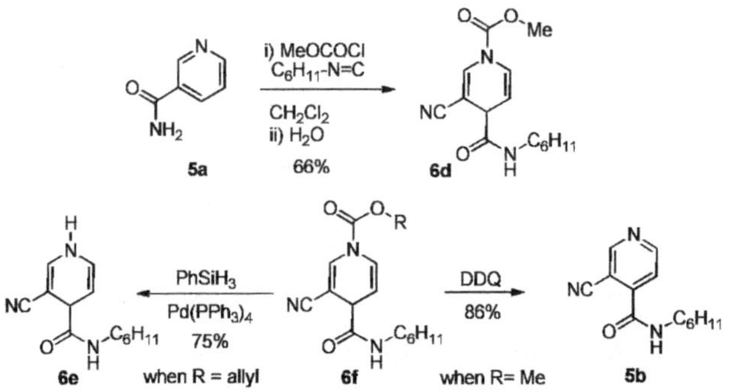

Scheme 39. Reissert-Ugi MCRs with nicotinamide and post-condensation transformations.

In our search for a direct functionalization of pyridines in isocyanide-MCRs, a challenge that has eluded a general solution and, taking into account the relevance of this heterocycle in organic and medicinal chemistry, we studied new approaches involving more aggressive activating agents. Inspired by a report by Corey,[99] triflic anhydride was tested in Reissert-Ugi processes (Scheme 40). Although the expected connectivity patterns were established, the corresponding products were isolated in reduced yields, with low regioselectivity and the oxidation compounds were also detected. However, the main problem was the massive decomposition (probably *via* polymerization) of the isocyanide in the presence of the triflic anhydride. We then turned our attention towards milder activating agents. Thus, trifluoroacetic anhydride (TFAA) was used in the MCR with pyridine and cyclohexylisocyanide. To our great surprise, pyridine was reactive but, instead of leading to Reissert-Ugi-type products, the process afforded the stable acid fluoride dipole **72a**.[100] An analogous reaction with isoquinoline was also productive in higher yields and the structural assignment of the isolated adduct **73a** was performed by spectroscopic means and X-ray analysis.

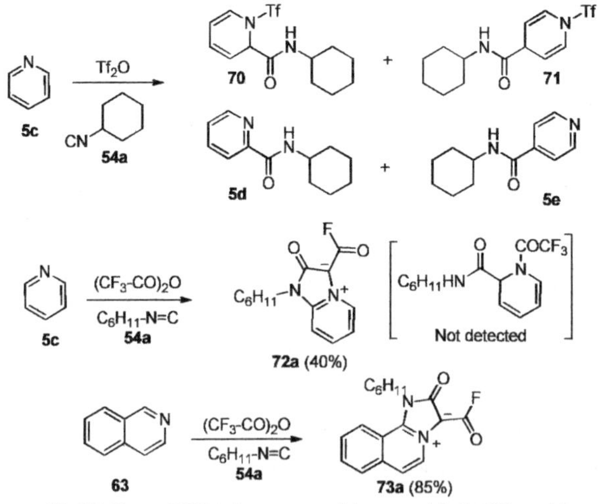

Scheme 40. Tf$_2$O- and TFAA-promoted isocyanide MCRs with azines.

50

The scope of the reaction was then explored and the reactivity range on the three components was analyzed (Scheme 41). A wide range of isocyanides was tested under the conditions previously described. The MCR was successful with alkyl, benzyl and aromatic isocyanides, with few exceptions (*tert*-butyl and TOSMIC). However, a competitive reaction pathway was noted in the case of electron-rich isocyanides and Arndtsen-type dipoles were formed[101] (Scheme 42). With respect to the azine component, isoquinolines, pyridine and phthalazine afforded the desired acid fluoride adducts. However, methyl isonicotinate, 4-methoxypyridine and pyridazines formed the corresponding Arndtsen dipoles. Interestingly, structural variations affecting the fluorinated anhydride were tolerated and it was determined that the minimum requirement for this input was the presence of two fluorine atoms at the carbonyl α-position. Trichloroacetic anhydride was also reactive, but the process was more complex.[100] In contrast, the dichloro derivative did not afford dipole-type adducts.

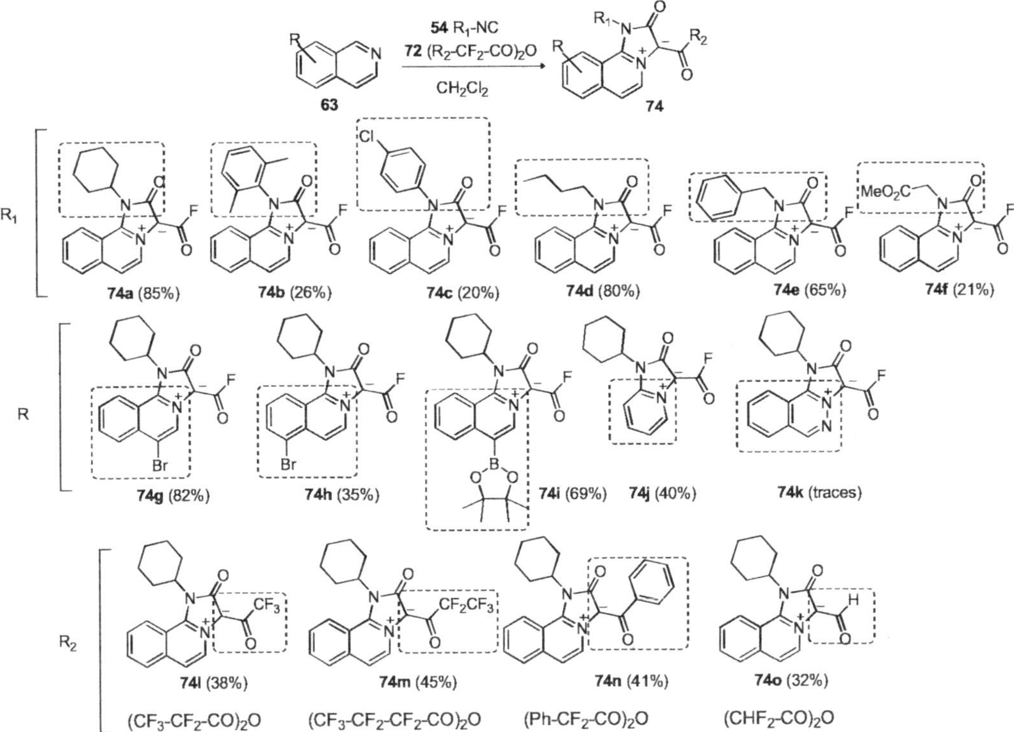

Scheme 41. Range of isocyanides, azines and anhydrides in the TFAA-Isocyanide-Azine MCR.

The mechanistic proposal for this complex reaction is shown in Scheme 42. El Kaïm described the activation of isocyanides by halogenated anhydrides,[102,103] a process involving the formation of intermediate **W**. This species may undergo nucleophilic attack by the isoquinoline component to generate adduct **X**, which may evolve by intramolecular addition of the imine moiety to the α-position of the azinium ion and fluoride loss, thereby resulting in the formation of the difluoroepoxide **Y**. This neutral intermediate would expel a fluoride anion and rearrange to form the stable acyl fluoride dipole **74**. Alternatively, the

TFAA-activated azines (**Z**) may suffer the isocyanide attack on the carbonyl group to yield species analogous to **X** and then follow the domino sequence up to the end.

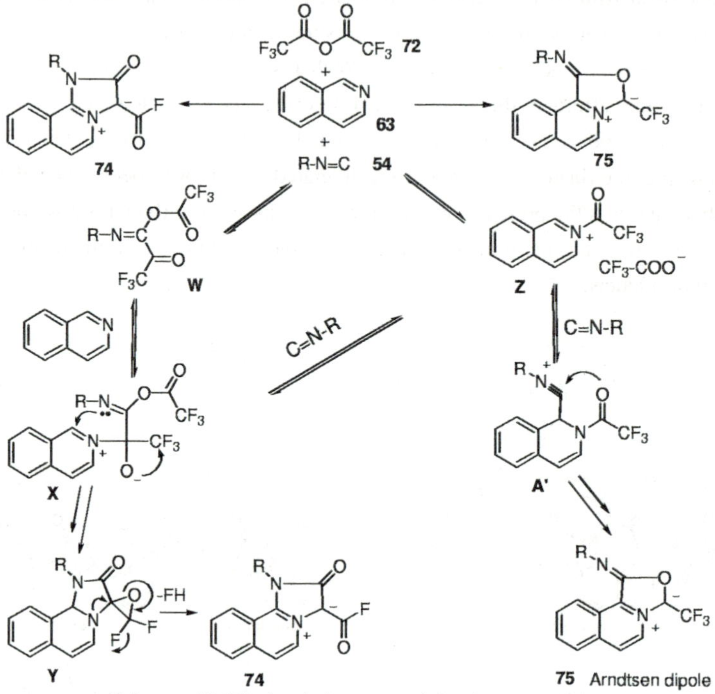

Scheme 42. Mechanistic proposal for the new MCR.

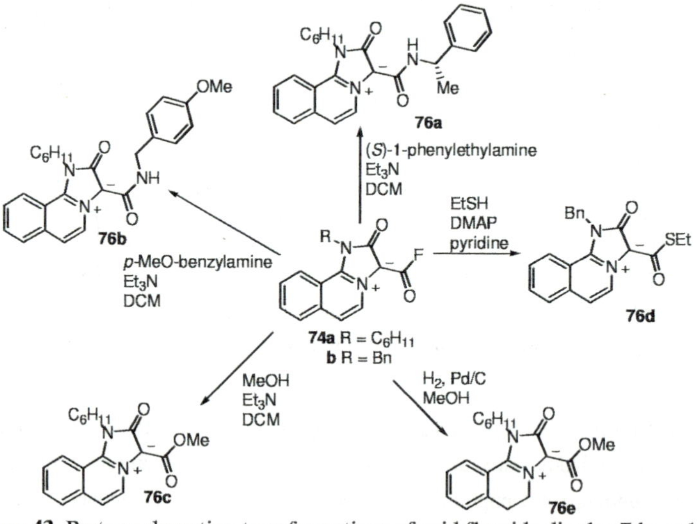

Scheme 43. Post-condensation transformations of acid fluoride dipoles **74a** and **74b**.

Moreover, this nucleophilic addition may take place upon the azine ring to furnish a nitrilium ion (**A'**), which in turn can be stabilized by the neighboring carbonyl leading to the Arndtsen-type dipoles **75**.

Interestingly some isocyanides (such as isocyanoacetates) afford compounds from both routes in comparable yields. This observation indicates competitive rates between the two pathways.

The acid fluoride dipole, although stable enough to be properly purified (by standard chromatography and crystalization techniques), displayed the expected reactivity for an activated carbonyl derivative towards nucleophilic reagents and thus it was further functionalized with alcohols, amines and thiols (Scheme 43). Interestingly, these protocols could be extended to biopolymers and DNA sequences were labeled with these dipolar tags. The structural features and the biological activities of these conjugates are currently being investigated.

4. Conclusions

The work performed in our group in recent years has advanced the description of new processes involving the participation of common heterocyclic structures in MCRs. In particular, the rich chemistry of azines and cyclic enol ethers has been merged with complex domino sequences of Povarov and isocyanide-MCRs to yield a broad range of scaffolds with high degree of structural diversity (Scheme 44). The mechanistic hypotheses, used as a starting point in this research, are the main tool to rationalize and develop new transformations from experimental observations. It is expected that the introduction of heterocycles in these domino reactions will increase their synthetic usefulness in the straightforward preparation of complex molecules, especially bioactive compounds and drugs.

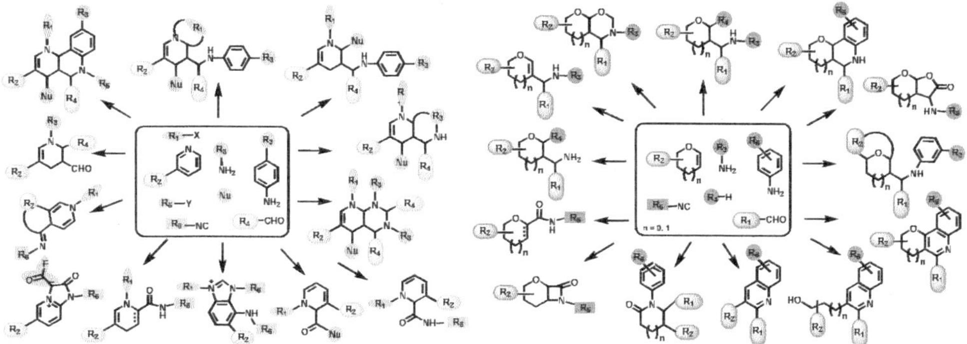

Scheme 44. Scaffold variability accessible from Povarov- and isocyanide-MCRs involving heterocycles.

Acknowledgments

We warmly thank all the members of our research group involved in this project for their enthusiasm and dedication. Financial support from the DGICYT (projects CTQ 2003-00089, 2006-03794 and 2009-07758) and IRB Barcelona is gratefully acknowledged. We are especially thankful to Laboratorios Almirall and Grupo Ferrer (Barcelona) for their support. Prof. Fernando Albericio (IRB Barcelona) and Prof. Jieping Zhu (CNRS Gif-sur-Yvette) are thanked for useful suggestions.

References

1.	Zhu, J.; Bienaymé, H. *Multicomponent reactions*; Wiley-VCH, 2005.
2.	Tietze, L. F.; Brasche, G.; Gericke, K. M. *Domino Reactions in Organic Synthesis*; Wiley-VCH, 2006.
3.	Dömling, A.; Ugi, I. *Angew. Chem. Int. Ed.* **2000**, *39*, 3168–3210.

4. Ugi, I. *Pure Appl. Chem.* **2001**, *73*, 187–191.

5. Ramón, D. J.; Yus, M. *Angew. Chem. Int. Ed.* **2005**, *44*, 1602–1634.

6. Bienaymé, H.; Hulme, C.; Oddon, G.; Schmitt, P. *Chem. Eur. J.* **2000**, *6*, 3321–3329.

7. Weber, L. *Drug Discovery Today* **2002**, *7*, 143–147.

8. Dömling, A. *Curr. Opin. Chem. Biol.* **2002**, *6*, 306–313.

9. Hulme, C.; Gore, V. *Curr. Med. Chem.* **2003**, *10*, 51–80.

10. Renner, S.; van Otterlo, W. A. L.; Dominguez Seoane, M.; Mocklinghoff, S.; Hofmann, B.; Wetzel, S.; Schuffenhauer, A.; Ertl, P.; Oprea, T. I.; Steinhilber, D.; Brunsveld, L.; Rauh, D.; Waldmann, H. *Nat. Chem. Biol.* **2009**, *5*, 585–592.

11. Wetzel, S.; Klein, K.; Renner, S.; Rauh, D.; Oprea, T. I.; Mutzel, P.; Waldmann, H. *Nat. Chem. Biol.* **2009**, *5*, 581–583.

12. Burke, M. D.; Schreiber, S. L. *Angew. Chem. Int. Ed.* **2004**, *43*, 46–58.

13. Walsh, D. P.; Chang, Y.-T. *Chem. Rev.* **2006**, *106*, 2476–2530.

14. Schreiber, S. L. *Chem. & Eng. News* **2003**, *81*, 51–61.

15. Sunderhaus, J. D.; Martin, S. F. *Chem. Eur. J.* **2009**, *15*, 1300–1308.

16. Ugi, I.; Werner, B.; Dömling, A. In *Targets in Heterocyclic Systems-Chemistry and Properties*; Attanasi, O. A.; Spinelli, D., Eds.; Società Chimica Italiana: Rome, 2000; Vol. 4, pp. 1–24.

17. Isambert, N.; Lavilla, R. *Chem. Eur. J.* **2008**, *14*, 8444–8454.

18. For an extensive overview, see: Heathcock, C. H.; Schreiber, S. L. In *Comprehensive Organic Synthesis*; Trost, B. M.; Fleming, I., Eds.; Pergamon Press: Oxford, 1991; Vol. 1 and 2.

19. Povarov, L. S. *Russ. Chem. Rev.* **1967**, *36*, 656–670.

20. Kouznetsov, V. V. *Tetrahedron* **2009**, *65*, 2721–2750.

21. Glushkov, V. A.; Tolstikov, A. G. *Russ. Chem. Rev.* **2008**, *77*, 137–159.

22. Ma, Y.; Qian, C.; Xie, M.; Sun, J. *J. Org. Chem.* **1999**, *64*, 6462–6467.

23. Kobayashi, S.; Ishitani, H.; Nagayama, S. *Synthesis* **1995**, 1195–1202.

24. Kobayashi, S.; Ishitani, H.; Nagayama, S. *Chem. Lett.* **1995**, 423–424.

25. Sridharan, V.; Perumal, P. T.; Avendaño, C.; Menéndez, J. C. *Org. Biomol. Chem.* **2007**, *5*, 1351–1353.

26. Maiti, G.; Kundu, P. *Tetrahedron Lett.* **2006**, *47*, 5733–5736.

27. For a general review, see: Kobayashi, S.; Sugiura, M.; Kitagawa, H.; Lam, W. W.-L. *Chem. Rev.* **2002**, *102*, 2227–2302.

28. Kobayashi, S.; Nagayama, S. *J. Am. Chem. Soc.* **1998**, *120*, 2985–2986.

29. Li, M.; Sun, E.; Wen, L. R.; Sun, J.; Li, Y.; Yang, H. *J. Comb. Chem.* **2007**, *9*, 903–905.

30. Savitha, G.; Perumal, P. T. *Tetrahedron Lett.* **2006**, *47*, 3589–3593.

31. Yadav, J. S.; Reddy, B. V. S.; Sadasiv, K.; Reddy, P. S. R. *Tetrahedron Lett.* **2002**, *43*, 3853–3856.

32. Yadav, J. S.; Reddy, B. V. S.; Uma Gayathri, K.; Prasad, A. R. *Synthesis* **2002**, 2537–2541.

33. Liu, H.; Dagousset, G.; Masson, G.; Retailleau, P.; Zhu, J. *J. Am. Chem. Soc.* **2009**, *131*, 4598–4599.

34. Akiyama, T.; Morita, H.; Fuchibe, K. *J. Am. Chem. Soc.* **2006**, *128*, 13070–13071.

35. For mechanistic studies, see: Loh, T. P.; Koh, K. S.-V.; Sim, K.-Y.; Leong, W.-K. *Tetrahedron Lett.* **1999**, *40*, 8447–8451.

36. Hermitage, S.; Howard, J. A. K.; Jay, D.; Pritchard, R. G.; Probert, M. R.; Whiting, A. *Org. Biomol. Chem.* **2004**, *2*, 2451–2460.

37. Lavilla, R. *J. Chem. Soc., Perkin Trans. 1* **2002**, 1141–1156, and references cited therein.

38. Lavilla, R.; Bernabeu, M. C.; Carranco, I.; Díaz, J. L. *Org. Lett.* **2003**, *5*, 717–720.

39. Brewster, M. E.; Simay, A.; Czako, K.; Winwood, D.; Farag, H.; Bodor, N. *J. Org. Chem.* **1989**, *54*, 3721–3726.

40. Kobayashi, S.; Manabe, K. *Acc. Chem. Res.* **2002**, *35*, 209–217.

41. Kobayashi, S.; Nagayama, S. *J. Am. Chem. Soc.* **1997**, *119*, 10049–10053.

42. Dondoni, A.; Massi, A. *Tetrahedron Lett.* **2001**, *42*, 7975–7978.

43. Carranco, I.; Díaz, J. L.; Jiménez, O.; Vendrell, M.; Albericio, F.; Royo, M.; Lavilla, R. *J. Comb. Chem.* **2005**, *7*, 33–41.

44. Powell, D. A.; Batey, R. A. *Tetrahedron Lett.* **2003**, *44*, 7569–7573.
45. Huang, T.; Li, C.-J. *Tetrahedron Lett.* **2000**, *41*, 6715–6719.
46. Kobayashi, S.; Akiyama, R.; Kitagawa, H. J. *Comb. Chem.* **2001**, *3*, 196–204.
47. For precedent in the field, see: Kiselyov, A. S.; Smith, L. I.; Armstrong, R. W. *Tetrahedron* **1998**, *54*, 5089–5096.
48. Lavilla, R.; Carranco, I.; Díaz, J. L.; Bernabeu, M. C.; de la Rosa, G. *Mol. Diversity* **2003**, *6*, 171–175.
49. For a related *N*-acylation using a functionalized aldehyde, see: Khadem, S.; Udachin, K. A.; Enright, G. D.; Prakesch, M.; Arya, P. *Tetrahedron Lett.* **2009**, *50*, 6661–6664.
50. An enantioselective version of this reaction has been disclosed; see: Takamura, M.; Funabashi, K.; Kanai, M.; Shibasaki, M. *J. Am. Chem. Soc.* **2001**, *123*, 6801–6808.
51. Reissert, A. *Chem. Ber.* **1905**, *38*, 1603.
52. Carranco, I.; Dìaz, J. L.; Jiménez, O.; Lavilla, R. *Tetrahedron Lett.* **2003**, *44*, 8449–8452.
53. For the related diamination of DHPs, see: Lavilla, R.; Kumar, R.; Coll, O.; Masdeu, C.; Spada, A.; Bosch, J.; Espinosa, E.; Molins, E. *Chem. Eur. J.* **2000**, *6*, 1763–1772.
54. For the use of this term, see the work on biomimetic cyclization of alkenes: Johnson, W. S.; Chen, Y. Q.; Kellogg, M. S. *J. Am. Chem. Soc.* **1983**, *105*, 6653–6656, and related papers in this field.
55. Lucchini, V.; Prato, M.; Scorrano, G.; Stivanello, M.; Valle, G. *J. Chem. Soc., Perkin Trans. 2* **1992**, 259–266.
56. Borrione, E.; Prato, M.; Scorrano, G.; Stivanello, M.; Lucchini, V. *J. Heterocycl. Chem.* **1988**, *25*, 1831–1833.
57. Zhang, J.; Li, C.-J. *J. Org. Chem.* **2002**, *67*, 3969–3971.
58. Chen, L.; Li, C.-J. *Green Chemistry* **2003**, *5*, 627–629.
59. Batey, R. A.; Simoncic, P. D.; Lin, D.; Smyj, R. P.; Lough, A. J. *Chem. Commun.* **1999**, 651–652.
60. Batey, R. A.; Powell, D. A. *Chem. Commun.* **2001**, 2362–2363.
61. Powell, D. A.; Batey, R. A. *Org. Lett.* **2002**, *4*, 2913–2916.
62. Ghosh, A. K.; Kawahama, R.; Wink, D. *Tetrahedron Lett.* **2000**, *41*, 8425–8429.
63. Ghosh, A. K.; Kawahama, R. *Tetrahedron Lett.* **1999**, *40*, 4751–4754.
64. Baudelle, R.; Melnyck, P.; Déprez, B.; Tartar, A. *Tetrahedron* **1998**, *54*, 4125–4140.
65. Jiménez, O.; de la Rosa, G.; Lavilla, R. *Angew. Chem. Int. Ed.* **2005**, *44*, 6521–6525.
66. Katritzky, A.; Rachwak, B.; Rachwal, S. *J. Org. Chem.* **1995**, *60*, 3993–4001.
67. Talukdar, S.; Chen, C.-T.; Fang, J.-M. *J. Org. Chem.* **2000**, *65*, 3148–3153.
68. Dagousset, G.; Drouet, F.; Masson, G.; Zhu, J. *Org. Lett.* **2009**, *11*, 5546–5549.
69. Bi, B.; Maurer, K.; Moeller, K. D. *Angew. Chem. Int. Ed.* **2009**, *48*, 5872–5874.
70. Camps, P.; Formosa, X.; Galdeano, C.; Muñoz-Torrero, D.; Ramírez, L.; Gómez, E.; Isambert, N.; Lavilla, R.; Badia, A.; Clos, M. V.; Bartolini, M.; Mancini, F.; Andrisano, V.; Arce, M. P.; Rodríguez-Franco, M. I.; Huertas, Ó.; Dafni, T.; Luque, F. J. *J. Med. Chem.* **2009**, *52*, 5365–5379.
71. Barluenga, J.; Mendoza, A.; Rodríguez, F.; Fañanás, F. J. *Angew. Chem. Int. Ed.* **2009**, *48*, 1644–1647.
72. Barluenga, J.; Mendoza, A.; Rodríguez, F.; Fañanás, F. J. *Angew. Chem. Int. Ed.* **2008**, *47*, 7044–7047.
73. Isambert, N.; Cruz, M.; Arévalo, M. J.; Gómez, E.; Lavilla, R. *Org. Lett.* **2007**, *9*, 4199–4202.
74. Stevenson, P. J.; Graham, I. *Arkivoc* **2003**, 139–144.
75. Vicente-García, E.; Catti, F.; Ramón, R.; Lavilla, R. *Org. Lett.* **2010**, *12*, 860–863.
76. Dömling, A. *Chem. Rev.* **2006**, *106*, 17–89.
77. Masdeu, C.; Díaz, J. L.; Miguel, M.; Jiménez, O.; Lavilla, R. *Tetrahedron Lett.* **2004**, *45*, 7907–7909.
78. Masdeu, C.; Gómez, E.; Williams, N. A. O.; Lavilla, R. *QSAR Comb. Sci.* **2006**, *25*, 465–473.
79. Kühle, E.; Anders, B.; Zumach, G. *Angew. Chem. Int. Engl. Ed.* **1967**, *79*, 663–680.
80. Ruppert, I. *Tetrahedron Lett.* **1980**, *21*, 4893–4896.
81. Lavilla, R. *Curr. Org. Chem.* **2004**, *8*, 715–737.
82. Masdeu, C.; Gómez, E.; Williams, N. A. O.; Lavilla, R. *Angew. Chem. Int. Ed.* **2007**, *46*, 3043–3046.
83. Kabbe, H. J. *Chem. Ber.* **1969**, *102*, 1404–1409.
84. Bez, G.; Zao, C.-G. *Org. Lett.* **2003**, *5*, 4991–4993.
85. Oshita, M.; Yamashita, K.; Tobisu, M.; Chatani, N. *J. Am. Chem. Soc.* **2005**, *127*, 761–766.

86. Korotkov, V. S.; Larionov, O. V.; de Meijere, A. *Synthesis* **2006**, 3542–3546.

87. Tarragó, T.; Masdeu, C.; Gómez, E.; Isambert, N.; Lavilla, R.; Giralt, E. *ChemMedChem* **2008**, *3*, 1558–1565.

88. Williams, N. A. O.; Masdeu, C.; Diaz, J. L.; Lavilla, R. *Org. Lett.* **2006**, *8*, 5789–5792.

89. Janvier, P.; Sun, X.; Bienaymé, H.; Zhu, J. *J. Am. Chem. Soc.* **2002**, *124*, 2560–2567.

90. Pirali, T.; Tron, G. C.; Zhu, J. *Org. Lett.* **2006**, *8*, 4145–4148, and references therein.

91. Bonne, D.; Dekhane, M.; Zhu, J. *Org. Lett.* **2005**, *7*, 5285–5288.

92. Zhu, J. *Eur. J. Org. Chem.* **2003**, 1133–1144.

93. For an earlier example of this transformation, see: Behnke, D.; Taube, R.; Illgen, K.; Nerdinger, S.; Herdtweck, E. *Synlett* **2004**, 688–692.

94. For a related process on an *N*-alkylquinolinium salt, see: Ugi, I.; Boettner, E. *Ann.* **1963**, *670*, 74–80.

95. Díaz, J. L.; Miguel, M.; Lavilla, R. *J. Org. Chem.* **2004**, *69*, 3550–3553.

96. Popp, F. D.; Takeuchi, I.; Kant, J.; Hamada, Y. *J. Chem. Soc., Chem. Commun.* **1987**, 1765–1766.

97. Lorsbach, B. A.; Bagdanoff, J. T.; Miller, R. B.; Kurth, M. J. *J. Org. Chem.* **1998**, *63*, 2244–2250.

98. Chen, C.; McDonald, I. A.; Munoz, B. *Tetrahedron Lett.* **1998**, *39*, 217–220.

99. Corey, E. J.; Tian, Y. *Org. Lett.* **2005**, *7*, 5535–5537.

100. Arévalo, M. J.; Kielland, N.; Masdeu, C.; Miguel, M.; Isambert, N.; Lavilla, R. *Eur. J. Org. Chem.* **2009**, 617–625.

101. Cyr, D. J. S.; Martin, N.; Arndtsen, B. A. *Org. Lett.* **2007**, *9*, 449–452.

102. El Kaïm, L. *Tetrahedron Lett.* **1994**, *35*, 6669–6670.

103. El Kaïm, L.; Pinot-Périgord, E. *Tetrahedron* **1998**, *54*, 3799–3806.

SYNTHESIS AND CHEMISTRY OF 3(2*H*)-FURANONES

Timm T. Haug and Stefan F. Kirsch*

Department Chemie, Technische Universität München, Lichtenbergstr. 4, D-85748 Garching, Germany
(e-mail: stefan.kirsch@ch.tum.de)

Abstract. *Heterocycles probably represent the largest and most varied class of chemical compounds currently registered. This review highlights the chemistry of 3(2H)-furanones, a relatively small class of naturally occurring heterocycles which have a wide range of biological activities in medicinal chemistry. Selected natural products are followed by pharmaceuticals and other biologically active compounds that contain the basic heterocyclic structure. Then, a special emphasis is placed on the key strategies that have been developed for the construction of the 3(2H)-furanone motif. Additionally, the use of 3(2H)-furanones in total synthesis and in selected synthetic transformations is discussed in a detailed way.*

Contents

1. Introduction

Furan-3(2*H*)-ones are five-membered heterocycles that play an important role as key structural elements of many naturally occurring compounds. For instance, they are found as flavor compounds in fruits[1] and the motif is found in various natural products extracted from fungi and plants. Due to promising medicinal activity in a number of cases, many different approaches to multiply substituted furan-3(2*H*)-ones have been developed, many of which differ only slightly in the type of bond formation.

This account aims to summarize the numerous strategies for 3(2*H*)-furanone synthesis by discussing both classical condensation strategies and transition-metal-catalyzed approaches. Another focus of this account is the synthetic value of such synthetic strategies in the construction of complex natural products. To this end, representative examples of natural products containing the 3(2*H*)-furanone moiety as well as structures that are significant from a medicinal point of view are presented first in order to provide a better understanding for the relevance of the heterocyclic core. This review covers the literature until 2009 and does not include patent literature.

In parallel to the isolation of the parent compound 3(2*H*)-furanone,[2] many natural products comprising this core structure have been identified as illustrated by bullatenone (**1**) (Figure 1).[3–6] The heterocyclic system has the spectroscopic characteristics of an enolic ether with an adjacent carbonyl group, as it is shown by its NMR and IR spectroscopic data:

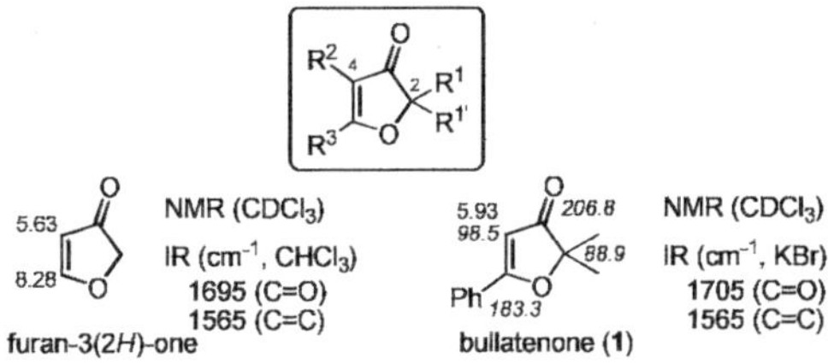

Figure 1. Representative spectroscopic data for 3(2*H*)-furanones.

2. Structural motif in natural compounds

In 1954, the first 3(2*H*)-furanone was isolated from a natural source by Brandt *et al.*.[4] Although the structure was originally misassigned as 3-dihydro-3-methyl-6-phenyl-4-pyrone, Parker *et al.* revised the structure in 1958 to be 2,3-dihydro-2,2-dimethyl-3-oxo-5-phenylfuran, bullatenone (**1**).[5]

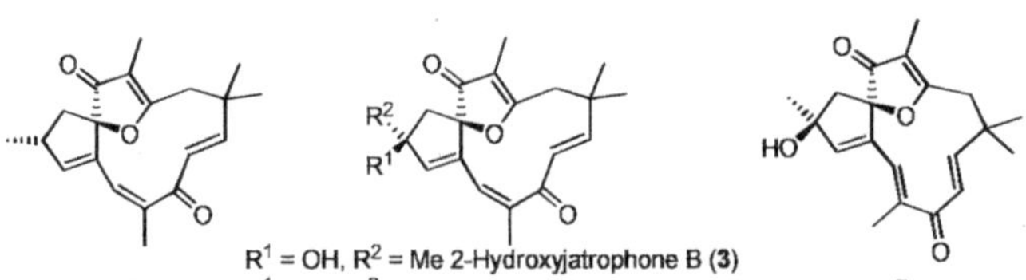

R^1 = OH, R^2 = Me 2-Hydroxyjatrophone B (**3**)
2 R^1 = Me, R^2 = OH 2-Hydroxyjatrophone A (**4**) **5**

Figure 2

In 1970, Kupchan *et al.* examined extracts of *Jatropha gossyplifolia* L. and identified therein a novel macrocyclic diterpenoid that was named jatrophone (**2**) (Figure 2).[6,7] Of particular interest was the spirocyclic core motif that soon became a challenging element of structure for synthetic chemists (see below), especially as this compound exhibited antileukemic activity in biological tests.

Following the discovery of jatrophone and its derivatives (**3–5**) (Figure 2),[6,7] a variety of different natural products with a 3(2*H*)-furanone unit has been identified.

2.1. Spirocyclic furanones

A novel class of 3(2*H*)-furanones including various representative compounds contains a spirocyclic core structure with a furan heterocycle. Illustrative of this class are, *e.g.* inoscavin A (**6**),[8] trachyspic acid (**7**)[9] and longianone (**8**) (Figure 3).[10] Trachyspic acid (**7**) was isolated from *Talaromyces trachyspermus* and showed inhibitory activity against heparinase.[9] Longianone (**8**), which has been isolated from the fungus *Xylaria longiana* in 1999,[10] is supposed to be a structural isomer of the fungal toxin patulin although its absolute configuration has not been determined so far. This similitude in structure suggests a related biosynthetic pathway.[11] Additionally, hyperolactones A-C (**9–11**) were isolated from *Hypericum chinese* L. (Figure 3).[12]

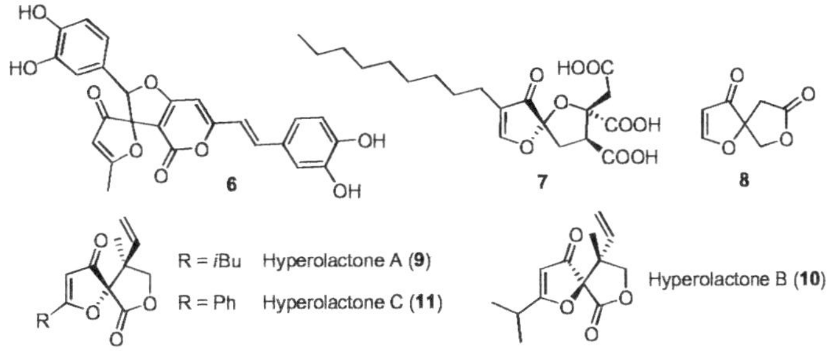

Figure 3

Various spirocyclic 3(2*H*)-furanones with a pyrrolidinone ring have been identified amongst natural metabolites. This unprecedented core structure has been revealed in a wide range of compounds including azaspirene (**12**) (isolated from the fungus *Neosartorya* sp),[13] pseurotin A (**13**) and pseurotin F$_2$ (**14**),[14,15] synerazol (**15**) (isolated from *Aspergillus fumigatus* SANK 10588)[16] and FD-838 (**16**) (isolated from *A. fumigatus fresenius* F-838) (Figure 4).[17] FD-838, for example, was shown to inhibit the growth of Gram-positive bacteria and fungi.

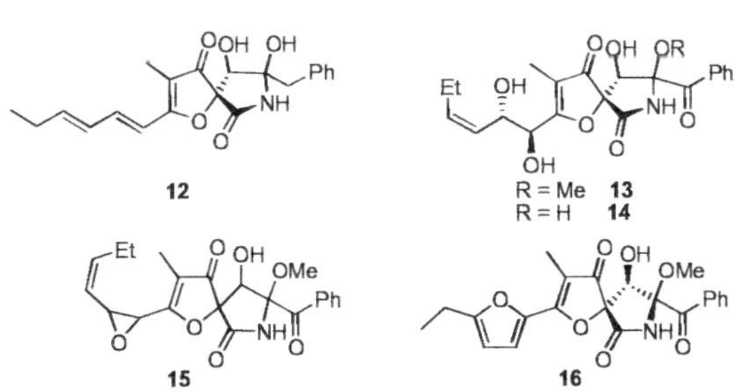

Figure 4

59

Some members of the recently found phelligridins also contain a spirocyclic 3(2*H*)-furanone motif. Phelligridins E (**17**)[18] and G (**18**)[19] were isolated by Shi from the mushroom *Phellinus igniarius* (Figure 5).

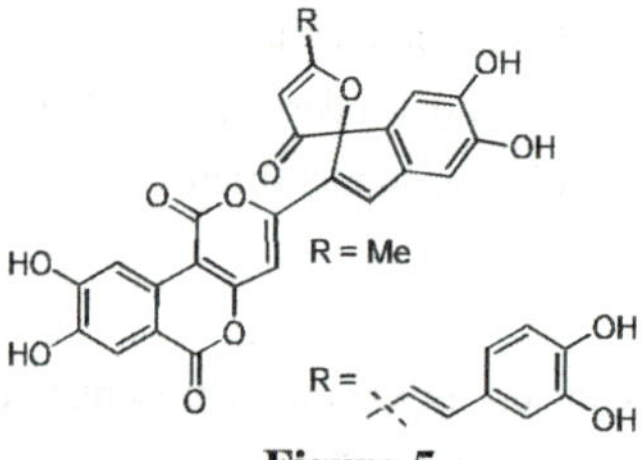

Figure 5

2.2. Germacranolides

Apart from the spirocyclic core, a manifold of other 3(2*H*)-furanone containing compounds were isolated from different natural sources. Two structurally related 3(2*H*)-furanones were assigned by Vichnewski *et al.* in 1976: goyazensolide (**19**) (from *Eremanthus goyazensis*)[20] and 15-deoxygoyazensolide (**20**) (from *Vanillosmopsis erythropappa*) (Figure 6).[21]

Several other sesquiterpenes of the germacranolide-type were found that incorporate a 3(2*H*)-furanone core, the most important one being budlein A (**21**). Budlein A has been isolated from *Viguiera buddleiaeformis* in 1976 by Jiménez and coworkers.[22] Lychnophorolide A (**22**), the C8-epimer of budlein A, was isolated by Le Quesne in 1982 from *Lychnophora affinis*.[23] Another representative of this substrate class is ladibranolide (**24**) from *Viguiera ladibractate*,[24] whereas the saponified variant, atripliciolide (**25**), was found in *Isocarpha* species in 1978.[25] Ortega *et al.* described the structures of zexbrevin (**26**) and calaxin (**27**) (Figure 6).[26]

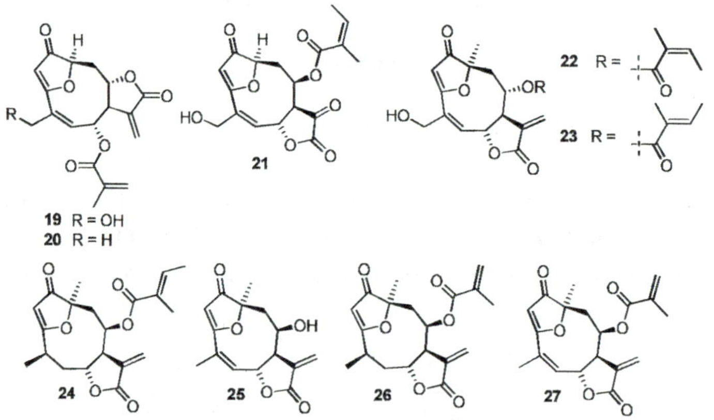

Figure 6

The eremantholides (**28–30**) (Figure 7), isolated in 1978 by Le Quesne *et al.* from *Eremanthus elaeagnus*,[27a] are modified germacranolides that possess an additional tetrahydrofuran ring. Only recently, Lopes *et al.* could isolate two novel sesquiterpene lactones of the germacranolide-type (**31** and **32**) (Figure 7) from *Lychnophora ericoides*.[28] Eremantholide A (**28**) exhibits significant levels of *in vitro* antitumor activity against a variety of tumor cell lines.[27b,c]

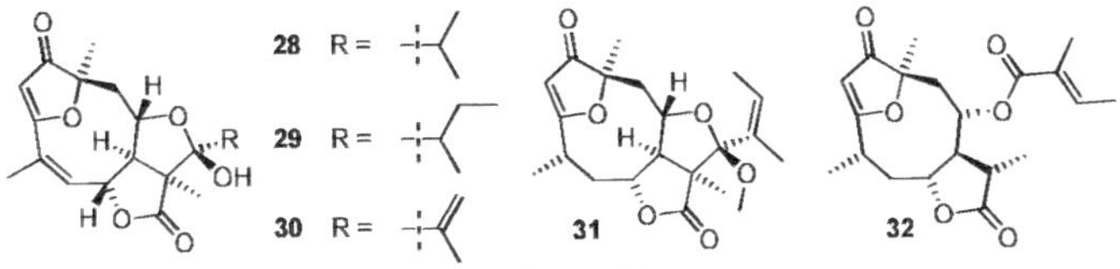

Figure 7

2.3. Non-spirocyclic 3(2*H*)-furanones

In 1967, Lahey and MacLeod were able to locate geiparvarin (**33**) (Figure 8) in the leaves of *Geijera parviflora* Lindl.[29,30] *In vitro* screenings showed significant anti-proliferative effects for this compound. Tachrosin (**34**) is a 3(2*H*)-furanone that was isolated in 1972 from *Tephrosia polystachyoides* Bak. f. (Figure 8).[31]

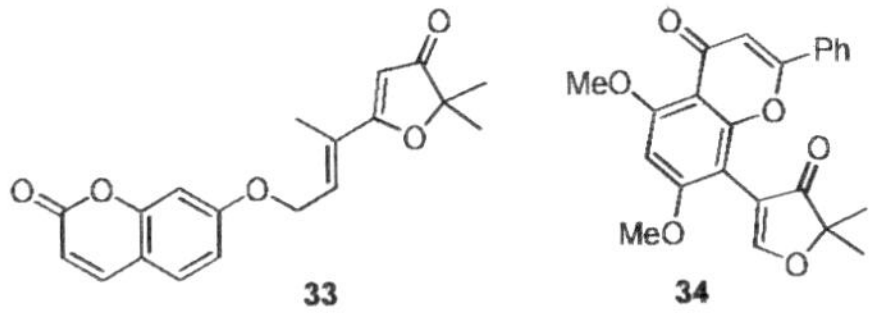

Figure 8

A whole class of siphonarienfuranones, members of the polypropionates, consist of a 3(2*H*)-furanone part containing a hydroxyl group at C2. For example, *E*- and *Z*-siphonarienfuranones (**35**) (Figure 9) were isolated by Faulkner *et al.* in 1984.[32] A C2-epimer and a deoxysiphonarienfuranone were found in *Siphonaria pectinata*.[33] Neovasifuranones A (**36**) and B (**37**) (Figure 9) were isolated in 1995 by Fukuyama *et al.* from the fungus *Neocosmospora vasinfecta* NHL2298.[34]

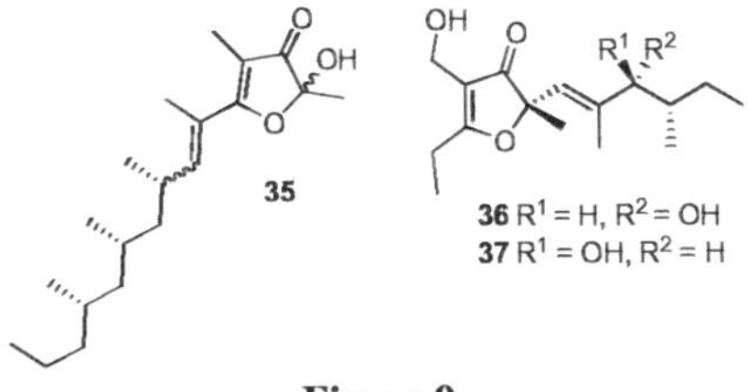

Figure 9

Further 3(2*H*)-furanone containing compounds are represented by glucoside derivatives, such as psydrin (**38**) from *Psydrax livia*,[35] phoeniceroside (**39**) from *Juniperus phoenicea* L.[36] and furaneol glucoside (**40**) that was found in strawberry juice (Figure 10).[37]

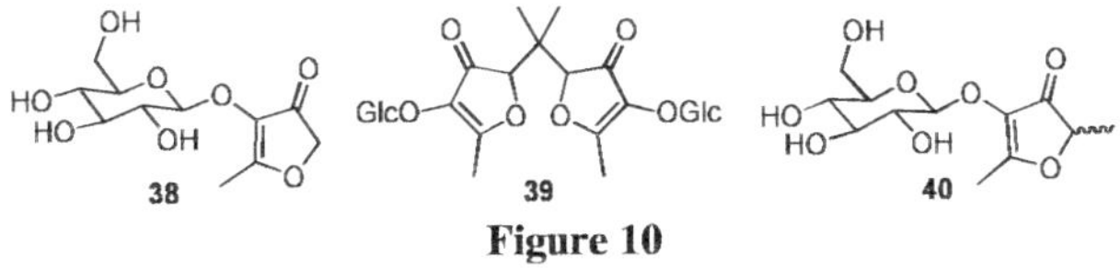

Figure 10

The dimeric chilenone A (**41**) was isolated as a racemic mixture.[38] Accordingly, chilenone B (**42**) is the trimer of 2-methyl-3(2*H*)-furanone (Figure 11).[39]

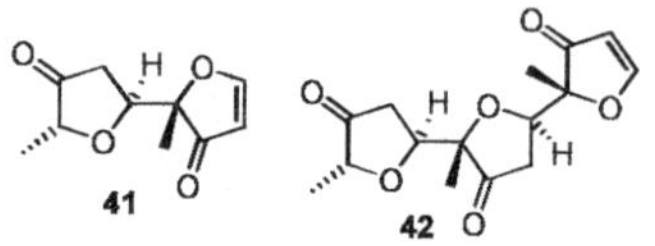

Figure 11

In 2004, terrefuranone (**43**) (Figure 12) was isolated from *Aspergillus terreus* by Gunatilaka *et al.*.[40] Biological testing against four cancer cell lines and human fibroblast cells revealed no activity. The phytotoxic metabolite gregatin B (**44**) (Figure 12) was isolated in 1975 from *Cephalosporium gregatum* and, subsequently, in 1980 from *Aspergillus panamensis* [41,42] First, its structure was suggested to be *O*-methyl-tetronic acid. Pattenden and Clemo revised the structure in 1982 showing the 3(2*H*)-furanone skeleton.[43,44]

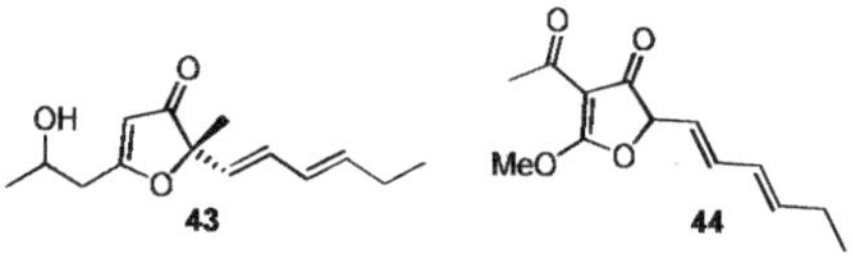

Figure 12

In food chemistry, several 3(2*H*)-furanones have been identified as important flavor compounds. A possible pathway for their ubiquitous appearance is through Maillard reaction between reducing sugars and amino acids.[1] Moreover, 3(2*H*)-furanones can be produced by yeast or are important for the flavor of different fruits. Notably, HDMF or Furaneol is a commerically available food additive.[1]

3. Medicinal chemistry

Bullatenone (**1**), the first 3(2*H*)-furanone that was isolated from a natural source, was found to show antiulcer activity that did not depend on inhibition of gastric acid secretion but on the strengthening of gastric and duodenal mucosal defenses. This seminal observation resulted in the preparation of several bullatenone derivatives with an increase in activity.[45]

Georgiev *et al.* investigated the antiallergic properties of 3(2*H*)-furanones in 1988. In this study, it was shown that compounds like **12** (Figure 13) can inhibit the actions of the mediators serotonin, histamine and bradykinin by 100% when administered intraperitoneally to rats at a dose of 100 mg/kg. In the active anaphylaxis assay in rats, compound **45** suppressed the edema by 45% at a dose of 100 mg/kg.[46]

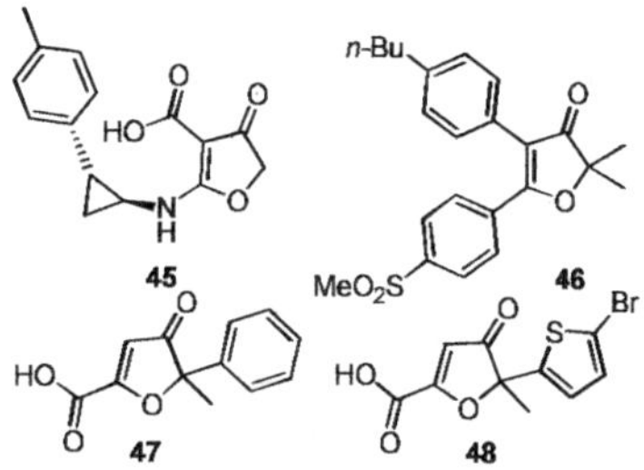

Figure 13

Chung *et al.* developed furanone **46** (Figure 13) that is one of the most potent cyclooxygenase-2 (COX-2) inhibitors known to date.[47] Since inhibition of cyclooxygenase-1 leads to gastro-intestinal toxicity, it is important to note that the 4,5-diaryl-3(2*H*)-furanone derivatives inhibit COX-2 selectively over COX-1, and most of the compounds tested possess COX-2/COX-1 selectivity over 100-fold.[48]

Acifran (**47**) (Figure 13) was developed by Ayerst Laboratories in the early 1980s as a lipid lowering agent without detailed knowledge of the molecular target.[49] With the discovery of a G-protein coupled receptor (GPCR) with which nicotinic acid interacts (GPR109a), a manifold of effort was initiated in the search of more potent and selective agonists of the receptor.[50,51] The high and low affinity niacin receptors GPR109a and GPR109b are expected to be ideal targets for high-density lipoprotein (HDL) elevating drugs for the treatment of atherosclerosis. A series of acifran analogs was prepared and evaluated as agonists for GPR109a and GPR109b, resulting in identification of compounds with improved activity at these receptors.[50,52] Substituted thienyl analogues were considerably more potent than **47**. In particular, 5-bromo-3-thienyl derivative **48** (Figure 13) was 0.11 μM at GPR109a and moderately selective over GPR109b.[50] Other 3(2*H*)-furanones that lowered serum triglycerides in rats were found earlier.[49]

Furan-3(2*H*)-ones were shown to be valuable in the inhibition of monoamine oxidase of type B (MAO-B), an enzyme that is involved in the neurotransmitter amine metabolism.[53] Selective MAO-B inhibitors have the potential to play a key role in the treatment of Parkinson's disease.[54] In this context, it is notable that geiparvarin (**33**) selectively inhibits MAO-B over MAO-A (pIC_{50} = 6.84 *vs* 4.57). Carotti *et al.* synthesized substituted geiparvarin derivatives and, thus, developed the demethyl congener of geiparvarin, furanone **49** (Figure 14), that has a significantly enhanced MAO-B inhibitory activity (pIC_{50} = 7.55) and a 850-fold selectivity for the MAO-B isoform.[53] Building up on those results, Chimichi *et al.* verified the importance of the coumarinyloxy moiety for pharmaceutical appliances.[55] Compounds **50–52** (Figure 14) represent the best compounds for antiproliferative activity that were found in this study. Without showing significant activity, these compounds were also tested against leukemia- and carcinoma-derived human cells. Of primary interest, geiparvarin derivative **53** (Figure 14) was recently shown to effectively induce apoptosis in a promyelocytic leukemia cell line (HL-60).[56] Manfredini *et al.* prepared geiparvarin analogues that demonstrated cytostatic activity against human tumor cells.[57–61] Re *et al.* synthesized xanthone analogues of Geiparvarin that were tested against lymphocytic leukemia.[62]

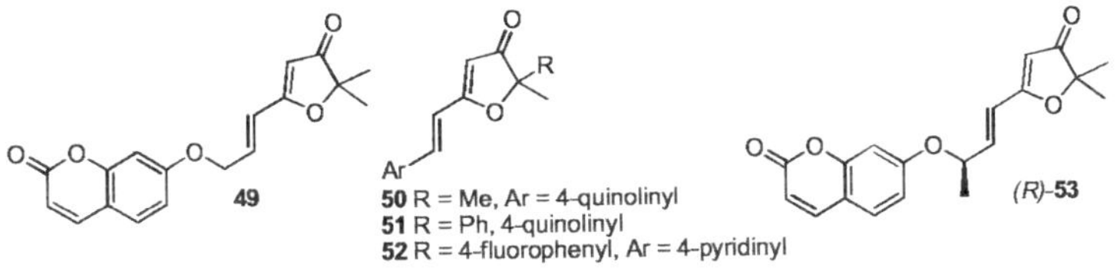

Figure 14

Marko *et al.* successfully used 3(2*H*)-furanones to modulate the growth of human tumor cells.[63] As a possible origin of action, the interaction of thiol groups in proteins with the conjugated carbonyl structure of the furanone was discussed.[64,65]

Simple molecules such as HDMF (4-hydroxy-2,5-dimethyl-3(2*H*)-furanone) and HEMF (4-hydroxy-2-ethyl-5-methyl-3(2*H*)-furanone) obtained from soy sauce and from Maillard reactions of hexoses and amino acids [66–68] demonstrated DNA-strand break activity through induction of active oxygen radicals (Figure 15).

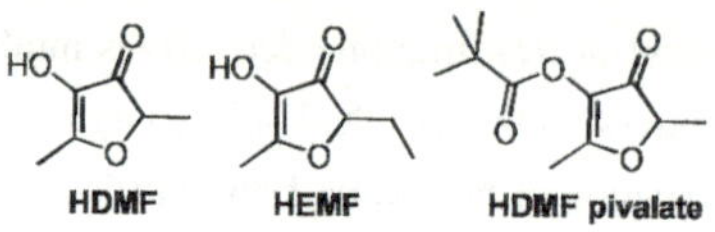

Figure 15

It was shown that the jatrophone derivatives hydroxyjatrophone B (**3**) and hydroxyjatrophone C (**5**) from *Jatropha gossypiifolia* have an antineoplastic activity comparable with that of the parent jatrophone (**2**).[6,69]

Mori *et al.* examined over 45 acylated hydroxyfuranone derivatives and discovered that 2,5-dimethyl-4-pivaloyloxy-3(2*H*)-furanone (HDMF pivalate) (Figure 15) is highly effective against cataracts in an *in vitro* model of a rat lens.[70,71]

4. Synthetic strategies

The 3(2*H*)-furanone syntheses that are the subject of this chapter can be generally divided into two major categories: the first, condensation and substitution strategies and the second, less studied, construction of the cyclic system through intramolecular addition reactions. In both cases, classical methods and transition-metal-catalyzed processes are discussed. Each section provides the underlying bond-forming strategies along with selected examples and variants. Specific approaches that do not fit into this general classification are subsequently detailed.

4.1. Cyclization/elimination approaches

A class of cascade reactions that result in the formation of 3(2*H*)-furanone heterocycles is based on a dehydration step proceeding from reaction intermediates of type **A** (Figure 16). As the preceding cyclization can be realized with diverse substrate classes, various bond-forming steps, discussed below, were employed.

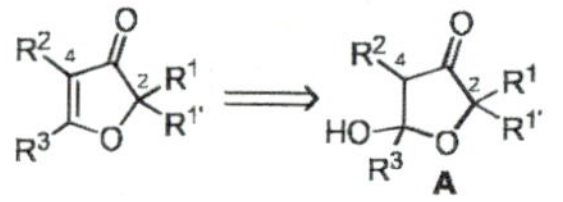

Figure 16. General dehydration strategy.

4.1.1. Bond formation between O1 and C5

As shown in Figure 17, the most simple way to afford intermediates of type **A** is by bond formation between O1 and C5. To this end, α'-hydroxy 1,3-diketone units are ideal substrates for the requisite cyclization that undergoes ring formation under both basic and acidic reaction conditions. Based on this fundamental reactivity, a variety of processes has been developed, through which 3(2*H*)-furanones are easily accessed. The synthetic value of these reactions is immense since the huge majority of synthetic strategies utilizes this bond-setting.

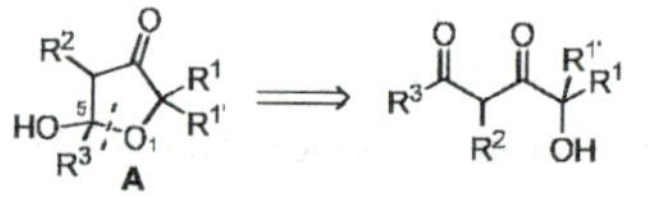

Figure 17. Bond formation between O1 and C5.

Amos B. Smith III *et al.* investigated the attractive potential of α'-hydroxy-1,3-diketones for the synthesis of 3(2*H*)-furanones in great detail (Scheme 1).[72] While the general approach to the heterocycle core involved the acid-catalyzed cyclization/dehydration discussed above, a focus was put on the efficient substrate synthesis. For instance, the intermolecular acylation of hydroxy ketones such as **54** was realized with acetyl imidazolide employing kinetic deprotonation (Scheme 1, eq. 1). Although this sequence led to clean formation of 3(2*H*)-furanone **55**, conversion was only in the range of 30–40%. A widely applicable strategy involved an initial aldol reaction followed by oxidation with Collins' reagent and treatment with mild aqueous acid to effect cyclization/dehydration (*e.g.*, **56**→**57**; Scheme 1, eq. 2). This two-step protocol was a general approach with wide applicability in the synthesis of numerous 3(2*H*)-furanones that underwent many variations.

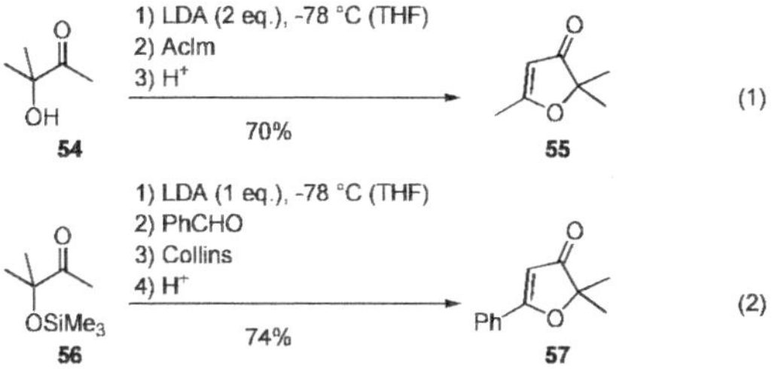

Scheme 1. Synthesis of 3(2*H*)-furanones according to Smith III *et al.*.

Extension of this synthetic strategy provides 3(2*H*)-furanones under basic rather than acidic conditions (Scheme 2, eq. 1).[73] This approach might be of particular value when α'-hydroxy-1,3-diketones carry functional groups that are sensitive to acid. For example, Tamm *et al.* used K$_2$CO$_3$ to cyclize **60** (Scheme 2, eq. 2). Nevertheless, in their route toward the total synthesis of the pseurotins, Tamm *et al.* relied on the acid-mediated ring-closure to give 3(2*H*)-furanone **63** in 75% yield (Scheme 2, eq. 3).[74]

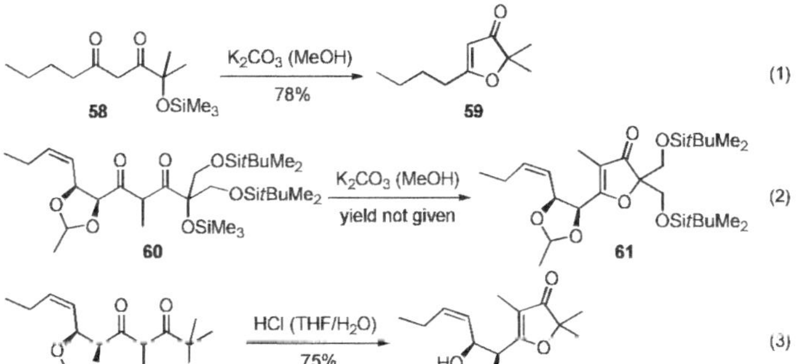

Scheme 2. Acid- and base-catalyzed ring closures in comparison.

Several substrate classes were investigated as synthetic equivalents for α'-hydroxy-1,3-diketones.[75,76] For example, allenic equivalents were described to be directly converted into 3(2*H*)-furanones under acidic conditions.[77] A more widespread equivalent for α'-hydroxy-1,3-diketones is the heterocyclic core of isoxazolines and isoxazoles that can be transformed into the intermediates prone to cyclization with hydrogenolysis. For instance, Torssell *et al.* showed that, upon condensation of silyl nitronates or nitrile oxides with vinyl ketones, 5-acyl-2-isoxazolines are formed that are effectively reduced with Raney nickel and H_2 in ethanol. Cyclization then proceeds in the presence of acetic acid to provide the corresponding 3(2*H*)-furanones *via* 1,4-diketo-2-ol intermediates such as **65** (Scheme 3, eq. 1).[78,79] A sequence based on the reduction of isoxazolines was also described by Curran *et al.* (Scheme 3, eq. 2).[80] Standard Raney-nickel mediated reduction produced **68**, which was not isolated, but directly acidified to yield the corresponding 3(2*H*)-furanone **69** in 51–77% yield. Tsuge *et al.* made use of nitriloxides that are easily converted into isoxazoles such as **70** by addition to substituted propargylic alcohols.[81] The isoxazoles were then treated with hydrogen and the resulting open-chain intermediate was cyclized under acidic conditions in good yields (Scheme 3, eq. 3). Importantly, the resulting phosphonates such as **71** could undergo subsequent Horner-Wadsworth-Emmons reactions. Simoni *et al.* constructed isoxazoles by exposure of ethyl 2,4-dioxo-alkanoates to hydrochloric hydroxylamine.[82,83] Here, the conditions for the successive transformation into 3(2*H*)-furanones remained unaltered. This strategy was also employed by Chimichi *et al.* to access novel geiparvarin analogues.[84] A Baylis-Hillman approach followed by hydrogenation was used by Batra *et al.* to obtain enaminone intermediates of type **68**.[85]

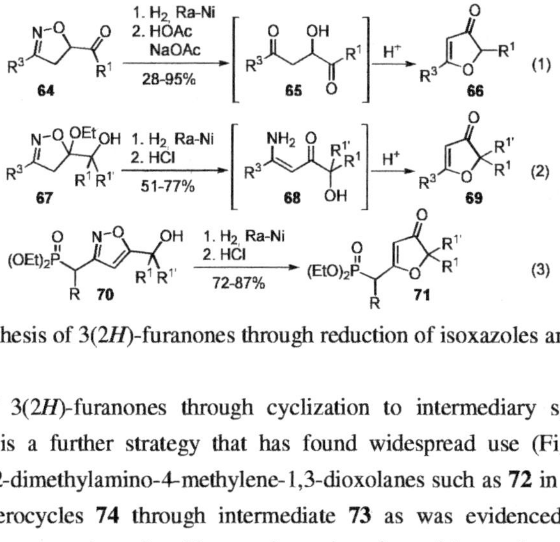

Scheme 3. Synthesis of 3(2*H*)-furanones through reduction of isoxazoles and isoxazolines.

The formation of 3(2*H*)-furanones through cyclization to intermediary species of type **A'** and subsequent elimination is a further strategy that has found widespread use (Figure 18). For example, Hoffmann *et al.* heated 2-dimethylamino-4-methylene-1,3-dioxolanes such as **72** in DMF at 60 °C to obtain the 3(2*H*)-furanone heterocycles **74** through intermediate **73** as was evidenced by NMR experiments (Scheme 4).[86] Yields were not given for this transformation. Stasevich *et al.* assembled 2-aminomethyl furanones such as **77** starting from epoxides and secondary amines.[87,88] The cyclization/elimination sequence as depicted in Scheme 4 (eq. 2) results in yields between 31% and 54%. Interestingly, addition of hydrobromic or hydrochloric acid instead of the secondary amine leads to the construction of 2-halomethyl-3(2*H*)-furanones in 40–56% yield.

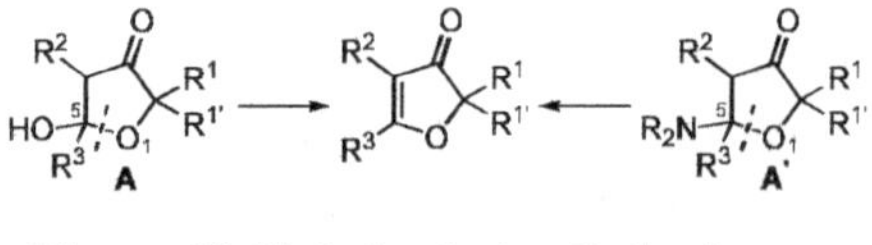

Figure 18. Variation in the elimination step.

Notably, instead of dimethylamine to be replaced, protocols were developed during the course of which Me$_2$S is displaced to give the desired 3(2H)-furanones.[89]

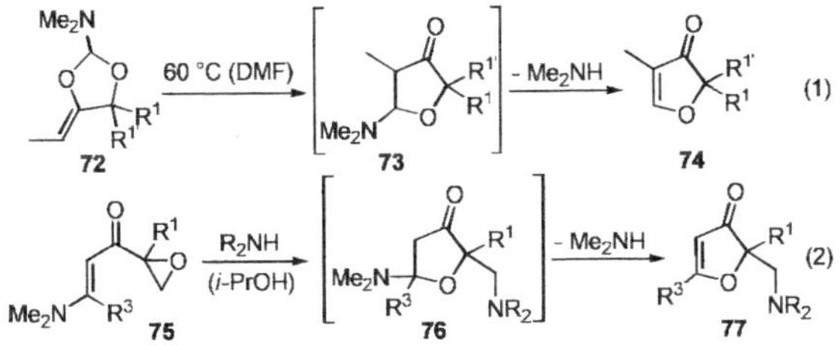

Scheme 4. Synthetic strategies with elimination of HNMe$_2$.

Since the strategies discussed in this section are based on a ring-forming cyclization to create the bond between O1 and C5 of the 3-furanone core, in most cases, α'-hydroxy-1,3-diketones are accessed as reactive intermediates for cyclization in the one or other way. An interesting variant was developed by Kutateladze *et al.* providing a rapid entry into flexibly substituted 3(2H)-furanones.[90] To this end, butylthiomethyl-1,3-dithiane (**78**) dimerized upon lithiation with n-BuLi followed by addition to aldehydes such as propanal to smoothly yield adduct **79** (Scheme 5). In the presence of an excess of HgCl$_2$, this bis-dithianes hydrolyzes in aqueous acetonitrile to directly give furanone **80**. Besides this efficient one-pot multicomponent approach to 3(2H)-furanones, aldehydes protected as dithianes were previously used in furanone syntheses by Carlson *et al.*.[77]

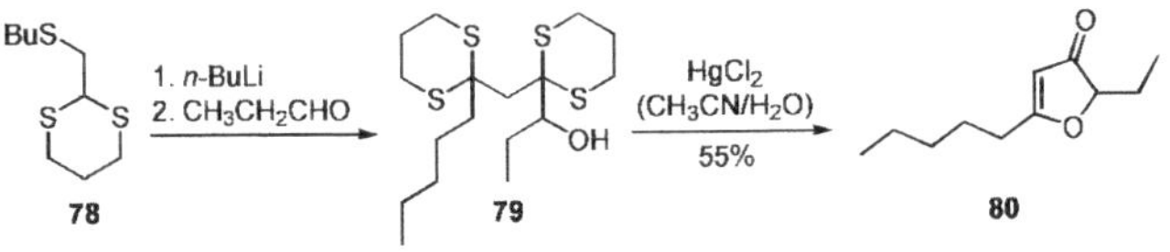

Scheme 5. Dithiane precursors.

Another way to form α'-hydroxy-1,3-diketones *in situ* is illustrated in Scheme 6. Inoue *et al.* converted 4-hydroxyalk-2-ynones such as **81** in a one-pot protocol into the corresponding 3(2H)-furanones.[91] Product formation in the presence of an alkyl halide and K$_2$CO$_3$ under an atmosphere of CO$_2$ can be explained by the intermediary formation of 1,3-dioxolan-2-one **82** followed by the degradation to **83** (Scheme 6). A related domino carboxylation/decarboxylation sequence *via* cyclic carbamates was also introduced for the synthesis of 3(2H)-furanones.[92] Substrates of type **81** were easily obtained through transition metal-catalyzed cross-coupling of propargylic alcohols with aryl iodides in the presence of CO.[93,94]

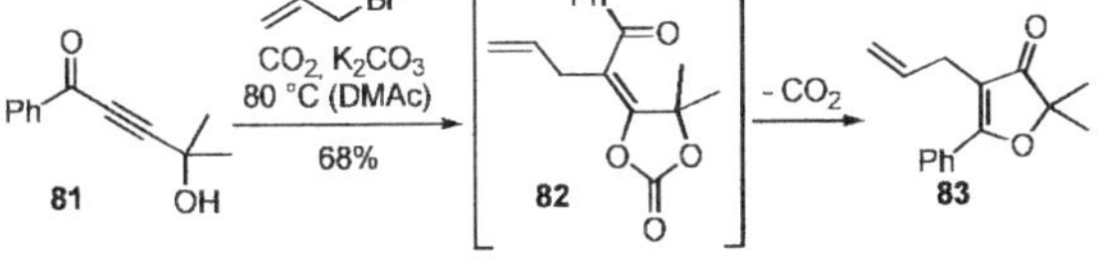

Scheme 6. Alkylative synthesis of 3(2H)-furanone according to Inoue *et al.*.

A class of reactions that involves a cyclization to create the bond between C5 and O1 of the 3-furanone moiety is based on the hydration of an acetylene moiety to result in α'-hydroxy 1,3-diketone species (Figure 19). According to this consideration, Williams *et al.* demonstrated that 3(2*H*)-furanone **85** is easily accessed from acetylenic alcohol **84** by utilizing $BF_3 \cdot OEt_2$ in the presence of catalytic amounts of mercuric oxide and trichloroacetic acid (Scheme 7, eq. 1).[95] Saimoto *et al.* disclosed a protocol to afford 3(2*H*)-furanones by utilizing a mixture of sulfuric acid and methanol.[96] Alternatively, the action of a polymer reagent Hg/Nafion-H was found to make the very same conversion possible (Scheme 7, eq. 2).[97] Schlessinger *et al.* followed this strategy for their studies on the synthesis of quassimarin.[98] 3(2*H*)-Furanone intermediate **89** was formed with sulfuric acid in THF from propargylic alcohol precursor **88** in 81% yield (as a mixture of anomers) (Scheme 7, eq. 3). A way to build up perfluoropropyl-3(2*H*)-furanones was discovered by Calas *et al.*.[99] They used 3-(perfluoro-butyl)alk-2-yn-1-ols such as **90** to produce carbonyl compound **91** with formic acid through rapid triple bond hydration (Scheme 7, eq. 4). Surprisingly, under these Rupe conditions, elimination to the unsaturated carbonyl compound was not observed. Instead, clean cyclization and elimination of HF led to the formation of the 3(2*H*)-furanone skeleton.

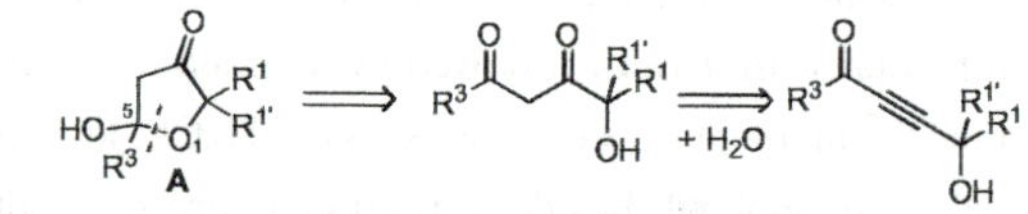

Figure 19. Ynone hydration strategy.

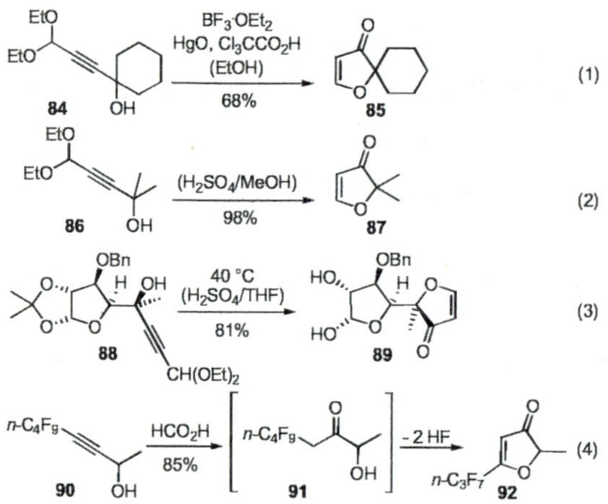

Scheme 7. Triple bond hydration approaches.

4.1.2. Bond formation between C4 and C5

Another conceivable pathway to access 3(2*H*)-furanones through a final dehydration is that key intermediate **A** results from a bond formation between C4 and C5. As depicted in Figure 20, α-acyloxy-ketones are suitable substrates, which are treated with a base to initiate cyclization. Following this consideration, in 1971, Margaretha reported that the sodium hydride mediated acylation of hydroxy ketone **93** with ethyl formate led to cyclization. Under acidic conditions, dehydration occurred to afford furanone **87** in 50% yield (Scheme 8, eq. 1).[100–102] Nevertheless, a more detailed study carried out by Smith III *et al.* showed that α-acetyloxy-ketones are of limited value as starting materials for the selective synthesis of 3(2*H*)-furanones.[72] In the case of R^3 = Ph, the desired 3(2*H*)-furanone **96** was obtained in 88%; on the other hand, the reaction conditions (NaH/DMSO) led to the formation of butenolide **97** in poor yield when R^3 is methyl. These results suggest that an equilibrium is established between the ketone and ester enolates, thus, limiting the applicability. Nevertheless, this cyclization mode was rendered feasible through enhancing the acidity of the acyl group. For example, spiro furanone derivatives such as **100** were formed in excellent yields through condensation of γ-acetoxy-β-ketoesters of type **99** under basic conditions.[103] These compounds were easily accessed through hydrolysis of orthoesters **98**, which were afforded in a palladium-catalyzed oxidative cyclization/carbonylation of propargylic esters. In a similar fashion, α-hydroxy-ketones undergo condensation with diethyl oxalate under basic conditions.[49] Subsequent 5-*exo*-trig cyclization and saponification leads to the carboxylic acids **101** containing the 3(2*H*)-furanone moiety. With specific substrates, a related cyclization step was induced with CsF.[104,105]

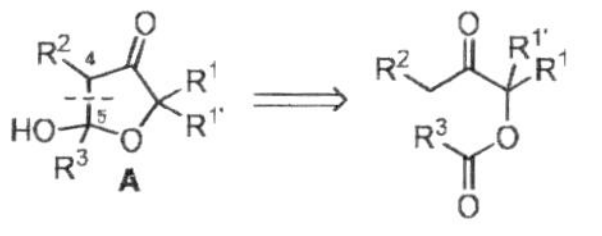

Figure 20. Bond formation between C4 and C5.

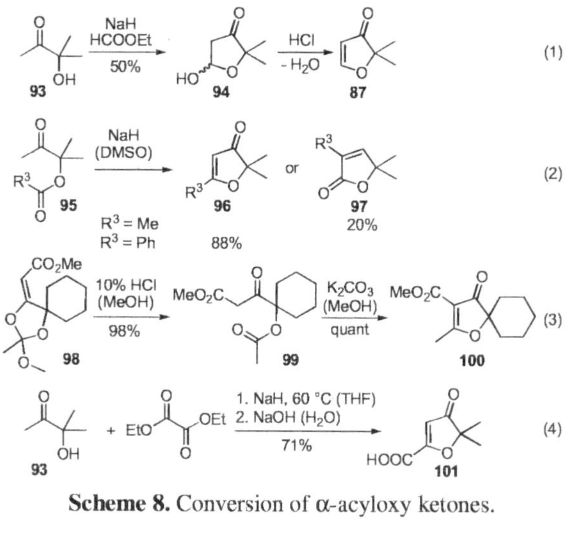

Scheme 8. Conversion of α-acyloxy ketones.

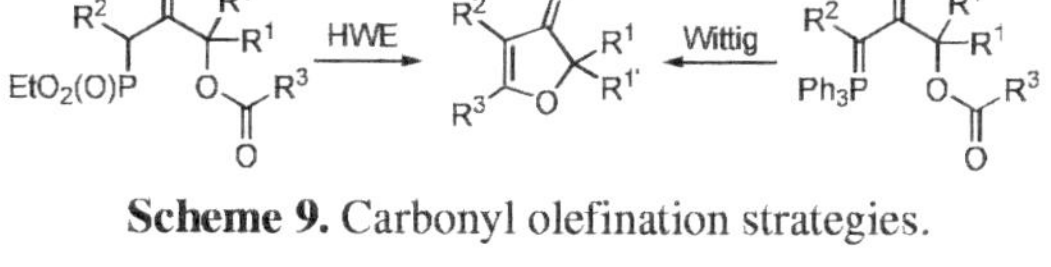

Scheme 9. Carbonyl olefination strategies.

An alternative entry into 3(2*H*)-furanones through bond formation between C4 and C5 was realized with carbonyl olefination reactions. Both a Wittig[106] and Horner-Wadsworth-Emmons reactions[107–110] were employed to directly install the C=C double bond of the heterocyclic system. The underlying strategies of this approach are illustrated in Scheme 9.

4.1.3. Bond formation between C2 and C3

A particular access to a variety of 3(2*H*)-furanones is through bond forming cyclization between C2 and C3 (Figure 21). In these processes, an intramolecular condensation leads to the construction of the ring system. As exemplified for the reaction of diester **102**, treatment with NaOEt results in furanone formation using a Dieckmann cyclization as key step (Scheme 10). Notably, the double bond geometry changes during the course of the reaction, presumably due to the basic reaction conditions.[111]

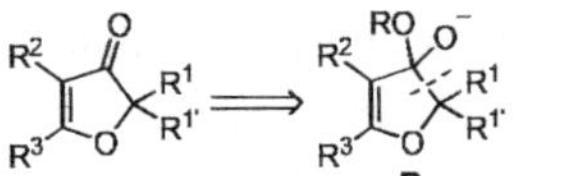

Figure 21. Bond formation between C2 and C3 through Dieckmann condensation.

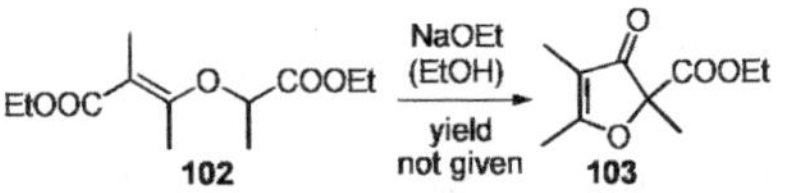

Scheme 10. Example for cyclization through Dieckmann condensation.

4.2. Substitution approaches

A general method for synthesizing substituted 3(2*H*)-furanones by bond formation between O1 and C2 is based on an intramolecular nucleophilic substitution in the α-position of the C3 carbonyl group in **C** (Figure 22). In most cases, this strategy is utilized for the synthesis of 3(2*H*)-furanone skeletons that are monosubstituted at C2 ($R^{1'}$ = H; R^1 = alkyl, aryl) due to the ease of cyclization for these systems.

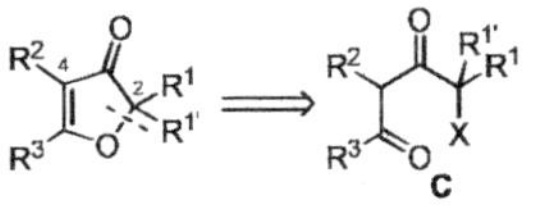

Figure 22. General substitution strategy.

In a seminal work, Fröhlisch *et al.* demonstrated that γ-halo-β-keto-carboxylic esters can undergo an intramolecular cyclization reaction upon treatment with triethylamine to yield 3(2*H*)-furanones (Scheme 11, eq. 1).[112] A related intramolecular alkylation was used by Yamaguchi *et al.* to cyclize 6-halo-3,5-dioxoalkanoates such as **106**.[113] This entry is particularly attractive for the preparation of a wide range of functionalized 3(2*H*)-furanones since the starting 6-halo-1,3,5-trioxo compounds are easily accessed from various α-chloroacetic acid chlorides and 1,3-bis(trimethylsilyloxy)-1,3-butadienes.[114,115] In a powerful two-component cyclocondensation, isocyanate **108** reacted with ethyl 4-bromo-3-oxopentanoate (**109**) to the corresponding 3(2*H*)-furanone.[116]

Further extension of this heterocycle synthesis was accomplished by combining a C-acylation of an enolate with a cyclization in a domino process that represents a flexible entry into 4,5-diaryl furanones **112** (Scheme 12).[117] While the corresponding acyl bromides mainly funrnished O-acylation products, the selective C-acylation is due to the softness of the cyanide leaving group. With an excess of base, the one-pot sequence results in yields between 33% and 77%. Importantly, the resulting furanones possess two methyl substituents at C2.

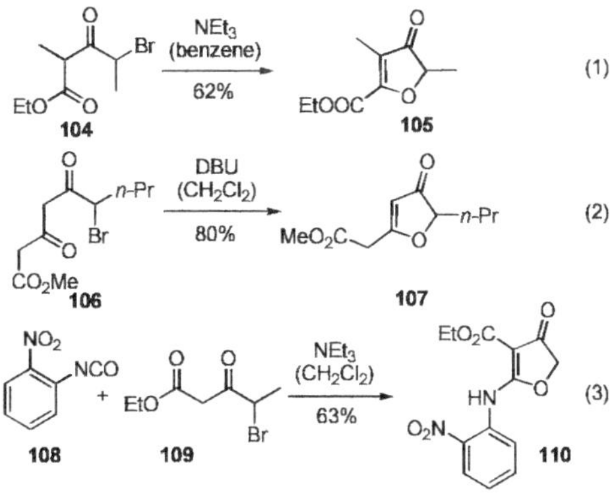

Scheme 11. Selected furanone syntheses with an alkylative cyclization.

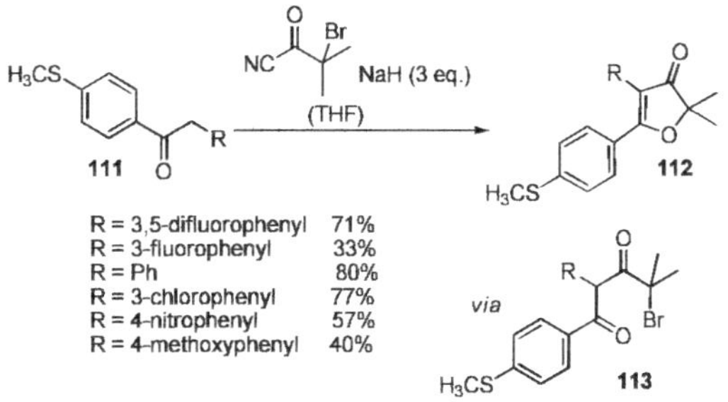

Scheme 12. Domino acylation/cyclization sequence.

While most of the substitutions that follow the general synthetic scheme illustrated in Figure 22 take use of bases,[118] an interesting alternative was developed by Hoffmann et al.. Here, the conversion of dibromoketone **114** was achieved with a zinc-copper couple in DMF (Scheme 13).[119–121]

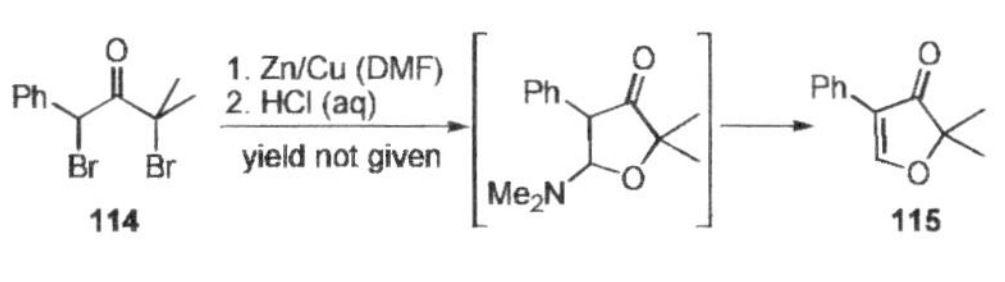

Scheme 13. Cyclization with metals.

It should be noted that a multitude of other reactions toward 3(2*H*)-furanones with a condensation or substitution step have been reported; due to their specificity, such reactions are not discussed in detail although they are useful in distinct cases. For example, Meister et al. investigated several approaches to 3(2*H*)-furanones using halogens as leaving groups.[2,122] Greenhill et al. found the unexpected formation of 3(2*H*)-furanones when investigating the Mannich reaction of butane 2,3-dione with dimethylamine hydrochloride and *para*-formaldehyde.[123] Skvortsov *et al.* produced 5-amino-3(2*H*)-furanones from nitriles such as **116** with LiOH in moderate yields (Scheme 14).[124] Nevertheless, the scope of this reaction appears to be limited.

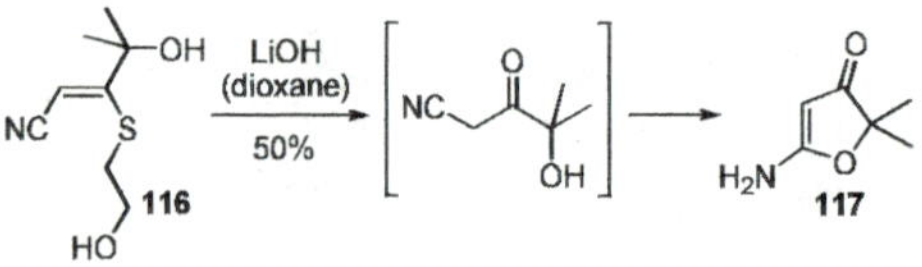

Scheme 14. Nitriles as cyclization precursors.

4.3. Addition approaches

The addition of an oxygen nucleophile onto a π-system as illustrated in Figure 23 easily generates 3(2*H*)-furanones. When alkynes are attacked (see **D**), the heterocycles are formed directly without need for an additional oxidation. On the other hand, substrates of type **E** lack the grade of unsaturation that leads to 3(2*H*)-furanone products, thus, requiring an oxidative cyclization step. Independent of the type of reagent employed, these cyclizations are 5-*endo* processes creating a new bond between O1 and C5.

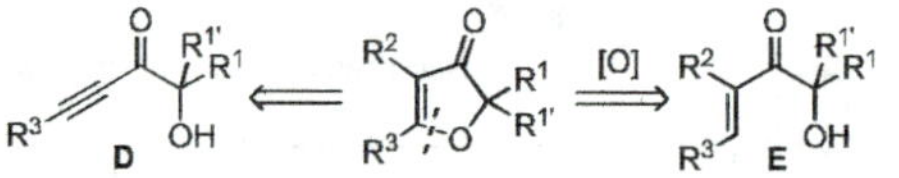

Figure 23. Modes of heterocyclization.

As an early example of reacting linear ynones in a 5-*endo*-dig closure, Baldwin *et al.* simply refluxed **118** in methanol to effect cyclization in the presence of NaOMe (Scheme 15).[125] In the absence of the base, formation of bullatenone (**1**) was not observed. This cyclization is also induced under acidic conditions (*e.g.*, trifluoroacetic acid)[126] albeit with a reduced rate. Marson *et al.* prepared 3(2*H*)-furanones in a catalytic asymmetric domino process consisting of Sharpless dihydroxylation of enone **119** and subsequent mercury-catalyzed cyclization.[127] The corresponding furanones were obtained in moderate to high yields and good enantioselectivity.

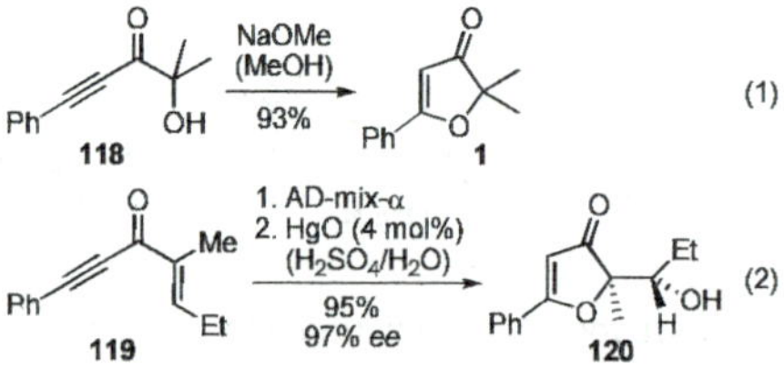

Scheme 15. 5-*Endo*-dig cyclization.

A fascinating domino cyclization reaction for the synthesis of 3(2*H*)-furanones was developed by Gouverneur *et al.*.[128] The cyclization of α-hydroxyynones with alkyl or aryl substituents at the alkyne gave the corresponding furanones such as **122** in the presence of an excess of ethyl acrylate (Scheme 16). The mechanism of the palladium-catalyzed cascade is believed to proceed through an initial coordination of

Pd(II) to the alkyne resulting in a Wacker-type 5-*endo* oxypalladation. Since the vinylpalladium species **123** cannot undergo β-hydride elimination, a subsequent Heck-type carbopalladation with ethyl acrylate gives **124**. After reductive elimination, Pd(0) is reoxidized by molecular oxygen to complete the catalytic cycle.

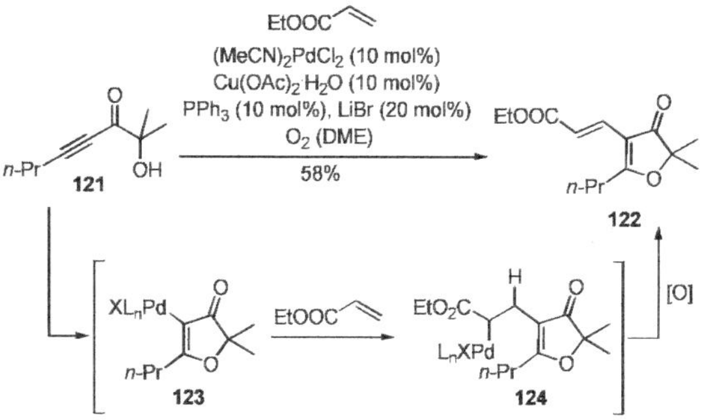

Scheme 16. Domino Wacker/Heck reaction.

A general method for synthesizing substituted 3(2*H*)-furanones by trapping a stabilized oxonium ion intermediate with various alcohol nucleophiles was introduced first by Liu *et al.* (Scheme 17).[129] In this reaction, the onium ion **126** is formed by the nucleophilic attack of the carbonyl oxygen onto the gold-activated alkyne.

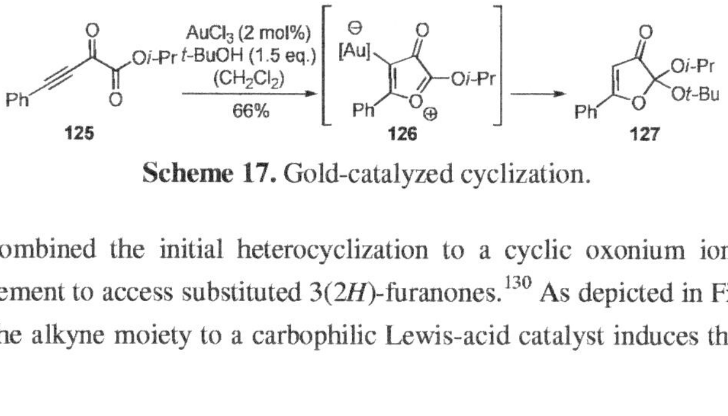

Scheme 17. Gold-catalyzed cyclization.

Kirsch *et al.* combined the initial heterocyclization to a cyclic oxonium ion intermediate with a pinacol-type rearrangement to access substituted 3(2*H*)-furanones.[130] As depicted in Figure 24, it is believed that coordination of the alkyne moiety to a carbophilic Lewis-acid catalyst induces the 5-*endo* attack of the carbonyl group.

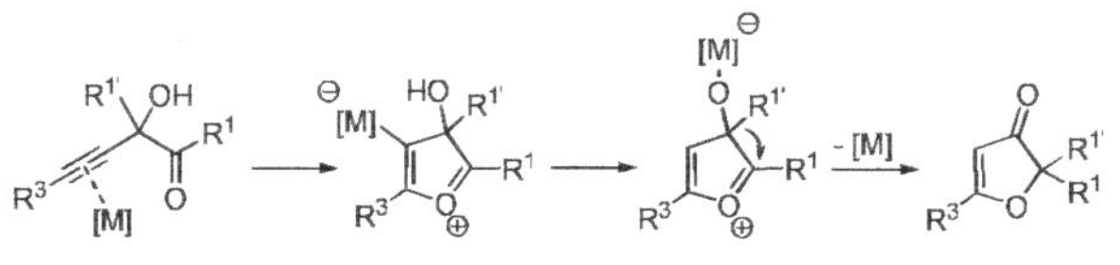

Figure 24. Proposed pathway for the synthesis of 3(2*H*)-furanones.

The putative oxonium ion then undergoes a 1,2-alkyl migration analogous to a formal α-ketol rearrangement. In the course of this cascade, a bond between O1 and C5 is formed accompanied by the construction of the quaternary center at C2. A representative example of this synthesis of 3(2*H*)-furanones is shown in Scheme 18.[131,132] Spirocyclic furanones are easily accessible through ring-contracting cyclization (*e.g.*, **128**→**129**). Acyclic systems also reacted, albeit with a significantly reduced scope. Shortly after, a closely related protocol to access spirocyclic and open-chained 3(2*H*)-furanones was devised by Sarpong *et* *al.*.[133] A valuable extension of this heterocycle synthesis was accomplished by using *N*-iodosuccinimide (NIS) in an electrophile-induced route to 4-iodo-3-furanones.[134] In a variant, AuCl$_3$ catalyzes the very same domino reaction in the presence of NIS to provide the iodo-substituted heterocycles in moderate to excellent yields. In combination with a subsequent cross coupling reaction, this transformation provides a convenient and flexible approach to fully substituted 3(2*H*)-furanones.

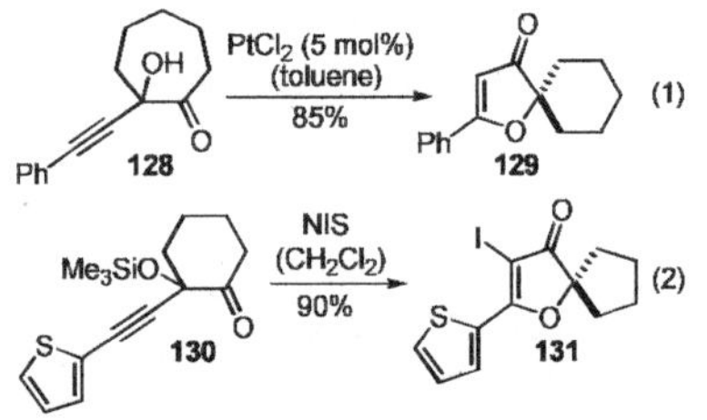

Scheme 18. Furanone syntheses following a heterocyclization/1,2-migration pathway.

One can find only a few reactions in literature that take use of a cyclization onto an enone moiety of type **E** under bond formation between O1 and C5. The most impressive example of this type was developed by Gouverneur *et al.* while investigating palladium-catalyzed oxidative cyclizations (Scheme 19).[135] Here, a Wacker-type oxypalladation is terminated by a β-hydride elimination that gives rise to the furanone core structure. Subsequent oxidation of the palladium species with molecular oxygen regenerates the palladium(II) that is required for the double bond activation. These oxidative heterocyclizations take place without affecting the stereochemical integrity of the starting substrates such as **132**. As a singular example, Tius *et al.* reported a rapid ring closure of diketone **134** under acidic conditions followed by double-bond migration to afford furanone **135** in 78% yield.[136]

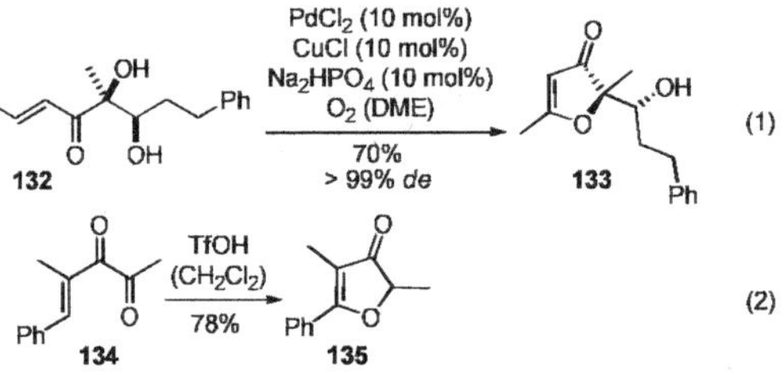

Scheme 19. Cyclization with enones.

The formation of a 3(2*H*)-furanone was also observed when reacting acetylacetone with 3-methyl-furan-2,5-dione **136**. This reaction most likely proceeds *via* a 5-*exo* Michael-addition as shown in Scheme 20 (eq. 1).[137,138] A related ring forming reaction is responsible for the one-pot conversion of 2-(acylmethyl)-furans **139** into 3(2*H*)-furanones **141** (Scheme 20, eq. 2).[139] By treatment with *m*-chloroperbenzoic acid (MCPBA) such furans undergo a selective oxidation of the furan nucleus leading to an intermediate **140** that is not isolable and directly undergoes cyclization under the reaction conditions.

Heterocyclizations involving allenes are also described to access 3(2*H*)-furanones *via* an addition approach although the number of examples disclosed in literature are low. In this context, it was demonstrated that allenic ketones afford 3(2*H*)-furanones in the presence of Hg(OAc)$_2$ (Scheme 21, eq.

1).[140] An interesting 3(2H)-furanone synthesis that proceeds through allene species was developed by Hiyama *et al.* in a work that was aimed to overcome the lack of regiocontrol in the hydration of alkynes (Scheme 21, eq. 2).[141] A mechanism *via* Ag(I)-catalyzed cyclization under acetoxy migration followed by oxidation of the resulting enol acetates with DDQ was proposed to produce furanones such as bullatenone (1).

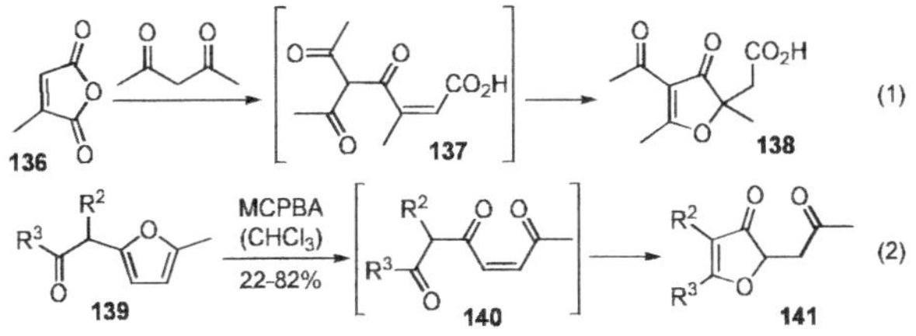

Scheme 20. Furanone synthesis according to Berner and Kolsaker.

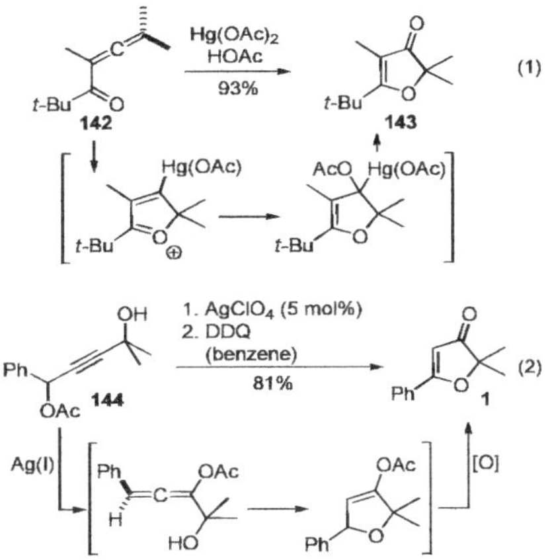

Scheme 21. Allenic species in 3(2H)-furanone syntheses.

4.4. Miscellaneous reactions

A variety of strategies that do not exactly fit into the sections discussed above have also been established as tools for the construction of 3(2H)-furanones. In some cases, the heterocyclic core of 3(2H)-furanones was unexpectedly formed *via* skeletal reorganization during isomerization attempts on various substrate classes.[142–145] In other cases,[146–157] the formation of 3(2H)-furanone structures is somewhat limited due to the characteristics of the employed substrates.

An interesting protocol was developed by Calter *et al.* by using a C-O insertion reaction.[158,159] This powerful sequence efficiently converts diazoacetals such as **145** into intermediates of type **146**, presumably through metal carbenoid formation in the presence of Rh$_2$(OAc)$_4$ (Scheme 22, eq. 1). Upon treatment with alumina, the instability of these compounds led to the clean formation of 3(2H)-furanone **147** in good overall yield. In a recent study, Winkler *et al.* demonstrated that 3-silyloxyfurans **148** undergo aldol reactions under furanone formation (Scheme 22, eq. 2).[160] The stereochemical outcome is influenced by the sterical hindrance of the employed aldehydes.

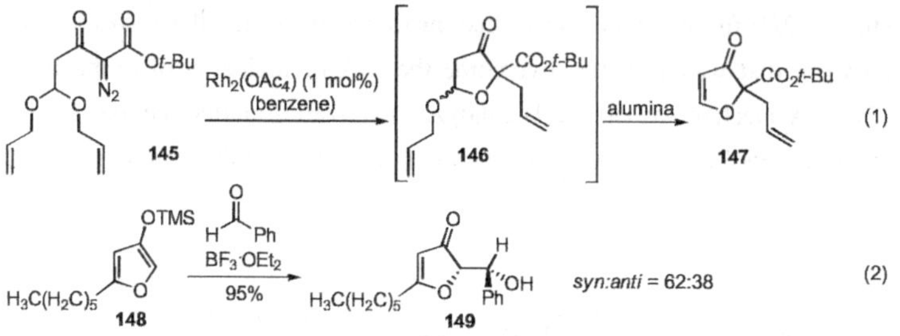

Scheme 22. C-O insertion sequence for the generation of 3(2*H*)-furanones.

Noteworthy, a number of recently developed methods for transition metal-catalyzed bond formations still remain to be applied to the synthesis of 3(2*H*)-furanones. For instance, the ring-closing metathesis reaction might be a potent method for the construction of the double-bond between C4 and C5 in the furanone core.

5. Selected applications in total synthesis

Total synthesis applications of the 3(2*H*)-furanone-forming reactions that are discussed in the previous sections are relatively rare due to the amount of natural products possessing this structural element (see Section 2). Although the following discussion puts the focus on the method used to construct the 3(2*H*)-furanone unit in the respective natural product, the general synthetic strategy is outlined as well.

It should be noted that, despite convenient *de novo* approaches towards gregatin B,[161,162] erigeronic acid,[163] tachrosin (**34**),[164] hyperolactone A (**9**),[165] hyperolactone C (**11**)[166] and wallemia C,[167] these naturally occurring 3(2*H*)-furanones are not detailed in the following part; the reader is directed to the original literature.

5.1. Bullatenone

In 1958, the total synthesis of bullatenone (**1**), the first 3(2*H*)-furanone isolated from natural sources, was successfully accomplished as outlined in Figure 25.[5] A Grignard addition of 2-methylbut-3-yn-2-ol onto benzaldehyde followed by oxidation and triethylamine-mediated cyclization gave crystalline **1**. Since then, a huge variety of bullatenone syntheses were reported that are not covered in this review.[3,73,80,82,97,168–173]

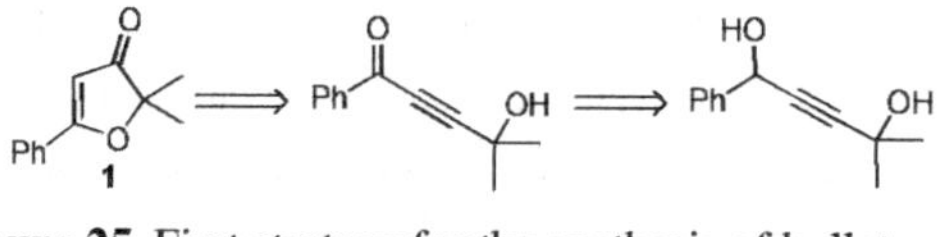

Figure 25. First strategy for the synthesis of bullatenone.

5.2. Geiparvarin

Apart from bullatenone, geiparvarin (**33**) is the naturally occurring 3(2*H*)-furanone with the most synthesis attempts. In 1980, Smith III *et al.* completed the first total synthesis of geiparvarin by direct alkylation of 5-ethyl-2,2-dimethylfuran-3(2*H*)-one (**150**) with aldehyde **151** followed by dehydration (Figure 26).[174,175] As the double bond configuration had not been determined before this *de-novo* approach, Smith's strategy had the major advantage to gain access to both diastereomers. Further strategies for the geiparvarin synthesis have a common key step that is the O-alkylation of phenol **152** with the furanone counterpart **153** (X = Br, OMs etc.). These strategies differ only in the way how the 3(2*H*)-furanone moieties are generated.

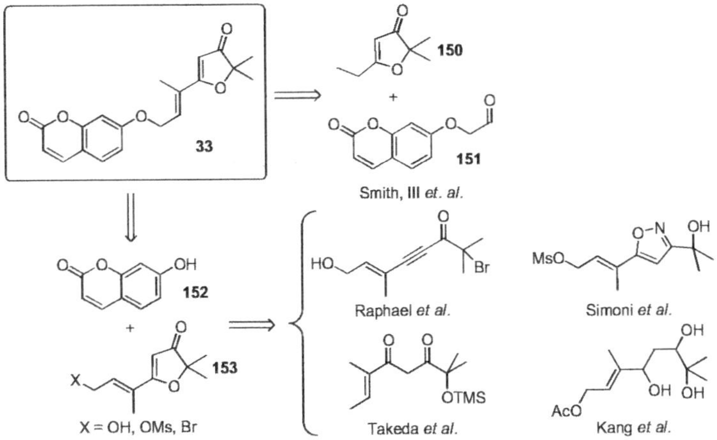

Figure 26. Strategies for the synthesis of geiparvarin.

For instance, Raphael *et al.* constructed the 3(2*H*)-furanone core by using a sequence of hydrolysis and cyclization of enamine **154** with aqueous acetic acid (Scheme 23).[169] The creation of **155** then follows a bond formation between O1 and C5 whereas **154** can be viewed as a synthetic equivalent for α'-hydroxy-1,3-diketones. Conversion of **155** to its mesylate and treatment with phenol **152** in refluxing acetone in the presence of K_2CO_3 produced geiparvarin. The same group developed alternative routes to geiparvarin with a late-stage construction of the 3(2*H*)-furanone moiety.[170]

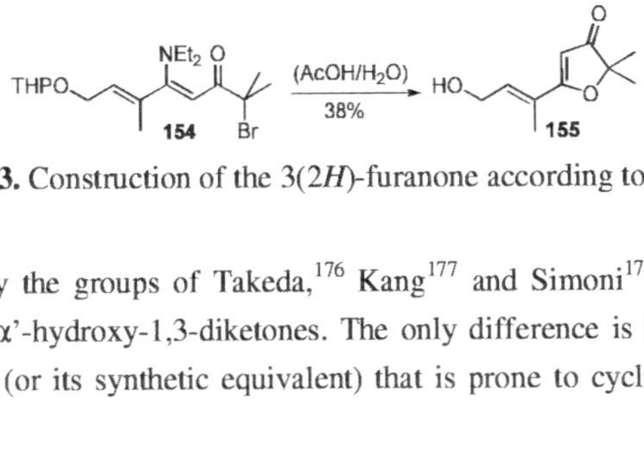

Scheme 23. Construction of the 3(2*H*)-furanone according to Raphael *et al.*.

Other approaches by the groups of Takeda,[176] Kang[177] and Simoni[178,179] also relied on the ease of furanone formation with α'-hydroxy-1,3-diketones. The only difference is the creation of the α'-hydroxy 1,3-diketone intermediate (or its synthetic equivalent) that is prone to cyclization under acidic conditions (Scheme 24).

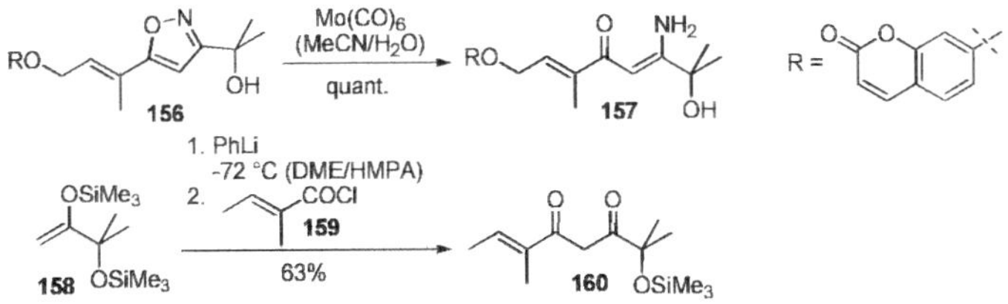

Scheme 24. α'-Hydroxy 1,3-diketone syntheses.

For example, Simoni *et al.* employed an isoxazole ring that, after unmasking with $Mo(CO)_6$ in wet acetonitrile, gave enaminoketone **157**.[178] Alternatively, Takeda *et al.* lithiated silylenol ether **158** to obtain 1,3-diketone **160** through addition of acid chloride **159**.[176] Further strategies that led to the successful synthesis of geiparvarin are not illustrated herein.[171,180]

5.3. Jatrophone

The seminal synthesis of jatrophone (**2**) developed by Smith III *et al.* is based on the assembly of the two intermediates **162** and **163** (Figure 27). The key step in the synthesis is an aldol reaction between both compounds followed by oxidation of the obtained alcohol with Collins' reagent. Exposure of α'-hydroxy-1,3-diketone **164** to HCl in THF resulted in the formation of 3(2*H*)-furanone **165**. It is important to note that the macrocyclization of key intermediate **161** was found to be sluggish yielding not more than 23%. Nevertheless, (+)-jatrophone, (+)-epijatrophone, (+)-hydroxyjatrophone A (**4**),[181a] (+)-hydroxyjatrophone B (**3**)[181a] and normethyljatrophone[181b] were synthesized in this way.

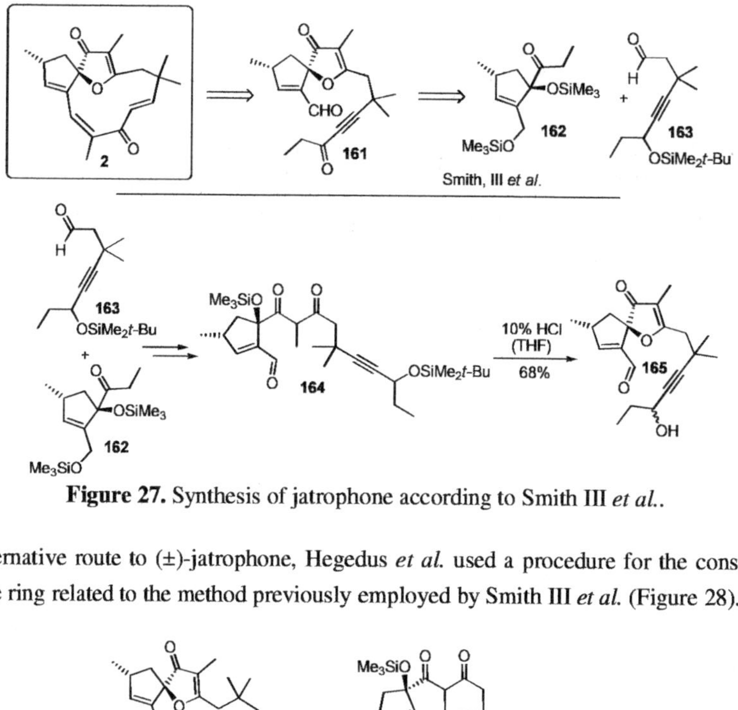

Figure 27. Synthesis of jatrophone according to Smith III *et al.*.

In an alternative route to (±)-jatrophone, Hegedus *et al.* used a procedure for the construction of the 3(2*H*)-furanone ring related to the method previously employed by Smith III *et al.* (Figure 28).[182]

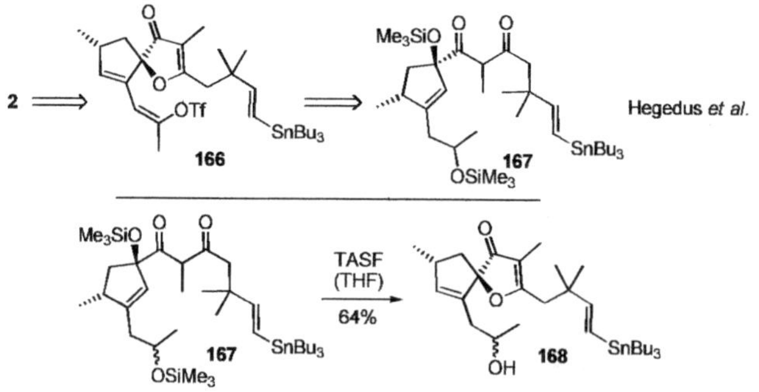

Figure 28. Synthesis of jatrophone according to Hegedus *et al.*.

To generate the spirofuranone **168** without protodestannylation, the cyclization of **167** required aprotic conditions. Such a requirement was perfectly fulfilled with tris(dimethylamino)sulfur (trimethylsilyl)-difluoride (TASF) as an anhydrous fluoride source. The synthesis of (±)-jatrophone and its epimer was completed by using an intramolecular palladium-catalyzed carbonylative coupling of an vinylic triflate with an organostannane (structure **166**). In the case of jatrophone, this macrocyclization yielded the natural product in 24%.

In 1992, the total synthesis of optically active (+)-jatrophone was described by Wiemer *et al.*.[110] A route was developed that used a chiral pool reagent to install the stereogenic center at C2 (Figure 29). The furanone ring system was generated in 62% through an intramolecular Horner-Wadsworth-Emmons reaction with key intermediate **171**. The resulting bicycle **173** was subjected to palladium-catalyzed cross-coupling with vinyl stannane **172** to afford **170**. The macrocyclization followed then the route developed by Smith III *et al.* to provide the common key intermediate **169**. The convergent synthesis affords ynone **169** in 11 steps from (*R*)-(+)-3-methyladipic acid in an overall yield of 15%.

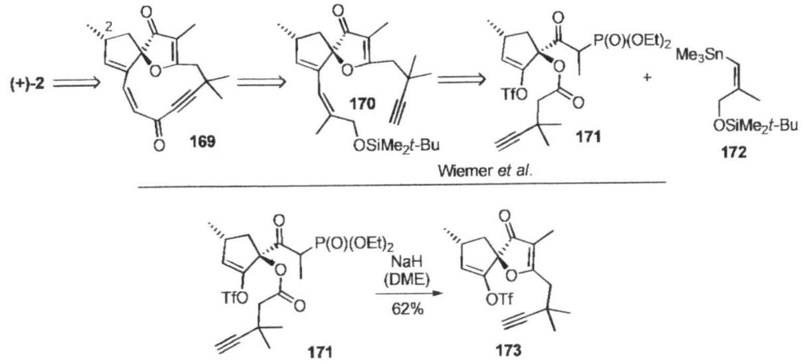

Figure 29. Synthesis of jatrophone according to Wiemer *et al.*.

5.4. Longianone

As outlined in Scheme 25, longianone (**8**) was synthesized by Steel in 1999 by using a radical cyclization reaction of **174** to create the spirocyclic compound **175**.[183] The 3(2*H*)-furanone unit was then generated through late-stage introduction of the α,β-unsaturation with PhSeCl followed by oxidation.

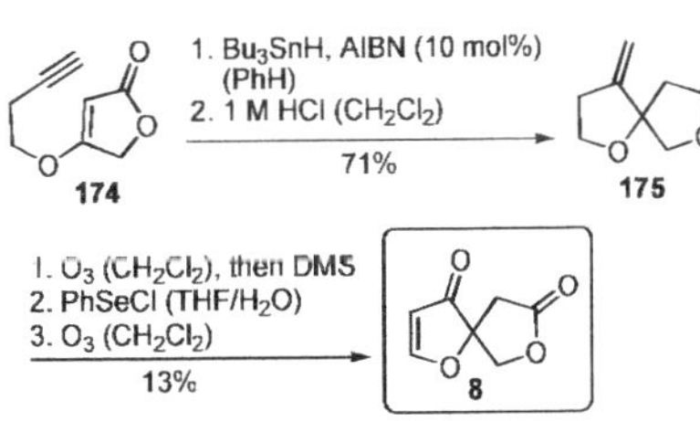

Scheme 25. Synthesis of longianone.

5.5. Trachyspic acid

The first total synthesis of (±)-trachyspic acid (**7**) was accomplished by Hatakeyama *et al.* based on a Cr(II)/Ni(II)-mediated key step to connect aldehyde **178** containing the citric acid moiety and the long-chain triflate **177** (Figure 30).[184] Thus, **176** was obtained in 83% yield in the presence of 10 mol% $NiCl_2$ and $CrCl_2$ in DMF. After Swern oxidation followed by desilylation, treatment of advanced intermediate **179** with 3 M $HClO_4$ in THF afforded spiroketal **180**. The $3(2H)$-furanone moiety was constructed through ozonolysis of **181** and exposure of the resulting ketone to TFA gave (±)-trachyspic acid. To access enantiopure trachyspic acid, Rizzacasa *et al.* developed a quite similar route that generates the $3(2H)$-furanone core *via* the very same synthetic scheme.[185]

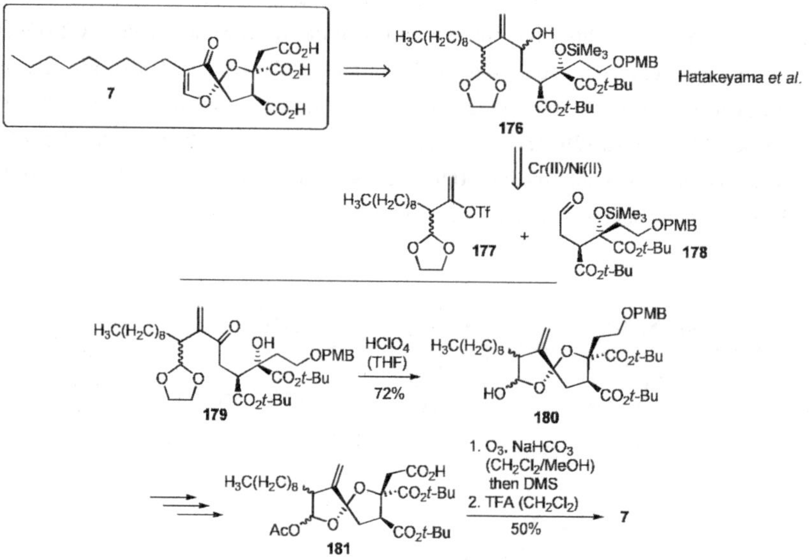

Figure 30. Synthesis of trachyspic acid according to Hatakeyama *et al.*.

5.6. Azaspirene, pseurotin A and pseurotin F₂

As illustrated in Figure 31, compound **182** was regarded as a versatile intermediate from which (−)-azaspirene (**12**) could be easily assembled.[186] The conversion of **182** into α'-hydroxy 1,3-diketone **183** was accomplished by using an aldol condensation with $(2E,4E)$-hepta-2,4-dienal followed by Dess-Martin oxidation.

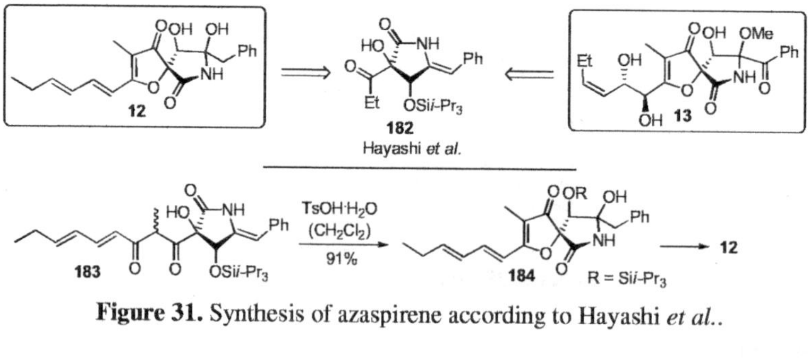

Figure 31. Synthesis of azaspirene according to Hayashi *et al.*.

Upon treatment of **183** with a catalytic amount of TsOH·H₂O, complete formation of the $3(2H)$-furanone moiety and hydration of the benzylidene group occurred to achieve furanone **184**. Deprotection of the triisopropylsilyl group with NH$_4$F then gives synthetic azaspirene that exhibited properties identical to those of the natural product.

For the total synthesis of pseurotin A (**13**) by Hayashi *et al.*, γ-lactam **182** was again key intermediate.[187] The furanone structure was introduced as discussed for the azaspirene synthesis. In 2004, Tadano *et al.* described the syntheses of pseurotin A, pseurotin F$_2$ and azaspirene whereas the furanone construction follows the assembly in the Hayashi syntheses.[188]

5.7. Eremantholide A

An elegant total synthesis of (+)-eremantholide A (**28**) confirming the assignment of its absolute configuration was accomplished by Boeckman, Jr. *et al.* in 1991.[189] The strategy features a Ramberg-Bäcklund sequence to effect the crucial ring contraction of key intermediate **185**. For the synthesis of the furanone ring, a classical cyclization of α'-hydroxy 1,3-diketone **186** was utilized (Figure 32).

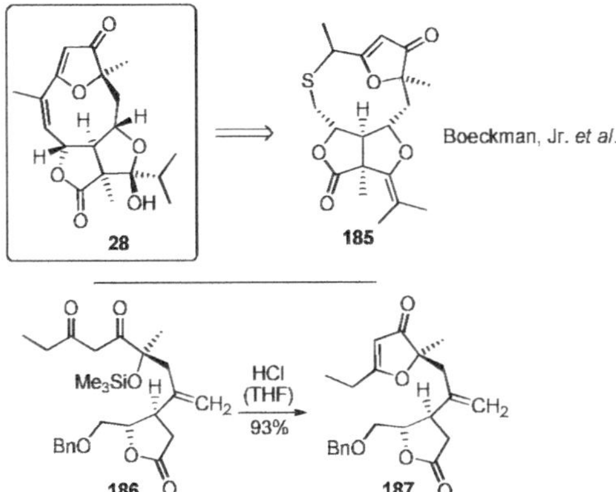

Figure 32. Synthesis of eremantholide A according to Boeckman, Jr. *et al.*.

As illustrated in Figure 33, an alternative route toward (+)-eremantholide A was described by Tadano *et al.* based on the enolate C-alkylation of simple 5-ethyl-2-methylfuran-3(2*H*)-one with triflate **188**.[190,191]

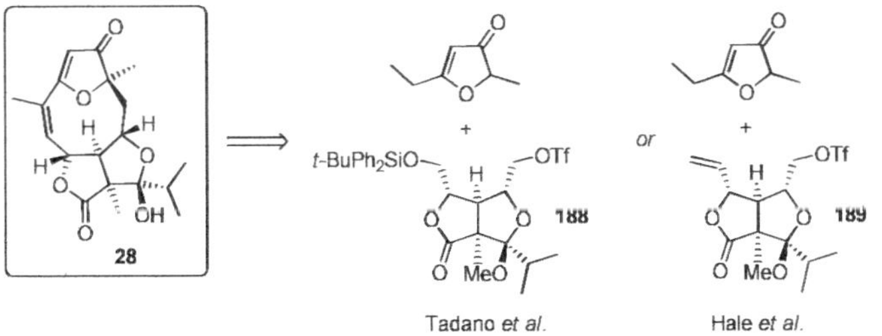

Figure 33. Synthesis of eremantholide A according to Tadano and Hale *et al.*.

The enolate alkylation was thoroughly investigated since the choice of base, solvent and additive was found to be critical for the diastereoselectivity. Best results for the desired diastereoisomer were obtained when using an equimolar amount of 18-c-6 and KHMDS in toluene. Macrocyclization was then achieved with an intramolecular vinylogous aldol reaction. Hale *et al.* mainly followed this approach to eremantholide A.[192] As alkylating reagent, triflate **189** was employed rather than **188**, thus enabling a ring-closing metathesis for macrocyclization.

6. Reactivity of 3(2*H*)-furanones

Since a huge variety of reactions with 3(2*H*)-furanones have been reported, this section is a personal selection that only summarizes the most intriguing transformations. As such, selected alkylations, ring modifications, photochemical reactions, and cycloadditions are briefly discussed in the following part.

6.1. Alkylations and related reactions

As discussed in the previous section, 3(2*H*)-furanones can be lithiated to form the corresponding enolates that react with various alkyl halides under predominant C-alkylation at C2 (Scheme 26).[72,84,193–195] In the case of a disubstituted C2-position, alkylation takes place at the γ-position.[72,196] From a synthetic point of view, it is interesting that γ-functionalization with phenylselenyl chloride followed by oxidative elimination affords valuable furadienones.[72]

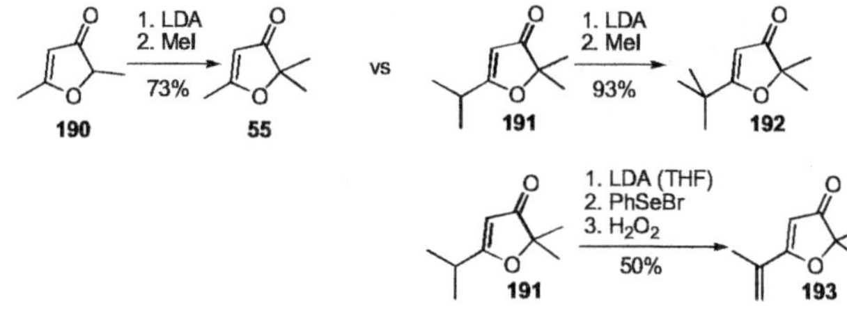

Scheme 26. Regioselective alkylations.

The reaction of Schiff bases with 3(2*H*)-furanones was also examined with the result that clean addition occurs at the C2-position of the furanone.[197] If present, a C5-methyl group is attacked in a second addition step thereafter.

Organocuprates add to simple 3(2*H*)-furanones in a conjugate addition.[72] In Baldwin's total synthesis of racemic methyl nonactate, such a conjugate addition of a Grignard reagent in the presence of CuBr·Me₂S followed by methylation yielded **194** in a *trans/cis*-mixture of 10/1 and 56% yield (Scheme 27).[198] Csuk *et al.* developed a related 1,4-addition of silyl ketene acetals such as **195** to 2,2-dimethyl-3(2*H*)-furanone.[199–201]

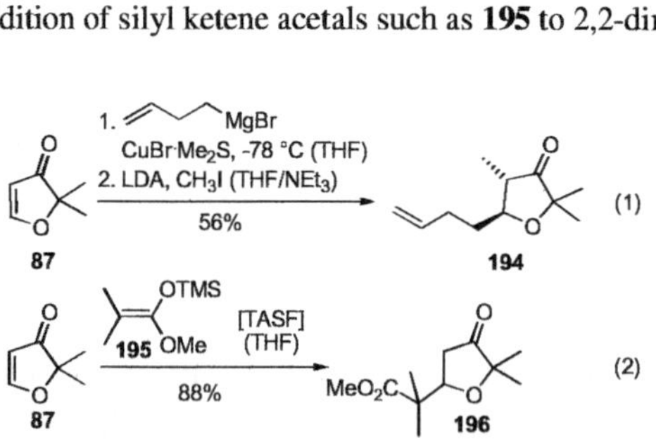

Scheme 27. Conjugate addition to 3(2*H*)-furanones.

6.2. Ring modifications

Ring modifying reactions are quite common to transform the 3(2*H*)-furanone moiety into other heterocycles of value. For example, pyrazolones are easily accessed by treating the furanone skeleton with diazonium salts (Scheme 28, eq. 1).[202] Another heterocyclic system that can be generated from furanone starting materials are pyrazoles. To realize this transformation ethyl 2-hydrazinylacetate might be used under basic conditions (Scheme 28, eq. 2).[203] On the other hand, 3(2*H*)-furanones are converted into isoxazoles such as **201** in the presence of hydroxylamine.[204] A variety of other heterocycle syntheses that rely on the nucleophilic ring opening of the furanone core and subsequent heterocyclization under formation of a novel ring system were reported as well.[116,205–211]

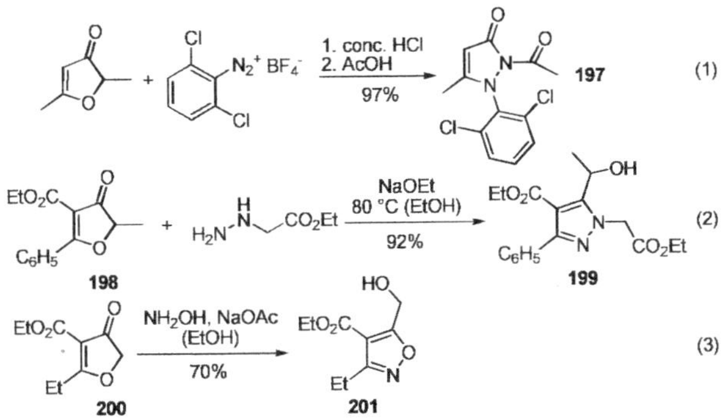

Scheme 28. Selected heterocycle syntheses.

In the presence of primary amines or ammonia, a simple ring opening of the 3(2*H*)-furanone unit was observed as exemplified in Scheme 29.[212,213]

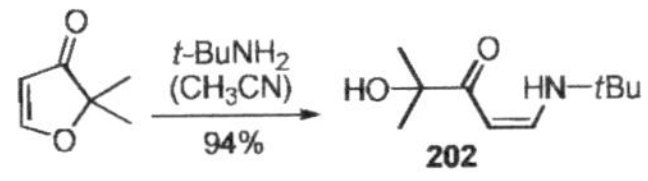

Scheme 29. Furanone opening with primary amines.

An interesting tetrahydrofuran synthesis was described by Baldwin *et al.* (Scheme 30).[214] After addition of dimethyloxosulfonium methylide to 2,2-dimethyl-3(2*H*)-furanone, a 1:1 mixture of the two products **203** and **204** was obtained. Unfortunately, it was not possible to alter the product distribution. In a similar way, treatment with dimethyl dioxirane led to the formation of corresponding epoxide.[215]

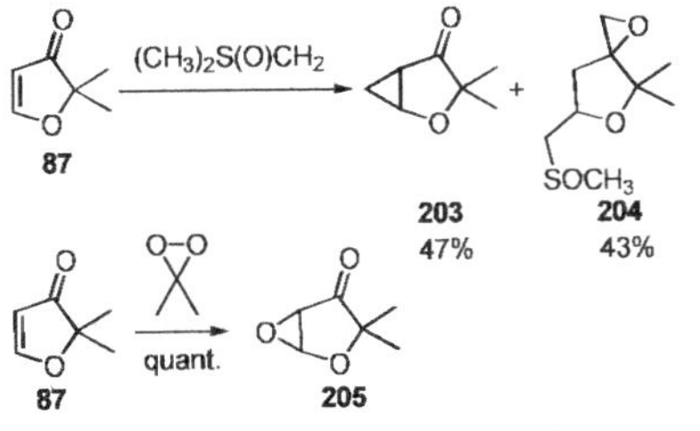

Scheme 30. Tetrahydrofuran formation with 3(2*H*)-furanones.

6.3. Photochemical reactions

The primary photochemical reaction of 3(2*H*)-furanones is the [2+2]-photocycloaddition. While investigating the intramolecular version of this reaction,[216] Bach *et al.* demonstrated that the regioselectivity can be controlled through the tether length. As depicted in Scheme 31, the intramolecular [2+2]-photocycloaddition allows for a rapid and high-yielding access to complex skeletons such as **207**. Notably, the regioselectivity was found to be completely reversed when employing substrate **208** under identical reaction conditions. Additionally, numerous examples on intermolecular [2+2]-photocycloaddition with 3(2*H*)-furanones are described.[141,217–224]

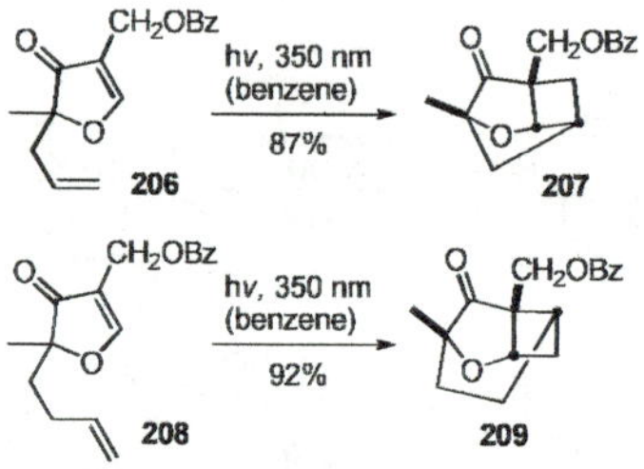

Scheme 31. Intramolecular [2+2]-photocycloaddition.

6.4. Cycloadditions

A rare example for the use of 3(2*H*)-furanones in cycloadditions was disclosed by Avery *et al.* as the first step in their total synthesis of laurenditerpenol.[225] It is noteworthy that this reaction only produced a single diastereomer in 69% yield as shown in Scheme 32. In the presence of a strong base, enolate **211** is formed that undergoes a [4+2]-cycloaddition process to provide **210**.[226,227] Nevertheless, a sequential Michael-addition/cyclization pathway provides an alternative explanation for the product formation.

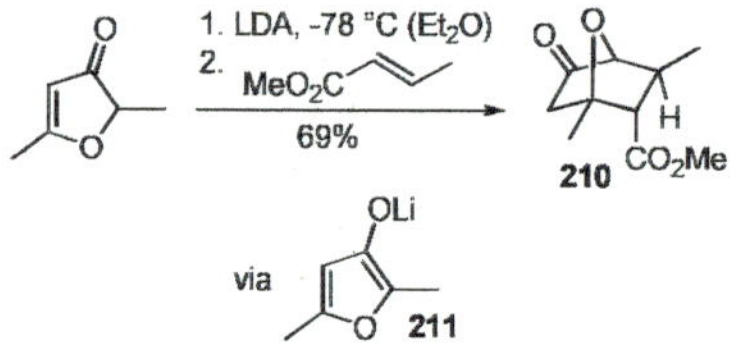

Scheme 32. Anion-assisted Diels-Alder reaction.

7. Conclusions

This brief overview was intended to demonstrate the value of 3(2*H*)-furanones, a class of five-membered heterocycles that can be broadly found as structural elements of many natural products and pharmaceutically important substances. Natural products possessing this structural element frequently exhibit significant biological activity. Although, as highlighted herein, a tremendous number of synthetic methods is known to approach the heterocyclic core, the synthesis of simple and more complex 3(2*H*)-furanones remains an area of ongoing interest. Moreover, 3(2*H*)-furanones are valuable building blocks for the construction of further functionalities as exemplified in Scheme 33.[228]

It is certain that many more protocols for the construction of polysubstituted 3(2*H*)-furanones will be described in the future.[229] One future trend is easy to predict: a number of methods will be applied to the target-oriented synthesis of complex structures. Currently, the methods to introduce the 3(2*H*)-furanone moiety into a complex target molecule are mainly based on the cyclization of α'-hydroxy-1,3-diketones.

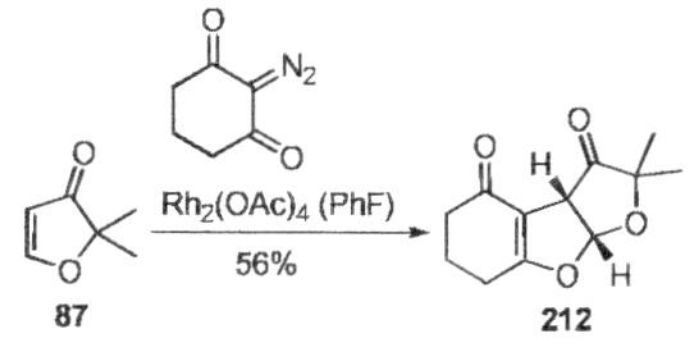

Scheme 33. Application in the synthesis of pseudosemiglabrin.

Acknowledgments

Research at Technische Universität München in the area of this review has been supported by the Fonds der Chemischen Industrie and DFG (Deutsche Forschungsgemeinschaft). Generous support from Prof. Thorsten Bach is gratefully acknowledged. I thank all co-workers involved in the development of noble-metal catalyzed cascade reactions for valuable contributions, effort and enthusiasm.

References

1. Slaughter, J. C. *Biol. Rev.* **1999**, *74*, 259–276.
2. Meister, C.; Scharf, H. D. *Synthesis* **1981**, 737–739.
3. Reffstrup, T.; Boll, P. M. *Acta Chem. Scand.* **1977**, *31B*, 727–729.
4. Brandt, C. W.; Taylor, W. I.; Thomas, B. R. *J. Chem. Soc.* **1954**, 3245–3254.
5. Parker, W.; Raphael, R. A.; Wilkinson, D. I. *J. Chem. Soc.* **1958**, 3871–3873.
6. Kupchan, S. M.; Sigel, C. W.; Matz, M. J.; Renauld, J. A. S.; Haltiwanger, R. C.; Bryan, R. F. *J. Am. Chem. Soc.* **1970**, *92*, 4476–4477.
7. Kupchan, S. M.; Sigel, C. W.; Matz, M. J.; Gilmore, C. J.; Bryan, R. F. *J. Am. Chem. Soc.* **1976**, *98*, 2295–2300.
8. Kim, J.; Yun, B.; Shim, Y. K.; Yoo, I. *Tetrahedron Lett.* **1999**, *40*, 6643–6644.
9. Shiozawa, H.; Takahashi, M.; Takatsu, T.; Kinoshita, T.; Tanzawa, K.; Hosoya, T.; Furuya, K.; Takahashi, S.; Furihata, K.; Seto, H. *J. Antibiot.* **1995**, *48*, 357–362.
10. Edwards, R. L.; Maitland, D. J.; Oliver, C. L.; Pacey, M. S.; Shields, L.; Whalley, A. J. S. *J. Chem. Soc., Perkin Trans. 1* **1999**, 715–719.
11. Goss, R. J. M.; Fuchser, J.; O'Hagan, D. *Chem. Commun.* **1999**, 2255–2256.
12. Aramaki, Y.; Chiba, K.; Tada, M. *Phytochemistry* **1995**, *38*, 1419–1421.
13. Asami, Y.; Kakeya, H.; Onose, R.; Yoshida, A. *Org. Lett.* **2002**, *4*, 2845–2848.
14. Bloch, P.; Tamm, C.; Bollinger, P.; Petcher, T. J.; Weber, H. P. *Helv. Chim. Acta* **1976**, *59*, 133–137.
15. Bloch, P.; Tamm, C. *Helv. Chim. Acta* **1981**, *64*, 304–315.
16. Ando, O.; Satake, H.; Nakajima, M.; Sato, A.; Nakamura, T.; Kinoshia, T.; Furuya, K.; Haneishi, T. *J. Antibiot.* **1991**, *44*, 382–389.
17. Mizoue, K.; Okazaki, T.; Hanada, K.; Amamoto, T.; Yamagishi, M.; Omura, S. *Eur. Pat. Appl.* EP216607.
18. Mo, S.; Wang, S.; Zhou, G.; Yang, Y.; Li, Y.; Chen, X.; Shi, J. *J. Nat. Prod.* **2004**, *67*, 823–828.
19. Wang, Y.; Mo, S.; Wang, S.; Li, S.; Yang, Y.; Shi, J. *Org. Lett.* **2005**, *7*, 1675–1678.
20. Vichnewski, W.; Sarti, S. J.; Gilbert, B.; Herz, W. *Phytochemistry* **1976**, *15*, 191–193.
21. Vichnewski, W.; Lopes, J. N. C.; Filho, D. D. S.; Herz, W. *Phytochemistry* **1976**, *15*, 1775–1776.
22. de Vivar, A. R.; Guerrero, C.; Díaz, E.; Bratoeff, E. A.; Jiménez, L. *Phytochemistry* **1976**, *15*, 525–527.
23. Quesne, P. W. L.; Menachery, M. D.; Pastore, M. P.; Kelley, C. J.; Brennan, T. F.; Onan, K. D.; Raffauf, R. F. *J. Org. Chem.* **1982**, *47*, 1519–1521.

24. Gao, F.; Wang, H.; Mabry, T. J. *Phytochemistry* **1987**, *16*, 779–781.
25. Bohlmann, F.; Mahanta, P. K.; Natu, A. A.; King, R. M.; Robinson, H. *Phytochemistry* **1978**, *17*, 471–474.
26. de Vivar, A. R.; Guerrero, C.; Diaz, E.; Ortega, A. *Tetrahedron* **1970**, *26*, 1657–1664.
27. (a) Quesne, P. W. L.; Levery, S. B.; Menachery, M. D.; Brennan, T. F.; Raffauf, R. F. *J. Chem. Soc., Perkin Trans. 1* **1978**, 1572–1580. (b) Geran, R. I.; Greenberg, N. H.; Macdonald, M. M.; Schumacher, A. M.; Abbott, B. J. *Cancer Chemother. Rep.* **1972**, *3*, 1–17. (c) Hladon, B.; Twardowski, T. *Pol. J. Pharmacol. Pharm.* **1979**, *31*, 35–43.
28. Sakamoto, H. T.; Flausino, D.; Castellano, E. E.; Stark, C. B. W.; Gates, P. J.; Lopes, N. P. *J. Nat. Prod.* **2003**, *66*, 693–695.
29. Lahey, F. N.; Macleod, J. K. *Aust. J. Chem.* **1967**, *20*, 1943–1955.
30. Dreyer, D. L.; Lee, A. *Phytochemistry* **1972**, *11*, 763–767.
31. Vleggaar, R.; Smalberger, T. M.; de Waal, H. L. *Tetrahedron Lett.* **1972**, 703–704.
32. Capon, R. J.; Faulkner, D. J. *J. Org. Chem.* **1984**, *49*, 2506–2508.
33. Paul, M. C.; Zubia, E.; Ortega, M. J.; Salvá, J. *Tetrahedron* **1997**, *53*, 2303–2308.
34. Furumoto, T.; Fukuyama, K.; Hamasaki, T.; Nakajima, H. *Phytochemistry* **1995**, *40*, 745–752.
35. Nahrstedt, A.; Rockenbach, J.; Wray, V. *Phytochemistry* **1995**, *39*, 375–378.
36. Comte, G.; Allais, D. P.; Chulia, A. J. *Tetrahedron Lett.* **1996**, *37*, 2955–2958.
37. Mayerl, F.; Näf, R.; Thomas, A. F. *Phytochemistry* **1989**, *28*, 631–633.
38. Martin, A. S.; Rovirosa, J.; Muiior, O.; Chen, M. H. M.; Guneratne, R. D.; Clardy, J. *Tetrahedron Lett.* **1983**, *24*, 4063–4066.
39. Martin, A. S.; Rovirosa, J.; Xu, C.; Lu, H. S. M.; Clardy, J. *Tetrahedron Lett.* **1987**, *28*, 6013–6014.
40. He, J.; Wijeratne, E. M. K.; Bashyal, B. P.; Zhan, J.; Seliga, C. J.; Liu, M. X.; Pierson, E. E.; Pierson, L. S.; VanEtten, H. D.; Gunatilaka, A. A. L. *J. Nat. Prod.* **2004**, *67*, 1985–1991.
41. Kobayashi, K.; Ui, T. *Tetrahedron Lett.* **1975**, *47*, 4119–4122.
42. Anke, H.; Schwab, H.; Achenbach, H. *J. Antibiot.* **1980**, *33*, 931–939.
43. Clemo, N. G.; Pattenden, G. *Tetrahedron Lett.* **1982**, *23*, 589–592.
44. Clemo, N. G.; Pattenden, G. *J. Chem. Soc., Perkin Trans. 1* **1985**, 2407–2412.
45. (a) Felman, S. W.; Jirkovsky, I.; Memoli, K. A.; Borella, L.; Wells, C.; Russell, J.; Ward, J. *J. Med. Chem.* **1992**, *35*, 1183–1190. (b) Rappai, J. P.; Raman, V.; Unnikrishnan, P. A.; Prathapan, S.; Thomas, S. K.; Paulose, C. S. *Bioorg. Med. Chem. Lett.* **2009**, *19*, 764–765.
46. Mack, R. A.; Zazulak, W. I.; Radov, L. A.; Baer, J. E.; Stewart, J. D.; Elzer, P. H.; Kinsolving, C. R.; Georgiev, V. S. *J. Med. Chem.* **1988**, *31*, 1910–1918.
47. Shin, S. S.; Byun, Y.; Lim, K. M.; Choi, J. K.; Lee, K.; Moh, J. H.; Kim, J. K.; Jeong, Y. S.; Kim, J. Y.; Choi, Y. H.; Koh, H.; Park, Y.; Oh, Y. I.; Noh, M.; Chung, S. *J. Med. Chem.* **2004**, *47*, 792–804.
48. Shin, S. S.; Noh, M.; Byun, Y. J.; Choi, J. K.; Kim, J. Y.; Lim, K. M.; Ha, J.; Kim, J. K.; Lee, C. H.; Chung, S. *Bioorg. Med. Chem. Lett.* **2001**, *11*, 165–168.
49. Jirkovsky, I.; Cayen, M. N. *J. Med. Chem.* **1982**, *25*, 1154–1156.
50. Jung, J.; Johnson, B. R.; Duong, T.; Decaire, M.; Uy, J.; Gharbaoui, T.; Boatman, P. D.; Sage, C. R.; Chen, R.; Richman, J. G.; Connolly, D. T.; Semple, G. *J. Med. Chem.* **2007**, *50*, 1445–1448.
51. Boatman, P. D.; Richman, J. G.; Semple, G. *J. Med. Chem.* **2008**, *51*, 7653–7662.
52. Mahboubi, K.; Witman-Jones, T.; Adamus, J. E.; Letsinger, J. T.; Whitehouse, D.; Moorman, A. R.; Sawicki, D.; Bergenhem, N.; Ross, S. A. *Biochem. Biophys. Res. Comm.* **2006**, *340*, 482–490.
53. Carotti, A.; Carrieri, A.; Chimichi, S.; Boccalini, M.; Cosimelli, B.; Gnerre, C.; Carotti, A.; Carrupt, P.; Testa, B. *Bioorg. Med. Chem. Lett.* **2002**, *12*, 3551–3555.
54. Cesura, A. M.; Pletscher, A. *Prog. Drug Res.* **1992**, *38*, 171–297.
55. (a) Chimichi, S.; Boccalini, M.; Cosimelli, B.; Dall'Acqua, F.; Viola, G. *Tetrahedron* **2003**, *59*, 5215–5223. (b) Chimichi, S.; Boccalini, M.; Cravotto, G.; Rosati, O. *Tetrahedron Lett.* **2006**, *47*, 2405–2408. (c) Viola, G.; Vedaldi, D.; Dall'Acqua, F.; Basso, G.; Disarò, S.; Spinelli, D.; Cosimelli, B.; Chimichi, S. *Chem. Biodiversity* **2004**, *1*, 1265–1280.

56. Chimichi, S.; Boccalini, M.; Salvador, A.; Dall'Acqua, F.; Basso, G.; Viola, G. *ChemMedChem* **2009**, *4*, 769–779.

57. Baraldi, P. G.; Guarneri, M.; Manfredini, S.; Simoni, D.; Balzarini, J.; Clercq, E. D. *J. Med. Chem.* **1989**, *32*, 284–288.

58. Simoni, D.; Manfredini, S.; Tabrizi, M. A.; Bazzanini, R.; Baraldi, P. G.; Balzarini, J.; Clercq, E. D. *J. Med. Chem.* **1991**, *34*, 3172–3176.

59. Baraldi, P. G.; Manfredini, S.; Simoni, D.; Tabrizi, M. A.; Balzarini, J.; Clercq, E. D. *J. Med. Chem.* **1992**, *35*, 1877–1882.

60. Manfredini, S.; Baraldi, P. G.; Bazzanini, R.; Guarneri, M.; Simoni, D.; Balzarini, J.; Clercq, E. D. *J. Med. Chem.* **1994**, *37*, 2401–2405.

61. Manfredini, S.; Simoni, D.; Ferroni, R.; Bazzanini, R.; Vertuani, S.; Hatse, S.; Balzarini, J.; Clercq, E. D. *J. Med. Chem.* **1997**, *40*, 3851–3857.

62. Valenti, P.; Recanatini, M.; Re, P. D.; Galimbeni, W.; Avanzi, N.; Filipeschi, S. *Arch. Pharm.* **1985**, *318*, 923–926.

63. Marko, D.; Habermeyer, M.; Kemény, M.; Weyand, U.; Niederberger, E.; Frank, O.; Hofmann, T. *Chem. Res. Toxicol.* **2003**, *16*, 48–55.

64. Fujita, E.; Nagao, Y. *Bioorg. Chem.* **1977**, *6*, 287–309.

65. Ishibashi, K.; Takahashi, W.; Takei, H.; Kakinuma, K. *Agric. Biol. Chem.* **1987**, *51*, 1045–1049.

66. Hiramoto, K.; Sekiguchi, K.; Aso-o, R.; Ayuha, K.; Ni-Iyama, H.; Kato, T.; Kikugawa, K. *Fd. Chem. Toxic.* **1995**, *33*, 803–814.

67. Li, X.; Hiramoto, K.; Yoshida, M.; Kato, T.; Kikugawa, K. *Fd. Chem. Toxic.* **1998**, *36*, 305–314.

68. Hiramoto, K.; Kato, T.; Takahashi, Y.; Yugi, K.; Kikugawa, K. *Mutation Res.* **1998**, *415*, 79–83.

69. Taylor, M. D.; Smith III, A. B.; Furst, G. T.; Gunasekara, S. P.; Bevelle, C. A.; Cordell, G. A.; Farmworth, N. R.; Kupchan, S. M.; Uchida, H.; Branfman, A. R.; Dailey, R. G.; Sneden, A. T. *J. Am. Chem. Soc.* **1983**, *105*, 3177–3183.

70. Sasaki, T.; Yamakoshi, J.; Saito, M.; Kasai, K.; Matsudo, T.; Koga, T.; Mori, K. *Biosci. Biotechnol. Biochem.* **1998**, *62*, 1865–1869.

71. Sasaki, T.; Yamakoshi, J.; Saito, M.; Kasai, K.; Matsudo, T.; Kikuchi, M.; Koga, T.; Mori, K. *Biosci. Biotechnol. Biochem.* **1998**, *62*, 2145–2154.

72. Smith III, A. B.; Levenberg, P. A.; Jerris, P. J.; Scarborough Jr., R. M.; Wovkulich, P. M. *J. Am. Chem. Soc.* **1981**, *103*, 1501–1513.

73. Shao, X.; Tamm, C. *Tetrahedron Lett.* **1991**, *32*, 2891–2892.

74. Dolder, M.; Shao, X.; Tamm, C. *Helv. Chim. Acta* **1990**, *73*, 63–68.

75. Arimoto, H.; Nishiyama, S.; Yamamura, S. *Tetrahedron Lett.* **1990**, *31*, 5619–5620.

76. Arimoto, H.; Ohba, S.; Nishiyama, S.; Yamamura, S. *Tetrahedron Lett.* **1994**, *35*, 4581–4584.

77. Sher, F.; Isidor, J. L.; Taneja, H. R.; Carlson, R. M. *Tetrahedron Lett.* **1973**, *14*, 577–580.

78. Andersen, S. H.; Das, N. B.; Jorgensen, R. D.; Kjeldsen, G.; Knudsen, J. S.; Sharma, S. C.; Torssell, K. B. G. *Acta Chem. Scand.* **1982**, *36B*, 1–14.

79. Andersen, S. H.; Sharma, K. K.; Torssell, K. B. G. *Tetrahedron* **1983**, *39*, 2241–2245.

80. Curran, D. P.; Singleton, D. H. *Tetrahedron Lett.* **1983**, *24*, 2079–2082.

81. Tsuge, O.; Kanemasa, S.; Suga, H. *Chem. Lett.* **1987**, 323–326.

82. Baraldi, P. G.; Barco, A.; Benetti, S.; Manfredini, S.; Pollini, G. P.; Simoni, D. *Tetrahedron* **1987**, *43*, 235–242.

83. Baraldi, P. G.; Barco, A.; Benetti, S.; Manfredini, S.; Pollini, G. P.; Simoni, D. *Tetrahedron Lett.* **1984**, *38*, 4313–4316.

84. Chimichi, S.; Boccalini, M.; Cosimelli, B.; Viola, G.; Vedaldi, D.; Dall'Acqua, F. *Tetrahedron Lett.* **2002**, *43*, 7473–7476.

85. Saxena, R.; Singh, V.; Batra, S. *Tetrahedron* **2004**, *60*, 10311–10320.

86. Carpenter, B. K.; Clemens, K. E.; Schmidt, E. A.; Hoffmann, H. M. R. *J. Am. Chem. Soc.* **1972**, *94*, 6213–6214.

87. Stasevich, G. Z.; Bubel', O. N.; Tishchenko, I. G.; Musa, Y. *Chem. Heterocycl. Compd.* **1988**, *24*, 1321–1324.

88. Zvonok, A. M.; Kuz'menok, N. M.; Stanishevskii, L. S. *Chem. Heterocyc. Comp.* **1988**, *24*, 839–844.

89. Yamamoto, M. *J. Chem. Soc.; Chem. Commun.* **1975**, 289–290.

90. Valiulin, R. A.; Halliburton, L. M.; Kutateladze, A. G. *Org. Lett.* **2007**, *9*, 4061–4063.

91. Kawaguchi, T.; Yasuta, S.; Inoue, Y. *Synthesis* **1996**, 1431–1432.

92. Inoue, Y.; Ohuchi, K.; Imaizumi, S. *Tetrahedron Lett.* **1988**, *29*, 5941–5942.

93. Inoue, Y.; Taniguchi, M.; Hashimoto, H.; Ohuchi, K.; Imaizumi, S. *Chem. Lett.* **1988**, 81–82.

94. Inoue, Y.; Ohuchi, K.; Yen, I.; Imaizumi, S. *Bull. Chem. Soc. Jpn.* **1989**, *62*, 3518–3522.

95. Williams, D. R.; Abbaspour, A.; Jacobsen, R. M. *Tetrahedron Lett.* **1981**, *22*, 3565–3568.

96. Saimoto, H.; Shinoda, M.; Matsubara, S.; Oshima, K.; Hiyama, T.; Nozaki, H. *Bull. Chem. Soc. Jpn.* **1983**, *56*, 3088–3092.

97. Saimoto, H.; Hiyama, T.; Nozaki, H. *Bull. Chem. Soc. Jpn.* **1983**, *56*, 3078–3087.

98. Schlessinger, R. H.; Wong, J. W.; Poss, M. A. *J. Org. Chem.* **1985**, *50*, 3950–3951.

99. Gomez, L.; Calas, P.; Commeyras, A. *J. Chem. Soc.; Chem. Commun.* **1985**, *21*, 1493–1494.

100. Gebel, R.; Margaretha, P. *Helv. Chim. Acta* **1992**, *75*, 1633–1638.

101. Margaretha, P. *Tetrahedron Lett.* **1971**, *12*, 4891–4892.

102. Lehmann, H.-G. *Angew. Chem. Int. Ed.* **1965**, *4*, 783–784.

103. Kato, K.; Nouchi, H.; Ishikura, K.; Takaishi, S.; Motodate, S.; Tanake, H.; Okudaira, K.; Mochida, T.; Nishigaki, R.; Shigenobu, K.; Akita, H. *Tetrahedron* **2006**, *62*, 2545–2554.

104. Sebti, S.; Foucaud, A. *Tetrahedron* **1986**, *42*, 1361–1367.

105. Villemin, D.; Jaffrès, P.; Hachémi, M. *Tetrahedron Lett.* **1997**, *38*, 537–538.

106. Babin, P.; Dunoguès, J.; Duboudin, F.; Petraud, M. *Bull. Soc. Chim. Fr.* **1982**, 125–128.

107. Drtina, G. J.; Sampson, P.; Wiemer, D. F. *Tetrahedron Lett.* **1984**, *25*, 4467–4470.

108. Sampson, P.; Roussis, V.; Drtina, G. J.; Koerwitz, F. L.; Wiemer, D. F. *J. Org. Chem.* **1986**, *51*, 2525–2529.

109. Roussis, V.; Gloer, K. B.; Wiemer, D. F. *J. Org. Chem.* **1988**, *53*, 2011–2015.

110. Han, Q.; Wiemer, D. F. *J. Am. Chem. Soc.* **1992**, *114*, 7692–7697.

111. Brettle, R.; Hydes, P. C.; Seddon, D.; Sutton, J. R.; Tooth, R. *J. Chem. Soc., Perkin Trans. 1* **1973**, 345–348.

112. Föhlisch, B.; Lutz, D.; Gottstein, W.; Dukek, U. *Liebigs Ann. Chem.* **1977**, 1847–1861.

113. Yamaguchi, M.; Shibato, K.; Nakashima, H.; Minami, T. *Tetrahedron* **1988**, *44*, 4767–4775.

114. Langer, P.; Krummel, T. *Chem. Commun.* **2000**, 967–968.

115. Langer, P.; Krummel, T. *Chem. Eur. J.* **2001**, *7*, 1720–1727.

116. Mack, R. A.; Georgiev, V. S.; Wu, E. S. C.; Matz, J. R. *J. Org. Chem.* **1993**, *58*, 6158–6162.

117. Lee, K.; Choi, Y. H.; Joo, Y. H.; Kim, J. K.; Shin, S. S.; Byun, Y. J.; Kim, Y.; Chung, S. *Bioorg. Med. Chem.* **2002**, *10*, 1137–1142.

118. Inokuchi, T.; Matsumoto, S.; Tsuji, M.; Torii, S. *J. Org. Chem.* **1992**, *57*, 5023–5027.

119. Barrow, M.; Richards, A. C.; Smithers, R. H.; Hoffmann, H. M. R. *Tetrahedron Lett.* **1972**, *13*, 3101–3102.

120. Noyori, R.; Hayakawa, Y.; Makino, S.; Hayakawa, N.; Takaya, H. *J. Am. Chem. Soc.* **1973**, *95*, 4103–4105.

121. Noyori, R.; Hayakawa, Y. *Tetrahedron* **1985**, *41*, 5879–5886.

122. Meister, H. *Liebigs Ann. Chem.* **1971**, *752*, 163–174.

123. Greenhill, J. V.; Ingle, P. H. B.; Ramli, M. *J. Chem. Soc., Perkin Trans. 1* **1972**, 1667–1669.

124. Skvortsov, Y. M.; Moshchevitina, E. I.; Mal'kina, A. G. *Chem. Heterocycl. Compd.* **1988**, *24*, 1069.

125. Baldwin, J. E.; Thomas, R. C.; Kruse, L. I.; Silberman, L. *J. Org. Chem.* **1977**, *42*, 3846–3852.

126. Brennan, C. M.; Johnson, C. D.; McDonnell, P. D. *J. Chem. Soc., Perkin Trans. 2* **1989**, 957–961.

127. Marson, C. M.; Edaan, E.; Morrell, J. M.; Coles, S. J.; Hursthouse, M. B.; Davies, D. T. *Chem. Commun.* **2007**, 2494–2496.

128. Silva, F.; Reiter, M.; Mills-Webb, R.; Sawicki, M.; Klär, D.; Bensel, N.; Wagner, A.; Gouverneur, V. *J. Org. Chem.* **2006**, *71*, 8390–8394.

129. Liu, Y.; Liu, M.; Guo, S.; Tu, H.; Zhou, Y.; Gao, H. *Org. Lett.* **2006**, *8*, 3445–3448.

130. Crone, B.; Kirsch, S. F. *Chem. Eur. J.* **2008**, *14*, 3514–3522.

131. Kirsch, S. F.; Binder, J. T.; Liébert, C.; Menz, H. *Angew. Chem. Int. Ed.* **2006**, *45*, 5878–5880.

132. Binder, J. T.; Crone, B.; Kirsch, S. F.; Liébert, C.; Menz, H. *Eur. J. Org. Chem.* **2007**, *38*, 1636–1647.

133. Bunnelle, E. M.; Smith, C. R.; Lee, S. K.; Singaram, S. W.; Rhodes, A. J.; Sarpong, R. *Tetrahedron* **2008**, *64*, 7008–7014.

134. Crone, B.; Kirsch, S. F. *J. Org. Chem.* **2007**, *72*, 5435–5438.

135. Reiter, M.; Turner, H.; Mills-Webb, R.; Gouverneur, V. *J. Org. Chem.* **2005**, *70*, 8478–8485.

136. Batson, W. A.; Sethumadhavan, D.; Tius, M. A. *Org. Lett.* **2005**, *7*, 2771–2774.

137. Berner, E. *Acta Chem. Scand.* **1952**, *6*, 646–653.

138. Berner, E.; Kolsaker, P. *Tetrahedron* **1968**, *24*, 1199–1203.

139. Antonioletti, R.; Bonadies, F.; Scettri, A. *Tetrahedron Lett.* **1987**, *28*, 2297–2298.

140. Wolff, S.; Agosta, W. C. *J. Org. Chem.* **1985**, *50*, 4707–4711.

141. Saimoto, H.; Hiyama, T.; Nozaki, H. *J. Am. Chem. Soc.* **1981**, *103*, 4975–4977.

142. Eicher, T.; Born, T. *Liebigs Ann. Chem.* **1972**, *762*, 127–153.

143. Caine, D. S.; Samuels, W. D. *Tetrahedron Lett.* **1980**, *21*, 4057–4060.

144. Amemiya, S.; Kojima, K.; Sakai, K. *Chem. Pharm. Bull.* **1984**, *32*, 457–462.

145. Baumann, M.; Hoffmann, W. *Synthesis* **1981**, 709.

146. Nishinaga, A.; Nakamura, K.; Matsuura, T. *Tetrahedron Lett.* **1978**, 3597–3600.

147. Nishinaga, A.; Nakamura, K.; Matsuura, T. *Tetrahedron Lett.* **1980**, *21*, 1269–1272.

148. Nishinaga, A.; Nakamura, K.; Matsuura, T. *J. Org. Chem.* **1983**, *48*, 3700–3703.

149. Nishinaga, A.; Nakamura, K.; Matsuura, T. *J. Org. Chem.* **1982**, *47*, 1431–1435.

150. Benjamin, B. M.; Raaen, V. F.; Hagaman, E. W.; Brown, L. L. *J. Org. Chem.* **1978**, *43*, 2986–2991.

151. Yoshioka, M.; Funayama, K.; Hasegawa, T. *J. Chem. Soc., Perkin Trans. 1* **1989**, 1411–1414.

152. Yoshioka, M.; Oka, M.; Ishikawa, Y.; Tomita, H.; Hasegawa, T. *J. Chem. Soc.; Chem. Commun.* **1986**, 639–640.

153. Venturello, C.; D'Aloisio, R. *Synthesis* **1977**, 754–755.

154. McCarthy, D. G.; Collins, C. C.; O'Driscoll, J. P.; Lawrence, S. E. *J. Chem. Soc., Perkin Trans. 1* **1999**, 3667–3675.

155. Saberi, S. P.; Salter, M. M.; Slawin, A. M. Z.; Thomas, S. E.; Williams, D. J. *J. Chem. Soc., Perkin Trans. 1* **1994**, 167–172.

156. Besenyei, G.; Neszmélyi, A.; Holly, S.; Simándi, L. I.; Párkányi, L. *Can. J. Chem.* **2001**, *79*, 649–654.

157. Athanasellis, G.; Detsi, A.; Prousis, K.; Igglessi-Markopoulou, O.; Markopoulos, J. *Synthesis* **2003**, 2015–2022.

158. Tolochko, L. A.; Sipyagin, A. M.; Kartsev, V. G. *Chem. Heterocycl. Compd.* **1986**, *22*, 225.

159. Ceccherelli, P.; Curini, M.; Marcotullio, M. C.; Rosati, O.; Wenkert, E. *Synth. Commun.* **1994**, *24*, 891–897.

160. Winkler, J. D.; Oh, K.; Asselin, S. M. *Org. Lett.* **2005**, *7*, 387–389.

161. Takaiwa, A. Yamashita, K. *Agric. Biol. Chem.* **1982**, *46*, 1721–1722.

162. Takaiwa, A. Yamashita, K. *Agric. Biol. Chem.* **1984**, *48*, 2061–2065.

163. Gogoi, S.; Argade, N. P. *Tetrahedron* **2006**, *62*, 2999–3003.

164. Antus, S.; Farkas, L.; Mógrádi, M.; Boross, F. *J. Chem. Soc., Perkin Trans. 1* **1977**, 948–953.

165. Ichinari, D.; Ueki, T.; Yoshihara, K.; Kinoshita, T. *Chem. Commun.* **1997**, 1743.

166. Kraus, G. A.; Wei, J. *J. Nat. Prod.* **2004**, *67*, 1039–1040.

167. Ito, M.; Tsukida, K.; Toube, T. P. *J. Chem. Soc., Perkin Trans. 1* **1981**, 3255–3257.

168. Takeda, A.; Tsuboi, S.; Sakai, T. *Chem. Lett.* **1973**, 425–426.

169. Jackson, R. F. W.; Raphael, R. A. *Tetrahedron Lett.* **1983**, *24*, 2117–2120.

170. Jackson, R. F. W.; Raphael, R. A. *J. Chem. Soc., Perkin Trans. 1* **1984**, 535–539.

171. Inoue, Y.; Ohuchi, K.; Imaizumi, S.; Hagiwara, H.; Uda, H. *Synth. Commun.* **1990**, *20*, 3063–3068.

172. Adams, J.; Poupart, M.; Grenier, L.; Schaller, C.; Ouimet, N.; Frenette, R. *Tetrahedron Lett.* **1989**, *30*, 1749–1752.

173. Tarnchompoo, B.; Thebtaranonth, Y. *Tetrahedron Lett.* **1984**, *25*, 5567–5570.

174. Smith III, A. B.; Jerris, P. J. *Tetrahedron Lett.* **1980**, *21*, 711–714.

175. Jerris, P. J.; Smith III, A. B. *J. Org. Chem.* **1981**, *46*, 577–585.

176. Sakai, T.; Ito, H.; Yamawaki, A.; Takeda, A. *Tetrahedron Lett.* **1984**, *25*, 2987–2988.

177. Kang, S. H.; Hong, C. Y. *Tetrahedron Lett.* **1987**, *28*, 675–678.

178. Baraldi, P. G.; Barco, A.; Benetti, S.; Guarneri, M.; Manfredini, S.; Pollini, G. P.; Simoni, D. *Tetrahedron Lett.* **1985**, *26*, 5319–5322.

179. Baraldi, P. G.; Barco, A.; Benetti, S.; Casolari, A.; Manfredini, S.; Pollini, G. P.; Simoni, D. *Tetrahedron* **1988**, *44*, 1267–1272.

180. Chen, K.; Joullié, M. M. *Tetrahedron Lett.* **1984**, *25*, 393–394.

181. (a) Smith III, A. B.; Lupo, A. T.; Ohba, M.; Chen, K. *J. Am. Chem. Soc.* **1989**, *111*, 6648–6656. (b) Smith III, A. B.; Guaciaro, M. A.; Schow, S. R.; Wovkulich, P. M.; Toder, B. H.; Hall, T. W. *J. Am. Chem. Soc.* **1981**, *103*, 219–222.

182. Gyorkos, A. C.; Stille, J. K.; Hegedus, L. S. *J. Am. Chem. Soc.* **1990**, *112*, 8465–8472.

183. Steel, P. G. *Chem. Commun.* **1999**, 2257–2258.

184. Hirai, K.; Ooi, H.; Esumi, T.; Iwabuchi, Y.; Hatakeyama, S. *Org. Lett.* **2003**, *5*, 857–859.

185. Zammit, S. C.; Ferro, V.; Hammond, E.; Rizzacasa, M. A. *Org. Biomol. Chem.* **2007**, *5*, 2826–2834.

186. Hayashi, Y.; Shoji, M.; Yamaguchi, J.; Sato, K.; Yamaguchi, S.; Mukaiyama, T.; Sakai, K.; Asami, Y.; Kakeya, H.; Osada, H. *J. Am. Chem. Soc.* **2002**, *124*, 12078–12079.

187. Hayashi, Y.; Shoji, M.; Yamaguchi, S.; Mukaiyama, T.; Yamaguchi, J.; Kakeya, H.; Osada, H. *Org. Lett.* **2003**, *5*, 2287–2290.

188. Aoki, S.; Oi, T.; Shimizu, K.; Shiraki, R.; Takao, K.; Tadano, K. *Bull. Chem. Soc. Jpn.* **2004**, *77*, 1703–1716.

189. Boeckman, R. K.; Yoon, S. K.; Heckendorn, D. K. *J. Am. Chem. Soc.* **1991**, *113*, 9682–9684.

190. Takao, K.; Ochiai, H.; Yoshida, K.; Hashizuka, T.; Koshimura, H.; Tadano, K.; Ogawa, S. *J. Org. Chem.* **1995**, *60*, 8179–8193.

191. Takao, K.; Ochiai, H.; Hashizuka, T.; Koshimura, H.; Tadano, K.; Ogawa, S. *Tetrahedron Lett.* **1995**, *36*, 1487–1490.

192. Li, Y.; Hale, K. J. *Org. Lett.* **2007**, *9*, 1267–1270.

193. Smith III, A. B.; Scarborough, R. M. *Tetrahedron Lett.* **1978**, 4193–4196.

194. Smith III, A. B.; Jerris, P. J. *Synth. Commun.* **1978**, *8*, 421–426.

195. Caine, D. S.; Arant, M. E. *Tetrahedron Lett.* **1994**, *35*, 6795–6798.

196. Caine, D. S.; Arant, M. E. *Synlett* **2004**, 2081–2086.

197. Chantegrel, B.; Gelin, S. *Synthesis* **1981**, 45–47.

198. Baldwin, S. W.; McIver, J. M. *J. Org. Chem.* **1987**, *52*, 320–322.

199. Csuk, R.; Schaade, M. *Tetrahedron Lett.* **1993**, *34*, 7907–7910.

200. Csuk, R.; Schaade, M. *Tetrahedron* **1994**, *50*, 3333–3348.

201. Csuk, R.; Schaade, M.; Krieger, C. *Tetrahedron Lett.* **1996**, *52*, 6397–6408.

202. Venturello, C.; D'Aloisio, R. *Synthesis* **1979**, 283–288.

203. Gelin, S.; Chabannet, M. *Synthesis* **1978**, 448–450.

204. Deshayes, C.; Chabannet, M.; Gelin, S. *Synthesis* **1984**, 868–870.

205. Mack, R. A.; Georgiev, V. S. *J. Org. Chem.* **1987**, *52*, 477–478.

206. Georgiev, V. S.; Mack, R. A.; Walter, D. J.; Radov, L. A.; Baer, J. E. *Helv. Chim. Acta* **1987**, *70*, 1526–1530.

207. Gelin, S. *Synthesis* **1978**, 291–294.

208. Gelin, S.; Chabannet, M. *Synthesis* **1980**, 875–877.

209. Amato, C.; Reboul, J.; Finet, J.; Calas, P. *Synthesis* **2005**, 2527–2532.

210. Amato, C.; Calas, P. *Synthesis* **2003**, 2709–2712.

211. Almerico, A. M.; Mingoia, F.; Diana, P.; Barraja, P.; Lauria, A.; Montalbano, A.; Cirrincione, G.; Dattolo, G. *J. Med. Chem.* **2005**, *48*, 2859–2866.
212. Venturello, C.; D'Aloisio, R. *Tetrahedron Lett.* **1982**, *23*, 2895–2898.
213. Patjens, J.; Margaretha, P. *Synthesis* **1990**, 476–477.
214. Baldwin, S. W.; Blomquist, H. R. *Tetrahedron Lett.* **1982**, *23*, 3883–3886.
215. Adam, W.; Hadjiarapoglou, L. *Chem. Ber.* **1990**, *123*, 2077–2079.
216. Bach, T.; Kemmler, M.; Herdtweck, E. *J. Org. Chem.* **2003**, *68*, 1994–1997.
217. Baldwin, S. W.; Mazzuckelli, T. J. *Tetrahedron Lett.* **1986**, *27*, 5975–5978.
218. Baldwin, S. W.; Crimmins, M. T. *J. Am. Chem. Soc.* **1980**, *102*, 1198–1200.
219. Baldwin, S. W.; Crimmins, M. T. *Tetrahedron Lett.* **1978**, 4197–4200.
220. Baldwin, S. W.; Wilkinson, J. M. *Tetrahedron Lett.* **1979**, 2657–2660.
221. Padwa, A.; Ku, A.; Sato, E. *Tetrahedron Lett.* **1976**, 2409–2412.
222. Baldwin, S. W.; Fredericks, J. E. *Tetrahedron Lett.* **1982**, *23*, 1235–1238.
223. Baldwin, S. W.; Landmesser, N. G. *Tetrahedron Lett.* **1982**, *23*, 4443–4446.
224. Gebel, R.; Margaretha, P. *Chem. Ber.* **1990**, *123*, 855–858.
225. Chittiboyina, A. G.; Kumar, G. M.; Carvalho, P. B.; Liu, Y.; Zhou, Y.; Nagle, D. G.; Avery, M. A. *J. Med. Chem.* **2007**, *50*, 6299–6302.
226. Caine, D. S.; Collison, R. F. *Synlett* **1995**, 503–504.
227. Caine, D. S.; Paige, M. A. *Synlett* **1999**, 1391–1394.
228. Pirrung, M. C.; Lee, Y. R. *J. Am. Chem. Soc.* **1995**, *117*, 4814–4821.
229. During the preparation of this account, a fascinating asymmetric allylic alkylation of 3(2*H*)-furanones was employed for the synthesis of hyperolactone C and biyouyanagin A: Du, C.; Li, L.; Xie, Z. *Angew. Chem. Int. Ed.* **2009**, *48*, 7853–7856.

ETHYNYLATION OF PYRROLE NUCLEUS WITH HALOACETYLENES ON ACTIVE SURFACES

Boris A. Trofimov and Lyubov N. Sobenina

A. E. Favorsky Irkutsk Institute of Chemistry, Siberian Branch of the Russian Academy of Sciences, 1 Favorsky Str., Irkutsk RUS-664033, Russian Federation (e-mail: boris_trofimov@irioch.irk.ru)

Abstract. *Recent publications covering new cross-coupling reactions of substituted pyrroles, 4,5,6,7-tetra-hydroindoles and indoles with haloacetylenes on active surfaces (MgO, CaO, ZnO, BaO, Al_2O_3, TiO_2, ZrO_2, K_2CO_3, $CaCO_3$, $ZrSiO_4$) to regioselectively afford 2-ethynylpyrrole, -4,5,6,7-tetrahydroindoles and 3-ethynylindoles are discussed. These reactions proceed at room temperature without transition-metal catalysts and base under solvent-free conditions. The yields of ethynylation products most often range between 60 to 70%, in certain cases reaching 90–94%. Some reactivity and structure peculiarities ([2+2]-cycloaddition to the acetylene moiety and rotational isomerism along the triple bond in crystals) of novel C-ethynylpyrroles, -4,5,6,7-tetrahydroindoles and -indoles are considered.*

Contents

1. Introduction
2. Reactions of pyrroles, 4,5,6,7-tetrahydroindoles and indoles with haloacetylenes on active surfaces
    2.1. Discovery of the reaction of pyrroles with haloacylacetylenes on active surfaces
    2.2. A deeper insight into the reaction of 2-arylpyrroles with benzoylbromoacetylene
    2.3. The ethynylation of 1-vinylpyrroles with benzoylbromoacetylene on alumina
    2.4. Effect of active surfaces on the pyrrole ethynylation with haloacetylenes
    2.5. The ethynylation of 4,5,6,7-tetrahydroindoles with halopropynoates on active surfaces
    2.6. The ethynylation of indoles with benzoylbromoacetylene on alumina
3. A peculiarity of the C-ethynylpyrroles and -indoles reactivity
4. Rotational isomerism of 2-(2-benzoylethynyl)-5-phenylpyrrole in crystal state
5. Conclusions

Acknowledgments

References

1. Introduction

The development of efficient methodologies for the regioselective functionalization of the pyrrole nucleus is of great importance, since these ring units occur as structural motifs in numerous biologically active natural products and pharmaceuticals.[1] Among the functionalized pyrroles and indoles, the C-ethynyl derivatives attract major attention due to the rich chemistry of the triple bond.[2] Consequently, considerable efforts have been devoted to the development of new methodologies for efficient synthesis of C-ethynyl-pyrroles, -4,5,6,7-tetrahydroindoles and -indoles.

However, almost all the known syntheses of C-ethynylpyrroles, -4,5,6,7-tetrahydroindoles and -indoles require functionalized pyrroles and indoles as reactants.

Thus, a number of *C*-ethynylpyrroles were obtained by base-assisted elimination reactions of pyrrole derivatives such as 2- and 3-(2-chloroethenyl)pyrroles,[3] 2-(2,2-dihaloethenyl)pyrroles,[4] 2-pyrrolyl-benzylketones.[5] Some *C*-ethynylpyrroles were synthesized by pyrolysis of 2-functional substituted ethenylpyrroles[6] and pyrrolylaroylmethylenephoshoranes.[7] Also, the synthesis of 2-ethynyl-1-methylpyrrrole from 2-iodo-1-methylpyrrole and ethynylmagnesium bromide in the presence of $ZnBr_2$ and $Pd(PPh_3)_4$ has been reported.[8]

Copper derivatives of 1-methylpyrrole reacted with substituted iodopropyne to afford the corresponding propynylpyrrole.[9] Similarly, α-lithiated 1-methylpyrrole with FC≡CH gave 2-ethynyl-1-methylpyrrole.[10]

Some syntheses of *C*-ethynylpyrroles[11] and -indoles[12] involved simultaneous building up the heterocyclic ring. Today, the most frequently used methodology for the introduction of acetylenic substituent to the pyrrole nucleus is cross-coupling of halopyrroles or -indoles with terminal acetylenes (Sonogashira reaction and its modifications).[2,13–15]

However, as a rule, this reaction was limited to pyrroles with electron-withdrawing substituents[14] and *N*-substituted derivatives.[15] Also, this reaction seems to be inapplicable for electron-deficient acetylenes (at least no examples of the cross-coupling with acetylenes bearing electron-acceptor function was as yet reported). Instability of simple halogenated pyrroles,[16] their decreased reactivity in the cross-coupling[17] and a significant interference of the side reductive dehalogenation[18] also make additional difficulties.

Therefore, it would be of methodological importance to devise a new type of coupling, in which non-halogenated pyrroles could be cross-coupled with readily available halogenated acetylenes.

Meanwhile, the reaction of pyrroles with haloacetylenes can be complicated by Diels-Alder condensation[19] and nucleophilic addition[20] of pyrrole nucleus to the triple bond as well as by the reactions of functional groups (in the cases of functionalized acetylenes and pyrroles). Recently, it was found that the pyrrole nucleus is readily ethynylated with haloacetylenes having electron-deficient substituents, when the reaction was carried out on the active surfaces of metal oxides or salts. The results of the systematic study of this novel cross-coupling reaction are reviewed in this chapter.

2. Reactions of pyrroles, 4,5,6,7-tetrahydroindoles and indoles with haloacetylenes on active surfaces

2.1. Discovery of the reaction of pyrroles with haloacylacetylenes on active surfaces

In the first report[21] devoted to ethynylation of pyrroles with haloacetylenes on active surfaces, the reaction of unsubstituted pyrrole (**1a**), 2-phenylpyrrole (**1b**) and 4,5,6,7-tetrahydroindole (**1c**) with acylbromoacetylenes was briefly described. These pyrroles **1a–c** were coupled with benzoyl- (**2a**) and thienoylbromoacetylenes (**2b**) on the surface of alumina to give 2-(acylethynyl)pyrroles **3a–f** (Scheme 1).

The reaction proceeded at room temperature for 30–60 minutes and was slightly exothermic. Experimentally, the reactants were pulverized with a 10-fold mass excess of Al_2O_3 under solvent-free conditions, through some amounts of solvents (*n*-hexane, Et_2O) to take off the products from the reaction mixture were required.

The reaction was 100% regioselective: no isomeric 1- or 3-(acylethynyl)pyrroles were detected in the reaction mixture (^{1}H-NMR). Side products of the coupling were 2-acyl-1,1-di(pyrrol-2-yl)ethenes **4a–f**, which were first assumed to be formed by the addition of a starting pyrroles to the major products **3a–f** (Scheme 2).

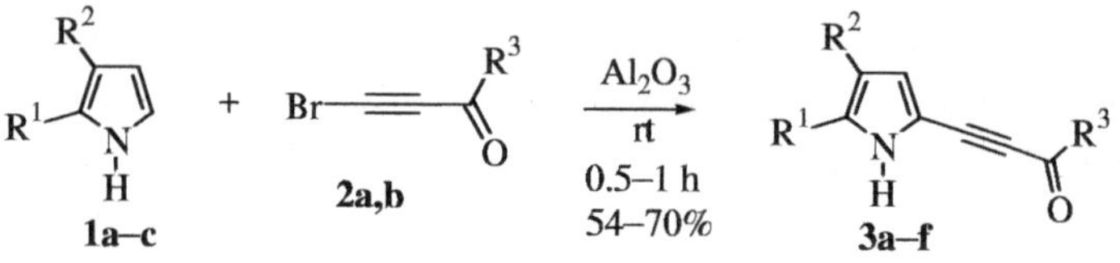

R^3 = Ph: R^1 = R^2 = H (**a**); R^1 = Ph, R^2 = H (**b**); R^1- R^2 = (CH$_2$)$_4$ (**c**);
R^3 = 2-thienyl: R^1 = R^2 = H (**d**); R^1 = Ph, R^2 = H (**e**); R^1- R^2 = (CH$_2$)$_4$ (**f**)

Scheme 1

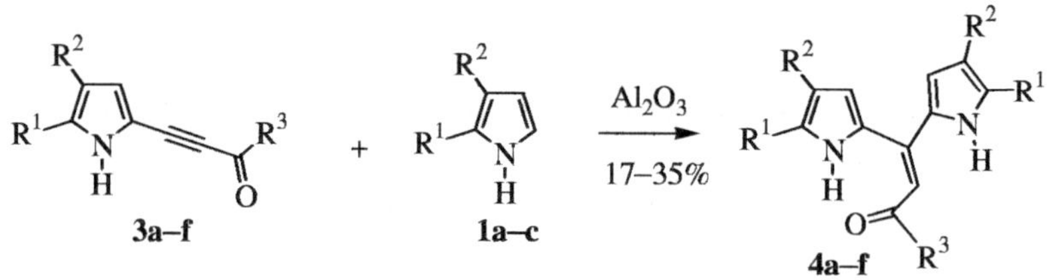

Scheme 2

The yields of **4** were normally up to 19%, expectedly increasing when excess pyrrole, higher temperatures or longer reaction times were employed. For example, the reaction of 2-fold molar excess of pyrrole **1c** with bromothienoylacetylene **2b** gave 55% and 35% yields of **3f** and **4f**, respectively (Schemes 1 and 2).

The mechanism of this new ethynylation is certainly completely different from the Sonogashira coupling[13–15] and probably involves an addition-elimination sequence (Scheme 3) promoted by the coordinatively unsaturated centres of Al$_2$O$_3$ (electrophilic assistance) and by grinding up the reactants with the solid phase.

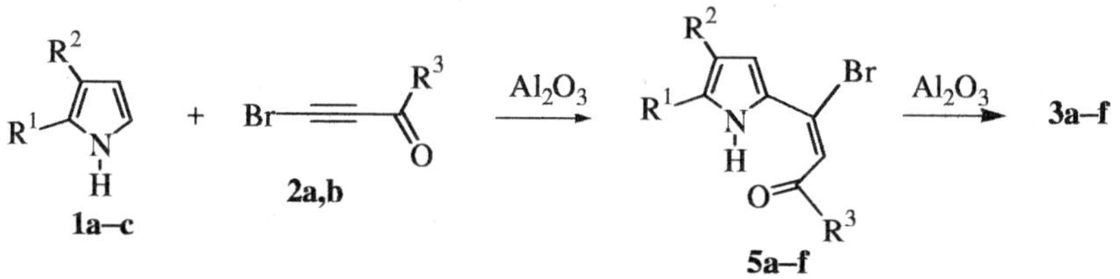

Scheme 3

In fact, *E*-2-(1-bromo-2-thienoylethenyl)-5-phenylpyrrole **5e**, a possible intermediate (from the reaction of **1b** with **2b**), has been identified in the CDCl$_3$ extract of the reaction mixture by ^{1}H-NMR technique.

However, upon chromatography (Al$_2$O$_3$) of the reaction mixture (1 hour after the reaction beginning), the pyrrole **5e** was not discernable (^{1}H-NMR) and the yield of the corresponding ethynylpyrrole **3e** was 60%. If silica, instead of alumina, was employed as the reaction medium, the adduct **5e** became the major product (the yield reached 60%) and the ethynylpyrrole **3e** was detectable (^{1}H-NMR) as traces only.

2.2. A deeper insight into the reaction of 2-arylpyrroles with benzoylbromoacetylene

A closer investigation of the pyrrole ethynylation with haloacetylenes was performed on 2-arylpyrroles with a benzoylbromoacetylene partner.[22] Cross-coupling of 2-arylpyrroles **1b,6a–c** with benzoylbromoacetylene **2a** on alumina at room temperature gave 2-aryl-5-(benzoylethynyl)pyrroles **3b,7a–c** in 45–94% yields (Scheme 4). In addition, small amounts of compounds **5b,8b,c** and **4b,9a–c** were formed.

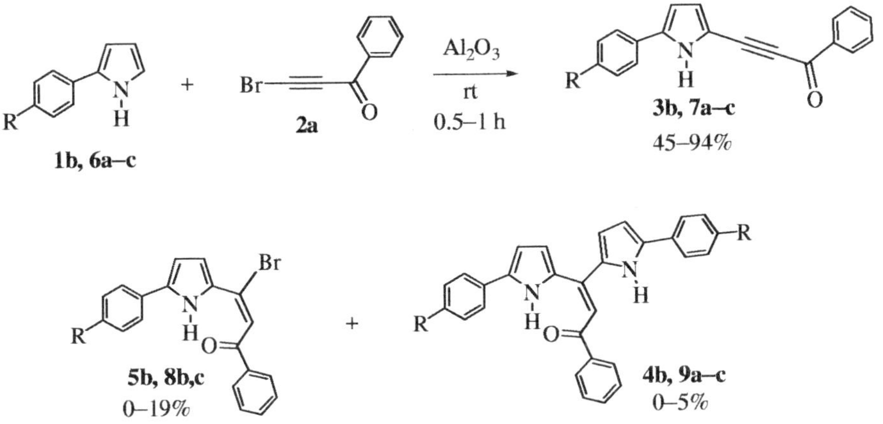

R = H (**1b, 3b, 4b, 5b**), Me$_2$N (**6a, 7a, 9a**), MeO (**6b, 7b, 8b, 9b**), Cl (**6c, 7c, 8c, 9c**)

Scheme 4

As it was previously assumed,[21] 2-(acylethynyl)pyrroles **3** were formed as a result of elimination of hydrogen bromide molecule from the primary pyrrole-acylbromoacetylene adducts, 2-(2-acyl-1-bromoethenyl)pyrroles **5**. To verify this assumption, the reaction of pyrroles **1b,6a–c** with acetylene **2a** were monitored by ^{1}H-NMR spectroscopy (samples were withdrawn from the reaction mixtures at definite time intervals and dissolved in CDCl$_3$ and their ^{1}H-NMR spectra were recorded). It was found that 2-arylpyrroles **1b,6b,c** reacted with acetylene **2a** to give the corresponding 2-aryl-5-(benzoylethynyl)pyrroles **3b,7b,c** and 2-aryl-5-(2-benzoyl-1-bromoethenyl)pyrroles **5b,8b,c**. Their ratio changed slightly with time (Table 1). No bromoethenylpyrrole **8a** was detected in the reaction mixture obtained from pyrrole **6a** and acetylene **2a**: even after 10 minutes, the mixture contained only pyrroles **7a** and **9a** together with the initial reactants (Table 1).

The fact that the product ratio does not change during the process may be rationalized as follows: pyrroles **1b,6a–c** add to acetylene **2a** according to the nucleophilic *trans*-addition pattern to give Z-2-(bromoethenyl)pyrroles **5b,8a–c**. The latter are transformed along two pathways: (1) elimination of hydrogen bromide leading to the formation of ethynylpyrroles **3b,7a–c** and (2) isomerization to *E*-isomers stabilized by strong intramolecular hydrogen bond like that found in structurally related compounds.[23] The absence of **8a** in the reaction mixture is likely due to higher rate of the elimination of hydrogen bromide than that of compounds **8b,c**.

Since *cis*-elimination is slower than *trans*-elimination,[24] some amount of the *E*-isomers of 2-(bromoethenyl)pyrroles **5b,8b,c** is present in the reaction mixture (Scheme 5).

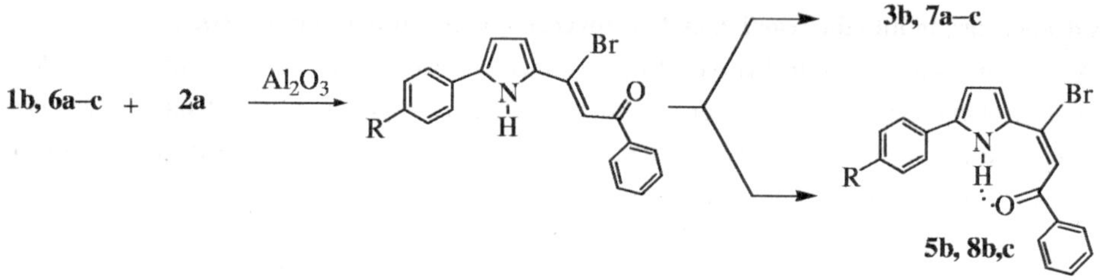

Scheme 5

The formation of ethynylpyrroles **3b,7b,c** from bromoethenylpyrroles **5b,8b,c** is supported by comparing the content of the latter in the crude product mixture with the yields of the isolated products. For example, pyrrole **6c** reacted with acetylene **2a** for 1 hour yielding (^{1}H-NMR) 81% of ethynylpyrrole **7c** and 19% of bromoethenylpyrrole **8c**. Chromatographic separation of the product mixture on alumina gave 94% of **7c** and 3% of compound **8c**. Moreover, bromoethenylpyrrole **8c** was found to undergo a readily transformation into ethynylpyrrole **7c** on alumina (conversion 92% in 1 hour). These results indicate that bromoethenylpyrrole **8c** is an intermediate product in the cross-coupling reaction. The *E*-isomer of bromoethenylpyrrole **8b** was more stable; its conversion into ethynylpyrrole **7b** in 50% yield was realized only by keeping the reaction mixture for 24 hours on alumina.

Table 1. Content of 2-aryl-5-(benzoylethynyl)pyrroles **3b,7a–c** and 2-aryl-5-(2-benzoyl-1-bromo-ethenyl)pyrroles **5b,8a–c** in the cross-coupling of the pyrroles **1b,6a–c** with benzoylbromoacetylene **2a** on Al$_2$O$_3$ (room temperature).

R	Reactions time (minutes)					
	10		30		60	
	Pyrrole, content		Pyrrole, content		Pyrrole, content	
H	**3b**, 75%	**5b**, 25%	**3b**, 74%	**5b**, 26%	**3b**, 78%	**5b**, 22%
Me$_2$N	**7a**, 51%	**8a**, 0%	**7a**, 64%	**8a**, 0%	**7a**, 70%	**8a**, 0%
MeO	**7b**, 53%	**8b**, 17%	**7b**, 59%	**8b**, 19%	**7b**, 64%	**8b**, 22%
Cl	**7c**, 72%	**8c**, 18%	**7c**, 79%	**8c**, 21%	**7c**, 81%[a]	**8c**, 19%[a]

[a]Yields of **7c** and **8c** after fractionation on Al$_2$O$_3$ are 94% and 3%, respectively.

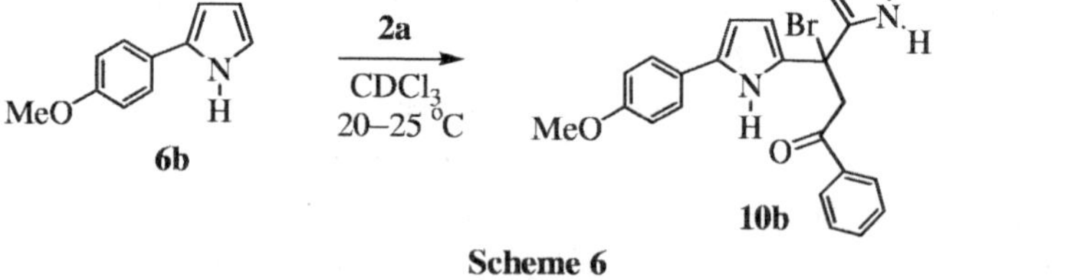

Scheme 6

Apart from the major products, ethynylpyrroles **3b,7a–c** and intermediate bromoethenylpyrroles **5b,8b,c**, the reaction mixtures contained small amounts of 2-benzoyl-1,1-dipyrrolylethenes **4b,9a–c**.

96

In some cases, the ^{1}H-NMR spectra of samples withdrawn from alumina on the initial stage (10 minutes) of the reactions contained a couple of broadened signals at δ 5.6 ppm and 12.1 ppm. According to the ^{1}H- and ^{13}C-NMR spectra, these signals were assigned to 2-benzoyl-1-bromo-1,1-di(pyrrol-2-yl)ethanes of type **10b** formed probably from unreacted pyrroles and benzoylbromoacetylene in chloroform (Scheme 6).

Dipyrrolylethane **10b** was also formed in an NMR ampoule (CDCl$_3$) from 2-(2-benzoyl-1-bromo-ethenyl)pyrrole **8b** and pyrrole **6b**. After treatment of a solution of dipyrrolylethane **10b** with aqueous potassium carbonate, dipyrrolylethene **9b** could be isolated.

2.3. The ethynylation of 1-vinylpyrroles with benzoylbromoacetylene on alumina

From theoretical and synthetic points of view it is intriguing to examine whether 1-vinylpyrroles, now accessible, obey to the ethynylation reaction with acylbromoacetylenes on active surfaces. The 1-vinyl group is known to deactivate α-position of the pyrrole ring[25] due to the negative inductive effect of the double bond[26] and its competitive conjugation with unshared nitrogen pair decreasing the pyrrole ring aromaticity. It also enhances the probability of Diels-Alder reactions.[27] Generally, the interaction of 1-vinylpyrroles with electron-deficient acetylenes represents a rewarding object for the investigation of competition of the two nucleophilic centres (α-position of the pyrrole ring and β-carbon of the vinyl group) with respect to electrophile.

In a recent paper,[28] it was shown that 1-vinyl- (**11a–c**) and 1-isopropenylpyrroles **11d–f** reacted with acetylenes **2a,b** on alumina regio- and chemoselectively to afford 1-vinyl- (**12a–c,g,h**) and 1-isopropenyl-2-(acylethynyl)pyrroles **12d–f,i** in 39–70% yields (Scheme 7).

The reactivity of pyrroles **11a–f** depends on the substituents in the pyrrole ring. So, under similar conditions the conversion of 1-isopropenylpyrroles **11d** and **11f** was 90 and 60%, respectively, while 1-iso-propenyl-4,5,6,7-tetrahydroindole **11e** reacted completely. During fractionating of the reaction mixtures on alumina the reaction was not stopped: as a result, neither pyrrole **11d** nor pyrrole **11f** were detected in any fractions.

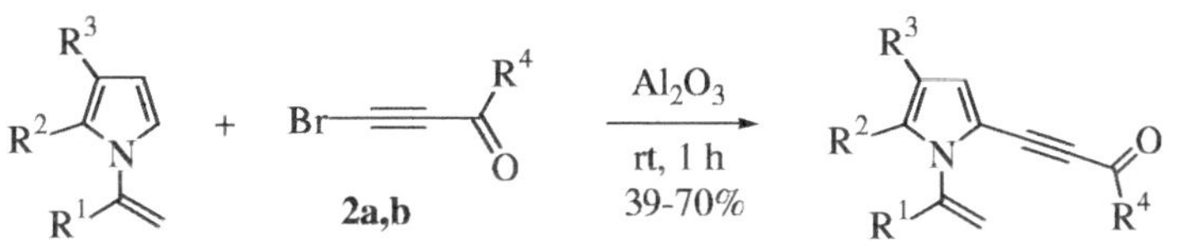

R^1 = H:

R^2 = R^3 = H (**11a, 12a**); R^2 - R^3 = (CH$_2$)$_4$ (**11b, 12b,g**); R^2 = Ph, R^3 = H (**11c, 12c,h**);

R^1 = Me:

R^2 = R^3 = H (**11d, 12d**); R^2 - R^3 = (CH$_2$)$_4$ (**11e, 12e,i**); R^2 = Ph, R^3 = H (**11f, 12f**);

R^4 = Ph (**12a–f**), 2-thienyl (**12g–i**)

Scheme 7

The ethynylation of 1-isopropenylpyrroles involves the formation of 1H-2-(acylethynyl)pyrroles **13d–f,i** as by-products.[28] They appear to be formed due to the hydrolysis of isopropenyl group catalyzed by the eliminated HBr (Scheme 8). This is supported by the fact that the starting 1-isopropenylpyrroles **11d–f** in the presence of Al_2O_3 without acetylenes **2a,b** under the reaction conditions are stable (it was shown using 1-isopropenylpyrrole **11e** as an example).

It seems likely that the hydrolysis begins with electrophilic addition of water molecule (contained in Al_2O_3) to the double bond of *N*-isopropenyl group and ends with decomposition of unstable α-pyrrolyl-alkanoles **14d–f,i** (Scheme 8).

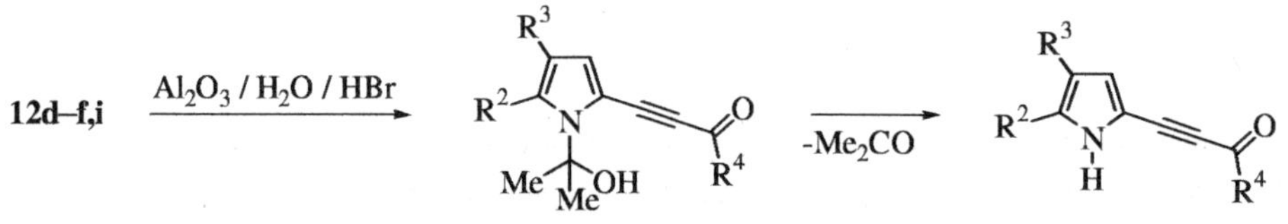

Scheme 8

Stability of the pyrroles **12d–f,i** depends on the substituents character in the pyrrole ring. It turned out that the least stable was 4,5,6,7-tetrahydroindole **12e**. In 1 hour after the reaction beginning, the content of 1*H*-2-ethynylpyrrole **13e** in the reaction mixture amounted to as much as 28% and this value still further increased during the fractioning on alumina (up to 40%). 2-Ethynylpyrroles **13d,f,i** were obtained (in 5, 3 and 10% yields, correspondingly) only when 1-isopropenyl-2-ethynylpyrroles **12d,f,i** were chromatographered on Al_2O_3.

Thus, the pyrroles having electron-donating substituents loose isopropenyl group more readily and that was in good agreement with the hydrolysis scheme (Scheme 8), which began with electophilic attack of proton to the double bond. This scheme was confirmed also with the fact that less protophile vinyl group was stable under the reaction conditions. Indeed, 1-vinyl-2-ethynylpyrroles **12a–c,g,h** did not form the corresponding 2-ethynylpyrroles at all.

The ethynylation of 1-vinyl- and 1-isopropenylpyrroles apparently proceeded through the stage of their nucleophilic addition to the activated triple bond with the formation of adducts **15**, which further eliminated HBr to afford the corresponding ethynylpyrroles **12a–i** (Scheme 9).[28] All attempts to detect the adducts **15** in the reaction mixtures were not met with success.

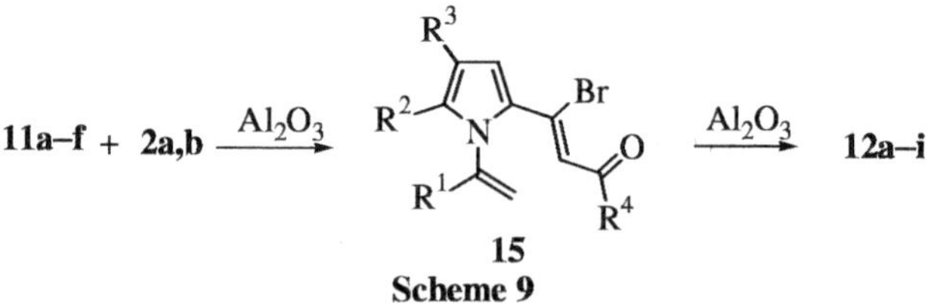

Scheme 9

2.4. Effect of active surfaces on the pyrrole ethynylation with haloacetylenes

As discussed above, the direct ethynylation of pyrroles with electrophilic haloacetylenes proceeded readily on the alumina active surface. Special emphasis should be made that the reaction was found to be facile and rapid at room temperature and required no transition-metal catalyst, base and solvent, thus closely corresponding to the green chemistry standards.[29] This reaction looked as opening a promising avenue to early inaccessible pyrroles with acetylenic substituents, useful building blocks to design drugs and new materials for molecular electronics. Consequently, it was reasonable to elucidate whether alumina was the unique metal oxide to affect the ethynylation and whether other active surfaces for this reaction could be found. Therefore, the room temperature interaction of 2-phenylpyrrole **1b** with benzoylbromoacetylene **2a** was investigated[30] upon their grinding with diverse metal oxides and salts (10-fold amount) as active surfaces. The reaction course [conversion of reactants **1b** and **2a** (Scheme 10) and the ratio of products] was determined by the ^{1}H-NMR spectra of the CDCl$_3$ extracts from the reaction mixture (Table 2). This reaction was chosen as a standard for the pyrrole ethynylation with haloacetylenes.

As reported earlier,[21,22] this reaction, when carried out on Al$_2$O$_3$, gave ethynylpyrrole **3b** as a major product, small amounts of the intermediate **5b** and the side product **4b**.

Apart from signals of these compounds, in the ^{1}H-NMR spectra of the samples taken from the reaction mixtures, two broad singlets at 5.70 ppm and 12.44 ppm appeared. According to ^{1}H- and ^{13}C-NMR spectroscopic data, these signals were assigned to unstable dipyrrolylbromoethane **16**, which can not be isolated (Scheme 10).

It has been shown (Table 2)[30] that certain metal oxides are active media for the interaction studied, particularly ZnO and BaO, the activity of which surpass that of Al$_2$O$_3$. The content of the ethynylation product in the reaction mixture was 81% (ZnO), 73% (BaO) and 71% (Al$_2$O$_3$) after 60 minutes.

No direct dependence of the basicity on the oxide efficiency in the cross-coupling reaction is observed as follows from the Table 2 (*i.e.*, ZnO is a weaker base than CaO but shows much higher activity in the cross-coupling). Moreover, alumina of different basicity, even with K$_2$CO$_3$ additive, gave practically the same results. One may suppose that transition metal contaminations of the above metal oxides and salts influence the reaction studied. However, this does not agree with the practically similar results obtained with different metal oxides (MgO, CaO, ZnO, BaO, Al$_2$O$_3$, Table 2). It is improbable that all the active surfaces are equally contaminated with the same transition metals.

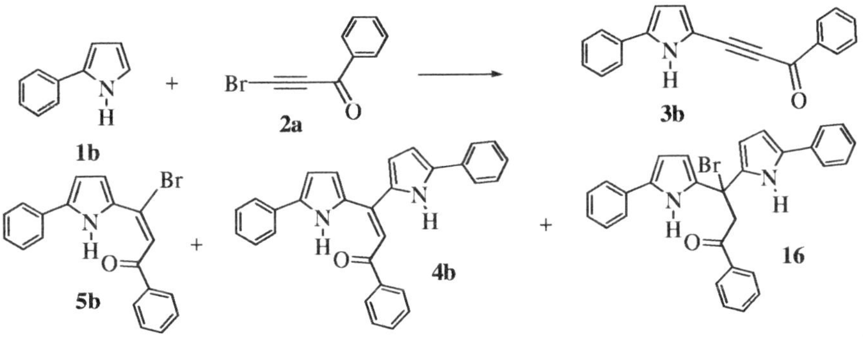

Scheme 10

As the case of Al$_2$O$_3$[21] the cross-coupling was accompanied by the formation of bromoethenylpyrrole **5b** and dipyrrolylethene **4b**, the former being an intermediate and the latter a side product (Scheme 10).

The metal oxides (TiO$_2$, ZrO$_2$) as well as the salts (CaCO$_3$, ZrSiO$_4$) were found to be inactive in the cross-coupling. Instead, they were highly active in the nucleophilic addition of pyrrole **1b** to the triple bond of benzoylbromoacetylene **2a** to afford the adduct **5b** (the content of which in the reaction mixture ranged from 81% up to 87%).[30] Thus, these metal oxides and salts are specific active surfaces for effecting chemo-, regio- and stereoselective addition of pyrrole **1b** (by its C2-position) to the triple bond of acetylene **2a**.

In all cases, dipyrrolylethene **4b** was formed in much smaller amounts (0–19%). This was obviously the product of substitution of the bromine atom in the intermediate **5b** by the pyrrolyl moiety.

Table 2. ^{1}H-NMR spectroscopic monitoring of the reaction of 2-phenylpyrrole **1b** with benzoyl-bromoacetylene **2a** (room temperature) on active surfaces.

Active surface	Reaction times (minutes)									
	10					60				
	Composition of reaction mixture (%)									
	1b	**3b**	**5b**	**4b**	**16**	**1b**	**3b**	**5b**	**4b**	**16**
MgO	54	41	5	0	0	25	69	6	0	0
CaO	71	19	7	3	0	39	50	9	2	0
ZnO	0	78	18	4	0	0	81	15	4	0
BaO	7	69	20	4	0	0	73	20	6	0
Al$_2$O$_3$ (pH 7.5)	17	62	21	0	0	5	71	23	0	0
Al$_2$O$_3$ (pH 9.3)	17	62	21	0	0	6	68	26	0	0
Al$_2$O$_3$^{a} (pH 9.3)	14	61	20	5	0	12	68	20	0	0
TiO$_2$	9	0	70	0	21	0	0	87	13	0
ZrO$_2$	0	0	73	0	27	0	0	83	0	17
CaCO$_3$	0	0	59	0	41	0	0	81	19	0
ZrSiO$_4$	7	0	61	0	32	0	0	85	0	15

^{a}K$_2$CO$_3$ (5% relative to alumina) was used.

Apparently, the cross-coupling, resulting in the formation of the ethynylpyrrole **3**, proceeded *via* the intermediate **5** which further eliminates HBr. This was supported by the observation that, when the intermediate **5** (prepared on ZrO$_2$) was passed through an Al$_2$O$_3$ column, the ethynylpyrrole **3** was isolated.

Obviously, the route to ethynylpyrrole **3** *via* the charge-transfer complex **17** and further the salt **18** avoiding the adduct **4**, could not be excluded (Scheme 11). Such a reaction may be realized as a parallel or complementary channel of the above HBr elimination from the adduct **5**.

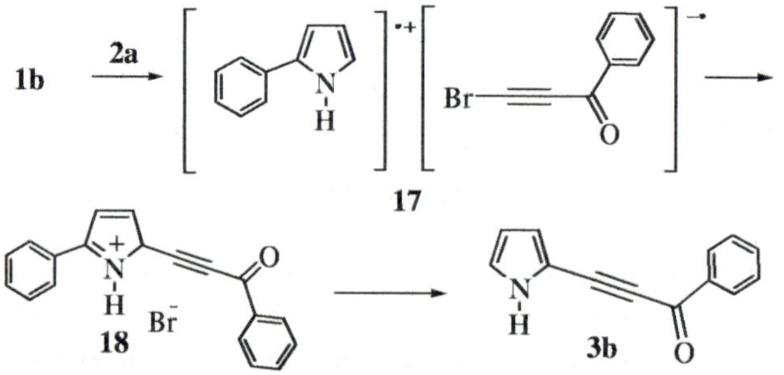

Scheme 11

100

It is known that reactions on the metal oxide surfaces often proceed by one electron transfer and hence for an investigation of their mechanisms, the ESR technique is appropriate.[31]

The reaction between the pyrrole **1b** and acetylene **2a** on active surfaces of the oxides: ZnO, Al_2O_3 (pH 9.5), TiO_2, ZrO_2 and the salts: $CaCO_3$, $ZrSiO_4$ was monitored by the ESR technique.[30] Singlets with g-factors in the region of free radicals were observed (Table 3).

The value of g-factors changed depending on the surface nature and varied within 0.0006. The asymmetric signals detected for $ZrSiO_4$, TiO_2 and ZrO_2 under microwave power saturation split in two lines. The line with g-factor 2.0023 was assigned to the 2-phenylpyrrole cation radical. The second line with g-factors 2.0032–2.0044 (depending on active surface nature) was likely to be attributed to the complex of benzoylbromoacetylene anion-radical with active surface. At the same time, microwave power saturation applied to the reaction mixtures on oxides Al_2O_3 and ZnO did not cause the ESR signals splitting. Probably, the components of the reaction mixture activated by the crystalline lattice first underwent charge transfer to deliver the ion-radical pair. Depending on acid-base properties of the active surface the ion-pair components behaved diversely: on the acidic surfaces, the radical cation should be stabilized, while on the basic ones it was deprotonated to a neutral radical, thus leading to different reaction products **3** and **5**, respectively.

Thus, the experimental data point to the fact that ethynylation of pyrroles with benzoylbromoacetylene may involve a one electron transfer stage delivering ion-radicals stabilized by the crystalline lattice of the active surface.[30]

Table 3. ESR signal characteristics for the reaction of 2-phenylpyrrole **1b** with benzoylbromoacetylene **2a** on active surfaces.

Active surface	$N \cdot 10^{16}$, spin/g	g-factor	ΔH, mT
ZnO	35	2.0030	0.660
Al_2O_3 (pH 9.5)	10	2.0026	0.600
$CaCO_3$	6.9	2.0030	0.533
TiO_2	11	2.0027	0.710
ZrO_2	4.0	2.0024	0.622
$ZrSiO_4$	8.7	2.0024	0.580

The active surface participation in the ion-radical generation was also confirmed by appearance of the corresponding but very weak signals in the ESR spectra of the separated pairs: Al_2O_3-2-phenylpyrrole and Al_2O_3-benzoylbromoacetylene. Thus, approximately 0.5 hour after preparation, 2-phenylpyrrole on Al_2O_3 turned light brown. In the ESR spectrum appeared a weak ($3 \cdot 10^{11}$ spin/g) broad (1.1 mT) singlet with g-factor 2.0024, narrowing and increasing with time. Benzoylbromoacetylene on Al_2O_3 just right after the preparation, became pink and its ESR spectrum showed very weak (10^{14} spin/g) narrow (0.33 mT) singlet with g-factor 2.0032, which did not practically change with time.

On silica gel, the pyrroles **1a–c** reacted with acylbromoacetylenes **2a,b** under mild conditions (room temperature, moderate self-heating, 15–30 minutes) to form 2-acyl-1,1-di(pyrrol-2-yl)ethenes **4a–f** as major products (in yields of up to 60%, based on a starting pyrrole) (Scheme 12).[32]

The expected primary adducts, 2-(2-acyl-1-bromoethenyl)pyrroles **5a–f**, could be isolated (by chromatography on Al_2O_3) in 3–11% yields only.

Monitoring of the reaction of pyrrole **1a** with acetylene **2b** by ^{1}H-NMR confirmed that the primary adduct was 2-(1-bromo-2-thienoylethenyl)pyrrole **5d**, the content of which was increasing over the time studied. However, in this case the adduct **5d** was not isolated in pure form: during chromatography on alumina, it partially underwent HBr elimination to form the acetylene **3d** in 12% yield, while the major part of **5d** was polymerized. Still, 2-(2-acyl-1-bromoethenyl)pyrroles **5b,c,f** were stable enough to be isolated by chromatography on alumina in 3–11% yields.

The isolated yields of 2-acyl-1,1-di(pyrrol-2-yl)ethenes **4a–f** were also sensitive towards the nature of substituents in the pyrrole ring. Thus, from the equimolar mixtures of pyrroles **1b,c** and acetylene **2b**, after their short contact (15 minutes) with silica, the adducts **4e,f** were prepared in 57 and 60% yields, respectively. Meanwhile, in order to attain a preparatively acceptable yield (31%) of the dipyrrolylethene **4d**, it was necessary to use a two-fold molar excess of the pyrrole **1a** and increase the reaction time to 30 minutes. However, further prolonging the process resulted in resinification of the products.

In the above-mentioned cases, acetylenes **3e** and **3f** were also formed in low yields (3 and 2%, respectively). The yield of these products increased to 20–22% after chromatographic treatment of reaction mixtures on alumina, again supporting the HBr elimination from the primary adducts **5** to be the source of acetylenes **3**. Correspondingly, on silica 2-acyl-1,1-di(pyrrol-2-yl)ethenes **4a–f** were mainly formed as a result of the substitution of bromine atom in 2-(2-acyl-1-bromoethenyl)pyrroles **5a–f** by a second molecule of the pyrrole **1** (presumably, through an addition-elimination two-step process). Another probable route to **4a–f** was the addition of the pyrrole **1** to acetylenes **3a–f**. It was also possible that both of these routes took place.

2.5. The ethynylation of 4,5,6,7-tetrahydroindoles with halopropynoates on an active surfaces

Application of the above methodology to the synthesis of 4,5,6,7-tetrahydroindole-2-propynoates has been considered as justified since these compounds are promising protected ethynylpyrroles because the ester moiety can be easily removed through conventional decarboxylation procedures.[33] Furthermore, 4,5,6,7-tetrahydroindole-2-propynoates could be expected to undergo easy catalytic dehydrogenation (as in simpler cases[34]) to yield 2-substituted indoles, potential intermediates for many alkaloids and pharmacologically important targets.[35]

Although the methods for the preparation of 3-substituted indoles are well established, there is a need for easier access to 2-substituted indoles. Unlike 3-substituted derivatives, 2-substituted congeners, particularly 2-ethynylindoles, still represent a synthetic challenge, since most electrophilic aromatic substitution reactions of indoles occur at the position 3. Consequently, tetrahydroindole derivatives which are actually pyrroles (in terms their reactivity) can be considered as potential precursors of corresponding 2-substituted indoles.

Therefore, the cross-coupling of $1H$- and 1-vinyl-4,5,6,7-tetrahydroindoles with ethyl bromo- and iodopropynoates to give 3-(4,5,6,7-tetrahydroindol-2-yl)-2-propynoates has been studied.[36]

In contrast to benzoylbromoacetylene, which with 4,5,6,7-tetrahydroindole **1c** [Al$_2$O$_3$, pH 7.4, a 10-fold amount (by weight), room temperature, 0.5 hours] formed mainly 2-(benzoylethynyl)-4,5,6,7-tetrahydroindole, ethyl bromopropynoate **19a** under similar conditions reacted with **1c** to form a mixture of ethyl 3-(4,5,6,7-tetrahydroindol-2-yl)-2-propynoate **20** (20%), ethyl 3-bromo-3-(4,5,6,7-tetrahydroindol-2-yl)-2-propenoate **21** (62%) and ethyl 3,3-di(4,5,6,7-tetrahydroindol-2-yl)acrylate **22** (14%) (Scheme 13).

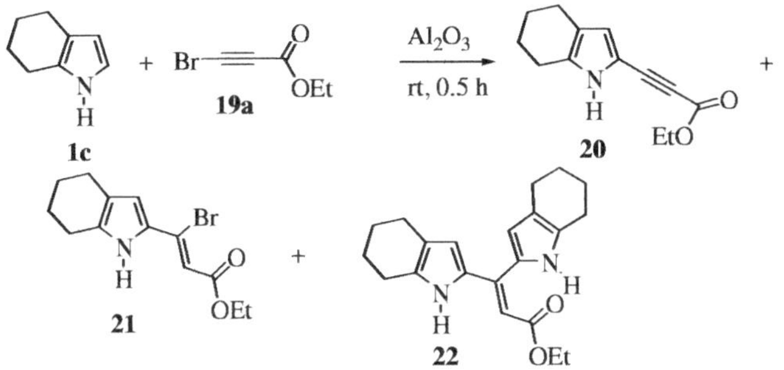

Scheme 13

With an increased alumina ratio (50-fold), product **21** was not detected and the only reaction products were indoles **20** (38%) and **22** (62%). A similar result was observed when a more basic sample of alumina (pH 9.5, a 50-fold amount) was employed.

When K$_2$CO$_3$ (10% relative to alumina) was added to the reaction mixture with a 50-fold amount of alumina (pH 9.5), the proportion of indolylpropynoate **20** in the reaction mixture increased to 58% (preparative yield 46%), while the content of di(indolyl)acrylate **22** dropped to 34%.

Unlike ethyl bromopropynoate **19a**, ethyl iodopropynoate **19b**[36] reacted with 4,5,6,7-tetrahydroindole **1c** (a ratios of **1c:19b**, 1:1, 2:1) on alumina of different pH values (7.4 and 9.5) and with different quantities (10- and 50-fold excess amount) to afford chemospecifically ethyl di(indolyl)acrylate **22** (yield 79%) (Scheme 14).

Upon mixing of equimolar quantities of the indole **1c** and ethyl bromopropynoate **19a** without alumina, a strong self-heating and bright violet coloration were observed. A caramel-like reaction product was formed consisting of ethyl di(indolyl)propanoate **24** and ethyl propynoate **19a** (Scheme 15).[36] With 2 mol equivalents of indole **1c** per 1 mol equivalent of ethyl bromopropynoate **19a**, the reaction furnished propanoate **24**, though accompanied by resinification.

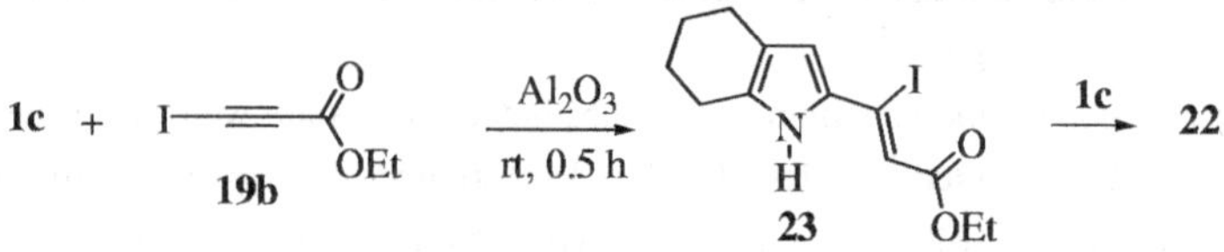

Scheme 14

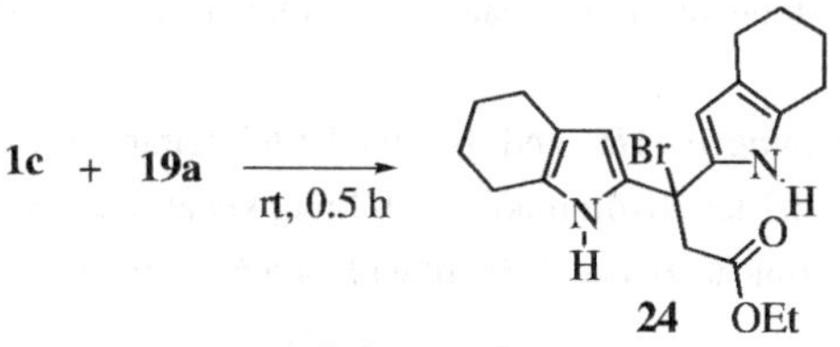

Scheme 15

As it was mentioned before, 1-vinyl-4,5,6,7-tetrahydroindole **11b** reacted with benzoylbromo-acetylene **2a** on Al_2O_3 selectively to give the corresponding ethynylindole **12b**. Under similar conditions, the indole **11b** and ethyl bromopropynoate **19a** formed indolylpropynoate **25** (77%) along with ethyl di(indolyl)acrylate **26** (23%) (Scheme 16). At a higher content of Al_2O_3 (a 50-fold amount), the selectivity of the reaction was greater and in 0.5 hours the ratio of **25:26** reached 92:8 (preparative yield of ethyl indolylpropynoate **25** in this case was 71%).

In the case of ethyl iodopropynoate **19b**, indole **11b** reacted slowly and in contrast to the indole **1c**, nonselectively: in 0.5 hours (Al_2O_3, a 10-fold amount) the starting material **11b** still remained (46%) with the ratio of **25:26** at 40:14.[36] To isolate ethyl di(indolyl)acrylate **26**, the reaction was carried out with **11b** and ethyl iodopropynoate **19b** (ratio of **11b:19b**, 1:1) on alumina (pH 9.5, a 5-fold amount) during 1 hour. In this case, the reaction gave ethyl indolylpropynoate **25** (53%) and ethyl di(indolyl)acrylate **26** (47%); the preparative yield of the latter was 31%.

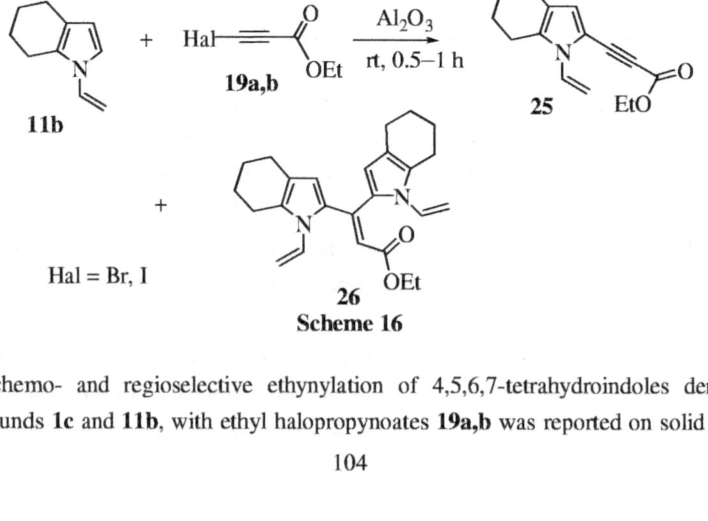

Scheme 16

Efficient chemo- and regioselective ethynylation of 4,5,6,7-tetrahydroindoles derivatives **27a–f**, including compounds **1c** and **11b**, with ethyl halopropynoates **19a,b** was reported on solid K_2CO_3 as novel active surface to afford the corresponding ethyl 3-(4,5,6,7-tetrahydroindol-2-yl)-2-propynoates **20,25,28a–f** in 62–90% yield (Scheme 17).[37]

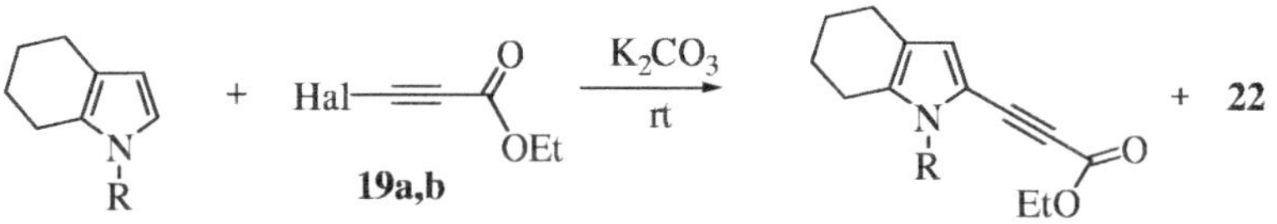

Hal = Br, I;
R = H (**20**), CH=CH$_2$ (**25**), Me (**28a**), Bn (**28b**), Me(CH)OPr-*i* (**28c**),
Me(CH)OBu-*n* (**28d**), CH$_2$CH$_2$SEt (**28e**), CH$_2$CH$_2$SPr-*n* (**28f**)

Scheme 17

The reaction proceeded smoothly at room temperature for about 60 minutes on grinding the reactants with solid K$_2$CO$_3$. The reaction was equally efficient both on air and under argon blanket. The products **20,25,28a–f** were isolated chromatographically by placing the solid reaction mixture on the top of an Al$_2$O$_3$-packed column and consequent eluting with *n*-hexane and diethyl ether.

In contrast to the similar reaction on the Al$_2$O$_3$ active surface,[36] in the case of ethyl bromopropynoate **19a**, no any side products were detected, whereas with ethyl iodopropynoate **19b**, 4,5,6,7-tetrahydroindole **1c** gave minor amounts of ethyl di(4,5,6,7-tetrahydroindol-2-yl)acrylate **22** (2–4%). Besides, the latter reaction appeared to be noticeably slower (41% conversion of pyrrole **1c** for 60 minutes).

The reaction is chemo- and regioselective. The expected addition products of the type 2-(1-bromo-2-ethoxycarbonylethenyl)-4,5,6,7-tetrahydroindole were not detectable in the reaction mixtures. Also, no traces of the corresponding 3-isomers contaminate the products; in spite of that it is the common knowledge that bulky and branched substitutions at position 1 in the pyrrole nucleus can direct the electrophilic attack to position 3.[38] In this reaction, even bulky substituents such as CH(Me)OPr-*i* and CH(Me)OBu-*n* did not change the high regioselectivity (actually specificity), though the reaction time was considerably increased (12 hours instead of 60 minutes) and portion-wise addition of acetylene **19a** to the reaction mixture was required. Obviously, the low reaction rate in this case resulted from the steric hindrance. Generally, the ethynylation tolerates the substituents R of quite a different nature (H, Alk, benzyl, vinyl, acetal and sulfide functions).

Astonishingly, the reaction did not occur in solution (diethyl ether, CHCl$_3$, CDCl$_3$) both with and without K$_2$CO$_3$, indicating a crucial role of the freshly ground potassium carbonate.

The key effect of the K$_2$CO$_3$ active surface on the ethynylation studied was also supported by the reactants/K$_2$CO$_3$ ratio influence on the product yield: at the 1:10 ratio the yield of **20** was close to quantitative, while at the higher concentrations of reactants in the K$_2$CO$_3$ medium (1:2, 1:5) the yield dropped to 50–55%. Meanwhile, the addition of K$_2$CO$_3$ (2.5 mol excess) to Al$_2$O$_3$ just insignificantly changed the usual (for Al$_2$O$_3$) products ratio: **20:22** for Al$_2$O$_3$ =2:3, for Al$_2$O$_3$+K$_2$CO$_3$=3:2.[36]

Formally, the ethynylation studied can be rationalized both as electrophilic substitution on the pyrrole ring or nucleophilic addition of the electron-rich pyrrole ring to the electron deficient triple bond. In both cases, the one-electron transfer step is plausible. Indeed, in the ESR spectrum of the reaction mixture, singlet (g=2.0023, ΔH=1.8 mT) was observed, thus confirming a one-electron transfer step from the pyrrole **1c** to the acetylene moiety, forming the ion-radical pair **29** which further transforms *via* intermediate salt **30** to ethyl propynoate **20** (Scheme 18).

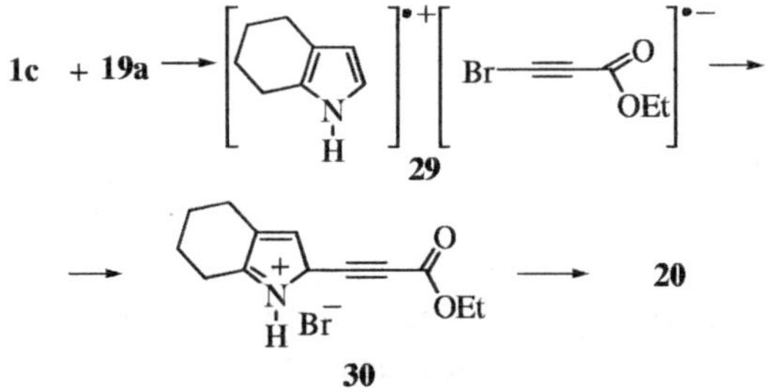

$$1c \ + \ 19a \ \longrightarrow$$

29

30

Scheme 18

To illustrate the easy aromatization of the ethyl 3-(4,5,6,7-tetrahydroindol-2-yl)-2-propynoates, the ethyl 3-(1-methyl-4,5,6,7-tetrahydroindol-2-yl)propynoate **28a** was refluxed in *o*-xylene with Pd/C (10% of Pd) for 12 hours to give ethyl 3-(indol-2-yl)propanoate **31** and ethyl 3-(indol-2-yl)acrylate **32** (4:1 ratio) resulting from the hydrogen redistribution between the cyclohexane ring and the triple bond (Scheme 19).[37]

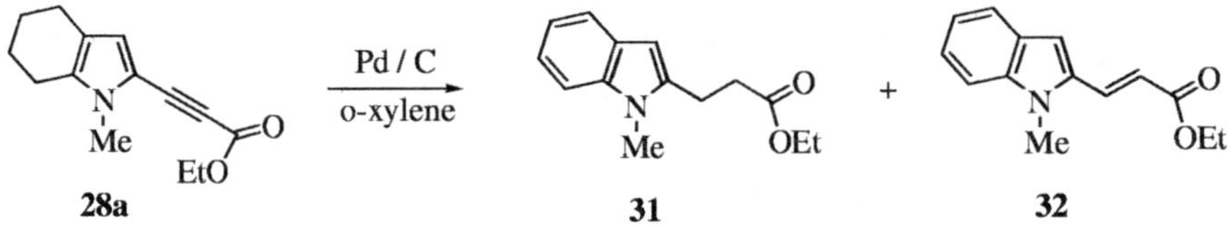

28a

31

32

Scheme 19

This confirmed that ethyl 3-(4,5,6,7-tetrahydroindol-2-yl)-2-propynoates **20,25,28a–f** synthesized were expectedly readily converted to 2-functionalized indoles.

2.6. The ethynylation of indoles with benzoylbromoacetylene on alumina

Indoles are important molecules, which exhibit a wide spectrum of biological activity.[1] Many naturally occurring products contain the indole skeleton.[39] The pharmacological activity of natural and synthetic indole derivatives has led to the extensive development of diverse methodologies for their synthesis. In particular, highly potent indole building blocks are those with acetylenic moieties. They are currently used in the design of numerous indole structures[40] due to the rich chemistry of the acetylenic function.[2] Certain syntheses of *C*-ethynylindoles are known (see introduction). Meanwhile, to our knowledge, there are no methods for introduction of an acetylenic moiety in place of the CH-indole hydrogen.

The successful development of the cross-coupling employing non-halogenated pyrroles and haloacetylenes stimulated the authors[41] to apply this methodology to the synthesis of acetylenic indole derivatives.

The experiments showed that indoles **33a,b** reacted smoothly with benzoylbromoacetylene **2a** under the following conditions: room temperature, 1 hour, when the reactants were ground with a 10-fold mass excess of Al_2O_3 under solvent-free conditions. The reaction proceeded chemo- and regioselectively to afford 3-(benzoylethynyl)indoles **34a,b** in 72 and 76% yields, respectively (Scheme 20).[41]

In the solid state but without Al_2O_3, the ethynylation did not take place (just ~1% of a mixture of unidentified products was formed).

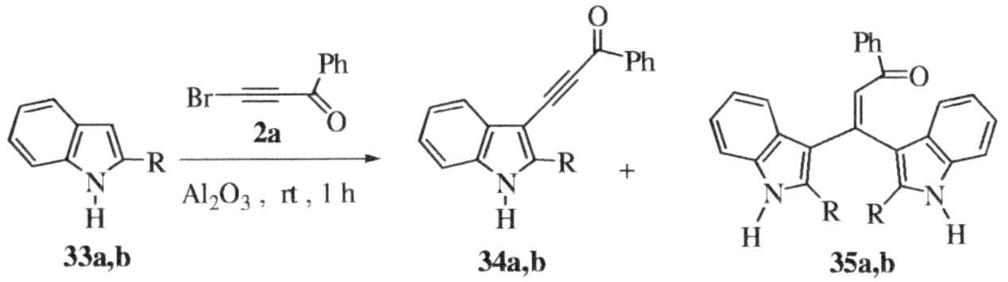

R = H (**a**) , Me (**b**)
Scheme 20

The only side products of the reaction were 1,1-di(indol-3-yl)-2-benzoylethenes **35a,b**, detectable in all cases (^{1}H-NMR) in small amounts (5–8%). The isolated yield of **35b** was 6%. The adducts **35a,b** were most probably formed by the substitution of bromine in the intermediates of this reaction, namely 3-(2-benzoyl-1-bromoethenyl)indoles **36a,b** since ethynylindoles **34a,b** could not add indoles **33a,b** under the reaction conditions (Scheme 21).

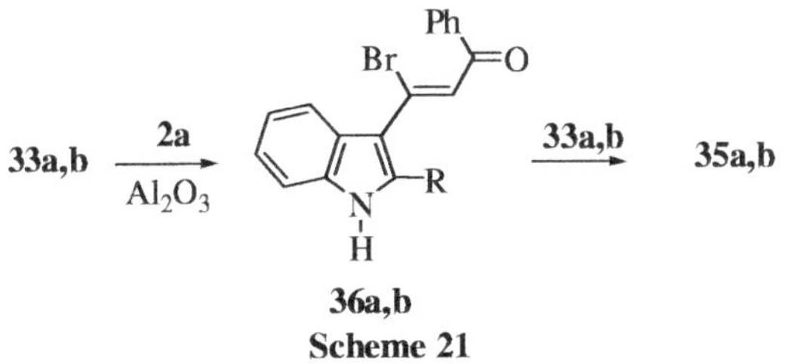

36a,b
Scheme 21

The condensed indole, benz[g]indole **37**, under the same conditions, was coupled with acetylene **2a** to furnish a mixture of 3- (**38**) and 2-ethynylindoles (**39**) in 45% overall yield (based on **37** conversion was 66%) (Scheme 22).[41] In this case, 2-(2-benzoyl-1-bromoethenyl)benz[g]indole **40**, stabilized by strong intramolecular H-bonding between the NH and C=O groups (δNH 14.59 ppm), was also observed.

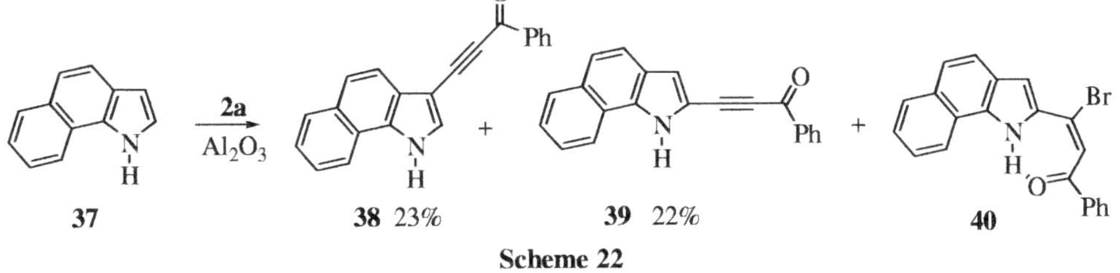

38 23% **39** 22% **40**
Scheme 22

107

In accordance with previous results,[21,22] 4,5-dihydrobenz[*g*]indole **41**, being actually a 2-phenyl-pyrrole, on reaction with benzoylbromoacetylene **2a** on alumina was readily converted to the corresponding 2-ethynylated derivative **42** in 68% yield. A side product of coupling, 2-benzoyl-1,1-di(4,5-dihydro-benz[*g*]indol-2-yl)ethene **43**, was isolated in 9% yield (Scheme 23).

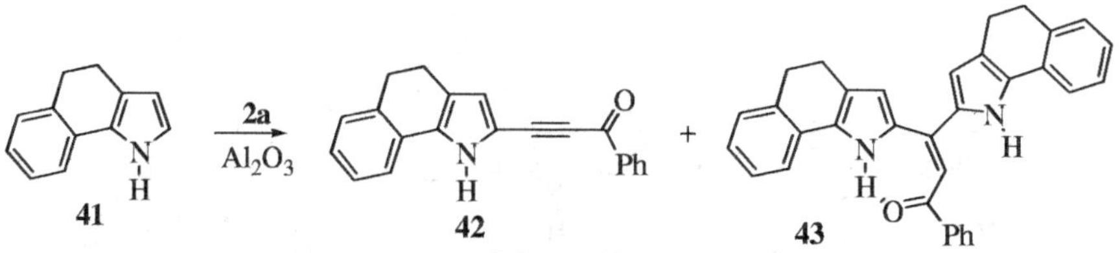

Scheme 23

By contrast to benzoylbromoacetylene **2a**, ethyl bromopropynoate **19a** reacted (room temperature, Al₂O₃) with indoles **33a,b** to give mainly the corresponding ethyl 3,3-di(indol-3-yl)acrylates **44a,b** in up to 57% yield (Scheme 24), whereas ethynylindoles **45a,b** were isolated in no more than 17% yield.[42] It should be noted that this reaction had a considerably lower rate than with benzoylbromoacetylene: indole **33a** with an equimolar amount of acetylene **19a** in 2 hours at room temperature gave only ~4% of ethynylindole **45a** and ~6% of acrylate **44a**. By heating the reactant mixture (52–54 °C, 1 hour), the concentration of acrylate **44a** increased to 18–19%, the fraction of ethynylindole **45a** being 4–5%. Apart these products, the reaction mixture contained unreacted indole **33a** and acetylene **19a**. In the presence of 2 equivalents of the latter (room temperature), the amount of ethynylindole **45a** increased to 20% in 1 hour and to 26% in 3 hours and it was isolated in 17% yield (after 3 hours). In this case, the concentration of acrylate **44a** did not exceed 10%. When the reaction was carried out with 2 equivalents of indole **33a**, the yield of acrylate **44a** reached 41%.

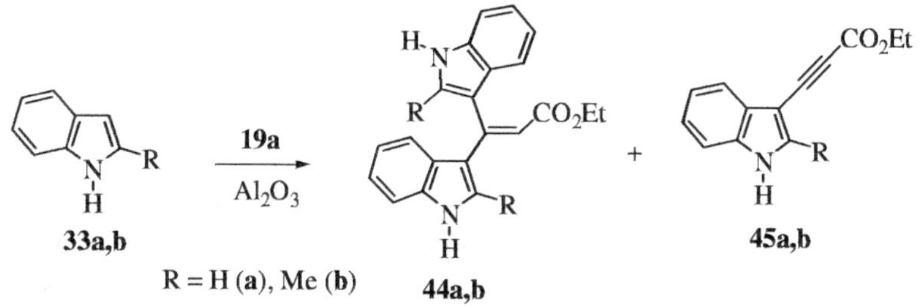

Scheme 24

2-Methylindole **33b** turned out to be more reactive than indole **33a** towards ethyl bromopropynoate **19a**: after 2 hours, only traces of **33b** were detected. The corresponding ethynylindole **45b** and ethyl di(indolyl)acrylate **44b** were formed in approximately equal amounts. However, the product ratio changed after the fractionation of the mixture on alumina column and compounds **44b** and **45b** were isolated in 57 and 12% yields, respectively.[42] The reactions of indoles **33a,b** with acetylene **19a** in the presence of other oxides, namely SiO₂, BaO, CaO, MgO, ZnO, were examined. Among these, only MgO showed a weak catalytic activity (~3–4% of ethynylindole **45a** and traces of acrylate **44a** were present in the reaction mixture after 1 hour at room temperature).[42]

108

3. A peculiarity of the *C*-ethynylpyrroles and -indoles reactivity

The above transition metal- and solvent-free ethynylation of pyrrole nucleus with electrophilic haloacetylenes on alumina[21] or other active surfaces[30] makes 2-ethynylpyrroles and 3-ethynylindoles, bearing electron withdrawing substituents at the triple bond, readily accessible. In particular, this new methodology has led to 2-acyl- and 2-alkoxycarbonylethynyl-4,5,6,7-tetrahydroindoles using large-scale 4,5,6,7-tetra-hydroindole[43] as starting material. These compounds, due to their possible easy aromatization, might be suitable intermediates to synthesize 2-functionalized indoles.[44]

2,3-Dichloro-5,6-dicyano-1,4-benzoquinone (DDQ) is especially effective for aromatization and it is often the reagent of choice to effect facile dehydrogenation of both simple and complex hydroaromatic compounds.[45] For example, fully aromatic derivatives were obtained in high yields by dehydrogenation of the corresponding tetrahydro-heteroaromatic compounds using DDQ.[46] It is known that the treatment of 4,7-dihydroindol-2-yl-fumarate with DDQ gives the expected indole in 94% yield.[47] Consequently, an attempt to apply this methodology to the synthesis of indoles with acetylenic substituents from 2-ethynyl-4,5,6,7-tetrahydroindoles **3c,20,28a,b** was recently undertaken.[48]

Surprisingly, the anticipated aromatization did not occur. Instead, [2+2]-cycloaddition to the triple bond to afford 1:1-adducts, bicyclo[4.2.0]octadienes **46a–d**, in 81–93% isolated yields was observed (Scheme 25).[48] The bright colored (cherry, deep cherry) adducts started precipitating instantly just after the reactants (pyrroles and DDQ) were mixed.

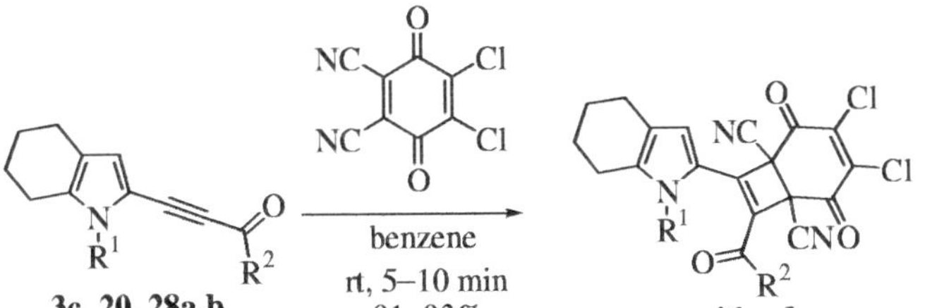

$R^2 = OEt; R^1 = H$ (**a**), Me (**b**), Bn (**c**); $R^2 = Ph, R^1 = H$ (**d**)

Scheme 25

In the ^{1}H-NMR spectra of the adducts **46a–d**, signals of indole protons were absent and the signals of four CH_2 groups of the cyclohexano moiety along with signals of pyrrole H-3 remained, the NH, H-3 couplings equaling to 1.7–2.0 Hz, that was typical for 2,4,5-trisubstituted pyrroles.[49] In the ^{13}C-NMR spectra, the signals of acetylenic carbons (δ 81.8–87.9 ppm) disappeared and the cyclobutene sp$_3$-signals (δ 53.8–52.2 ppm) were observed.

As to the question, whether the chlorine-substituted C=C bond of DDQ or the one with the cyano groups undergoes the cycloaddition, the comparison of the ^{13}C-NMR spectrum of the adducts with that of DDQ[50] unambiguously evidences in favor of structures **46a–d**. Indeed, the chemical shifts of the olefinic carbons C-3 and C-4 of quinone cycle (δ 142.9–144.0 ppm and δ 140.1–141.1 ppm, respectively) were similar to those of the chlorine substituted carbons (δ 141.0 ppm) in DDQ, being significantly different from carbon atoms bound with cyano groups (δ 125.1 ppm).

Generally, DDQ is a powerful oxidizing agent and it has proved to be a versatile reagent for various organic transformations, *e.g.* it is an active dienophile.[51] However, its participation in [2+2]-cycloadditions to an acetylenic moiety so far was not reported. Moreover, DDQ was used to oxidize phenylalkynes to the corresponding enynes,[52] *i.e.* with an intact triple bond. Also unknown remained any other [2+2]-cyclo-addition reactions at the triple bond of ethynylpyrroles. Thus, the observed [2+2]-cycloaddition of DDQ to the triple bond of 2-ethynyl-4,5,6,7-tetrahydroindoles was the first example of such reactions for acetylenes and DDQ.

Commonly, cycloaddition reactions of benzoquinones with acetylenes are mostly photochemical leading (depending on the quinone structure) either to the adduct at the carbonyl group[53] or at the double bond.[54] The ¹H-NMR monitoring of the reaction of 2-ethynyl-4,5,6,7-tetrahydroindoles **3c,20,28a,b** with DDQ in the dark confirmed that the reaction was not a photochemical one: in this case, the [2+2]-cyclo-addition was as efficient as in daylight.

The reaction was proved to be of a general character: apart from the 2-ethynyl-4,5,6,7-tetra-hydroindoles **3c,20,28a,b** other substituted 2-ethynylpyrroles, *e.g.* pyrroles **3b,7b,c**, reacted with DDQ in the same [2+2]-cycloaddition manner, though the process required a longer time (1 hour) and/or a polar solvent (acetone, acetonitrile or acetone-benzene) (Scheme 26).

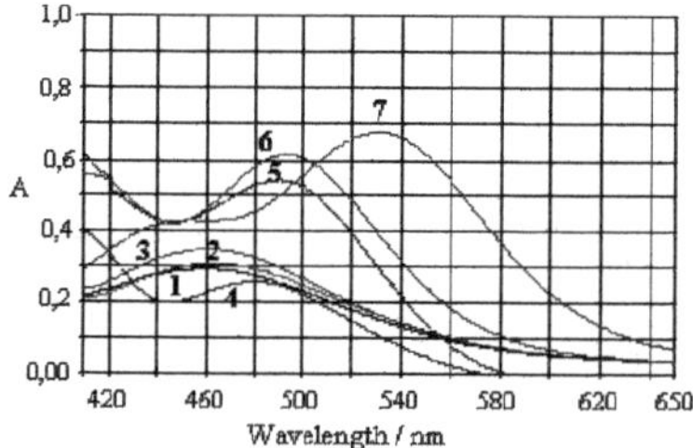

$R^2 = Ph; R^1 = Ph$ (**a**), $4\text{-}ClC_6H_4$ (**b**); $R^2 = 4\text{-}O_2NC_6H_4$, $R^1 = 4\text{-}MeOC_6H_4$ (**c**)

Scheme 26

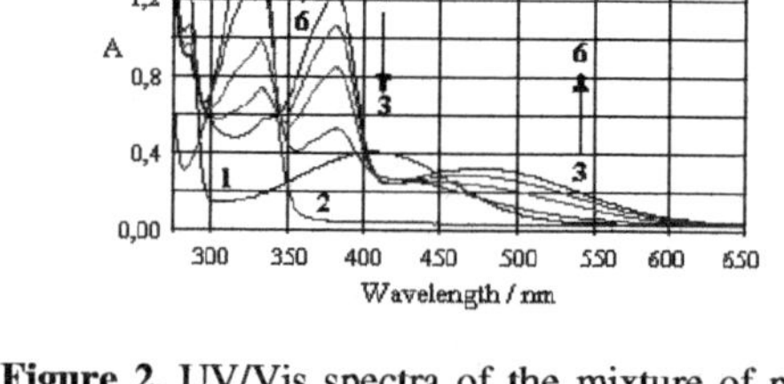

Figure 1. CT bands in the UV/Vis spectra of cycloadducts: 1-**46a**; 2-**46b**; 3-**46c**; 4-**46d**; 5- **47a**; 6-**47b**; 7-**47c**.

Figure 2. UV/Vis spectra of the mixture of pyrrole **20** (0.025 mM) and DDQ (0.025 mM) in benzene, room temperature; 1-DDQ; 2-pyrrole **20**; 3-after 2 min; 4-after 10 min; 5-after 25 min; 6-after 60 min.

The structure of the adducts **47a–c** was in agreement with the NMR spectra, showing that pyrrole C-3 and C-4 carbons and the corresponding protons H-3 and H-4 remained intact. A large difference (up to 22 ppm) between ¹³C-chemical shifts of the CN-substituted carbons in the tetrahydroindoles **46a–d** and

5-arylsubstituted series **47a–c** called for intramolecular charge transfer (CT) from arylpyrrole moiety (through conjugation mechanism) onto the dicyanocyclobutene counterpart of the molecule that was convincingly evidenced by the large red shifts (up to 70 nm) and hypochromic effect of the CT band in UV/Vis spectra of compounds **46a–d** and **47a–c** (Figure 1). This long-range charge transfer was also manifested by the ^{13}C- down-field shifts (1–3 ppm) of the most remote C_{para} carbons of the aryl substituents in the adducts **47a–c** as compared to the starting acetylenes **3b,7b,c**.

The reaction of 3-ethynylindoles **34a,b** with DDQ obeyed the same [2+2]-cycloaddition scheme to give cycloadducts **48a,b** in 84–87% yields, proceeding as smoothly as with 2-ethynylpyrroles **3b,7b,c** (acetone-benzene or acetonitrile, rt, 5–10 minutes) (Scheme 27).

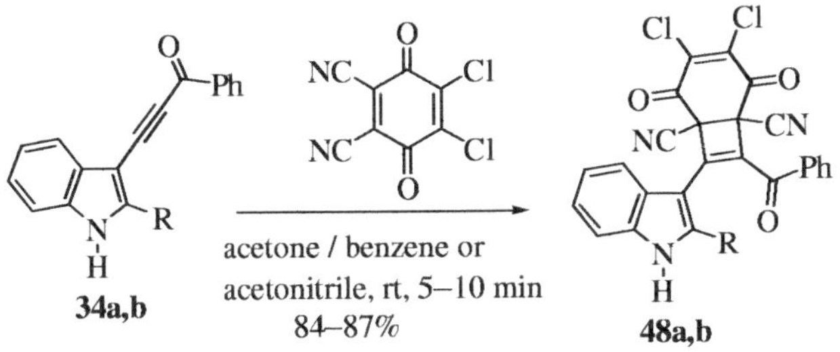

Scheme 27

Reactions of quinones often proceed with electron transfer and hence can be examined by UV/Vis[55] and ESR[55a,c,56] spectroscopy. Indeed, the experimental data (UV/Vis and ESR) obtained may be tentatively rationalized as an evidence for the electron transfer for the cycloaddition reaction under study. When monitored by UV/Vis spectroscopy (Figure 2), just after mixing of reagents in benzene, two sharp intense peaks at 330, 381 nm along with a broad absorption band in the region 400–600 nm were observed. Further evolution of the spectra within 60 minutes showed that the band at 330 nm was decreasing while the intensity of the band at 381 nm and the broad absorption in the region 400–600 nm were increasing and the maximum at 477 nm was appearing. The final spectrum corresponded to that of the adduct **46a**, isobestic point at 431 nm indicating the conversion **3c+DDQ→46a**.

In MeCN, at ~−30 °C the absorption band at 1082 nm (which disappeared at room temperature) was registered. According to reference,[55a] this band was assignable to the radical cation species like **49a**. It is known that in polar solvents like MeCN CT-complexes or ion-radical pairs such as **49** are dissociated, in this case, to radical cation **49a** and radical anion **49b** (Scheme 28).

Thus, the UV/Vis monitoring of the cycloaddition course showed that the starting acetylenes **3c,20,28a,b** upon contacting with DDQ were transformed rapidly to the cycloadducts **46a–d**, for which the strong intramolecular charge transfer was a major feature, while at low temperature (−30 °C), the intermediate radical cation became likely discernible (1082 nm). Notably, as mentioned above, for the adducts **47a–c** derived from 2-aryl-5-ethynylpyrroles, the CT band was significantly red-shifted (up to 70 nm) with simultaneous hypochromic effect (Δlg ε 0.53), thus indicating deep charge transfer with the participation of aryl substituents (Figure 1). Correspondingly, this was accompanied by the down-field shifts of C_{para} carbons of the aryl groups (see above) and a pronounced shielding of the CN-substituted carbons.

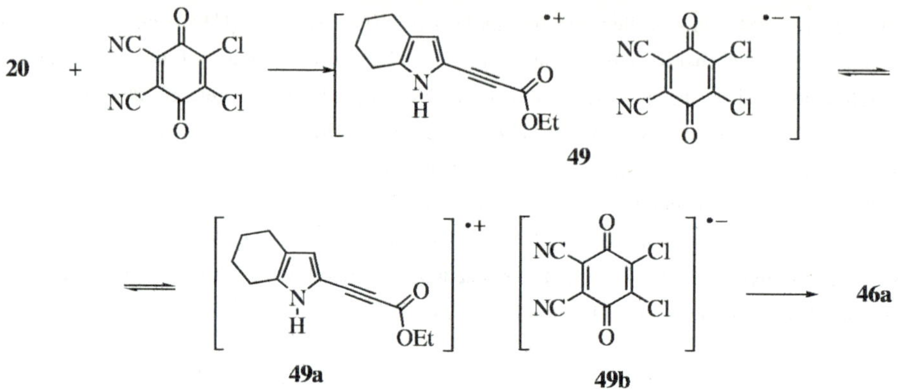

The scheme depicts the reaction of **20** + DDQ giving intermediate **49**, via radical ion pairs **49a** and **49b**, leading to **46a**.

Scheme 28

In the ESR spectrum of the mixture of pyrrole **20** and DDQ (1:1 molar ratio, room temperature, benzene), two overlapped signals (g-factors 2.0052 and 2.0048, respectively) with no hyperfine structure were observed. Judging from g-factor value, the first signal was assigned to the DDQ radical anion. To check this assignment, the reaction was monitored at a low temperature (the ampoule with reaction mixture under argon was frozen in the liquid nitrogen and was then placed in the ESR spectrometer). In the spectrum, a signal with hyperfine structure (five lines, aN=0.58 G, g=2.0052), attributable to the DDQ radical anion[57] was registered. This is in agreement with a tentative mechanism which involves electron transfer from the ethynylpyrrole moiety to DDQ to form ion-radical pair including the DDQ radical anion and the ethynylpyrrole radical cation, the latter being undetectable by the ESR technique here employed due to its lesser stability (separately, both reactants, pyrrole **20** and DDQ, are non paramagnetic in benzene).

In the solid state, adducts **46a–d**, **47a–c**, **48a,b** are paramagnetic ($\sim 10^{18}$ spin/g) as exemplified by the ESR spectrum of adduct **46a** where two overlap singlets (g_1=2.0052, ΔH_1=8.7 G; g_2=2.0042, ΔH_2=2.9 G) are observed. Upon dissolving of the adducts (*e.g.* in benzene) the paramagnetism vanishes. It may be assumed, that the paramagnetism of the solid sample results from the electron transfer in the associated molecules.

4. Rotational isomerism of 2-(2-benzoylethynyl)-5-phenylpyrrole in crystal state

The 2-(2-benzoylethynyl)-5-phenylpyrrole **3b** was found to crystallize in two visually distinctive forms: prisms **3b-1** and needles **3b-2** with different melting points.[58]

The prisms **3b-1** were less stable and upon repeated recrystallization transformed to the needles **3b-2**, which did not give the prisms when further recrystallized. The latter can only be isolated by the instant extraction of the alumina used in the reaction.

The X-ray analysis (Figures 3 and 4) revealed that the crystals represented the *cis-* (**3b-1**) and *trans-*(**3b-2**) rotamers with respect to mutual disposition of the nitrogen atom and the carbonyl group (Scheme 29). For **3b-1**, the torsion angle between the pyrrole and 5-phenyl planes was $\sim 10°$ and that between the benzoyl was $\sim 17°$, whereas in the molecule of **3b-2** these structural units were essentially planar ($\sim 7°$ and $0°$, respectively), that likely meant a stronger conjugation in the latter. The crystal packing of the two forms was entirely different: the prisms **3b-1** consisted of H-bonded macrocyclic dimers (Figure 5), while the needles **3b-2** were H-bonded chains (Figure 6).

112

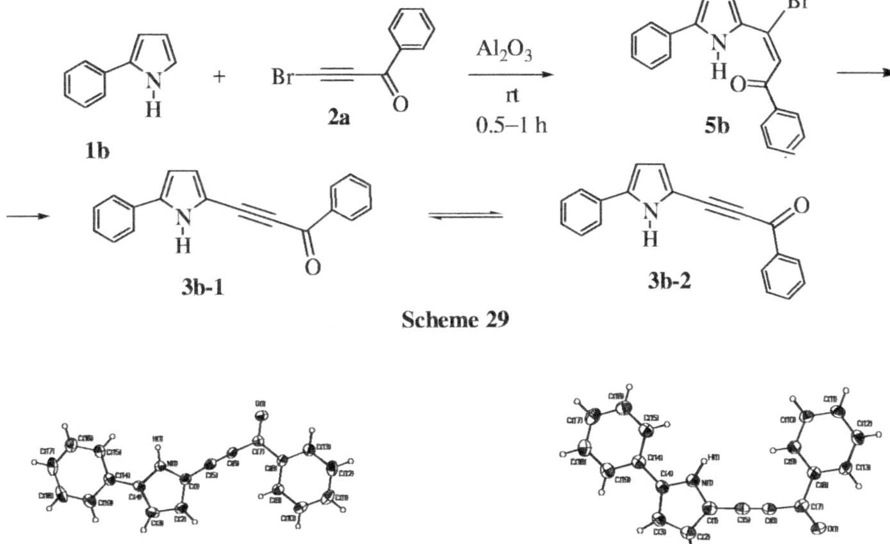

Scheme 29

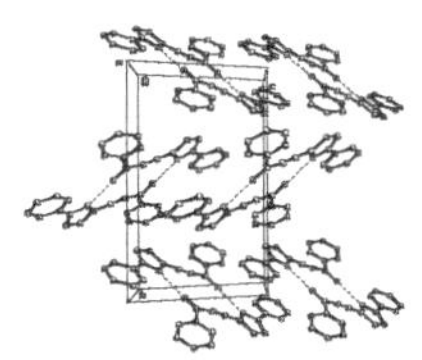

Figure 3. The general view of **3b-1**.

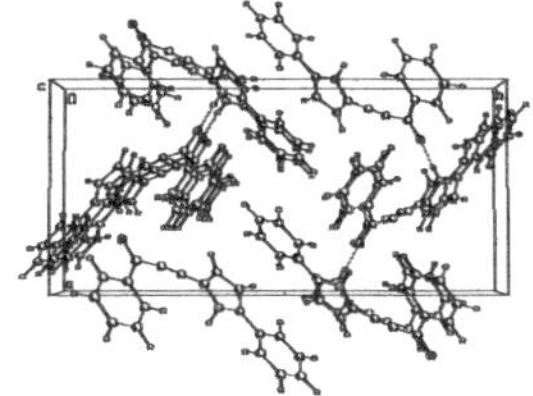

Figure 4. The general view of **3b-2**.

Figure 5. The H-bonded macrocyclic dimers in the crystal structure of **3b-1**.

Figure 6. The H-bonded chains in the crystal structure of **3b-2**.

The crystals **3b-1** and **3b-2** had different IR spectra (in KBr pellets, cm^{-1}), major differences being observed in the regions: (νNH) 3311 for **3b-1** and 3332 for **3b-2**; (νCO) 1630 (s), 1617 (sh, m) for **3b-1**, 1623 (s), 1610 (s) for **3b-2**; (δ ≡C-) 650 (s) for **3b-1**, 651 and 647 (m) for **3b-2**.

However, in a solution (CCl$_4$, 50^{-2} mmol/l), these differences disappeared and only a higher intensity of the C=O absorption [1637 (s) cm^{-1} for both **3b-1** and **3b-2**] and a slight lower frequency shift of the bound NH were observed for **3b-1** as compared to those of **3b-2** [3306 (s) cm^{-1} for **3b-1** and 3312 (w) cm^{-1} for **3b-2**]. In the UV/Vis spectra (in cyclohexane), the absorption maxima were the same for both crystals, though they were all systematically more intensive for **3b-2** [(λ$_{max}$/nm (lg ε for **3b-1**, **3b-2**)]: sh 218 (4.04, 4.13), 237 (3.91, 4.02), 269 (4.23, 4.31), sh 286 (4.14, 4.22), 385 (4.52, 4.58), reproducibility of the lg ε value being within 2–4% (from 3 different concentrations).

The NMR (^{1}H- and ^{13}C-) spectra were identical for both crystals in accordance with the expected low-rotation barrier between the two rotamers. Indeed, the B3LYP/6-311G(d) calculations of isolated molecules gave the total energy difference of 0.77 kcal/mol in favor of **3b-2**, with the **3b-2**→**3b-1** barrier equaling to 2.6 kcal/mol.

All the data were rationalized as follows: the *cis*-rotamer **3b-1** was likely originated from an alumina-induced *cis*-elimination in the intermediate **5b** with immediate stabilization of the *cis*-orientated N-H and C=O bonds in the macrocyclic dimer 2·**3b-1**, which then further transformed to the H-chained **3b-2** (Scheme 30).

Thus, the fixation of the *cis*- and *trans*-rotamers **3b-1** and **3b-2** was likely due to the strong H-bonding in the crystal state, which remained for a while in a solution.

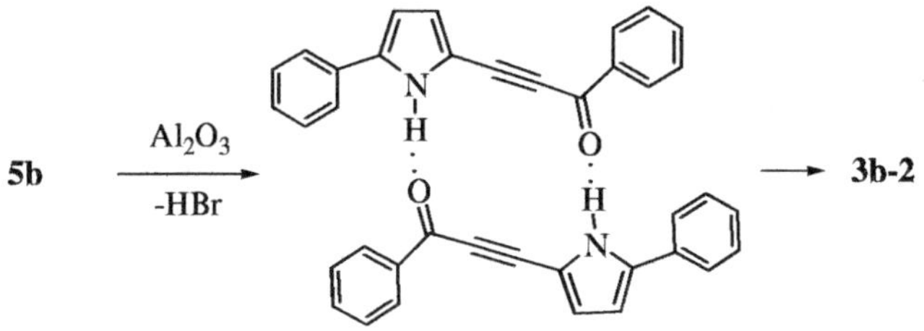

5b $\xrightarrow[\text{-HBr}]{\text{Al}_2\text{O}_3}$ **3b-1** ⟶ **3b-2**

Scheme 30

Just recently,[59] a crystalline diacetylene, 1,4-di(α-naphthyl)-1,3-butadiyne, was found to adopt a centrosymmetric conformation of the naphthalene rings around the -C≡C-C≡C- axis. However, the corresponding *cis*-rotamer was not detected.

5. Conclusions

The ethynylation of pyrroles and indoles with haloacetylenes on active surfaces, in spite of its recent discovery, has already attracted attention and been cited not only in original papers,[60] but also in reviews[61] and monographs.[62] The synthetic potential of this novel cross-coupling reaction is far from being fully understood and keeps developing. Additional possibilities of this methodology are to be expected from its modification using metal complex catalysis as it has been demonstrated by the successful ethynylation of fused nitrogen heterocycles by haloacetylenes in the presence of palladium catalyst in toluene.[63] Doping the active surfaces with catalytically active metals promises new adventures in this intriguing field.

Alkyl-, aryl- and heteroarylpyrroles, as well as tetrahydroindoles and hence indoles are now readily available *via* the two-step reaction of ketones (through ketoximes) with acetylene[64] and those haloacetylenes can be easily prepared by halogenation of terminal acetylenes.[65] Therefore, the ethynylation of pyrrole nucleus with haloacetylenes on active surfaces has a potential to become a useful economic and "green" methodology in pyrrole, indole and acetylene chemistry.

Acknowledgments

The work was carried out under financial support of leading scientific schools by the President of the Russian Federation (Grant NSh-3230.2010.3), the Russian Foundation of Basic Research (Grant No 09-03-00064a).

References

1. (a) Jones, R. A. *Pyrroles*; Wiley: New York, 1990; part 2. (b) Gribble, G. W. In *Comprehensive Heterocyclic Chemistry II*; Katritzky, A. R.; Rees, C. W.; Scriven, E. F. V., Eds.; Pergamon Press: Oxford, 1996; Vol. 2, pp. 207–258. (c) Sundberg, R. J. *Indoles*; Academic Press: London, 1996. (d) Katritzky, A. R.; Pozharskii, A. F. *Handbook of Heterocyclic Chemistry*; Pergamon Press: Oxford, 2000; Chapter 4. (e) Joule, J. A. In *Science of Synthesis (Houben-Weyl. Methods of Molecular Transformations)*; Thomas, E. J., Ed.; Georg Thieme: Stuttgart, 2000; Vol. 10, pp. 361–652.

2. (a) Barluenga, J.; Vazquez-Villa, H.; Ballesteros, A.; Gonzalez, J. M. *Adv. Synth. Catal.* **2005**, *347*, 526. (b) Barluenga, J.; Vazquez-Villa, H.; Merino, I.; Ballesteros, A.; Gonzalez, J. M. *Chem. Eur. J.* **2006**, *12*, 5790. (c) Kitagaki, S.; Okumura, Y.; Mukai, C. *Tetrahedron* **2006**, *62*, 10311. (d) Asao, N.; Aikawa, H. *J. Org. Chem.* **2006**, *71*, 5249. (e) Tiano, M.; Belmont, P. *J. Org. Chem.* **2008**, *73*, 4101. (f) Hellal, M.; Bourguignon, J.-J.; Bihel, F. J.-J. *Tetrahedron Lett.* **2008**, *49*, 62.

3. (a) Mironov, A. F.; Akimenko, L. V.; Rumyantseva, V. D.; Evstigneeva, R. P. *Khim. Geterosikl. Soed.* **1975**, 423; *Chem. Abstr.* **1975**, *83*, 28041b. (b) Kozhich, D. T.; Akimenko, L. V.; Mironov, A. F.; Evstigneeva, R. P. *Zh. Org. Khim.* **1977**, *13*, 2604. (c) Mironov, A. F.; Kozhich, D. T.; Vasilevskii, V. I.; Evstigneeva, R. P. *Synthesis* **1979**, 513. (d) Nizhnik, A. N.; Mironov, A. F. *Zh. Org. Khim.* **1979**, *15*, 1739. (e) Kozhich, D. T.; Vasilevskii, V. I.; Mironov, A. F.; Evstigneeva, R. P. *Zh. Org. Khim.* **1980**, *16*, 849; *Chem. Abstr.* **1980**, *93*, 46309f. (f) Mironov, A. F.; Nizhnik, A. N.; Kozhich, D. T.; Kozyrev, A. N.; Evstigneeva, R. P. *J. Gen. Chem. USSR (Engl.Transl.)* **1981**, *51*, 565; *Zh. Obshch. Khim.* **1981**, *51*, 699. (g) Nizhnik, A. N.; Mironov, A. F. *J. Gen. Chem. USSR (Engl.Transl.)* **1984**, *54*, 565; *Zh. Obshch. Khim.* **1984**, *54*, 2316. (h) Barrero, A. F.; Sanchez, J. F.; Oltra, J. E.; Teva, D. *J. Heterocycl. Chem.* **1991**, *28*, 939. (i) Diziere, R.; Savignac, P. *Tetrahedron Lett.* **1996**, *37*, 1783. (j) Jiang, X.; Smith, K. M. *J. Chem. Soc., Perkin Trans. 1* **1996**, 2072.

4. (a) Tietze, L. F.; Kettschau, G.; Heitmann, K. *Synthesis* **1996**, 851. (b) Kitano, Y.; Suzuki T.; Kawahara, E.; Yamazaki, T. *Bioorg. Med. Chem.* **2007**, *17*, 5863.

5. Soley, R.; Albericio, F.; Álvarez, M. *Synthesis* **2007**, 1559.

6. (a) Wentrup, C.; Winter, H. W. *Angew. Chem.* **1978**, *90*, 643. (b) Comer, M. C.; Despinoy, X. L. M.; Gould, R. O.; McNab, H.; Parsons, S. *J. Chem. Soc., Chem. Commun.* **1996**, 1083.

7. Brittain, J. M.; Jones, R. A.; Taheri, S. A. N. *Tetrahedron* **1992**, *48*, 7609.

8. Negishi, E.; Xu, C.; Tan, Z.; Kotora, M. *Heterocycles* **1997**, *46*, 209.

9. Brain, E. G.; Eglington, A. J.; James, B. G.; Nayler, J. H. C.; Osborne, N. F.; Pearson, M. J.; Smale, T. C.; Southgate, R.; Tolliday, P.; Basker, M. J.; Mizzen, L. W.; Sutheland, R. *J. Med. Chem.* **1977**, *20*, 1086.

10. Sauvêtre, R.; Normant, J. F. *Tetrahedron Lett.* **1982**, *23*, 4325.

11. (a) Vo-Quang, L.; Vo-Quang, Y. *J. Heterocycl. Chem.* **1982**, *19*, 145. (b) Kusumoto, T.; Hiyama, T.; Ogata, K. *Tetrahedron Lett.* **1986**, *27*, 4197. (c) Dell'Erba, C.; Giglio, A.; Mugnoli, A.; Novi, M.; Petrillo, G.; Stagnaro, P. *Tetrahedron* **1995**, *51*, 5181. (d) Novikov, M. S.; Khlebnikov, A. F.; Sidorina, E. S.; Kostikov, R. R. *J. Chem. Soc., Perkin Trans. 1* **2000**, 231. (e) Takeda, M.; Matsumoto, S.; Ogura, K. *Heterocycles* **2001**, *55*, 231. (f) Nedolya, N. A.; Brandsma, L.; Tolmachev, S. V. *Chem. Heterocycl. Compd.* **2002**, *38*, 745; *Khim.Geterotsikl. Soed.* **2002**, *38*, 843. (g) Matsumoto, S.; Kobayashi, T.; Ogura, K. *Heterocycles* **2005**, *66*, 319. (h) Knight, D. W.; Rost, H. C.; Sharland, Ch. M.; Singkhonrat, J. *Tetrahedron Lett.* **2007**, *48*, 7906.

12. Furstner, A.; Ernst, A.; Krause, H.; Ptock, A. *Tetrahedron* **1996**, *52*, 7329.

13. (a) Vasilevsky, S. F.; Sundukova, T. A.; Shvartsberg, M. S.; Kotlyarevsky, I. L. *Bull. Acad. Sci. USSR, Div. Chem. Sci. (English Transl.)* **1979**, *28*, 1536; *Izv. AN SSSR, Ser. Khim.* **1979**, *28*, 1661. (b) Mueller, T. J. J.; Ansorge, M.; Lindner, H. J. *Chem. Ber.* **1996**, *129*, 1433. (c) Liu, J.-H.; Yang, Q.-C.; Mak, T. C. W.; Wong, H. N. C. *J. Org. Chem.* **2000**, *65*, 3587. (d) Bergauer, M.; Gmeiner, P. *Synthesis* **2001**, 2281. (e) Bergauer, M.; Gmeiner, P. *Synthesis* **2002**, 274. (f) Bergauer, M.; Huebner, H.; Gmeiner, P. *Bioorg. Med. Chem. Lett.* **2002**, *12*, 1937. (g) Vasilevsky, S. F.; Klyatskaya, S. V.; Elguero, J. *Tetrahedron* **2004**, *60*, 6685. (h) Rohand, T.; Qin, W.; Boens, N.; Dehaen, W. *Eur. J. Org. Chem.* **2006**, 4658.

14. (a) Jux, N.; Koch, P.; Schmickler, H.; Lex, J.; Vogel, E. *Angew. Chem.* **1990**, *102*, 1429. (b) Jacobi, P. A.; DeSimone, R. W. *Tetrahedron Lett.* **1992**, *33*, 6239. (c) Jacobi, P. A.; Rajeswari, S. *Tetrahedron Lett.* **1992**, *33*, 6235. (d) Martire, D. O.; Jux, N.; Aramendia, P. F.; Negri, R. M.; Lex, J.; Braslavsky, S.; Schaffner, K.; Vogel, E. *J. Am. Chem. Soc.* **1992**, *114*, 9969. (e) Artis, D. R.; Cho, I.-S.; Jaime-Figueroa, S.; Muchowski, J. M. *J. Org. Chem.* **1994**, *59*, 2456. (f) Jacobi, P. A.; Guo, J. *Tetrahedron Lett.* **1995**, *36*, 2717. (g) Jacobi, P. A.; Guo, J.; Zheng, W. *Tetrahedron Lett.* **1995**, *36*, 1197. (h) Jacobi, P. A.; Guo, J.; Hauck, S. I.; Leung, S. H. *Tetrahedron Lett.* **1996**, *37*, 6069. (i) Jacobi, P. A.; Guo, J.; Rajeswari, S.; Zheng, W. *J. Org. Chem.* **1997**, *62*, 2907. (j) Jacobi, P. A.; Coutts, L. D.; Guo, J.; Hauk, S. I.; Leung, S. H. *J. Org. Chem.* **2000**, *65*, 205. (k) Jacobi, P. A.; DeSimone, R. W.; Ghosh, I.; Guo, J.; Leung, S. H.; Pippin, D. *J. Org. Chem.* **2000**, *65*, 8478. (l) Liu, J. H.; Chan, H.-W.; Wong, H. N. C. *J. Org. Chem.* **2000**, *65*, 3274. (m) Mitsui, T.; Kimoto, M.; Sato, A.; Yokoyama, S.; Hirao, I. *Bioorg. Med. Chem. Lett.* **2003**, *13*, 4515. (n) Tu, B.; Ghosh, B.; Lightner, D. A. *J. Org. Chem.* **2003**, *68*, 8950. (o) Tu, B.; Lightner, D. A. *J. Heterocycl. Chem.* **2003**, *40*, 707. (p) Wan, C.-W.; Burghart, A.; Chen, J.; Bergstroem, F.; Johansson, L. B.-A.; Wolford, M. F.; Kim, T, G.; Topp, M. R.; Hochstrasser, R. M.; Burgess, K. *Chem. Eur. J.* **2003**, *9*, 4430. (q) Tu, Bi.; Ghosh, B.; Lightner, D. A. *Monatsh. Chem.* **2004**, *135*, 519. (r) Nikitin, E. B.; Nelson, M. J.; Lightner, D. A. *J. Heterocycl. Chem.* **2007**, *44*, 739.

15. (a) Ortaggi, G.; Scarsella, M.; Scialis, R.; Sleiter, G. *Gazz. Chim. Ital.* **1988**, *118*, 743. (b) Alvarez, A.; Guzman, A.; Ruiz, A.; Velarde, E.; Muchowski, J. M. *J. Org. Chem.* **1992**, *57*, 1653. (c) Vasilevsky, S. F.; Verkruijsse, H. D.; Brandsma, L. *Rec. Trav. Chim. Pays-Bas,* **1992**, *111*, 529. (d) Sugihara, Y.; Miyatake, R.; Murata, I.; Imamura, A. *J. Chem. Soc., Chem. Commun.* **1995**, 1249. (e) Mal'kina, A. G.; Brandsma, L.; Vasilevsky, S. F.; Trofimov, B. A. *Synthesis* **1996**, 589. (f) Chan, H.-W.; Chan, P.-C.; Liu, J.-H.; Wong, H. N. C. *Chem. Commun.* **1997**, 1515. (g) Haubmann, C.; Huebner, H.; Gmeiner, P. *Bioorg. Med. Chem. Lett.* **1999**, *9*, 3143. (h) Wright, J. L.; Gregory, T. F.; Boxer, P. A.; Meltzer, L. T.; Serpa, K. A.; Wise, L. D.; Hong-Bae, S.; Huang, J. C.; Konkoy, C. S.; Upasani, R. B.; Whittemore, E. R.; Woodward, R. M.; Yang, K. C.; Zhou, Z.-L. *Bioorg. Med. Chem. Lett.* **1999**, *9*, 2815. (i) Brandsma, L.; Nedolya, N. A.; Verkruijsse, H. D.; Trofimov, B. A. *Chem. Heterocycl. Compd.* **2000**, *36*, 876; *Khim. Geterotsikl. Soed.* **2000**, 990. (j) Liao, J.-H.; Chen, C.-T.; Chou, H.-C.; Cheng, C.-C.; Chou, P.-T.; Fang, J. M.; Slanina, Z.; Chow, T. J. *Org. Lett.* **2002**, *4*, 3107. (k) Heynderickx, A.; Kaou, A. M.; Moustrou, C.; Samat, A.; Guglielmetti, R. *New J. Chem.* **2003**, *27*, 1425. (l) Fang, J. M.; Selvi, S.; Liao, J.-H.; Slanina, Z.; Chen, C.-T.; Chou, P.-T. *J. Am. Chem. Soc.* **2004**, *126*, 3559. (m) Hayes, M. E.; Shinokubo, H.; Danheiser, R. L. *Org. Lett.* **2005**, *7*, 3917. (n) Yasuda, T.; Imase, T.; Nakamura, Y.; Yamamoto, T. *Macromolecules* **2005**, *38*, 4687. (o) Laha, J. K.; Muthiah, Ch.; Taniguchi, M.; McDowell, B. E.; Ptaszek, M.; Lindsey, J. S. *J. Org. Chem.* **2006**, *71*, 4092. (p) Lu, S.-H.; Selvi, S.; Fang, J.-M. *J. Org. Chem.* **2007**, *72*, 117.

16. (a) Powers, J. C. *J. Org. Chem.* **1966**, *31*, 2627. (b) Gossauer, A. *Die Chemie der Pyrrole*; Springer: New York, 1974; pp. 326–332. (c) Brennan, M. R.; Erickson, K. L.; Szmalc, F. S.; Tansey, M. J.; Thornton, J. M. *Heterocycles* **1986**, *24*, 2879. (d) Bergman, J.; Venemalm, L. *J. Org. Chem.* **1992**, *57*, 2495. (e) Gossauer, A. In *Methoden der Organischen Chemie (Houben-Weil),* Bd. E6a; Hetarene, I. Hrsg; von Kreher, R. P., Eds.; Thieme: New York, 1994; Teil 1, S 750. (f) Brandsma, L.; Vasilevsky, S. F.; Verkruijsse, H. D. In *Application of Transition Metal Catalysts in Organic Synthesis*; Springer: Berlin, 1997; p. 34.

17. Vasilevsky, S. F.; Sundukova, T. A.; Shvartsberg, M. S.; Kotlyarevsky, I. L. *Izv. Akad. Nauk SSSR, Ser. Khim.* **1980**, 1871; *Chem. Abstr.* **1981**, *94*, 30464.

18. Sundukova, T. A.; Vasilevsky, S. F.; Shvartsberg, M. S.; Kotlyarevsky, I. L. *Izv. Akad. Nauk SSSR, Ser. Khim.,* **1980**, 726; *Chem. Abstr.* **1980**, *93*, 71446.

19. (a) Zhang, C.; Trudell, M. L. *J. Org. Chem.* **1996**, *61*, 7189. (b) Druzhinin, S. V.; Balenkova, E. S.; Nenajdenko, V. G. *Tetrahedron* **2007**, *63*, 7753. (c) Sun, H.; Lin, Y-J.; Wu, Y. L.; Wu, Y. *Synlett* **2009**, 2473.

20. (a) Sobenina, L. N.; D'yachkova, S. G.; Stepanova, Z. V.; Toryashinova, D.-S. D.; Albanov, A. I.; Ushakov, I. A.; Demenev, A. P.; Mikhaleva, A. I.; Trofimov, B. A. *Russ. J. Org. Chem. (Engl. Transl.)* **1999**, *35*, 916. (b) Trofimov, B. A. *Curr. Org. Chem.* **2002**, *6*, 1121.

21. Trofimov, B. A.; Stepanova, Z. V.; Sobenina, L. N.; Mikhaleva, A. I.; Ushakov, I. A. *Tetrahedron Lett.* **2004**, *45*, 6513.

22. Trofimov, B. A.; Sobenina, L. N.; Stepanova, Z. V.; Demenev, A. P.; Mikhaleva, A. I.; Ushakov, I. A.; Vakul'skaya, T. I.; Petrova, O. V. *Russ. J. Org. Chem. (Engl. Transl.)* **2006**, *42*, 1348.

23. Stepanova, Z. V.; Sobenina, L. N.; Mikhaleva, A. I.; Ushakov, I. A.; Elokhina, V. N.; Vorontsov, I. I.; Antipin, M. Yu.; Trofimov, B. A. *Russ. J. Org. Chem. (Engl. Transl.)* **2003**, *39*, 1636.

24. Norman, R. O. C.; Coxon, J. M. *Principles of Organic Synthesis; 3-rd Edition*; Blackie Academic & Professional: London, New York, 1993; pp. 96–97.

25. (a) Trofimov, B. A.; Mikhaleva, A. I. *N-Vinylpyrroles*; Nauka: Novosibirsk, 1984. (b) Trofimov, B. A. *Vinylpyrroles* In *Pyrroles* Chapter Two: *The synthesis, Reactivity and Physical Properties of Substituted Pyrroles*; Jones, A., Ed; Wiley, Inc: New York, 1992.

26. Zhdanov, Yu. A.; Minkin, V. I. *Correlation analysis in organic chemistry*; Rostov University Publishers: Rostov, 1966.

27. (a) Hania, M. M. *Asian J. Chem.* **2000**, *12*, 321. (b) Yoshida, T; Ito, A.; Ibusuki, K.; Murase, M.; Kotani, E. *Chem. Pharm. Bull.* **2001**, *49*, 1198.

28. Trofimov, B. A.; Sobenina, L. N.; Stepanova, Z. V.; Ushakov, I. A.; Petrova, O. V.; Tarasova, O. A.; Volkova, K. A.; Mikhaleva, A. I. *Synthesis* **2007**, 447.

29. Anastas, P.; Warner, J. In *Green Chemistry: Theory and Practice*; Oxford University Press: New York, 1998.

30. Trofimov, B. A.; Sobenina, L. N.; Stepanova, Z. V.; Vakul'skaya, T. I.; Kazheva, O. N.; Aleksandrov, G. G.; Dyachenko, O. A.; Mikhaleva, A. I. *Tetrahedron* **2008**, *64*, 5541.

31. (a) Rooney, J. J.; Pink, R. C. *Trans. Faraday Soc.* **1962**, *58*, 1632. (b) Corma, A.; Garcia, H. *Top. in Catal.* **1998**, *6*, 127. (c) Tanaka, M.; Kobayashi, K. *Chem. Commun.* **1998**, 1965. (d) Biancardo, M.; Argazzi, R.; Bignozzi, C. A. *Inorganic Chem.* **2005**, *44*, 9619.

32. Stepanova, Z. V.; Sobenina, L. N.; Mikhaleva, A. I.; Ushakov, I. A.; Elokhina, V. N.; Voronov, V. K.; Trofimov, B. A. *Synthesis* **2004**, 2736.

33. (a) Okamoto, S.; Ono, N.; Tani, K.; Yoshida, Y.; Sato, F. *J. Chem. Soc., Chem. Commun.* **1994**, 279. (b) Poulsen, T. B.; Bernardi, L.; Aleman J.; Overgaard, J.; Jorgensen, K. A. *J. Am. Chem. Soc.* **2007**, *129*, 441. (c) Kolarovic, A.; Faberova, Z. *J. Org. Chem.* **2009**, *74*, 7199.

34. (a) Lim, S.; Jabin, I.; Revial, G. *Tetrahedron Lett.* **1999**, *40*, 4177. (b) Nakagawa, T.; Kawada, A. *Jpn. Kokai Tokkyo Koho* **2004**; JP 2004010601: *Chem. Absr.* **2004**, *140*, 93921.

35. (a) Kuehne, M. E.; Podhorez, D. E.; Mulamba, T.; Bornmann, W. G. *J. Org. Chem.* **1987**, *52*, 347. (b) Hashimoto, C.; Husson, H.-P. *Tetrahedron Lett.* **1988**, *29*, 4563. (c) Modi, S. P.; Zayed, A.-H.; Archer, S. *J. Org. Chem.* **1989**, *54*, 3084. (d) Katritzky, A. R.; Li, J., Stevens, C. V. *J. Org. Chem.* **1995**, *60*, 3401. (e) Laronze, M.; Sapi, J. *Tetrahedron Lett.* **2002**, *43*, 7925.

36. Trofimov, B. A.; Sobenina, L. N.; Demenev, A. P.; Stepanova, Z. V.; Petrova, O. V.; Ushakov, I. A.; Mikhaleva, A. I. *Tetrahedron Lett.* **2007**, *48*, 4661.

37. Trofimov, B. A.; Sobenina, L. N.; Stepanova, Z. V.; Petrova, O. V.; Ushakov, I. A.; Mikhaleva, A. I. *Tetrahedron Lett.* **2008**, *49*, 3946.

38. (a) Belen'kii, L. I. *Heterocycles* **1994**, *37*, 2029. (b) Belen'kii, L. I.; Kim, T. G.; Suslov, I. A.; Chuvylkin, N. D. *Russ. Chem. Bull.* **2005**, *54*, 853.

39. (a) Kanekiyo, N.; Choshi, T.; Kuwada, T.; Sugino, E.; Hibino, S. *Heterocycles* **2000**, *53*, 1877. (b) Hibino, S.; Choshi, T. *Nat. Prod. Rep.* **2002**, *19*, 148.

40. (a) Passarella, D.; Giardini, A.; Martinelli, M.; Silvani, A. *J. Chem. Soc., Perkin Trans. 1* **2001**, 127. (b) Södenberg, B. C. G. *Coord. Chem. Rev.* **2003**, *241*, 147.

41. Sobenina, L. N.; Demenev, A. P.; Mikhaleva, A. I.; Ushakov, I. A.; Vasil'tsov, A. M.; Ivanov, A. V.; Trofimov, B. A. *Tetrahedron Lett.* **2006**, *47*, 7139.

42. Petrova, O. V.; Sobenina, L. N.; Ushakov, I. A.; Mikhaleva, A. I. *Russ. J. Org. Chem. (Engl. Transl.)* **2008**, *44*, 1512.

43. (a) Trofimov, B. A. In *Advances in Heterocyclic Chemistry*; Academic Press: New York, 1990; Vol. 51, p. 177. (b) Trofimov, B. A.; Mikhaleva, A. I.; Schmidt, E. Yu.; Ryapolov, O. A.; Platonov, V. B. Patent RF № RU 2297410, **2006**.

44. (a) Trofimov, B. A.; Mikhaleva, A. I.; Schmidt, E. Yu.; Ryapolov, O. A.; Platonov, V. B. Patent RF №
RU 2307830, **2006**. (b) Trofimov, B. A.; Mikhaleva, A. I.; Schmidt, E. Yu.; Ryapolov, O. A.; Platonov,
V. B. Patent 2345066, **2007**.

45. Bharate, S. B. *Synlett* **2006**, 496.

46. Begouin, A.; Hesse, S. In Proceedings of ECSOC-10 [on CD-ROM], The Tenth International
Electronic Conference on Synthetic Organic Chemistry, November 1–30, 2006; Seijas, J. A.; Tato, M.
P. V., Eds.; MDPI: Basel, **2006**, a007; also available online at http://193.144.75.244/congresos/
ecsoc/10/GOS/a007/a007.pdf).

47. Cavdar, H.; Saracoglu, N. *J. Org. Chem.* **2006**, *71*, 7793.

48. Trofimov, B. A.; Sobenina, L. N.; Stepanova, Z. V.; Ushakov, I. A.; Sinegovskaya, L. M.;
Vakul'skaya, T. I.; Mikhaleva, A. I. *Synthesis* **2010**, 470.

49. (a) Sobenina, L. N.; Drichkov, V. N.; Mikhaleva, A. I.; Petrova, O. V.; Ushakov, I. A.; Trofimov, B. A.
Tetrahedron **2005**, *61*, 4841. (b) Trofimov, B. A.; Petrova, O. V.; Sobenina, L. N.; Drichkov, V. N.;
Mikhaleva, A. I.; Ushakov, I. A.; Tarasova, O. A.; Kazheva, O. N.; Chekhlov, A. N.; Dyachenko, O. A.
Tetrahedron **2006**, *62*, 4146.

50. (a) Neidlein, R.; Kramer, W.; Leidholdt, R. *Helv. Chim. Acta* **1983**, *66*, 652. (b) Christl, M.; Brunn, E.;
Lanzendoerfer, F. *J. Am. Chem. Soc.* **1984**, *106*, 373.

51. (a) Nishida, S.; Murakami, M.; Oda, H.; Tsuji, T.; Mizuno, T.; Matsubara, M.; Kikai, N. *J. Org. Chem.*
1989, *54*, 3859. (b) Chiba, K.; Arakawa, T.; Tada, M. *J. Chem. Soc., Perkin Trans. 1* **1998**, 2939. (c)
Aly, A. A. *Z. Naturforsch, B: Chem. Sci.* **2005**, *60*, 106.

52. Montevecchi, P. C.; Navacchia, M. L. *J. Org. Chem.* **1998**, *63*, 8035.

53. (a) Zimmerman, H. E.; Craft, L. *Tetrahedron Lett.* **1964**, *5*, 2131. (b) Bryce-Smith, D.; Fray, G. I.;
Gilbert, A. *Tetrahedron Lett.* **1964**, *5*, 2137. (c) Bosch, E.; Hubig, S. M.; Kochi, J. K. *J. Am. Chem.
Soc.* **1998**, *120*, 386. (d) Wang, X. Y.; Yan, B. Z.; Wang, T. *Chin. Chem. Lett.* **2003**, *14*, 270.

54. (a) Pappas, S. P.; Portnoy, N. A. *J. Org. Chem.* **1968**, *33*, 2200. (b) Pappas, S. P.; Pappas, B. S.;
Portnoy, N. A. *J. Org. Chem.* **1969**, *34*, 520. (c) Kuehne, M. E.; Linde, H. *J. Org. Chem.* **1972**, *37*,
4031. (d) Kim, S. S.; Kim, A. R.; O, K. J.; Yoo, D. J.; Shim, S. C. *Chem. Lett.* **1995**, *24*, 787. (e) Kim,
S. S.; Kim, A. R.; Mah, Y. J.; Shim, S. C. *Chem. Lett.* **1999**, *28*, 395.

55. (a) Polyakov, N. E.; Konovalov, V. V.; Leshina, T. V.; Luzina, O. A.; Salakhutdinov, N. F.; Konovalova,
T. A.; Kispert, L. D. *J. Photochem. Photobiol. A: Chem.* **2001**, *141*, 117. (b) Pouretedal, H. R.;
Semnani, A.; Keshavarz, M. H.; Firooz, A. R. *Turk. J. Chem.* **2005**, *29*, 647. (c) Sun, D.; Rosokha, S.
V.; Kochi, J. K. *J. Phys. Chem. B* **2007**, *111*, 6655. (d) Rosokha, S. V.; Kochi, J. K. *Accounts Chem.
Res.* **2008**, *41*, 641. (e) Salem, H. *Eur. J. Chem.* **2009**, *6*, 332.

56. (a) Dworniczak, M. *React. Kinet. Catal. Lett.* **2003**, *78*, 65. (b) Lu, J-M.; Rosokha, S. V.; Neretin, I. S.;
Kochi, J. K. *J. Am. Chem. Soc.* **2006**, *128*, 16708.

57. Grampp, G.; Landgraf, S.; Rasmussen, K. *J. Chem. Soc., Perkin Trans. 2* **1999**, 1897.

58. Trofimov, B. A.; Stepanova, Z. V.; Sobenina, L. N.; Mikhaleva, A. I.; Sinegovskaya, L. M.; Potekhin,
K. A.; Fedyanin, I. V. *Mendeleev Commun.* **2005**, *6*, 229.

59. Rodriquez, J. G.; Tejedor, J. L. *Tetrahedron* **2005**, *61*, 3033.

60. Trofimov, A.; Chernyak, N.; Gevorgyan, V. *J. Am. Chem. Soc.* **2008**, *130*, 5636.

61. Banwell, M. G.; Goodwin, T. E.; Ng, S.; Smith, J. A.; Wong, D. A. *Eur. J. Org. Chem.* **2006**, *31*, 3043.

62. Tanaka, K. *Solvent-free Organic Synthesis*; Wiley-VCH: Weinheim, 2009; pp. 112–118.

63. Seregin, I. V.; Ryabova, V.; Gevorgyan, V. *J. Am. Chem. Soc.* **2007**, *129*, 7742.

64. (a) Gonzalez, F.; Sanz-Cervera, J. F.; Williams, R. M. *Tetrahedron Lett.* **1999**, *40*, 4519. (b) Varlamov,
A. V.; Voskresenskii, L. G.; Borisova, T. N.; Chernyshev, A. I.; Levov, A. N. *Chem. Heterocycl.
Compd.* **1999**, *35*, 613. (c) Chen, J.; Burghart, A.; Derecskei-Kovacs, A.; Burgess, K. *J. Org. Chem.*
2000, *65*, 2900. (d) Abele, E.; Lucevics, E. *Heterocycles* **2000**, *53*, 2285. (e) Abele, E.; Lucevics, E.
Chem. Heterocycl. Compd. **2001**, *37*, 141. (f) Gilchrist, T. L. *J. Chem. Soc., Perkin Trans. 1* **2001**,
2491. (g) Wang, L.-X.; Li, X.-G.; Yang, Y. Y.-L. *React. Funct. Polym.* **2001**, *47*, 125. (h)
Voskressensky, L. G.; Borisova, T. N.; Varlamov, A. V. *Chem. Heterocycl. Compd.* **2004**, *40*, 326. (i)
Williams, R. M. *Chem. Pharm. Bull.* **2002**, *50*, 711. (j) Ismail, M. A.; Boykin, D. W.; Stephens, C. E.;
Tetrahedron Lett. **2006**, *47*, 795. (k) Bellina, F.; Rossi, R. *Tetrahedron* **2006**, *62*, 7213. (l) Loudet, A.;
Burgess, K. *Chem. Rev.* **2007**, *107*, 4891. (m) Galenko, A. V.; Selivanov, S. I.; Lobanov, P. S.;

Potekhin, A. A. *Chem. Heterocycl. Compd.* **2007**, *43*, 1124. (n) Wang, Z. *Comprehensive Organic Name Reactions and Reagents*; Part 3; Wiley: London, 2009, p. 3824.

65. (a) Brandsma, L. *Preparative Acetylenic Chemistry*; Elsevier: New York, 1971; pp. 97–101. (b) Hopf, H.; Witulski, B. In *Modern Acetylene Chemistry*; Stang, P. J.; Diederich, F., Eds.; VCH: New York, 1995; pp. 48–53.

ACCESS TO SPIROCYCLIC PIPERIDINES, IMPORTANT BUILDING BLOCKS IN MEDICINAL CHEMISTRY

Yves Troin[*a] **and Marie-Eve Sinibaldi**[b,c]

[a]*Clermont Université, ENSCCF, EA 987, LCHG, BP 10448, F-63000 Clermont-Ferrand, France*
(e-mail: yves.troin@univ-bpclermont.fr)
[b]*Clermont Université, Université Blaise Pascal, SEESIB, BP 10448, F-63000 Clermont-Ferrand, France*
[c]*CNRS, UMR 6504, SEESIB, F-63177 Aubiere, France*

Abstract. Azaspirocyclic system and in particular spiropiperidines can be found in numerous natural products of relevant activities. This motif constitutes privileged substructures for drug discovery and has been also used as scaffolds for combinatorial libraries. Moreover, spiropiperidines themselves are useful tools for the mechanism of interaction of small nonpeptidic molecules. In this context, this article describes an overview of the synthetic attempts developed toward 3- and 4-spirocyclic piperidines.

Contents

1. Introduction

A great number of biological active compounds possess a heterocyclic substructure. Among them, the piperidine ring system is one of the most common motif found in numerous natural products, drugs and drugs candidates. In addition, when the piperidine ring is engaged in a spirocyclic system, it constitutes the privileged structural motifs of a great number of molecules of pharmaceutical and agrochemistry importance. Therefore, spiropiperidine derivatives are attractive molecules. The spiro centre could be found α, β or γ to the nitrogen atom leading to three different classes of piperidines that we can call 2-, 3- or 4-spiropiperidines.

Thus, 1-azaspiro[5.5]undecanes or 6-azaspiro[4.5]decanes (2-spiropiperidines) (Scheme 1) constitute the spirocyclic parts of a large number of natural compounds with remarkable biological activities such as the well-known alkaloid histrionicotoxin (HTX) which acts as potent non competitive blocker of nicotinic receptor-gated channels.[1] Variation in the length and degree of unsaturation of the side chain of HTX permitted an access to several derivatives possessing important neurophysiological properties.[1]

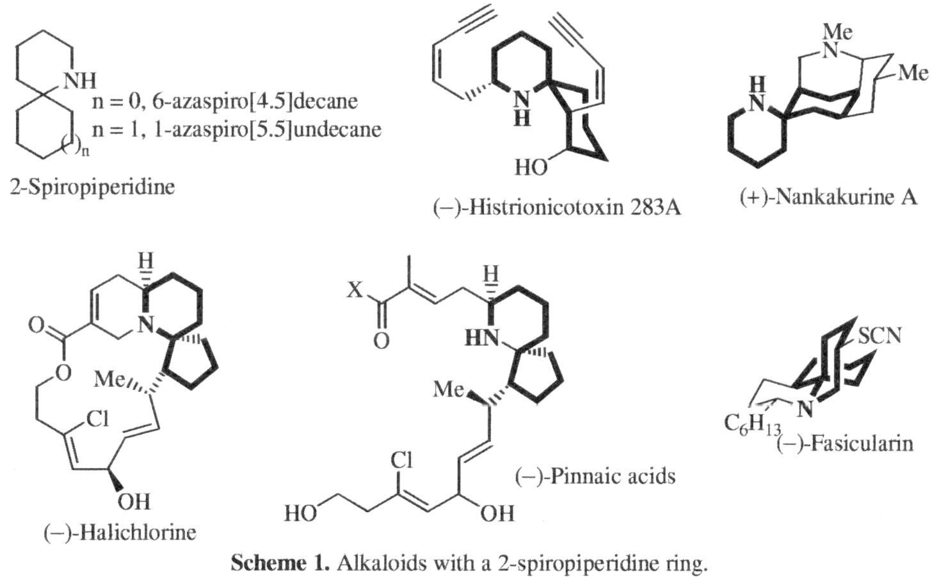

Scheme 1. Alkaloids with a 2-spiropiperidine ring.

More complex natural products possess azaspirocyclic skeletons. For instance, the marine alkaloid halichlorine, a vascular cell adhesion molecule-1 antagonist,[2] and the pinnaic acids which show specific inhibition of cytosolic phospholipase A_2 (IC_{50}=0.2 mM)[3] contain a 6-azaspiro[4.5]decane skeleton. The 1-azaspiro[5.5]undecane framework is found in the *lycopodium* alkaloid nankakurine A, a compound inducing secretion of neurotrophic factors recognized for their potential in combating neurodegenerative disorders which first enantioselective synthesis was described in 2008.[4] This system is also present in the tricyclic alkaloid fasicularin which is cytotoxic toward Vero cells (Scheme 1). The 3-spiropiperidines, in particular the 2-azaspiro[5.5]undecanes, are present in very simple natural compounds such as the γ-aminoalcohols nitramine, isonitramine and sibirine isolated in 1973. Moreover, more complex natural products exhibit this structural feature. Gymnodimine for example, a marine biotoxin which is responsible for a neurotoxic shellfish poisoning in 1993 of the North Island of New Zealand, is constituted by an unusual spirocyclic imine: in fact an azaspiro[5.5]undecadiene subunit, within a macrocarbocycle is responsible of the toxicity. Serratezomine A, with a seco-serratinine type skeleton, possesses also this typical core (Scheme 2). Other 3-spiropiperidines derivatives were recognized as biologically active substrate. Among them, we found fawcettidine, an acetylcholinesterase inhibitor (AchE) with a 7-azaspiro[4.5]decane system; the members of the manzamine A family can be mentioned, *e.g.* nakadomarin A, which acts as Cdk4 inhibitor and presents a cytotoxicity against murine lymphoma L1210 cells,[5] incorporating a 2,7-diazaspiro-[4.5]decane moiety (Scheme 2).

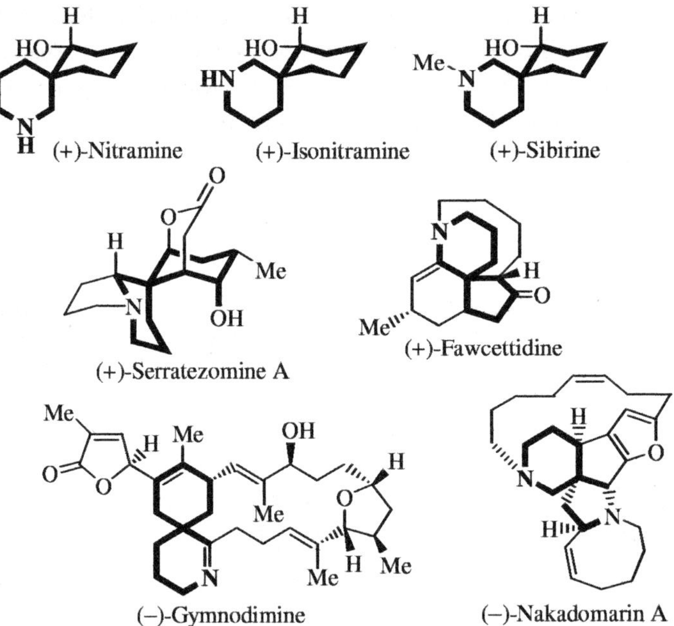

Scheme 2. Compounds incorporating a 3-spiropiperidine moiety.

Compounds with a 4-spiropiperidine motif like the spiroindolinone **I**, spirohydantoin **II** or spirotetraline **III** moieties possess interesting activities and most of them were synthetic derivatives. Thus, spiperone, also known as spiropitan, is an antipsychotic drug acting of both 5-HT and dopamine receptor used for the treatment of schizophrenia which has anti-inflammatory and neuroprotective effects in the central system by modulating glial activation.[6] MK-0667 has been detected as a potent peptidomimetic

Groth Hormone Secretagogues[7a,b] and various *N*-substituted analogs exhibited affinity to the nociceptin (NOP) receptor.[7c] L-387,384, a spirotetraline piperidine, possesses affinity for σ_1 ligand (IC$_{50}$=3.8 nM), whereas its oxygen analogue the spiro[[2]benzopyran-1,4'-piperidine] **A** is recognized as a selective σ_2 ligand (IC$_{50}$=0.90 nM);[8] L-366,509, a spiroindenepiperidine, acts as an oxytocin antagonist and BL1743, a reduced cyclohexadienone azaspirocycic compound, appeared to be an antiviral (Scheme 3).

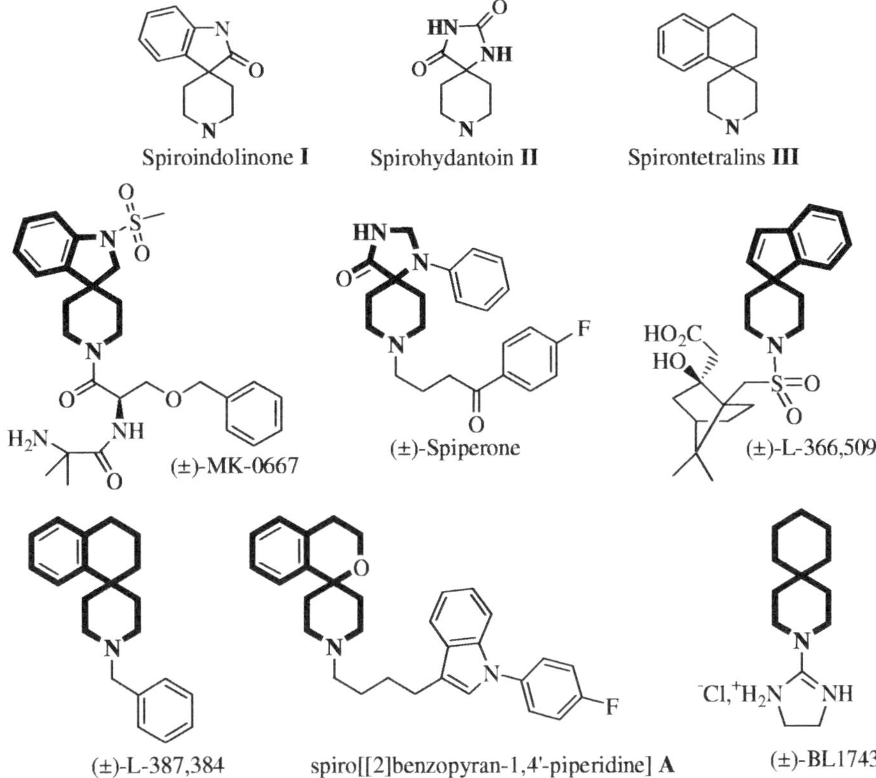

Scheme 3. Bioactive molecules with a 4-spiropiperidine ring.

As a lot of synthetic efforts have been aimed at the synthesis of the 2-spiropiperidine skeleton (in particular for the lead (–)-histrionicotoxin),[9] a large number of strategies were reported and recently summarized for this structure,[10] we will therefore focused on an overview of the diverse methods developed for the construction of the less studied but nevertheless attractive 3- and 4-azaspirocyclic frameworks.

2. 3-Spiropiperidines

2.1. General retrosynthetic pathways

The general pathways that have been developed to elaborate the 3-spiropiperidine motif involve two approaches depending of the order in which the piperidine nucleus and the carbocyclic ring were constructed.

In approaches **I**, the spirocycle is elaborated from a preformed piperidine ring. In most cases, a 2-piperidone derivative constitutes the point of anchorage from which the spirocyclic system was elaborated by formation of one C-C bond (*route I-A*) or two C-C bonds (*route I-C*). Enamines and their reactivity were also used as starting materials in a *route I-B* affording the formation of one C-C bond.

In approaches **II**, the carbocyclic ring was built first with or without the quaternary carbon centre in place and the piperidine was finally elaborated by ring closure classical reactions (*routes II-D*, formation of one C-C bond and *route II-E*) (Scheme 4).

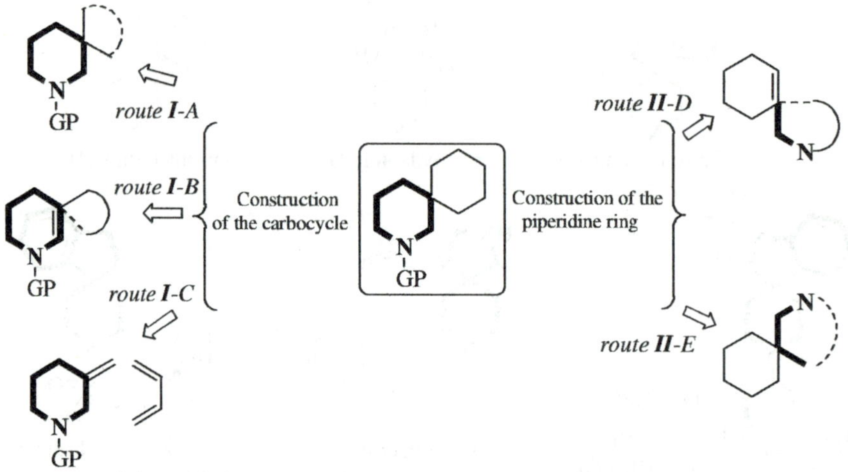

Scheme 4. Retrosynthetic general pathways to 3-spiropiperidines.

We described below the two approaches **I** and **II** giving details about the different routes developed in the literature for the edification of 3-spiropiperidines.

2.2. Approaches by construction of the spirocycle from a preformed piperidine ring
2.2.1. From 1,5-dicarbonyl compounds (*route I-A*)

This approach permitted the synthesis of racemic Nitramine, Isonitramine and Sibirine by Deyine *et al.*[11] in 2005. Thus, aminosilylenolether **1**, prepared from the commercially available 3-piperidinol, reacts with methyl vinyl ketone (MKV) (Michael-type addition) furnishing the key ketoaldehyde **2** in 73% yield.

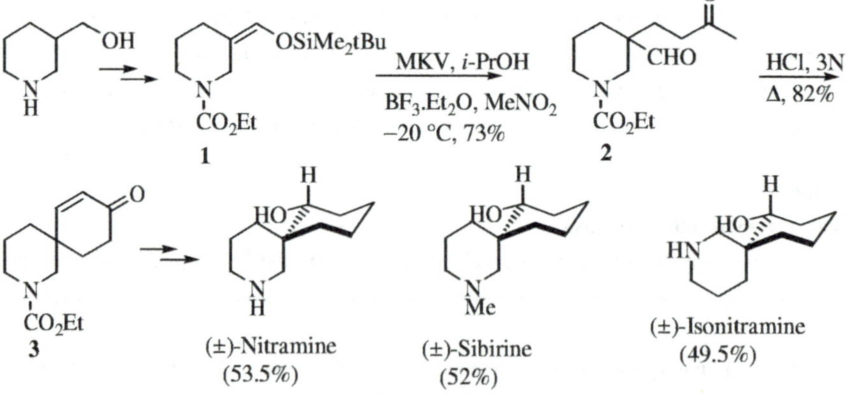

Scheme 5

124

The cyclization of **2** performed under acid conditions leads to the enone **3** which, after oxidation followed by hydrogenation, generates the azaspiranic alkaloids (Scheme 5).

2.2.2. Reductive cyclization (*route I-A*)

In the Nakadomarine A synthesis, the spirocyclic ring compound **5** was obtained in 44% yield by a classical reductive cyclization of the aldehyde-ester **4**, obtained in four steps and 41% yield from the methyl 4-oxo-3-piperidinecarboxylate, using 4-methoxybenzylamine followed by $NaBH_3CN$ reduction.[12] This spiro-γ-lactam **5** was then used to elaborate the tetracyclic A-B-C-D part of the natural product by functional transformation (Scheme 6).

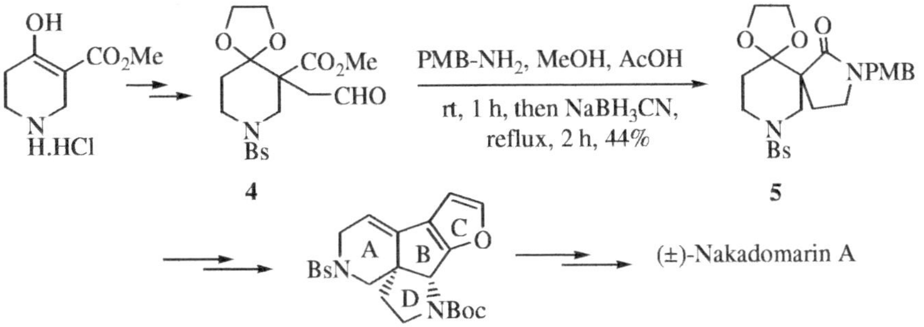

Scheme 6

2.2.3. RCM (*route I-A*)

The spiroimine ring system present in Gymnodimine (see Scheme 2 for structure) has attracted particular attention. Brimble and Trzoss[13] have developed a rapid method using a ring closing metathesis of the diene **7** as a key step.

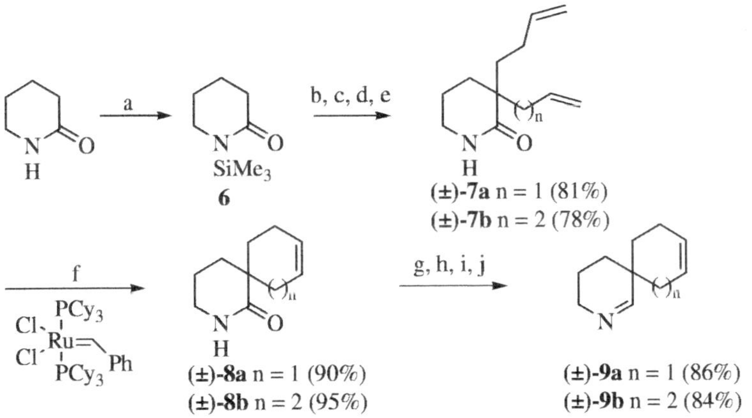

a. *n*-BuLi, THF, −78 °C, SiMe₃Cl, −78 °C–0 °C; b. LDA, −78 °C; c. 4-bromo-1-butene, −78 °C to 0 °C; d. followed when n=1 by LDA, −78 °C; e. allylbromide, −78 °C to 0 °C; f. Grubbs' catalyst, CH_2Cl_2, rt, 3 h; g. *n*-BuLi, THF, 78 °C; h. 2-(trimethylsilyl)ethyl-4-nitrophenyl carbonate, −78 °C to rt; i. LiEt₃H, −78 °C then work-up; j. Bu₄NF, THF, rt, 12 h.

Scheme 7

At first *N*-trimethylsilyl lactam **6** was prepared from 2-piperidone and, after a double alkylation, diene **7** was formed in good yield. Grubbs' catalyst $(Cl_2(PCy_3)_2RuCHPh)$-RCM gave spirolactams **8** in 90–95% yield. *N*-Protection of the lactams followed by reduction of the amido group using the conditions of Grieco and Kaufman (triethylborohydride)[14] and exposure of the resulting mixture to tetrabutylammonium fluoride furnished the desired spiroimines **9** (Scheme 7).

Using a similar approach, the chiral bicyclic lactams **10** served as a chiral template to prepare enantiopure spiropiperidines.[15] Thus, lactams **10** were doubly alkylated to **11** with good to high diastereoselectivity and were involved in a ring-closing metathesis using $Cl_2(PCy_3)_2RuCHPh$ catalyst to afford the corresponding spirocyclic lactams **12** in excellent yields. DIBAL-H reduction of the bicyclic lactam **12b**, for example, furnished finally the spiropiperidine **13** nearly quantitatively (Scheme 8).

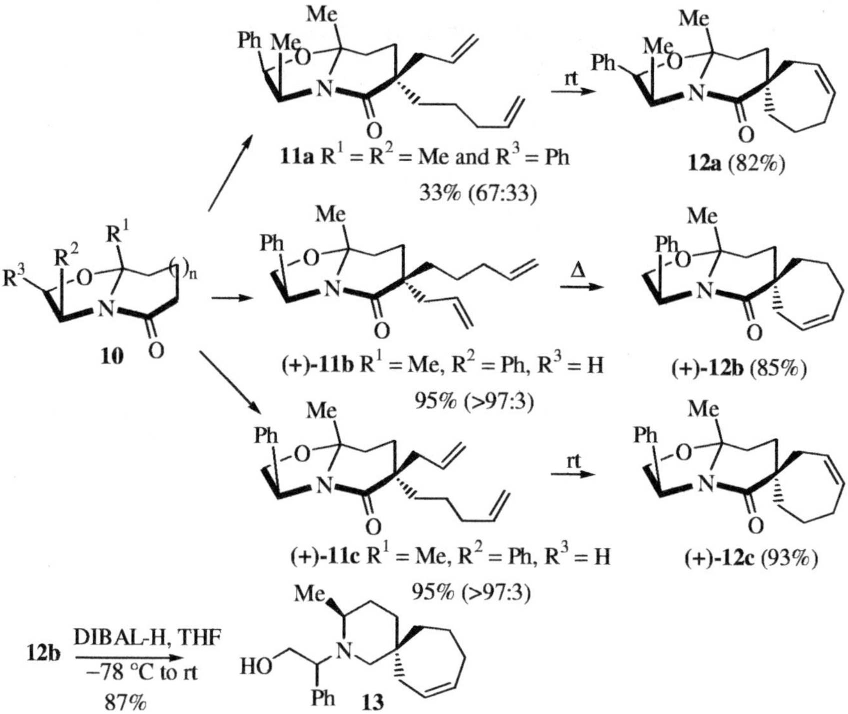

Scheme 8

2.2.4. Mitsunobu reaction (*route I-B*)

In the search directed toward the synthesis of useful modulators of neurokinin NK-1 receptor, Kulagowski *et al.*[16] chose to prepare 1-oxa-7-azaspiro[4.5]decane derivatives **17** starting from (2*S*,3*S*)-(+)-2-phenyl-3-piperidinol. This latter was stereospecifically transformed into **14** in 82% yield after protection of the amino group to a carbamate, a Swern oxidation of the hydroxy followed by a Grignard addition and a desilylation. A palladium(0)-mediated hydrostannylation of **14** provided a mixture of two vinyl stannanes isomers **15** which were cyclized under Mitsunobu conditions to furnish the spiropiperidine **16** in 63% yield. Further transformation *via* a Stille cross-coupling followed by a reduction of the double bond furnished lactam **17** (Scheme 9).

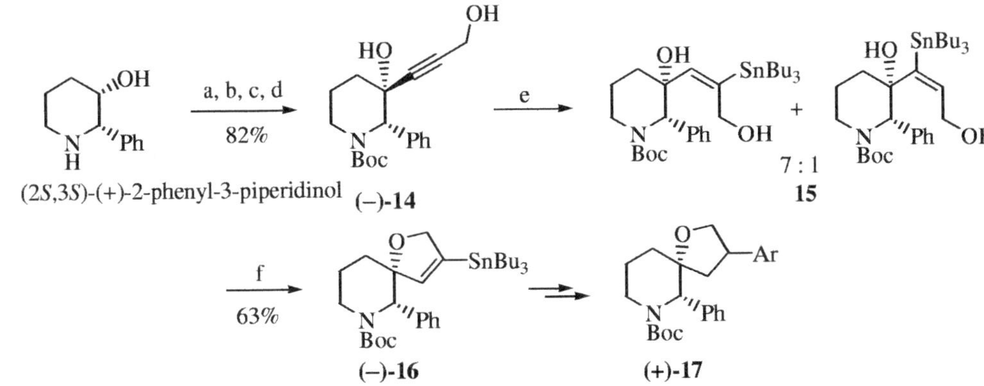

a. Boc_2O, Na_2CO_3, H_2O, CH_2Cl_2; b. Swern [O]; c. $BrMg-C{\equiv}C-CH_2OTMS$; d. Bu_4NF, THF;

e. n-Bu_3SnH/PhMe, $Pd(PPh_3)_4$; f. DEAD, PPh_3/THF.

Scheme 9

2.2.5. Metal-catalyzed cyclizations (*route I-B*)

Metal-catalyzed cyclizations involving an alkyne have emerged as a powerful methodology to construct nitrogen-containing heterocycles. Recently, this reaction process has been successfully applied to 1,6-enyne system **A** to form quaternary carbon centre leading to spiropiperidines or piperidones **B** (Scheme 10).

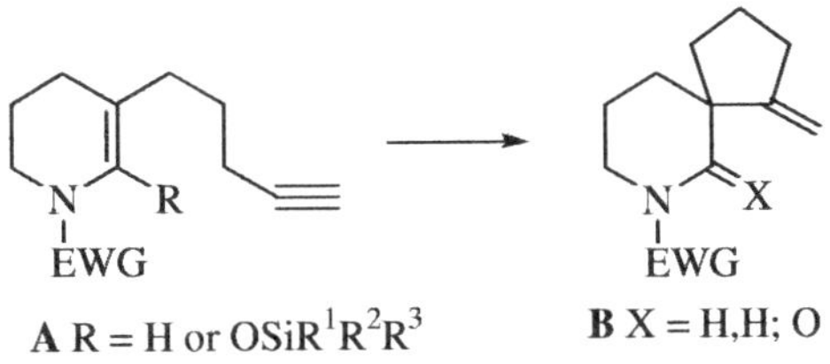

Scheme 10

Thus a selective mild 5-*exo-dig* cyclization of the ω-acetylenic silyl enol ether **18** occurred when it was treated in a 10:1 mixture of CH_2Cl_2/MeOH which is used as a proton source, with gold(I) catalyst and silver co-catalyst generating spirocycle **20**[17] (Scheme 11).

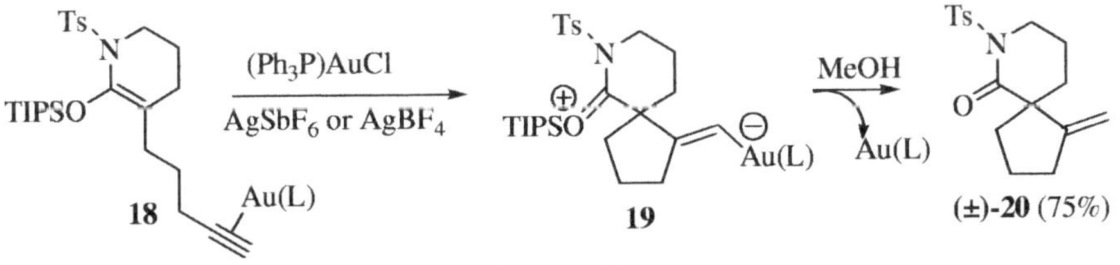

Scheme 11

In a similar manner, enesulfonamide **21** tethered to an alkyne cyclized in $PhCH_3$/MeOH giving the spirocyclic compounds **22** or **23** as a single isomer and 80% yield, differing about the possible migration of the double bond upon the choice of the catalyst (PtCl$_2$ or PtCl$_2$/AgOTf). The spiropiperidine **23** was then cleanly transformed in four steps in 29% overall yield to a 2:1 mixture of Nitramine and Isonitramine (Scheme 12).[18]

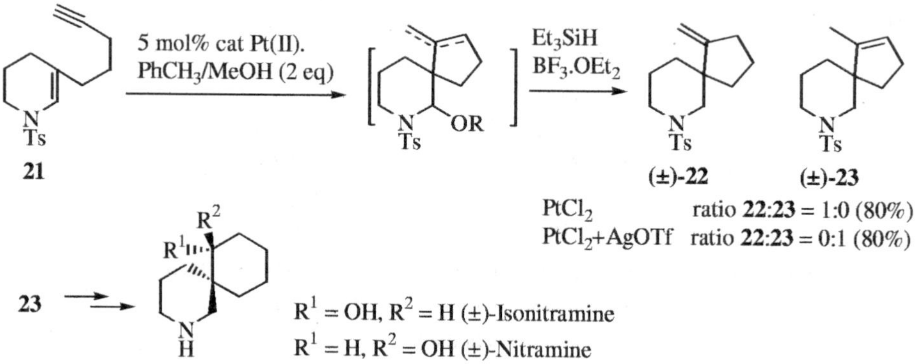

PtCl$_2$ ratio **22:23** = 1:0 (80%)
PtCl$_2$+AgOTf   ratio **22:23** = 0:1 (80%)

R^1 = OH, R^2 = H (±)-Isonitramine
R^1 = H, R^2 = OH (±)-Nitramine

Scheme 12

The powerful of palladium substrates in chemical transformations has been used to synthesize 3-spiropiperidines. In fact, the spirocyclic indoline ring system which presents numerous biological applications is considered as an important scaffold for the discovery of novel therapeutic agents because of its favourable physical properties for oral bioavailability. A few methods exist for the synthesis of the spiro[indoline-3,3'-piperidine] core **25**. However, Pfefferkorn and Choi[19] described in 2008 a simple approach to this skeleton using an α-arylation intramolecular reaction of **24** easily obtained from ethyl 3-piperidinecarboxylate (ethylnipecotate) (Scheme 13). Thus, the palladium-catalyzed cyclization of **24** at elevated temperature provided oxindole **25** in 54% yield. Further reduction with BH$_3$.SMe$_2$ followed by deprotection with chloroethylchloroformate furnished the corresponding 1,1'-*H*-spiro(indoline-3,3'-piperidine] in 49% yield.

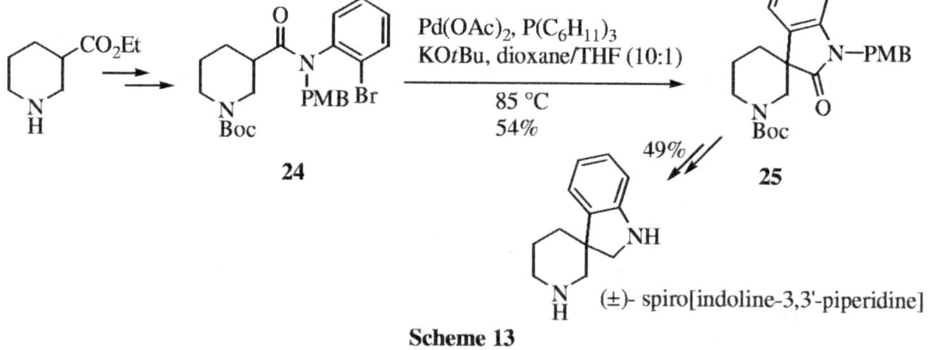

Scheme 13

Use of palladium(0) permitted the formation of spiroindenepiperidinedione moiety **27** and spiroindenepiperidones **28** from cyclic arylenamide **26** *via* a 5-*exo*-cyclization of the aryl-palladium intermediates **29** under an intramolecular Heck cyclization which allowed the insertion of the enamide double bond into the aryl-Pd bond.[20] Then, imide **27** provided through an oxidation of iminium ion **30** whereas amide **28** is issued of its reduction (Scheme 14).

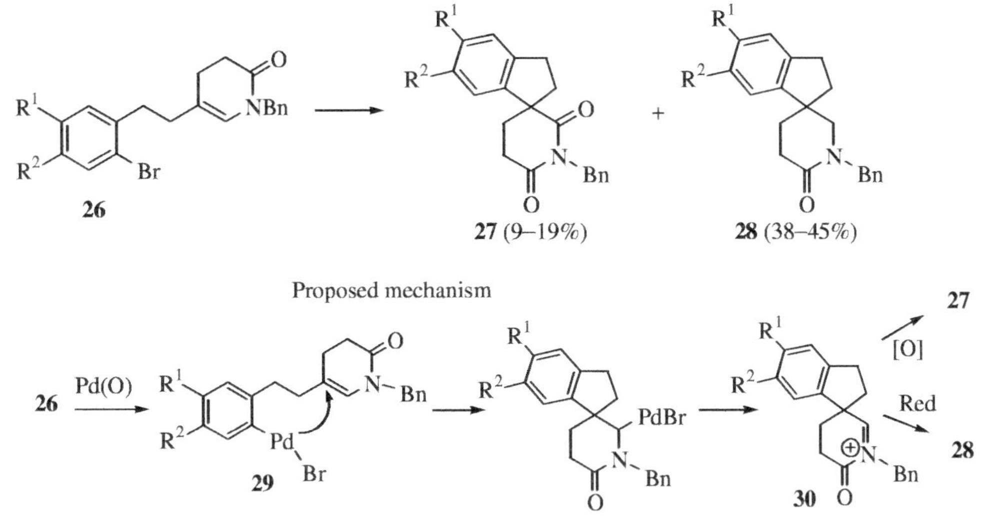

26

27 (9–19%)   +   **28** (38–45%)

Proposed mechanism

26 $\xrightarrow{Pd(0)}$ **29**   →   **30** $\xrightarrow{[O]}$ **27** $\xrightarrow{Red}$ **28**

Scheme 14

In 2008, Procter and coll.[21] used a Samarium(II)-mediated cyclization of γ,δ-unsaturated keto-lactams **31** bearing on the *N*-aryl moiety an electron-withdrawing substituent (Ar=4-C$_6$H$_4$-CF$_3$ or 4-C$_6$H$_4$-OMe) to obtain *syn*-spirolactams **32** with complete diastereoselectivity. The proposed mechanism consists of the SmI$_2$ reduction of the double bond of amide **31**, generating a radical-anion, which was further transformed into a samarium enolate which implies a chelation-controlled aldol cyclization affording the spirocyclic compound (Scheme 15).

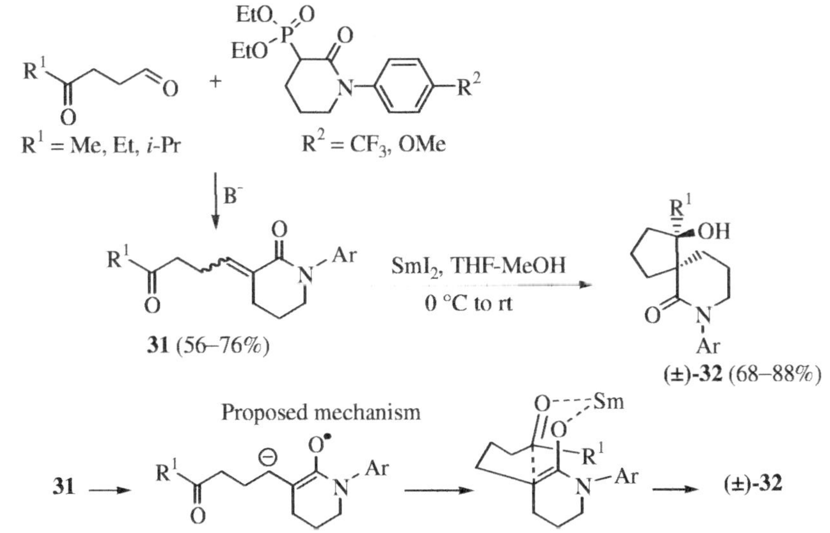

R^1 = Me, Et, *i*-Pr   R^2 = CF$_3$, OMe

31 (56–76%) $\xrightarrow{\text{SmI}_2,\ \text{THF-MeOH}}_{0\ °C\ to\ rt}$ (±)-**32** (68–88%)

Proposed mechanism

31 →   →   (±)-**32**

Scheme 15

2.2.6. Use of phenyloxazolopiperidine (*route I-B*)

The synthesis of natural (–)-Sibirine, (+)- and (–)-Isonitramine was efficiently achieved by Kunesch, Husson *et al.*[22] starting from the condensation of (*R*)-(–)-phenylglycinol and glutaraldehyde in the presence of sodium *p*-toluenesulfinate as a nucleophilic source. The key step is the formation of the spiro compound **33** *via* a stereocontrolled spiroaldolization process in which the enamine intermediate reacts on the aldehyde function in a chair-like transition state with the sulfone and the newly formed C-O bond in a diequatorial position (Scheme 16).

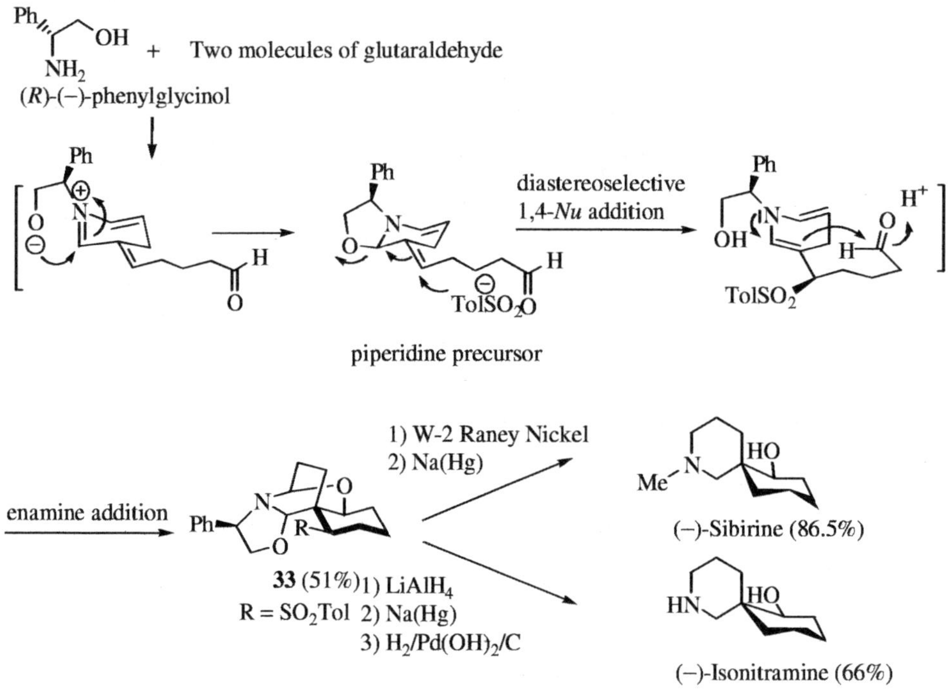

Scheme 16

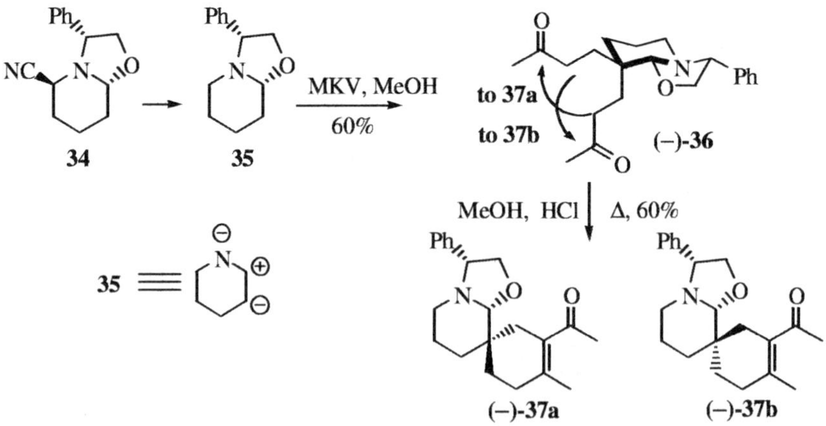

Scheme 17

The same authors developed in 2004 a rapid and efficient synthesis of 3-spiropiperidines using the key phenyloxazolopiperidine **35** as an equivalent nucleophilic synthon which reacts on methyl vinyl ketone in a Michael reaction to give the 3,3-bisubstituted oxazoline **36**.[23] Compound **36** underwent an acidic crotonization-dehydration reaction (*route A*) leading to a 1:1 mixture of the spirocyclic compounds **37a,b** (Scheme 17).

2.2.7. [1-3]-Dipolar cycloaddition (*route I-B*)

The [1,3]-dipolar cycloaddition of azomethine ylides is one of the most powerful methods for the formation of substituted 2,5-*cis*-pyrrolidines starting from metallo-stabilized ylide **38a** (*syn* orientation of substituents). However, if bulky azomethine ylide like **38b** is used, 2,5-*trans*-pyrrolidines are obtained due to the *anti*-orientation of the substituents.

Applying this methodology to the piperidinenone **39**, this reaction allowed the synthesis of the cycloadduct **40**, precursor of the ADE fragment of (−)-Nakadomarin A (see Scheme 2 for structure) which was isolated as a single diastereomer due to the cycloaddition of the ylide on the opposite face of the phenyl groups (Scheme 18).[24]

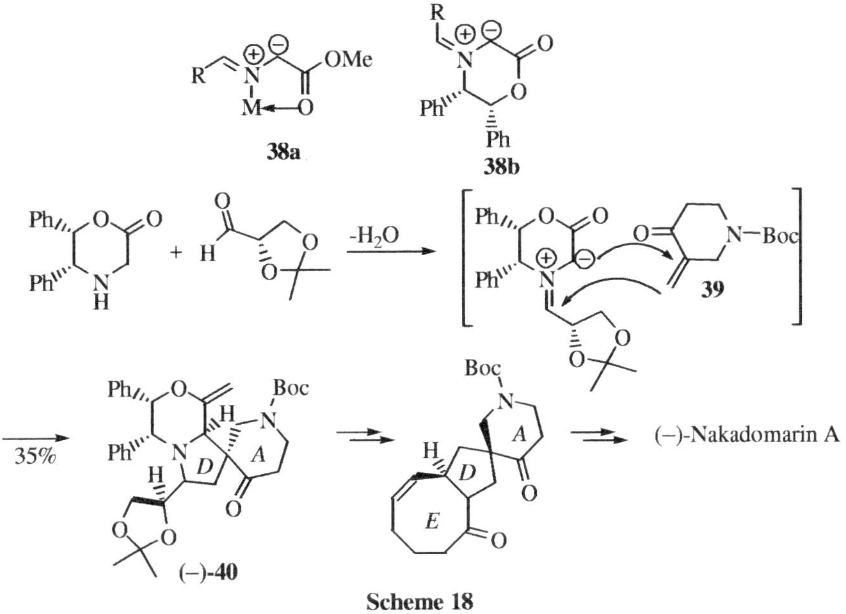

Scheme 18

2.2.8. Diels-Alder reaction (*route I-C*)

In their synthesis directed toward segments of (−)-Gymnodimine (see Scheme 2 for structure), Romo *et al.*[25a] prepared the spirocyclic imine core using an intermolecular Diels-Alder cycloaddition promoted by Et₂AlCl of an α-methylene-δ-lactam dienophile **41a** and the (*Z*)- or (*E*)- dienyne **42**. The reaction led to the spirocyclic compound **43** possessing the relative stereochemistry of the natural product.

The authors reported latter an asymmetric synthesis of the spirocyclic compound **43** starting from **41b** and (*E*)-isomer **42** *via*, this time, a diastereo- and enantioselective concerted Diels-Alder reaction catalyzed by an Evans' type copper-bis(oxazoline) complex (Scheme 19).[25b]

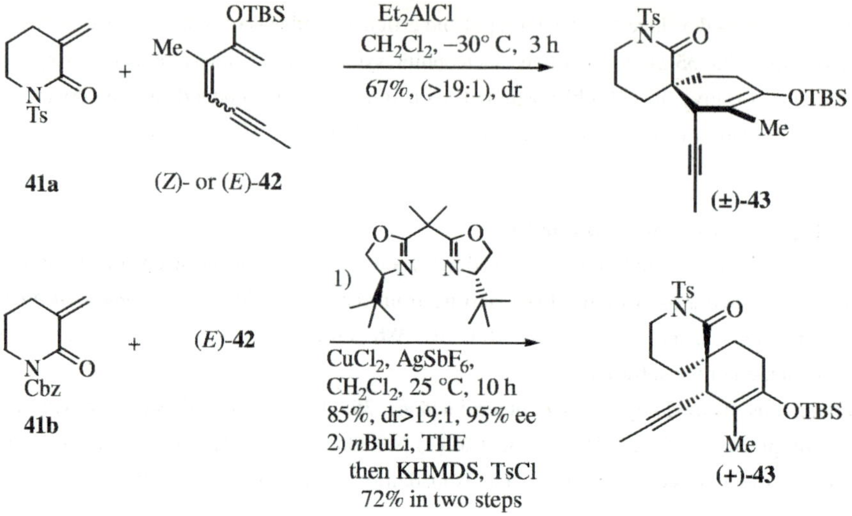

Similarly, Murai *et al.*[26] chose to obtain the spirocyclic part of Gymnodimine by highly *exo*-selective asymmetric Diels-Alder cycloaddition of the same methylene lactam **41b** and triene **44** in the presence of Ellman's copper *bis*(sulfinyl)-imidoamidine complex. The reaction proceeded with excellent regio- and diastereofacial selectivity allowing the formation of only one cycloadduct **45** in good yield (Scheme 20).

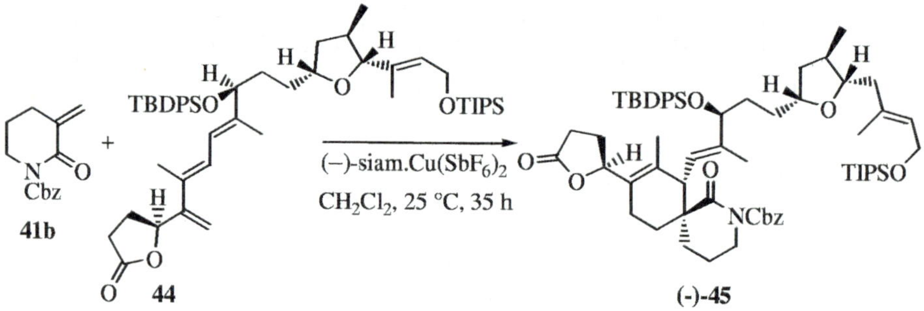

Scheme 19

Scheme 20

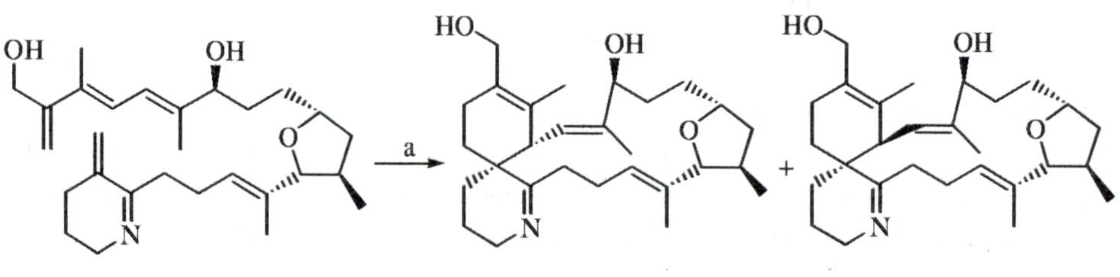

a: condition of cyclization = pH 6.5, sodium citrate/HCl buffer; H₂O, 36 °C, 48 h, conc = 60mM

Scheme 21

In another synthesis of Gymnodimine, Kishi and coll.[27] developed a Diels-Alder macrocyclization of the α,β-unsaturated imine **46** in water at pH 6.5 under dilute conditions and a moderate temperature of 36 °C. The reaction is not selective and a mixture of *endo-* and *exo-*cycloadduct (stable natural diastereomer) **47a** and **47b** was obtained in a 1:1 *ratio* (Scheme 21).

Starting from α-methylene lactam **41c**, reaction with 1-trimethylsilyloxy-1,3-butadiene permitted the access to **48**, precursor of (±)-Sibirine in 50% yield (Scheme 22).[28]

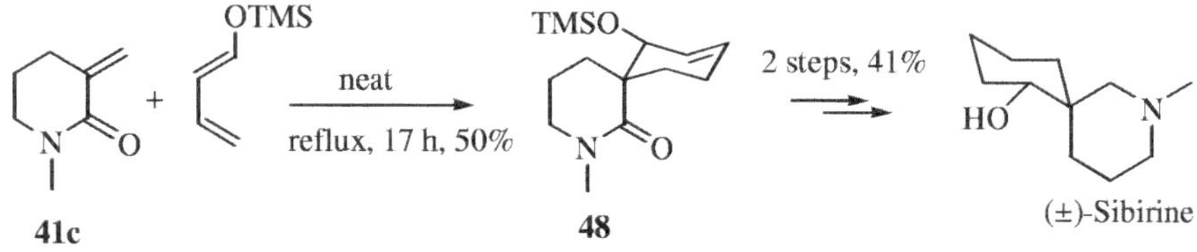

Scheme 22

2.3. Approaches by construction of the piperidine ring from a carbocyclic ring in place
2.3.1. Radical cyclization (*route II-D*)

Radical cyclization process is an effective method to allow the generation of quaternary centre in a regio- and stereocontrolled manner.

In this context, preparation of the phenyl selenide carbamate **49** as precursor for a radical intermediate is of particular interest.[29] Even if a 7-*endo*-cyclized product was expected in the cyclization step as well as the 6-*exo* adduct, the use of a sterically demanding group R² such as SO₂Ph, which acts also as a stabilizer on the radical intermediate, permitted the synthesis of the spirocyclic compounds **51a,b** through exclusive "path a" from **50**. Major compound **51a** was isolated and further transformed into (−)-Sibirine in 78% yield (Scheme 23).

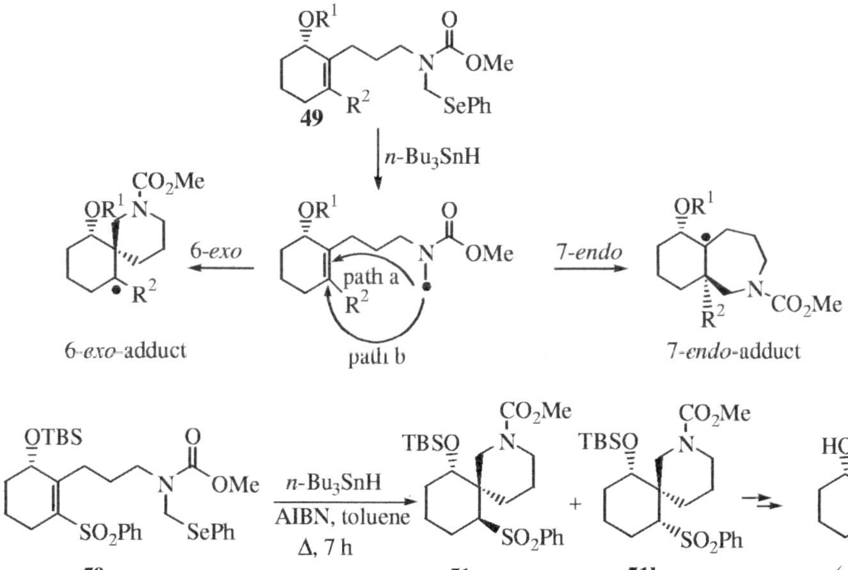

Scheme 23

2.3.2. Photocyclization (*route II-D*)

A general method for the preparation of spirolactam has been developed in 1999 by Piva *et al.*[30] following a strategy relying on the fragmentation of a [2+2] cyclobutane photoadduct by Lewis acid activation. Thus, the unsaturated oxoamide **52** was engaged in a cross-coupling metathesis with trimethylallylsilane which underwent a photocycloaddition giving compound **53**, in 84% yield, as a mixture of *syn*- and *anti*- isomers. Transformation of **53** into the spiranic vinyl compound **54** was accomplished with $BF_3.OEt_2$ in 79% yield (Scheme 24). The vinyl group was then easily cleaved by oxidation/decarbonylation procedure.

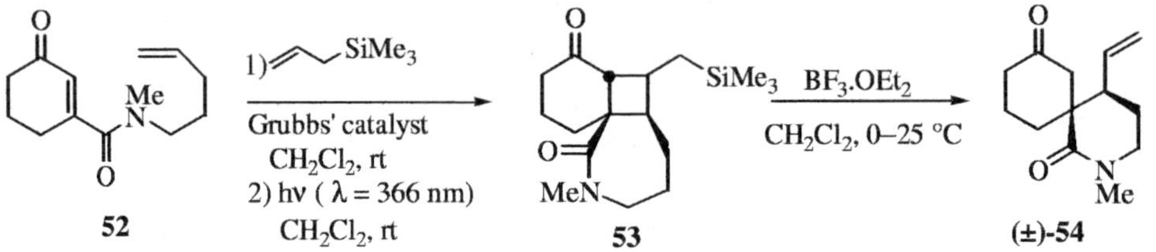

Scheme 24

2.3.3. *N*-Acyliminium ion cyclization (*route II-D*)

N-Acyliminium ion cyclization has been successfully applied to the synthesis of the spirocyclic part of Nakadomarin A (see Scheme 2 for structure).[31] Thus, treatment of amide **55** with $Sc(OTf)_3$ provided a mixture of two hemi-aminals **56α** and **56β** (54% yield) which resulted from a conjugate addition and subsequent capture of the *N*-acyliminium ion by adventitious water. Oxidation of the hemi-aminals afforded a spirocylic imide possessing the furan substituent in an equatorial position (Scheme 25).

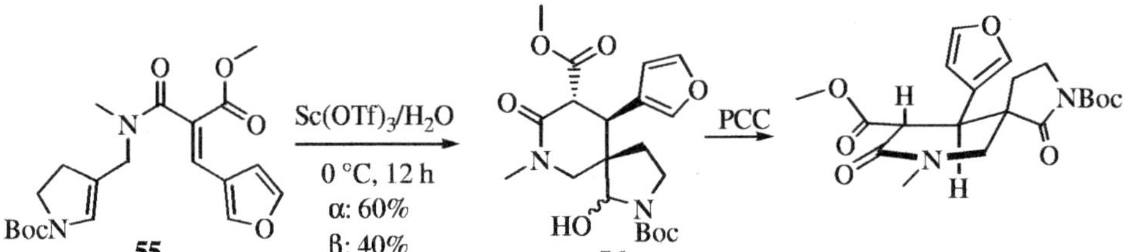

Scheme 25

2.3.4. Use of cyclohexadiene-Fe(CO)₃ (*route II-D*)

Intramolecular coupling between the cyclohexadiene-Fe(CO)$_3$ moiety of the *N*-butenyl amides **57** under a CO atmosphere led to a mixture of spiro-γ-lactams **58a,b**.[32] The presence of an electron-withdrawing group on the pendant olefin increases the stability of the starting complex **58**. Spirolactams **58a,b** were further transformed into spiroenone **59** in 70% yield by treatment with trimethylamine *N*-oxide (Me_3NO) followed by an acidic hydrolysis of the enol ether (Scheme 26).

2.3.5. Lactamization (*route II-E*)

Among the reactions developed to prepare heterocyclic compounds, halocyclization of unsaturated compounds is one of the most used, in particular, the iodolactonization process. In contrast, asymmetric iodolactamization has been less studied.

Li and Shen[33] investigated this reaction on a plethora of ω-unsaturated amides substituted with oxazolidines as chiral auxiliaries. Among them, hexenamides **60** were developed and led under basic BuLi treatment to δ-lactams **61** in 90–98% yields with excellent diastereoselectivities due to the most favour transition state placing the olefinic moiety and the *sec*-butyl group at the opposite faces of the imide to minimize steric repulsion (Scheme 27).

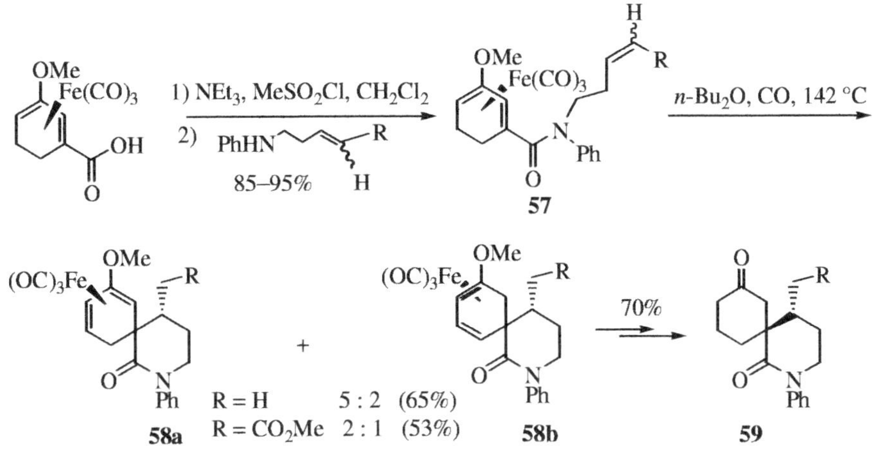

OMe
Fe(CO)3
OH
O
1) NEt3, MeSO2Cl, CH2Cl2
2) PhHN
R
H
85–95%
57
n-Bu2O, CO, 142 °C
(OC)3Fe
N
Ph
R = H 5 : 2 (65%)
R = CO2Me 2 : 1 (53%)
58a
58b
70%
59

Scheme 26

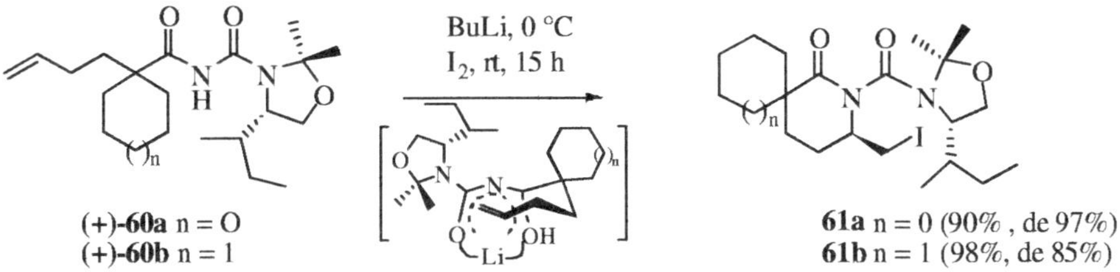

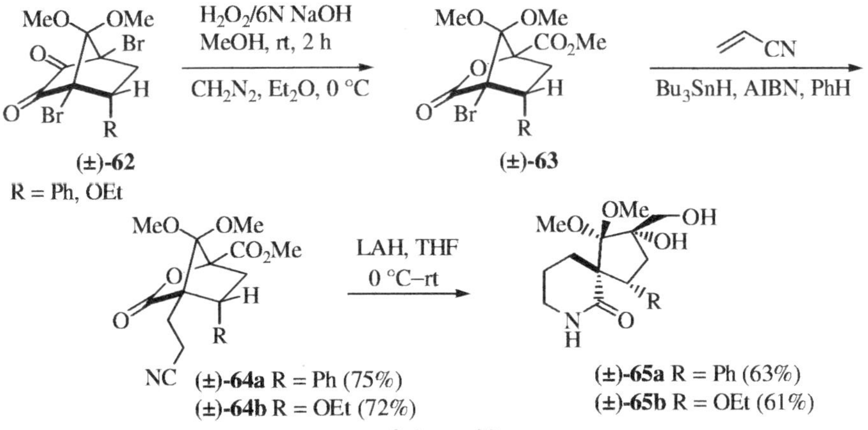

BuLi, 0 °C
I2, rt, 15 h
(+)-60a n = O
(+)-60b n = 1
61a n = 0 (90% , de 97%)
61b n = 1 (98%, de 85%)

Scheme 27

MeO
OMe
Br
H
R
(±)-62
R = Ph, OEt
H2O2/6N NaOH
MeOH, rt, 2 h
CH2N2, Et2O, 0 °C
CO2Me
(±)-63
CN
Bu3SnH, AIBN, PhH
LAH, THF
0 °C–rt
NC
(±)-64a R = Ph (75%)
(±)-64b R = OEt (72%)
OH
(±)-65a R = Ph (63%)
(±)-65b R = OEt (61%)

Scheme 28

A spirolactamization *via* radical-mediated intermolecular C-C bond formation of bridged lactones **63** with acrylonitrile followed by chemoselective reduction using LAH was reported for the preparation of the 7-azaspiro[4.5]decane **65**, a conformationally restricted pseudopeptide, which relative stereoche-mistry was deduced from X-ray analysis.[34] The bromo bicyclic lactone **63** was prepared as a single regioisomer by SNi2 alkaline H_2O_2 cleavage of the diketone **62** (Scheme 28).

2.3.6. Hydroamination (*route II-E*)

Intramolecular addition of N-H bond of an amine or amide on unactivated alkenes is an interesting reaction permitting an access to heterocyclic compounds. Nevertheless, this reaction suffers of the oxophilicity and /or basicity of the metal catalysts used and requires, sometimes, drastic conditions.

Recently, it was applied on primary and secondary ammonium salts using a mixture of Au(I) catalyst and AgOTf (5 mol%). As an example, spiropiperidine **66** was obtained in 91% yield (Scheme 29).[35] The mechanism of this cyclization was recently reported.[35b]

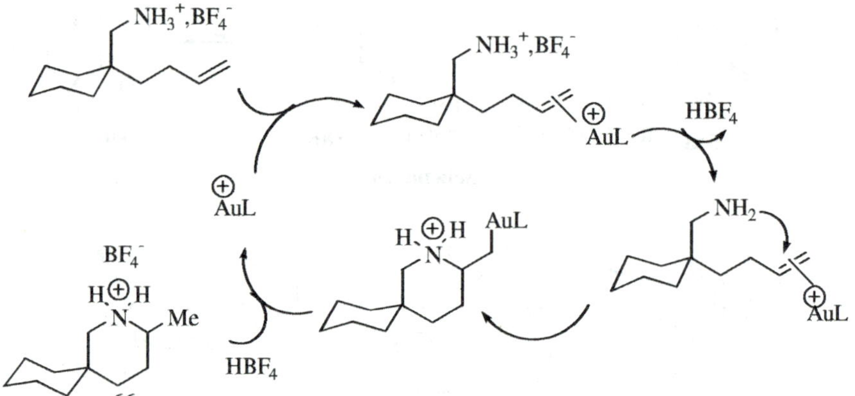

Scheme 29

3. 4-Spiropiperidines

3.1. General retrosynthetic pathways

The general approaches to elaborate the 4-spiropiperidine framework can be classified in two pathways, **III** and **IV**, depending if the piperidine cycle is preformed or not.

In approaches **III**, the piperidine nucleus served as an appendage to build the spiro ring. In this case, the spirocycle was elaborated from piperidine substituted at the C-4 position by a pendant olefine (*route III-F*) or from 4-piperidone (*route III-G*) by reaction on the ketone function which allowed the introduction of the element constituting the spirocycle.

In approaches **IV**, two routes were proposed. *Route IV-I* implicates the formation of the quaternary centre by a double alkylation step, whereas in *route IV-H*, the quaternary carbon is created at first and the piperidine ring is elaborated at the last step by reaction with amine-derivatives.

In addition, multi-component reactions (MCR) were developed allowing more complexity in the carbocyclic part of the synthesized 4-piperidine; in this case, either an amine or a piperidone was used to promote the piperidine part of the spiroheterocyclic compound (Scheme 30).

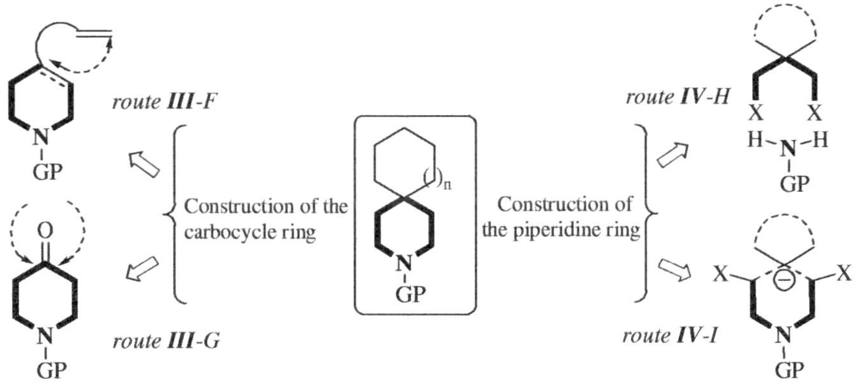

Scheme 30. General retrosynthetic pathways to 4-spiropiperidines.

3.2. Approaches by construction of the spirocycle from a preformed piperidine ring

3.2.1. Fischer indole reaction (*route III-F*)

Commercially isonipecotic acid was required in the preparation of MK-677 growth hormone secretagogue. This acid was readily transformed to aldehyde **67** which underwent a Fischer indole reaction by treatment with phenylhydrazine and TFA giving the spiroindoline **68** in 93% overall yield without other side-product.[36] Classical transformations furnished MK-677 in few steps with a good yield and in a synthetic process allowing the preparation of multikilogram quantities (Scheme 31). A solid phase synthesis of spiroindolines based upon this strategy was also reported.[37]

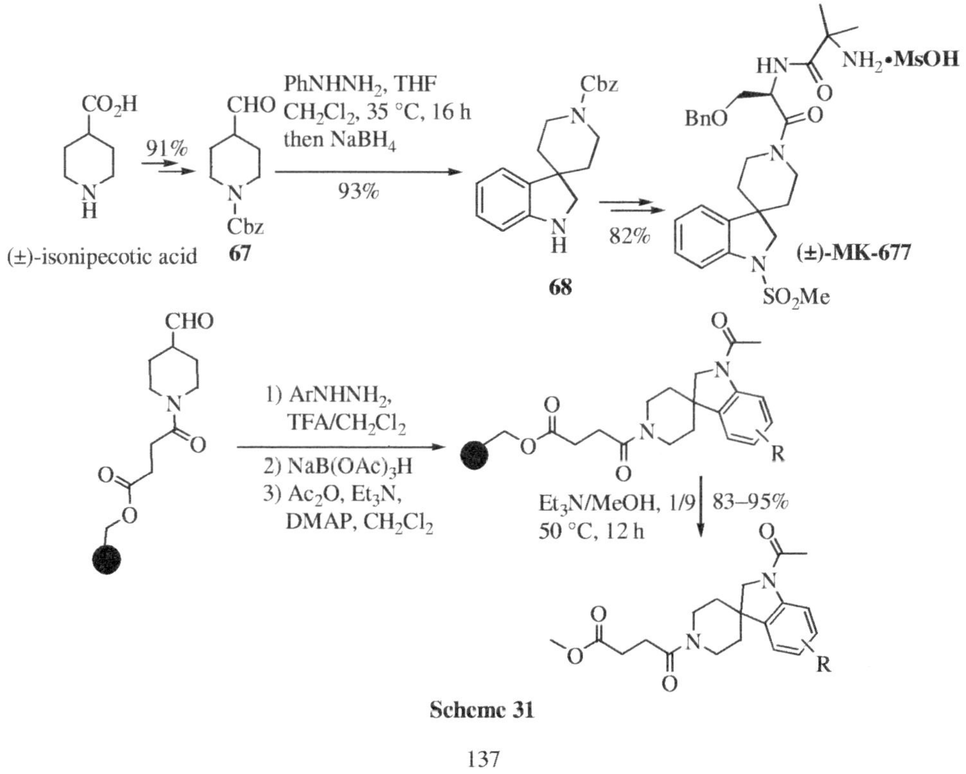

Scheme 31

3.2.2. Palladium-catalyzed cyclization (*route III-F*)

The intramolecular Heck reaction is a powerful reaction to create tetrasubstituted carbon centres and therefore it has been extensively used in the synthesis of spiroheterocycles. For example, this reaction was applied on bromophenylaminopiperidine **69** to produce a mixture of spiropiperidinequinolines **70** and **71** in 77% yield (9:1 ratio)[38] (Scheme 32).

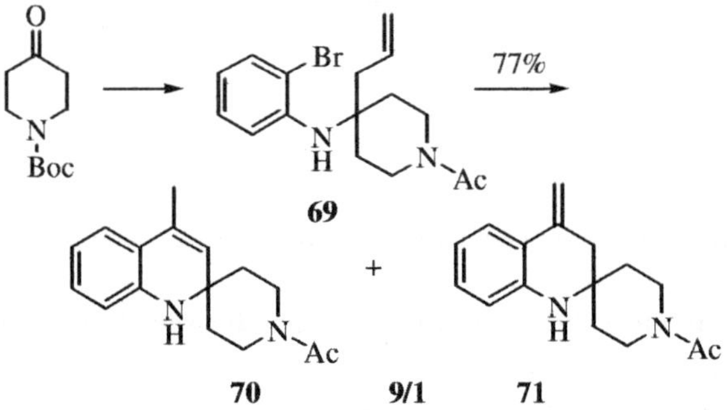

a. 2-Bromoaniline, Δ, toluene; b. allylMgBr, toluene; c. HCl, EtOH, Δ; d. Ac$_2$O; e. Pd(OAc)$_2$, PPh$_3$, NEt$_3$, CH$_3$CN.

Scheme 32

Spiroindolinones piperidines **73**[39a] were synthesized by palladium-catalyzed intramolecular α-arylation of amides **72** using Hartwig's method.[39b] After oxidative addition of halide amide **72** to palladium(0) complex, basic treatment induced the formation of an arylpalladium enolate complex which, after reductive elimination, furnished **73**. This method allowed an access through a five-steps sequence to a series of 3-spirocyclic indolin-2-ones with a variety of substituents on the phenyl group and piperidine rings, in 12–65% yield, which displayed affinity for ORL-1 receptor (Scheme 33).

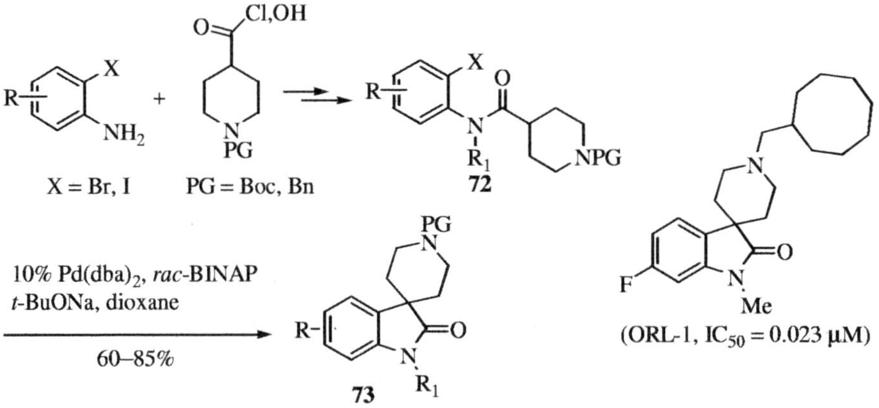

Scheme 33

3.2.3. Radical cyclization reaction (*route III-F*)

A radical approach was used to prepare spirobenzofuranpiperidines **74**. Among them, **74a** was detected as a potent tryptase inhibitor with oral efficacy in animal models of airway inflammation. Thus, *N*-Boc-4-piperidone was transformed into allyl chloride **75** in 52% yield. Coupling this unstable chloride with 2-bromo-4-cyanophenol furnished ether **76** in 67% yield. Radical-based cyclization with Bu$_3$SnH/AIBN in refluxing toluene afforded the spiropiperidine **77** in 85% yield. The synthesis of **74a** was then achieved from **77** in seven steps and 33% overall yield (Scheme 34).[40]

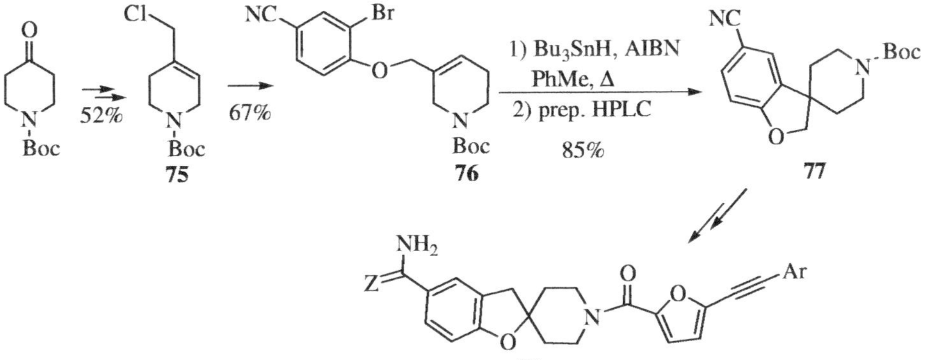

Scheme 34

Spirohexadienones were conveniently prepared by radical cyclizations in which the radical is generated by photolysis of iodides or reduction of diazonium salts.[41] More recently, an *ipso*-type oxidative radical cyclization process using (Me$_3$Si)$_3$SiH/Et$_3$B has been reported and applied to the synthesis of spirooxindoles and spirodihydroquinolones but in only low yields. Thus, the formal synthesis of aza-galanthamine was realized from precursor **79** in 40% yield, easily prepared by radical cyclization of **78** issued from available starting materials (Scheme 35).[42]

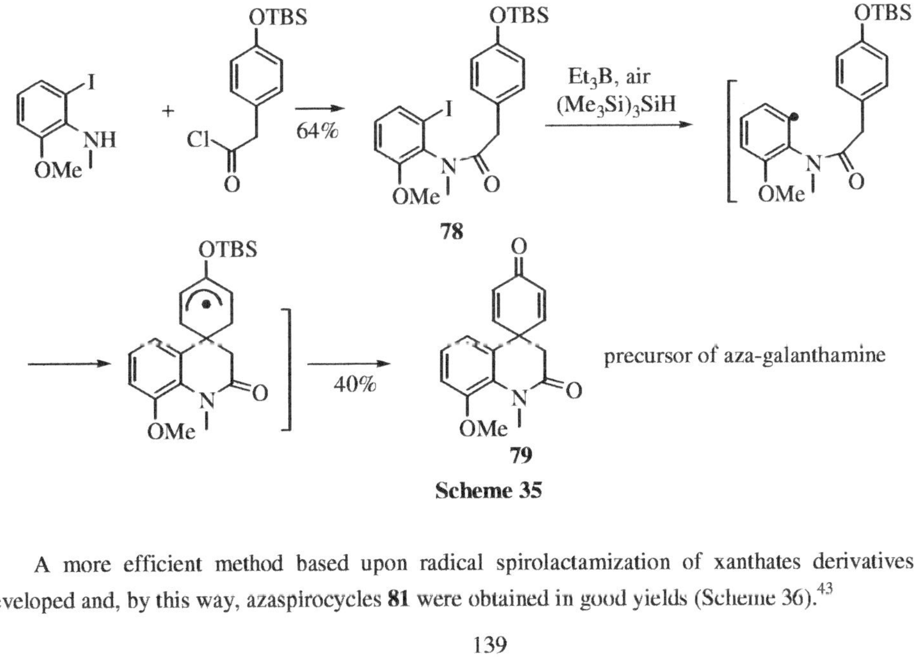

Scheme 35

A more efficient method based upon radical spirolactamization of xanthates derivatives **80** was developed and, by this way, azaspirocycles **81** were obtained in good yields (Scheme 36).[43]

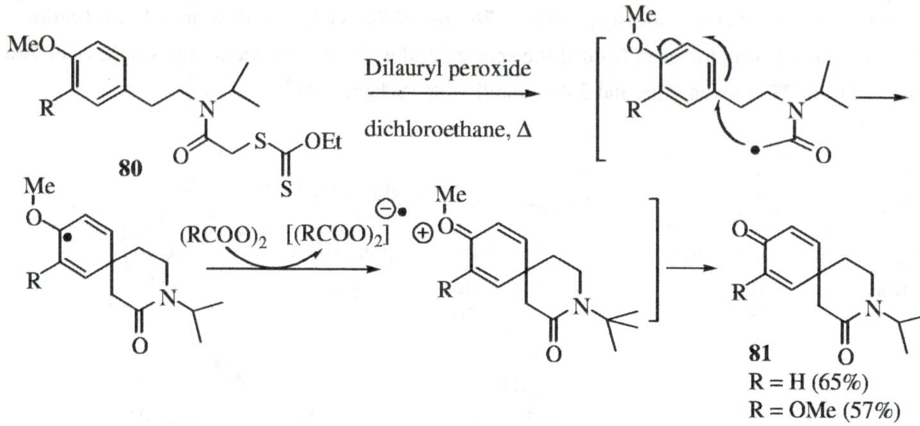

Scheme 36

3.2.4. From 4-piperidone (*route III-G*)

This path is the more general approach used. In this case, the ketone function of the 4-piperidone moiety constitutes the point of anchorage of the two carbon chains, precursor of the second cycle allowing the creation of the quaternary spirocentre by chemical classical reactions.

Spirohydantoins are considered privileged structures with potential for diverse biological activity. Among them, spirohydantoins-piperidines are attractive because they are recognized as antagonists of the melanin-concentrating hormone receptor and act as glycine transporter 1 (ClyT1) inhibitor. Screening of various substituents on the nitrogen atoms has been reported[44] and, in addition, a solution-phase parallel synthesis was developed.[44d] The skeleton of these interesting substrates was always prepared by the same method in a few steps. The synthesis started with *N*-substituted 4-piperidone which was used for the introduction of the hydantoine part through a Strecker reaction, employing a mixture of amines/TMSCN or amines/KCN. Cyclization of the aminonitrile **82** thus obtained into the spirocyclic structures **83** was then achieved, either with chlorosulfonylisocyanate followed by acid hydrolysis, or by a variety of substituted isocyanates and acid treatment in moderate to good yields (Scheme 37).

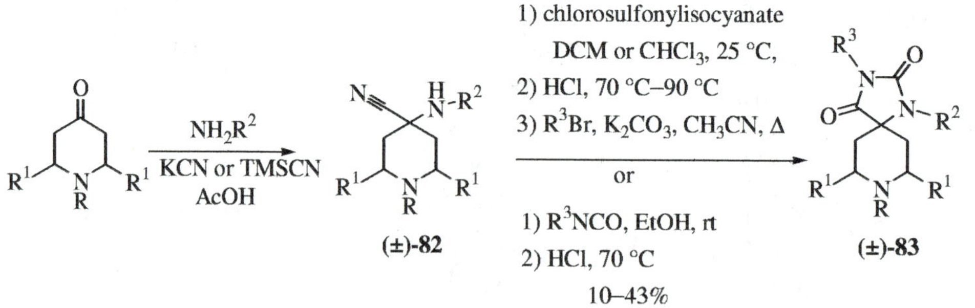

Scheme 37

Another strategy often used led upon a Knoevenagel-cyclocondensation reaction of *N*-substituted-4-piperidones with cyanoacetates furnishing intermediates **84**. These latter were further transformed either in spiropyrimidones **85** in 70–96% yield[45a] or, in piperidines **86** by treatment of **84** with acetate derivatives.

Compound **86** constituted then efficient precursors of 3,9-diazaspiro[5.5]undecanes **87**[45b] by simple reduction with $NaBH_4$-$CoCl_2$-MeOH (Scheme 38).

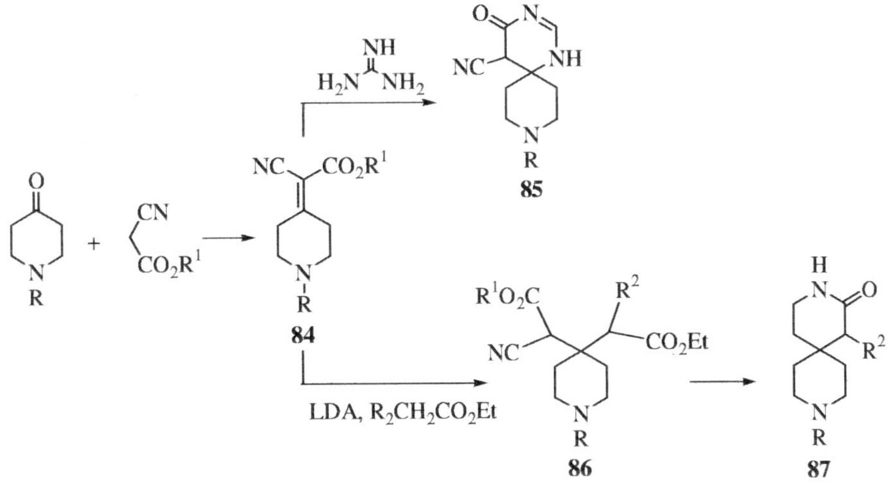

Scheme 38

3.3. Approaches by construction of the piperidine ring

3.3.1. Amine condensation (*route IV-H*)

In this approach, the quaternary centre is already formed and the piperidine ring was built by a reductive cyclization of an amine on an ester-aldehyde or a dialdehyde (**88**→**89**) (Scheme 39),[46] a condensation of an urea on a diester which was subsequently reduced (**90**→**91**) (Scheme 40)[47] or by addition of an allylmagnesium bromide to a ketimine issued from addition of an amine on the ketone function, followed by an acid catalyzed intramolecular Friedel-Crafts alkylation (**92**→**93**)[48](Scheme 41).

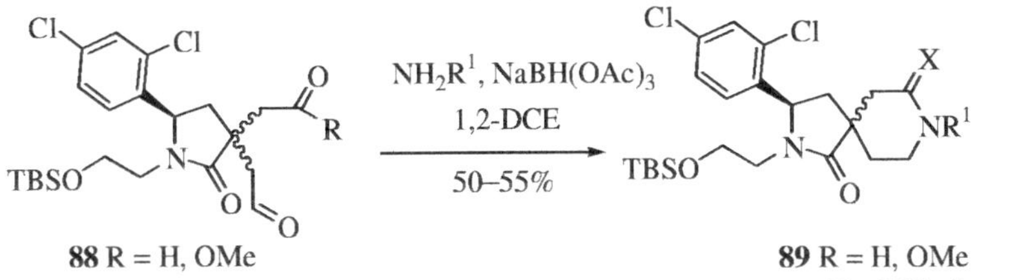

Scheme 39

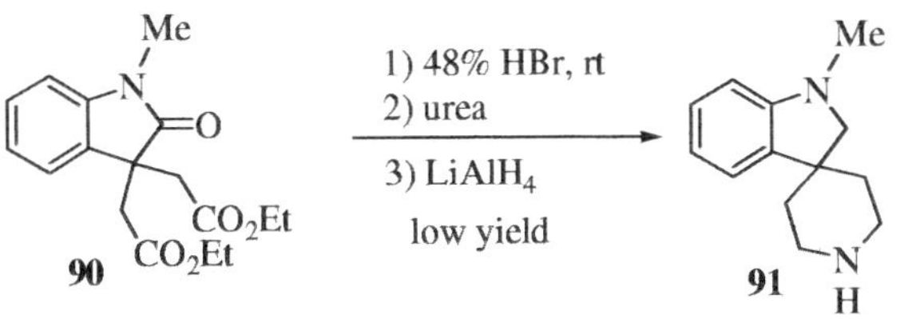

Scheme 40

141

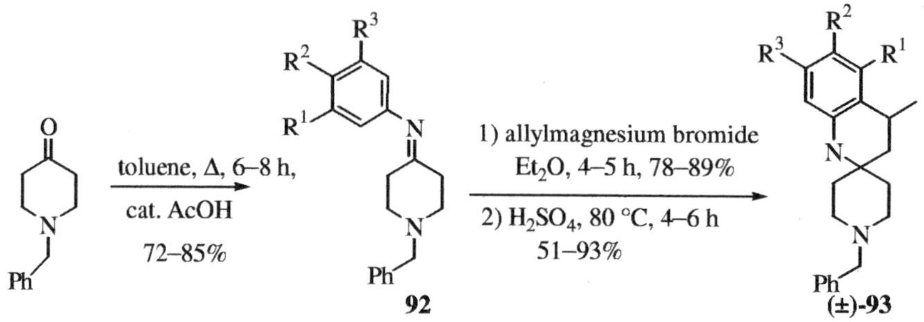

3.3.2. Nucleophilic substitution (*route IV-H*)

The classical route to prepare 4-spiropiperidines involves the dialkylation of a suitable activated compound by mechlorethanamine. Thus, spiroindolinone **94**[49a] and spiroindolylpiperidine **95**[49b] were obtained in good yields (Scheme 42).

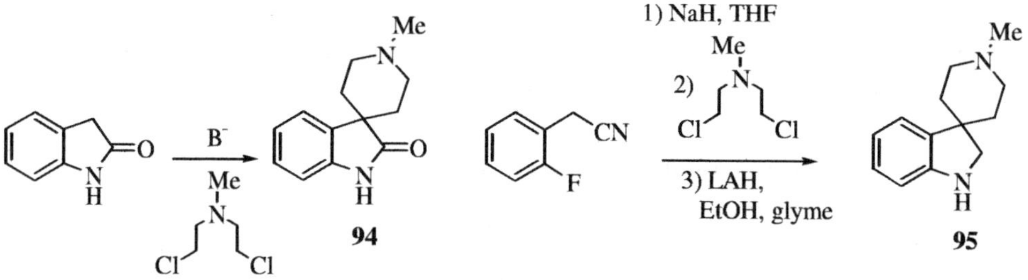

Scheme 42

3.3.3. Anionic cycloacylation of carbamate (*route IV-I*)

In this approach, a xanthen-9-yl anion was trapped by its carbamate side-chain situated at the same position.

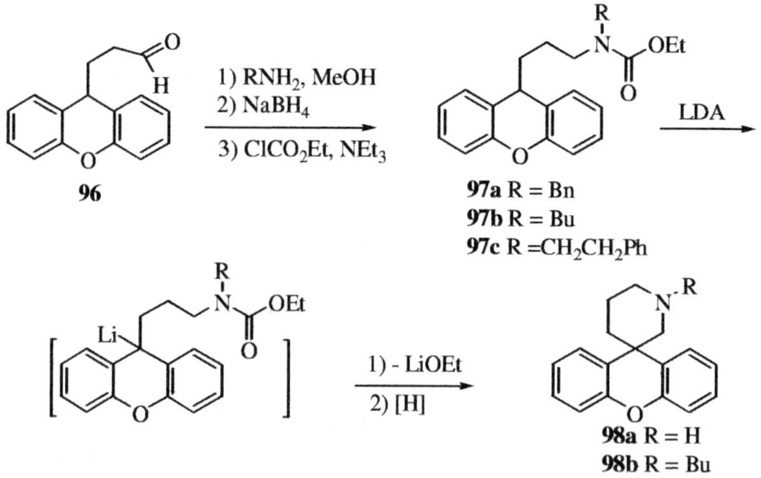

Scheme 43

Thus, carbamates **97** were easily prepared by reductive amination of **96** with primary amines/NaBH$_4$ in 60–76% yield. In the case of **97a** and **97b**, deprotonation with one equivalent of LDA afforded, after cyclization and reduction of the carbonyl group, the spiropiperidines **98a,b** in 55–85% yield. In the case of **97c**, two equivalents of LDA were necessary to deprotect both benzylic positions leading, after cyclization and subsequent β-elimination, to the secondary lactam which was isolated and further reduced into spirolactam **98a** in good yield (Scheme 43).[50]

3.4. Multi-component reaction

Multi-component reactions (MCRs) are procedures allowing the formation of complex molecules with high efficiency and stereoselectivity. For instance, functionalized spiroquinolines **99** were obtained through a one-pot three component cascade reaction between alkynoyls **100**, aldehydes **101** and aniline derivatives **102** using a Pt^{2+} (5 mol% of [PtMe$_2$(cod)] (cod=1,5-cyclooctadiene) and HBF$_4$) as the catalyst (Scheme 41).[51] Under these conditions, spiro[furan-2,4'-quinoline] derivatives **103** were synthezised in 72–87% yield in the most cases with a dr of 1:1 except for **103a–c** which were isolated as a single diastereomer (Scheme 44).

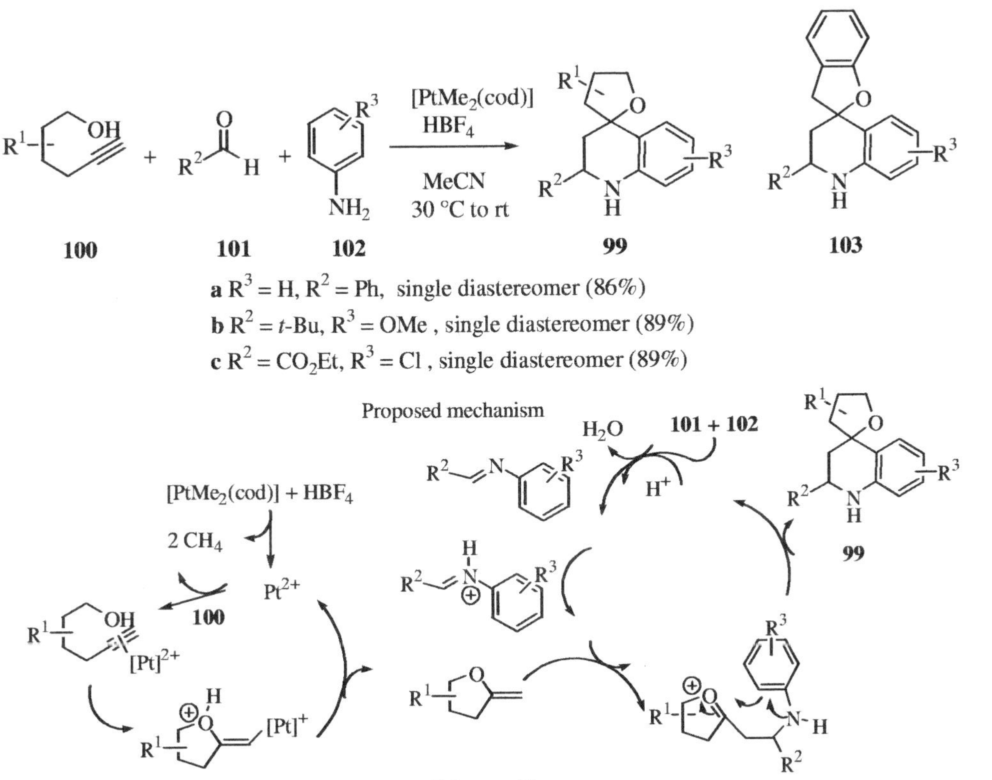

100   **101**   **102**   **99**   **103**

a R^3 = H, R^2 = Ph, single diastereomer (86%)

b R^2 = t-Bu, R^3 = OMe , single diastereomer (89%)

c R^2 = CO$_2$Et, R^3 = Cl , single diastereomer (89%)

Scheme 44

A library of spirodiketopiperazines **104** as potent CCR5 antagonists with anti-HIV activity, was synthesized on the basis of a solid phase Ugi multi-component reaction.[52] Thus, reaction of isonitrile resin with amines, 4-piperidones and carboxylic acids led to intermediate **105** which underwent a deprotection and a final cyclization-cleavage to give the spiro **104** in high purity (95.9–100%) and good yields (72–97%) (Scheme 45).

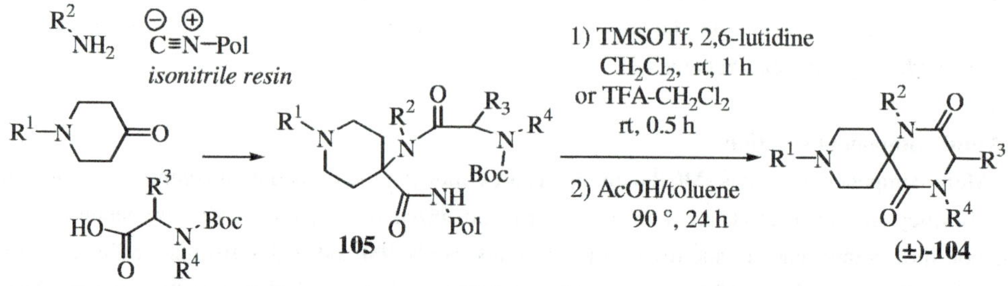

Scheme 45

4. Conclusion

Because of their diversified activities (antibacterial, antiviral, antigastrin, anticonvulsant...), 3- and 4-spiropiperidines were considered as important pharmacophores in medicinal chemistry and this has encouraged many chemists to develop a lot of synthetic approaches to obtain them.

References

1. (a) Spivak, C. E.; Maleque, M. A.; Oliveira, A. C.; Masukawa, L. M.; Tokuyama, T.; Daly, J. W.; Albuquerque, E. X. *Mol. Pharmacol.* **1982**, *21*, 351–361. (b) Daly, J. W.; Nishizawa, Y.; Edwards, J. A.; Waters, J. A.; Aronstam, R. S. *Neurochem. Res.* **1991**, *16*, 489–500. (c) Daly, J. W. *J. Nat. Prod.* **1998**, *61*, 162–172.

2. Kuramoto, M.; Tong, C.; Yamada, K.; Chiba, T.; Hayashi, Y.; Uemura, D. *Tetrahedron Lett.* **1996**, *37*, 3867–3870.

3. Chou, Y.; Kuramoto, M.; Otani, Y.; Shikano, M.; Yazawa, K.; Uemura, D. *Tetrahedron Lett.* **1996**, *37*, 3871–3874.

4. Nilsson, B. L.; Overman, L. E.; Read de Alaniz, J.; Rohde, J. M. *J. Am. Chem. Soc.* **2008**, *130*, 11297–11299, and references cited therein.

5. Kobayashi, J.; Watanabe, D.; Kawasaki, N.; Tsuda, M. *J. Org. Chem.* **1997**, *62*, 9236–9239. (b) Kobayashi, J.; Tsuda, M.; Ishibashi, M. *Pure Appl. Chem.* **1999**, *71*, 1123–1126.

6. Zheng, L. T.; Hwang, J.; Ock, J.; Lee, M. G.; Lee, W.-H.; Suk, K. *J. Neurochem.* **2008**, *107*, 1225–1235.

7. (a) Patchett, A. A.; Nargund, R. P.; Tata, J. R.; Chen, M.-H.; Barakat, K. J.; Johnston, D. B. R.; Cheng, K.; Chan, W. W.-S.; Butler, B.; Hickey, G.; Jacks, T.; Schleim, K.; Pong, S.-S.; Chaung, L.-Y. P.; Chen, H. Y.; Frazier, E.; Leung, K. H.; Chiu, S.-H. L.; Smith, R. G. *Proc. Natl. Acad. Sci. U.S.A.* **1995**, *92*, 7001–7005. (b) Palucki, B. L.; Feighner, S. D.; Pong, S.-S.; McKee, K. K.; Hreniuk, D. L.; Tan, C.; Howard, A. D.; Van der Ploeg, L. H. Y.; Patchett, A. A.; Nargund, R. P. *Bioorg. Med. Chem. Lett.* **2001**, *11*, 1955–1957. (c) Caldwell, J. P.; Matasi, J. J.; Zhang, H.; Fawzi, A.; Tulshian, D. B. *Bioorg. Med. Chem. Lett.* **2007**, *17*, 2281–2284.

8. (a) Maier, C. A.; Wünsch, B. *J. Med. Chem.* **2002**, *45*, 438–448. (b) Moltzen, E. K.; Perregaard, J.; Meier, E. *J. Med. Chem.* **1995**, *38*, 2009–2017.

9. Sinclair, A.; Stockman, R. A. *Nat. Prod. Rep.* **2007**, *24*, 298–326.

10. Dake, G. *Tetrahedron* **2006**, *62*, 3467–3492.

11. Deyine, A.; Poirier, J.-M.; Duhamel, L.; Duhamel, P. *Tetrahedron Lett.* **2005**, *46*, 2491–2493.

12. Nagata, T.; Nishida, A.; Nakagawa, M. *Tetrahedron Lett.* **2001**, *42*, 8345–8349.

13. Brimble, M. A.; Trzoss, M. *Tetrahedron* **2004**, *60*, 5613–5622.
14. Grieco, P. A.; Kaufman, M. D. *J. Org. Chem.* **1999**, *64*, 6041–6048.
15. Hughes, R. C.; Dvorak, C. A.; Meyers, A. I. *J. Org. Chem.* **2001**, *66*, 5545–5551.
16. Kulagowski, J. J.; Curtis, N. R.; Swain, C. J.; Williams, B. J. *Org. Lett.* **2001**, *3*, 667–670.
17. Minnihan, E. C.; Colletti, S. L.; Toste, F. D.; Shen, H. C. *J. Org. Chem.* **2007**, *72*, 6287–6289.
18. Harrison, T. J.; Patrick, B. O.; Dake, G. R. *Org. Lett.* **2007**, *9*, 367–370.
19. Pfefferkorn, J. A.; Choi, C. *Tetrahedron Lett.* **2008**, *49*, 4372–4373.
20. Satyanarayana, G.; Maier, M. E. *J. Org. Chem.* **2008**, *73*, 5410–5415.
21. Guazzelli, G.; Duffy, L. A.; Procter, D. J. *Org. Lett.* **2008**, *10*, 4291–4294.
22. François, D.; Lallemand, M.-C.; Selkti, M.; Tomas, A.; Kunesch, N.; Husson, H.-P. *Angew. Chem. Int. Ed.* **1998**, *37*, 104–105.
23. Poupon, E.; François, D.; Kunesch, N.; Husson, H.-P. *J. Org. Chem.* **2004**, *69*, 3836–3841.
24. Ahrendt, K. A.; Williams, R. M. *Org. Lett.* **2004**, *6*, 4539–4541.
25. (a) Yang, J.; Cohn, S. T.; Romo, D. *Org. Lett.* **2000**, *2*, 763–766. (b) Kong, K.; Moussa, Z.; Romo, D. *Org. Lett.* **2005**, *7*, 5127–5130.
26. (a) Tsujimoto, T.; Ishihara, J.; Horie, M.; Murai, A. *Synlett* **2002**, 399–402. (b) Ishihara, J.; Miyakawa, J.; Tsujimoto, T.; Murai, A. *Synlett* **1997**, 1417–1419.
27. Johannes, J. W.; Wenglowsky, S.; Kishi, Y. *Org. Lett.* **2005**, *7*, 3997–4000.
28. Uenoyama, Y.; Fukuyama, T.; Ryu, I. *Org. Lett.* **2007**, *9*, 935–937.
29. Koreeda, M.; Wang, Y.; Zhang, L. *Org. Lett.* **2002**, *4*, 3329–3332.
30. Faure, S.; Piva-Le Blanc, S.; Piva, O. *Tetrahedron Lett.* **1999**, *40*, 6001–6004.
31. Nilson, M. G.; Funk, R. L. *Org. Lett.* **2006**, *8*, 3833–3836.
32. Pearson, A. J.; Wang, X. *Tetrahedron Lett.* **2002**, *43*, 7513–7515.
33. Shen, M.; Li, C. *J. Org. Chem.* **2004**, *69*, 7906–7909.
34. Khan, F. A.; Dash, J. *J. Org. Chem.* **2003**, *68*, 4556–4559.
35. (a) Bender, C. F.; Widenhoefer, R. A. *Org. Lett.* **2006**, *8*, 5303–5305. (b) Bender, C. F.; Widenhoefer, R. A. *Chem. Commun.* **2008**, 2741–2743, and references cited therein.
36. Maligres, P. E.; Houpis, I.; Rossen, K.; Molina, A.; Sager, J.; Upadhyay, V.; Wells, K. M.; Reamer, R. A.; Lynch, J. E.; Askin, D.; Volante, R. P.; Reider, P. J.; Houghton, P. *Tetrahedron* **1997**, *53*, 10983–10992.
37. Cheng, Y.; Chapman, K. T. *Tetrahedron Lett.* **1997**, *38*, 1497–1500.
38. Cossy, J.; Poitevin, C.; Gomez Pardo, D.; Peglion, J. –L.; Dessinges, A. *J. Org. Chem.* **1998**, *63*, 4554–4557.
39. (a) Bignan, G. C.; Battista, K.; Connolly, P. J.; Orsini, M. J.; Liu, J.; Middleton, S. A.; Reitz, A. B. *Biorg. Med. Chem. Lett.* **2005**, *15*, 5022–5026. (b) Shaughnessy, K. H.; Hamann, B. C.; Hartwig, J. F. *J. Org. Chem.* **1998**, *63*, 6546–6553.
40. Costanzo, M. J.; Yabut, S. C.; Zhang, H.-C.; White, K. B.; de Garavilla, L.; Wang, Y.; Minor, L. K.; Tounge, B. A.; Barnakov, A. N.; Lewandowski, F.; Milligan, C.; Spurlino, J. C.; Abraham, W. M.; Boswell-Smith, V.; Page, C. P.; Maryanoff, B. E. *Bioorg. Med. Chem. Lett.* **2008**, *18*, 2114–2121.
41. (a) Hey, D. H.; Jones, G. H.; Perkins, M. J. *J. Chem. Soc. C* **1971**, 116–122. (b) Horii, Z.; Uchida, S.; Nakashita, Y.; Tsuchida, E.; Iwata, C. *Chem. Pharm. Bull.* **1974**, *22*, 583–586.
42. Gonzales-Lopez de Turiso, F.; Curran, D. P. *Org. Lett.* **2005**, *7*, 151–154.
43. Ibarra-Rivera, T. R.; Gámez-Montaño, R.; Miranda, L. D. *Chem. Commun.* **2007**, 3485–3487.
44. (a) Goebel, T.; Ulmer, D.; Projahn, H.; Kloeckner, J.; Heller, E.; Glaser, M.; Ponte-Sucre, A.; Specht, S.; Ramadam Sarite, S.; Hoerauf, A.; Kaiser, A.; Hauber, I.; Hauber, J.; Holzgrabe, U. *J. Med. Chem.* **2008**, *51*, 238–250. (b) Rowbottom, M. W.; Vickers, T. D.; Tamiya, J.; Zhang, M.; Dyck, B.; Grey, J.; Schwarz, D.; Heise, C. E.; Hedrick, M.; Wen, J.; Tang, H.; Wang, H.; Fisher, A.; Aparicio, A.; Saunders, J.; Goodfellow, V. S. *Bioorg. Med. Chem. Lett.* **2007**, *17*, 2171–2178. (c) Pinard, E.; Ceccarelli, S. M.; Stalder, H.; Alberati, D. *Bioorg. Med. Chem. Lett.* **2006**, *16*, 349–353. (d) Nieto, M. J.; Philip, A. E.; Poupaert, J. H.; McCurdy, C. R. *J. Comb. Chem.* **2005**, *7*, 258–263.

45. (a) Ramezanpour, S.; Hashtroudi, M. S.; Bijanzadeh, H. R.; Balalaie, S. *Tetrahedron Lett.* **2008**, *49*, 3980–3982. (b) Yang, H.; Lin, X.-F.; Padilla, F.; Rotstein, D. M. *Tetrahedron Lett.* **2008**, *49*, 6371–6374.
46. Hansen, J. D.; Newhouse, B. J.; Allen, S.; Anderson, A.; Eary, T.; Schiro, J.; Gaudino, J.; Laird, E.; Allen, A. C.; Chantry, D.; Eberhardt, C.; Burgess, L. E. *Tetrahedron Lett.* **2006**, *47*, 69–72.
47. Efange, S. M. N.; Kamath, A. P.; Khare, A. B.; Kung, M.-P.; Mach, R. H.; Parsons, S. M. *J. Med. Chem.* **1997**, *40*, 3905–3914.
48. (a) Kouznetsov, V. V.; Díaz, B. P.; Sanabria, C. M. M.; Vargas Méndez, L. Y.; Poveda, J. C.; Stashenko, E. E.; Bahsas, A.; Amaro-Luis, J. *Lett. Org. Chem.* **2005**, *2*, 29–32. (b) Vargas Méndez, L. Y.; Kouznetsov, V. V. *Tetrahedron Lett.* **2007**, *48*, 2509–2512.
49. (a) Goehring, R. R. *Org. Prep. Proceed. Int.* **1995**, *27*, 691–694. (b) Xie, J.-S.; Huang, C. Q.; Fang, Y.-Y.; Zhu, Y.-F. *Tetrahedron* **2004**, *60*, 4875–4878.
50. Quintás D.; Garciá, A.; Domínguez, D. *Tetrahedron Lett.* **2003**, *44*, 9291–9294.
51. Barluenga, J.; Mendoza, A.; Rodríguez, F.; Fañanás, F. J. *Angew. Chem. Int. Ed.* **2008**, *47*, 7044–7047.
52. Habashita, H.; Kokubo, M.; Hamano, S.-I.; Hamanaka, N.; Toda, M.; Shibayama, S.; Tada, H.; Sagawa, K.; Fukushima, D.; Maeda, K.; Mitsuya, H. *J. Med. Chem.* **2006**, *49*, 4140–4152.

MOST RELEVANT RECENT ENANTIOSELECTIVE SYNTHESIS OF PYRROLIDENES AND PIPERIDINES

Xavier Companyó, Andrea-Nekane Alba and Ramon Rios*

Departament de Química Orgànica, Universitat de Barcelona
Martí i Franqués 1-11, E-08028 Barcelona, Spain (e-mail: rios.ramon@icrea.cat)

Abstract. *In the past few years, much attention has been paid to the enantioselective synthesis of pyrrolidines and piperidines, given that they are extensively used as catalysts, ligands and constitute important structural motifs in medicinal chemistry. This manuscript has the aim to cover the principal organometallic and organocatalytic methodologies to synthesize them.*

Contents

1. Introduction
2. Enantioselective synthesis of piperidines
    2.1. Metal catalyzed approaches
    2.2. Organocatalytic approaches
3. Enantioselective synthesis of pyrrolidines
    3.1. Metal catalyzed approaches
    3.2. Organocatalytic approaches
4. Conclusions
References

1. Introduction

Pyrrolidines (**1**) and piperidines (**2**) are common substructures in many natural products and pharmacologically active compounds (Scheme 1). Both structures are nitrogen containing heterocycles and the great importance of these compounds is clearly stated by the efforts devoted by chemical community in the quest for new and highly effective methodologies that allow building of these privileged structures in an enantioselective fashion.

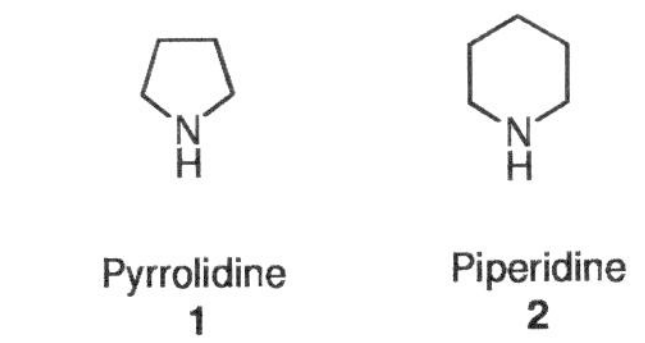

Pyrrolidine
1

Piperidine
2

Scheme 1. Nitrogen heterocyclic rings.

Quinolidines, proline derivatives, pipecolic acid derivatives, non-natural aminoacids, azasugars, and more, are only some few examples of the huge relevancy of these structures. From organometallic chemistry to organocatalysis, several methods have been developed for the enantioselective synthesis of these compounds. Some representative examples are illustrated in Scheme 2. CYB3 (**3**) is found in the seeds and leaves of Castanospermum austral,[1] Swainsonine (**4**) was isolated from Swainsona canescens in 1979,[2] Adenophorine (**5**) was discovered in plants in 2000,[3] (+)-241D (**6**) is a dendrobate alkaloid.[4]

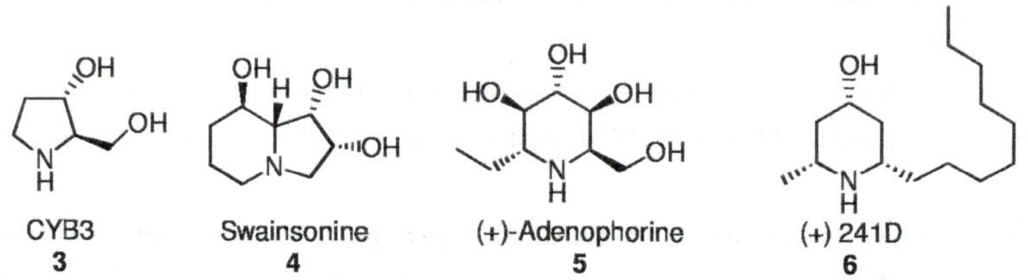

Scheme 2. Examples of natural products containing pyrrolidine or piperidine rings.

2. Enantioselective synthesis of piperidines

Piperidines (**2**) and their derivatives are prominent structural units of alkaloids and other natural compounds and are important intermediates in the synthesis of numerous bioactive pharmaceuticals, which often display interesting biological activities.[4] The stereoselective synthesis of piperidine derivatives is therefore important for organic and medicinal chemistry. Paroxetine (**7**), Nojirimicin (**8**) or Laccarin (**9**) (among others) have attracted much attention from the chemical community, which has developed a large number of enantioselective synthesis of them (Scheme 3).

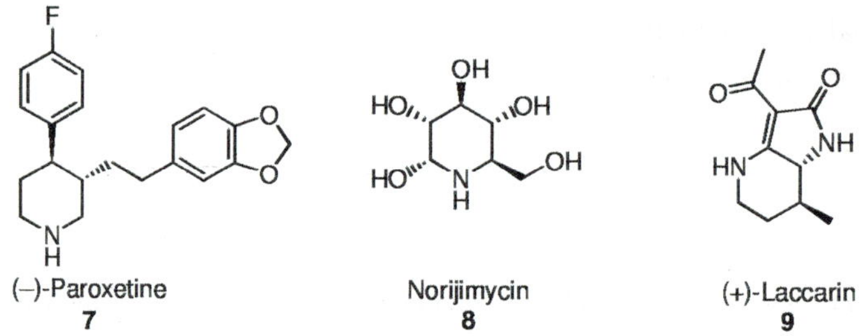

Scheme 3. Different biological active piperidines.

Piperidines (**2**) represent a class of heterocycles frequently associated with biologically active natural products and often embedded within scaffolds recognized as privileged by medicinal chemists.[5]

2.1. Metal catalyzed approaches

As discussed in the introduction, piperidines (**2**) constitute an interesting target for many chemists. The chemical community has devoted many efforts for the enantioselective synthesis of piperidines (**2**). One of the most common enantioselective approaches is the use of organometallic methodologies.

One of the most efficient enantioselective approaches was developed by Gallacher and co-workers. They have outlined a methodology based on the reactivity of 1,2- and 1,3-cyclic sulfamidates towards nucleophiles that provide a versatile approach to a range of enantiomerically pure *N*-heterocycles, including piperidines (**2**) and piperidones.[6]

Based in this strategy, Gallagher described the synthesis of (–)-paroxetine (**7**) and (+)-laccarin (**9**).[7] The approach to the synthesis of (–)-paroxetine (**7**) is shown in Scheme 4, being the key feature the intermediacy of the $C(3)$-aryl substituted 1,3-cyclic sulfamidate **14** and the high degree of stereochemical control exercised during the reaction of **14** with the enolate of dimethyl malonate (**15**). Commercially available keto-ester **10** was reduced using a [Ru][Cl-MeO-BIPHEP] catalyst system (**I**) to give alcohol **11** in 95% yield and 97% ee. Aluminium mediated amidation, followed by amide reduction, gave amino alcohol **13** and the two step cyclic sulfamidate formation proceeded smoothly to give the key intermediate **14** in excellent overall yield. After a subsequent nucleophilic displacement, the crude product was immediately subjected to mild acid hydrolysis and, afterwards, thermolysis, which gave lactam **16** as essentially a single diastereomer in 70% yield. Reduction and N-debenzylation gave (–)-paroxetine (**7**) in 52% overall yield (Scheme 4).

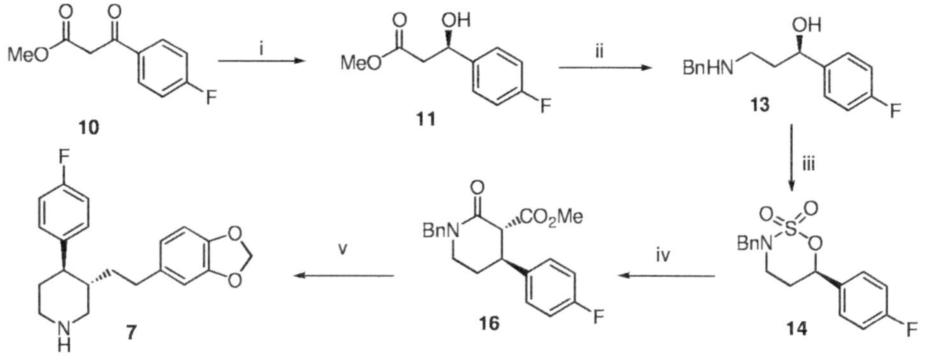

Reagents and conditions: i, [((S)-Cl-MeO-BIPHEP)Ru-(cymene)Cl]Cl (**I**) CH$_2$Cl$_2$ (0.5 mol%), H$_2$ (8 bar), MeOH, 60 °C (95%); ii, AlMe$_3$, BnNH$_2$ (**12**), PhMe, 0 °C to rt (100%), LiAlH$_4$, THF, reflux (98%); iii, SOCl$_2$, Et$_3$N, imidazole, CH$_2$Cl$_2$, 220 °C to 0 °C (95%); RuCl$_3$ (0.25 mol%), NaIO$_4$, MeCN–H$_2$O, 0 °C (87%); iv, dimethyl malonate (**15**), NaH, DMF, 60 °C; then 5 M HCl; then PhMe, reflux (70%); v, LiAlH$_4$,THF, reflux; MsCl, Et$_3$N, CH$_2$Cl$_2$; NaH, DMF, 90 °C, 10% Pd/C (35%), H$_2$ (6 bar), i-PrOH, AcOH, 50 °C, then aq. HCl, i-PrOH.

Scheme 4. Synthesis of (–)-Paroxetine reported by Gallagher.

In order to synthesize (+)-laccarin (**9**), commercially available ethyl (3R)-hydroxybutyrate (**17**) was converted to N-benzylamide (**18**), followed by reduction and conversion of the resulting aminoalcohol to the target sulfamidate (**19**) in 62% overall yield. The sodium enolate of diethylmalonate (**20**) reacted efficiently with cyclic sulfamidate (**19**) and a subsequent ring cleavage afforded the resulting adduct (**21**). The amine was Boc-protected in order to inhibit premature lactam formation (Scheme 5).

Amination at $C(2)$ was achieved using monochloramine (**22**) under basic conditions to give **23** in 75% yield. N-Acylation using diketene and cyclization under Claisen conditions, followed by ester hydrolysis and decarboxylation using aqueous acid, afforded **25**. N-Boc cleavage using TFA and then neutralization and thermolisys gave N-benzyl laccarin (**26**). Hydrogenation of N-benzyl laccarin in acetic acid and presence of TFA afforded, in an efficient and chemoselective way, (+)-laccarin (**9**).

In 2007, Hoveyda and co-workers developed an enantioselective synthesis of piperidines **27** based in a catalytic asymmetric ring opening metathesis.[8] Azabicycle **28** reacts with molybdenum-based catalyst (**IV**) to afford 2,6-disubstituted piperidines **27** with excellent yields and enantioselectivities (Scheme 6).

The same year, Rao and co-workers developed an asymmetric synthesis of piperidines with ring closing methatesis (RCM) as the key step.[9] In this paper, Rao used an asymmetric reduction of α-azido aryl ketone and consequently oxidative cleavage of an aryl group, RCM, and dihydroxylation reactions in a highly stereoselective and efficient manner. This new synthetic route has been demonstrated to be efficient for the synthesis of 1-aza sugars related to hydroxy piperidines that are highly potent and specific inhibitors of β-glycosidases. Soon after, Krishna and co-workers developed a similar approach in the synthesis of (−)-andrachcinidine.[10]

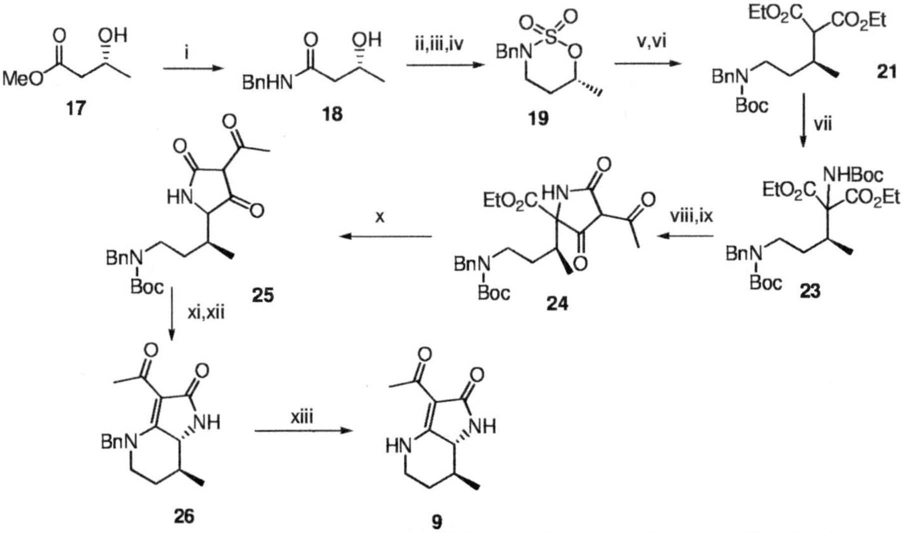

Reagents and conditions: i, AlMe$_3$, BnNH$_2$, PhMe, 0 °C to rt (88%); ii, LiAlH$_4$, THF, reflux (98%); iii, SOCl$_2$, Et$_3$N, imidazole, CH$_2$Cl$_2$, 220 °C to 0 °C (84%); iv, RuCl$_3$ (**II**) (0.25 mol%), NaIO$_4$, MeCN–H$_2$O, 0 °C (85%); v, diethyl malonate (**20**), NaH, DMF, 110 °C; then 5 M HCl; vi, Boc$_2$O, NaHCO$_3$, MeCN (81% from 3); vii, NH$_2$Cl (**22**) ca. 0.15 M in Et$_2$O, t-BuOK, THF, 0 °C to rt (75% + 23% recovered **13**); viii, diketene, cat. DMAP, THF, 60 °C; ix, NaOEt, EtOH, 60 °C; x, H$_2$O, then 5 M HCl; xi, TFA, CH$_2$Cl$_2$ then Et$_3$N; xii, PhMe, 80 °C (68% of **18a** and **18b** from **14**); xiii, 10% Pd/C (**III**) (65 wt%), H$_2$ (5.5 bar), TFA (93%).

Scheme 5. Sinthesis of Laccarin reported by Gallagher.

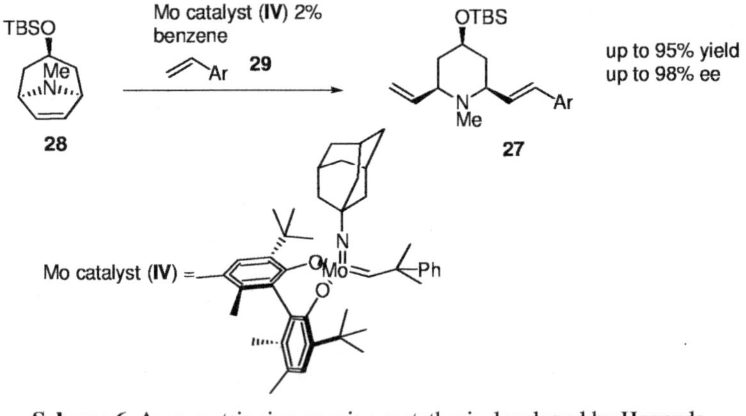

Scheme 6. Asymmetric ring opening metathesis developed by Hoveyda.

Helchem and co-workers developed in 2009 a new route for the enantioselective synthesis of piperidines based on an Ir-catalyzed allylic substitution as the key step.[11]

In order to show the power of this new methodology, the authors did the total synthesis of the dendrobate alkaloid (+)-241D (**6**) and (2*R*,4*S*,6*S*)-2-methyl-6-propylpiperidin-4-ol (**42**), a spruce alkaloid. The overall strategy is described in Scheme 7. Key steps are, as already mentioned, two asymmetric Ir-catalyzed allylic substitutions and an addition of Brown's chiral allylboron reagent Ipc$_2$B(allyl) to the intermediate aldehyde **33**. Given the availability of both enantiomers of all the chiral reagents, the configurational switch allows the preparation of each of eight stereoisomers with high selectivity.

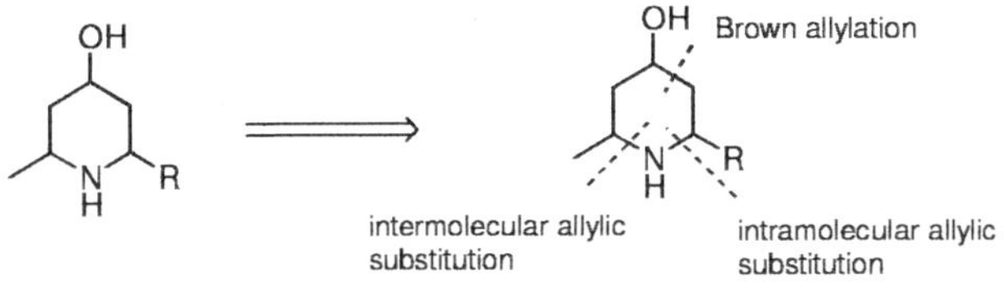

Scheme 7. Retrosynthetic analysis of 2,6-disubstituted hydroxypiperidines.

The synthesis started with the Ir-catalyzed amination of *trans*-crotyl methyl carbonate (**30**) with HN(CHO)Boc (**31**), an ammonia equivalent. This reaction was efficiently catalyzed by iridium catalyst **V**. Subsequent steps of deprotection, hydroboration/oxidation and Swern oxidation furnished aldehyde **33** in 89% yield (Scheme 8).

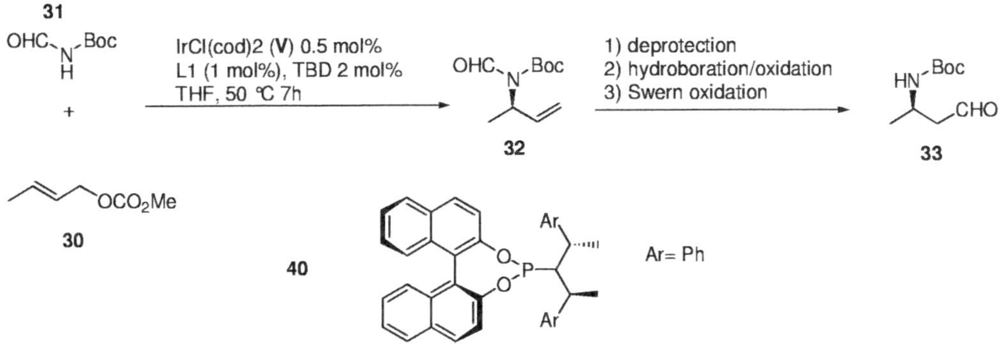

Scheme 8. Synthesis of **34** reported by Helmchen.

Then, **33** was treated with (+)-*B*-allyldisiopinocampheylborane (**34**) at low temperature. Finally, chain prolongation was carried out by cross metathesis (Grubbs' II catalyst **VI**) of the homoallylic alcohol **35** and biscarbonate **36**. Allylic carbonate **37** was produced in 87% yield as a 9:1 mixture of *E/Z* isomers. Cleavage of the Boc group under standard conditions gave the cyclization precursor **39** in 96% yield.

The Ir-catalyzed cyclization of amine **39** proceeded smoothly under standard conditions. The reaction was efficiently catalyzed by **41** as ligand, obtaining the piperidine **39** in 90% yield and total diastereoselectivity. It is noteworthy that only by choosing the opposite enantiomer of the ligand, the obtained diastereomer was the opposite (Scheme 9).

Next, compound **39** was modified to achieve (+)-241D (**6**) and (2*R*,4*S*,6*S*)-2-methyl-6-propylpiperidin-4-ol (**42**) (Scheme 10).

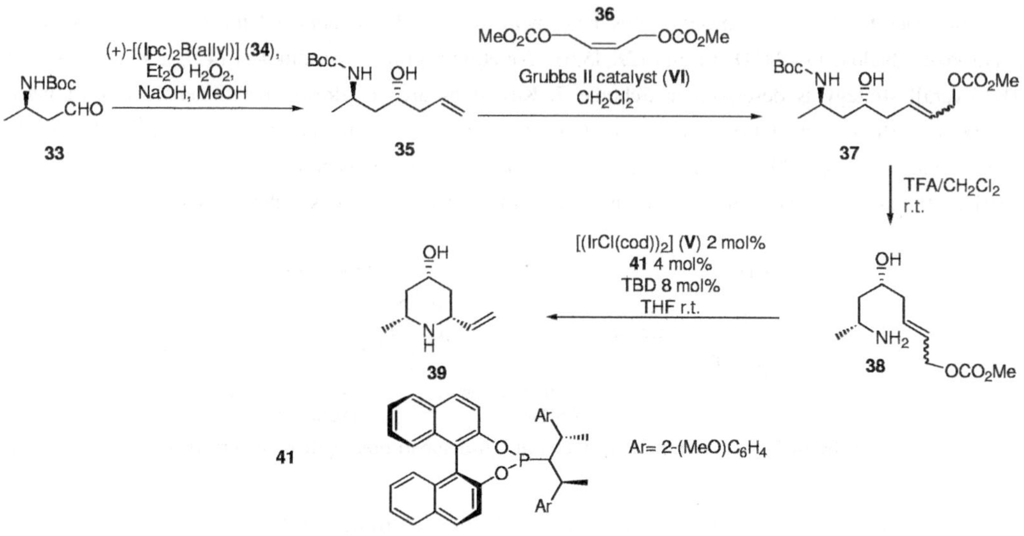

Scheme 9. Synthesis of compound 39.

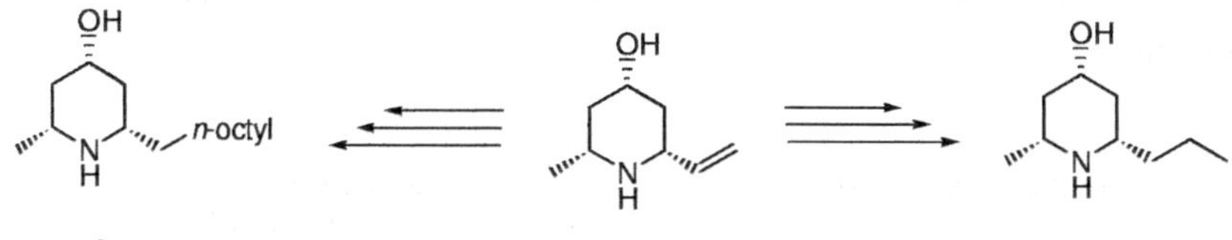

Scheme 10. Synthesis of (+)-241D (6) and Spruce alkaloid (41).

Very recently, Carretero and co-workers have described an elegant methodology which affords piperidine derivatives **43** *via* an inverse-electron-demand Diels-Alder.[12] *N*-(Heteroarylsulfonyl)-1-aza-1,3-dienes **44** react with ethyl vinyl ether (**45**) and this process is catalyzed by simple nickel salts (**VII**). The corresponding piperidines are obtained in high yields and enantioselectivities and with almost pure *endo* diastereoselectivity, as illustrated in Scheme 11.

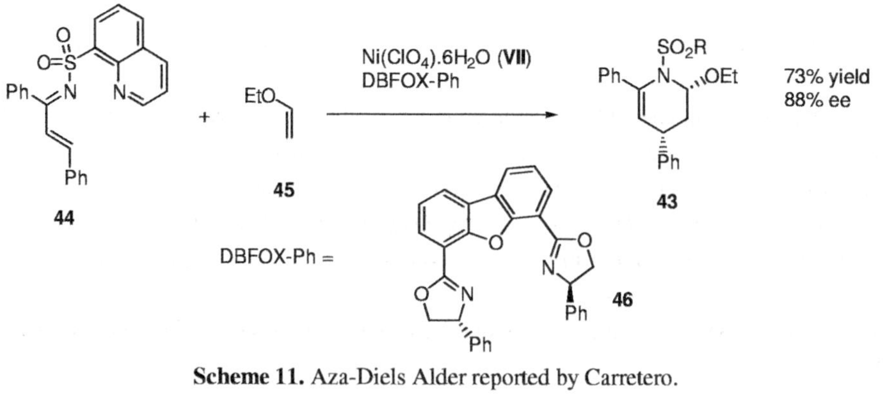

Scheme 11. Aza-Diels Alder reported by Carretero.

2.2. Organocatalytic approaches

The first researchers to synthesize a piperidine derivative through an organocatalytic methodology were Akiyama and co-workers[13] in 2006. They developed an enantioselective *aza*-Diels-alder reaction between Brassard's diene (1,3-dimethoxy-1-(trimethylsiloxy)butadiene, **47**) and aldimines **48** catalyzed by a chiral Brønsted acid **VIII** to afford piperidinones **49** (Scheme 12).

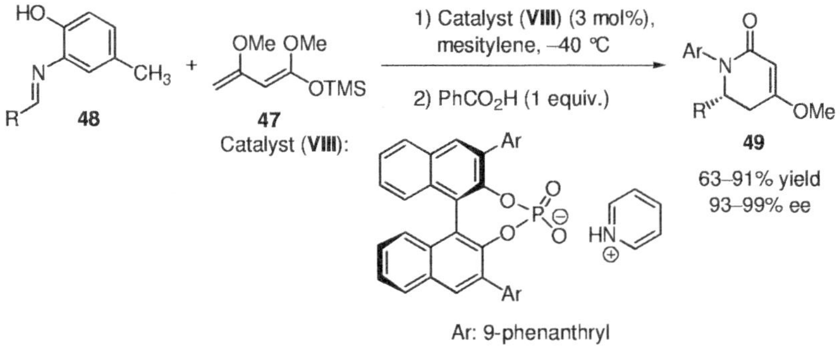

Scheme 12. Synthesis of piperidinone derivatives **49** reported by Akiyama.

Akiyama used the pyridinium salt of the chiral phosphoric acid derived from Binol (**VIII**) at low catalyst loadings (3 mol%) to obtain a wide set of functionalized piperidinones **49** in good yields (63–91%) and excellent enantioselectivities (93–99% ee). The improved results obtained with the catalyst in its salt form (**VIII**), instead of as free acid, can be explained due to its weaker acidity combined with the high reactivity and lability of Brassard's diene (**47**). When the reaction was carried out with the catalyst as free acid, although the high enantiocontrol was maintained, the cycloadduct yield decreased significantly.

It is noteworthy that the presence of the hydroxyl moiety on the *N*-aryl group is essential for attaining the high enantiocontrol. When the reaction was tested with an aldimine derived from *p*-methoxyaniline, the final product was obtained in low optical purity. This observation suggests a nine-membered transition state, wherein the phosphoryl oxygen atom forms a hydrogen bond with the hydrogen atom of the imine's OH moiety, allowing the preferential attack of the nucleophile by the *Re* face of the aldimine (Scheme 13).

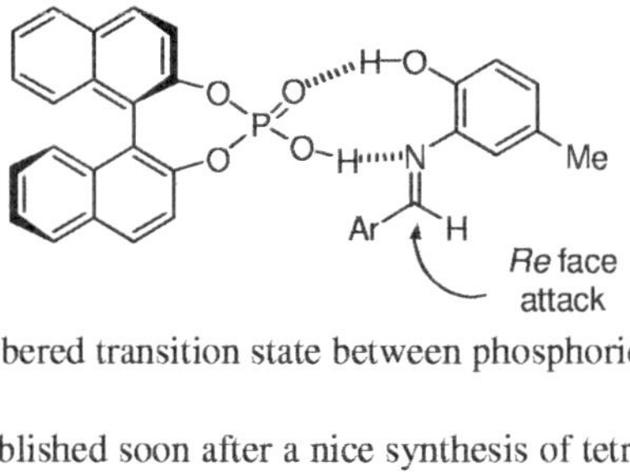

Scheme 13. Nine-membered transition state between phosphoric acid and an aldimine.

The same research group published soon after a nice synthesis of tetrahydroquinolidines **50** through an inverse-demand aza Diels-Alder reaction of aldimines **48** (azabutadienes) and enol ethers **45** (which act as electron-rich alkenes).[14] The process is efficiently catalyzed by the free phosphoric acid derived from BINOL (**IX**), affording schalemic tetrahydroquinolidines **50** with high to excellent enantioselectivities (87–96% ee) and in almost total *cis* diastereoselectivity (96:4–99:1 dr).

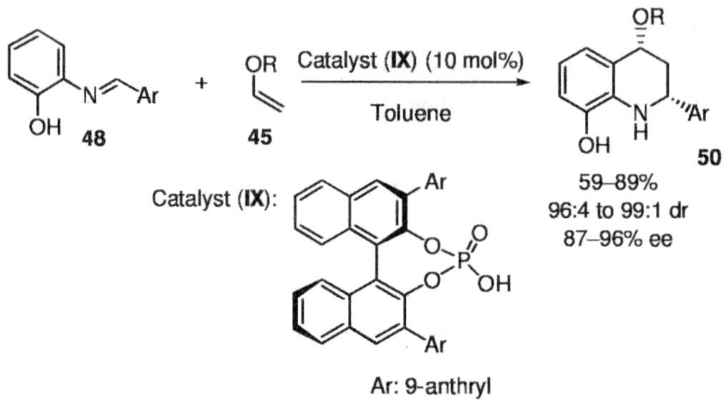

Scheme 14. Synthesis of tetrahydroquinoldinines **50** through an inverse-demand aza Diels-Alder.

In this reaction, aromatic and heteroaromatic aldimines **48** as well as different vinyl ethers **45** proved to be good substrates. The reaction even worked with cyclic vinyl ethers such as dihydrofuran and dihydropyran affording the tricycloadducts with excellent results.

Nearly at same time, Bode and co-workers[15] performed the first [NHC(N-Heterocyclic Carbene)]-catalyzed aza Diels-Alder reaction, allowing the obtaining of *cis*-3,4-disubstituted dihydropiridinones **51**.

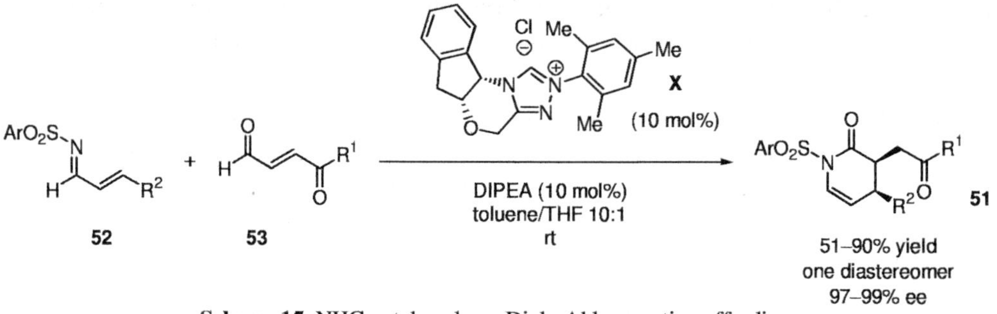

Scheme 15. NHC-catalyzed aza-Diels-Alder reaction affording
cis-3,4-disubstituted dihydropyridinones (**51**).

This reaction between N-sulfonyl-α,β-unsaturated imines (**52**) and β-activated enals (**53**) is efficiently catalyzed by a novel chiral triazolium salt (**X**).

As shown in Scheme 16, the enal (**53**) undergoes a nucleophilic addition of the carbene catalyst (**X**), forming the Breslow intermediate **54** or its homoenolate resonance structure **55**. Then, protonation of **55** affords the catalyst-bounded enol or enolate **56**, an exceptionally reactive dienophile which undergoes LUMO$_{\text{diene}}$-controlled Diels-Alder with the imine partner **52**, furnishing the dihydropyridinone derivatives **51** in excellent diastereo- and enantioselectivities. That exceptional diastereoselectivity can be rationalized by the high performance for an *endo* transition state and in the NHC-catalyzed system. This reaction mode is reinforced by the presence of the bulky triazolium moiety in the active dienophile **56** (Scheme 17). In addition, the *cis*-stereoselectivity would arise from a (Z)-enolate **56** reacting as the dienophile.

In 2007, Terada and co-workers[16] developed a domino aza-ene type reaction/cyclization cascade for one-pot entry to enantioenriched piperidines **57**.

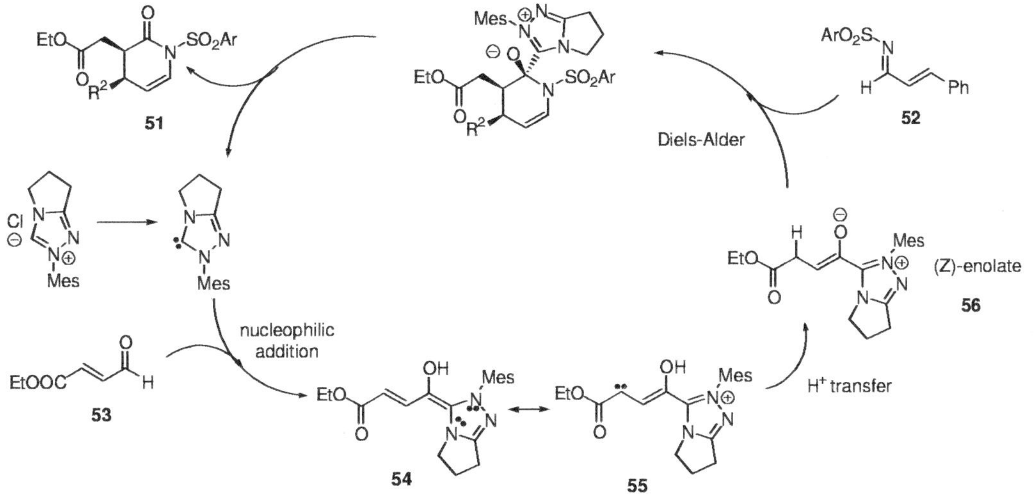

Scheme 16. Postulated catalytic cycle.

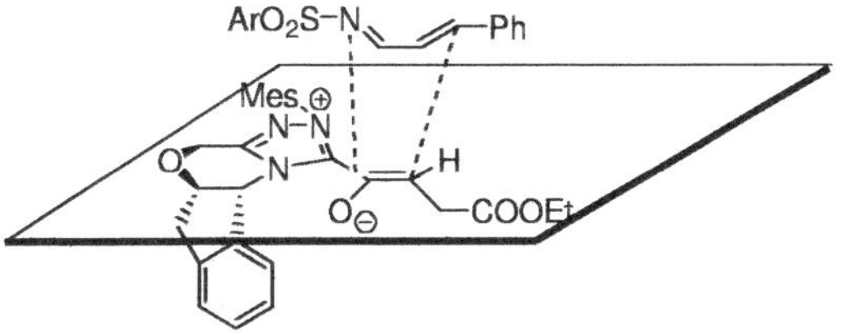

Scheme 17. Stereochemical model for *endo*-Diels-Alder cycloaddition.

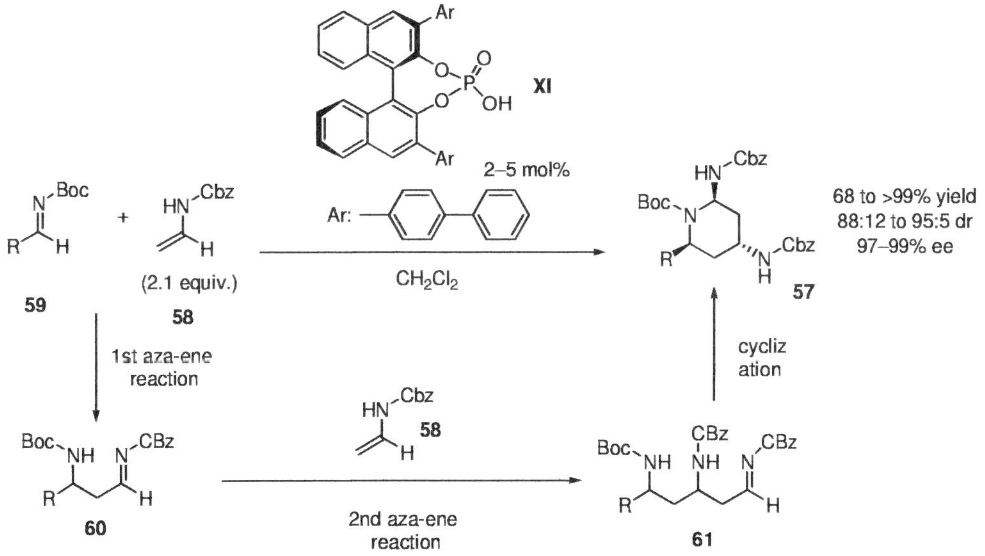

Scheme 18. Domino aza-ene type reaction/cyclization cascade for the construction
of enantioenriched piperidines (**57**).

The potential of such cascade transformations catalyzed by binol-derived phosphoric acid (**XI**) was highlighted through their ability to achieve a rapid increment in molecular complexity from simple enecarbamates **58** and a broad range of aldimines **59** (aromatic and aliphatic), affording the desired piperidine derivatives **57** with three new stereogenic centres in a highly diastereo (88:12–95:5 dr) and enantioselective (97–99% ee) manner.

Later on, Córdova and co-workers[17] reported a highly enantioselective synthesis of 1,2-dihydroquinolidine-3-carbaldehydes (**62**) through a domino aza-Michael/Aldol reaction between 2-amino-benzaldehydes **63** and enals **64** catalyzed by commercially available diphenylprolinol TMS protected ether (**XII**).

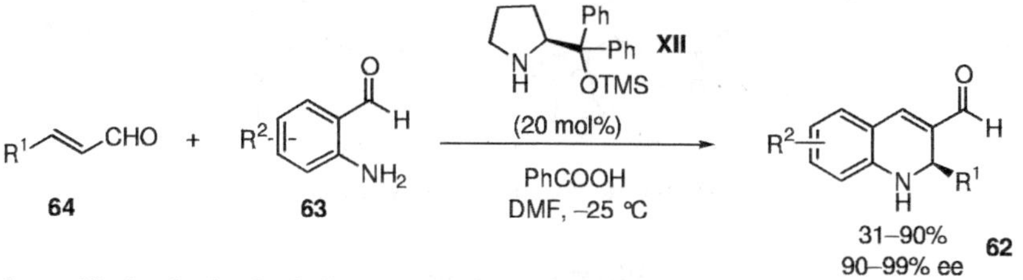

Scheme 19. Synthesis of 1,2-dihydroquinolidines (**62**) reported by Córdova and coworkers.

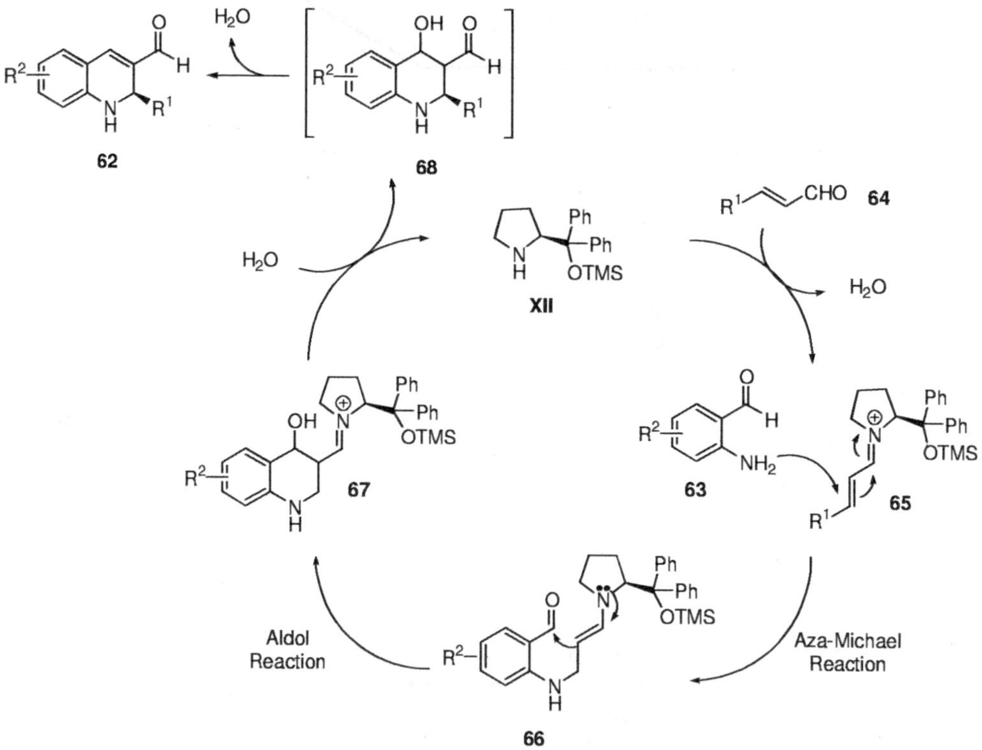

Scheme 20. Proposed catalytic cycle for the enantioselective formation
of 1,2-dihydroquinolidine-3-carbaldehydes (**62**).

The development of a conjugate addition of an amine to an electron-deficient α,β-unsaturated system represented an unprecedented organocatalytic process since, generally, an amine is a much weaker nucleophile than a thiol or an alcohol. Actually, this reaction represented the first asymmetric organocatalytic aza-Michael reaction of primary amines with α,β-unsaturated aldehydes.

Thus, a wide set of 1,2-dihydroquinolidines (**62**) can be synthesized through this domino sequence starting from both aromatic and aliphatic enals (**64**) as well as different substituted 2-aminobenzaldehydes (**63**), affording six-membered nitrogenated compounds with moderated to good yields (31–90%) and excellent enantioselectivities (90–99% ee) (Scheme 20).

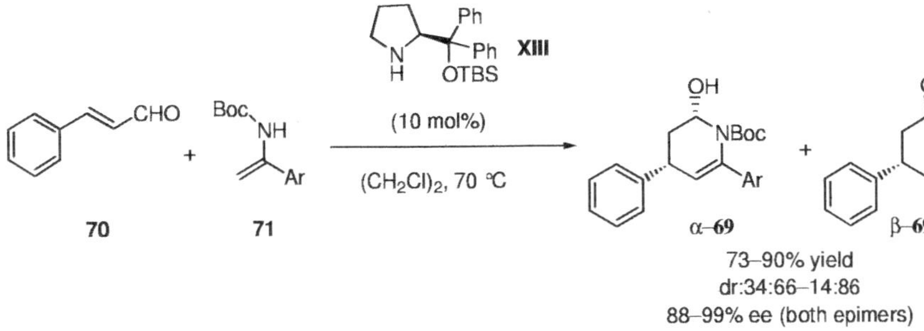

Scheme 21. Synthesis of piperidines (**69**) reported by Hayashi and co-workers.

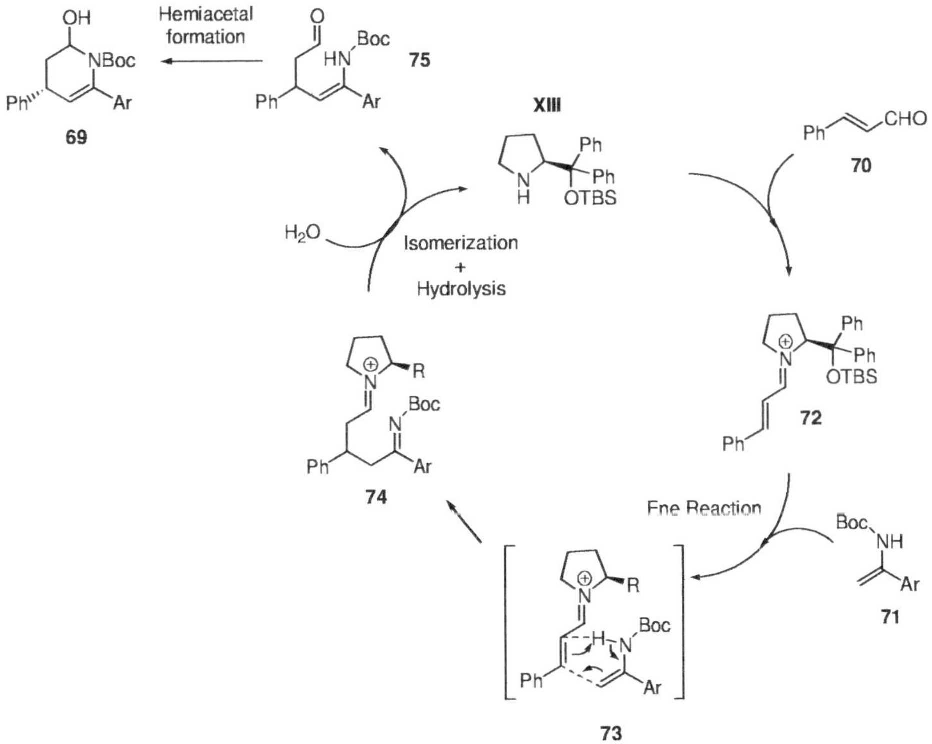

Scheme 22. A plausible reaction mechanism.

157

Some months later, Wang and co-workers[18] reported the same domino sequence employing 2-*N*-protected-aminobenzaldehydes in basic media, rendering also good results.

Hayashi and co-workers[19] published in 2008 a formal aza [3+3]-cycloaddition reaction for the formation of enantioenriched piperidines **69**. Serendipitously, they discovered that when cinnamaldehyde (**70**) and *N*-(1-phenylvinyl)acetamide (**71**) were reacted under the effect of a catalytic amount of diphenylprolinol TBS ether (**XIII**), the piperidine derivative **69** was furnished in good yields and excellent enantioselectivities, albeit in moderated diastereoselectivities, as depicted in Scheme 21.

This method consists on four consecutive reactions involving an asymmetric ene reaction to form the acylimine **74**, an isomerization to enamine (**75**), the catalyst hydrolysis and, finally, the hemiacetal cyclization to afford **69** (Scheme 22).

Notably, this methodology represents one of the few successful examples of asymmetric, catalytic ene reaction of α,β-unsaturated aldehydes as enophiles reported until that moment.

Soon after, Rueping and co-workers[20] published a spectacular highly enantioselective cascade reaction in which six steps were catalyzed by a chiral Brønsted acid (**IX**), providing enantioenriched tetrahydropyridines **76** (Scheme 23).

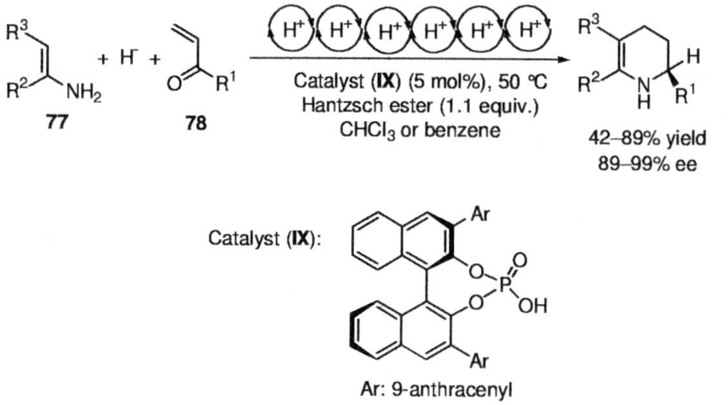

Scheme23. Cascade reaction reported by Rueping.

The exposition of a mixture of enamine **77** and vinyl ketone **78** to a catalytic amount of the Brønsted acid (**IX**) led to the formation of the corresponding Michael adducts **79a** and **79b**. The cyclization of **79a**, under acid equilibrium with **79b**, affords the hemiaminalic compound **80**, which, after fast water elimination, gives the dihidropyridine **81**. Afterwards, a subsequent Brønsted acid protonation induces the generation of iminium ion **82**, the chiral ion pair, which is activated for an enantioselective hydride transfer from Hantzsch ester **83** to render the desired product **76**. Notably, although the yields of the isolated products are quite low, in nearly all cases the cascade reaction proceeds with almost total enantiocontrol (Scheme 24).

Later on, Chen and co-workers [21] developed an organocatalytic asymmetric inverse-electron-demand aza-Diels-Alder reaction, providing an easy entry to piperidine derivatives **84** (Scheme 25). The reaction of *N*-sulfonyl-1-aza-1,3-butadienes **85** and the *in situ* generated chiral enamine as electron-rich alkene affords the nitrogenated compound in good yields (40–92%), excellent enantioselectivities (93–99%) and, in all cases, a total diastereocontrol (>99:1).

158

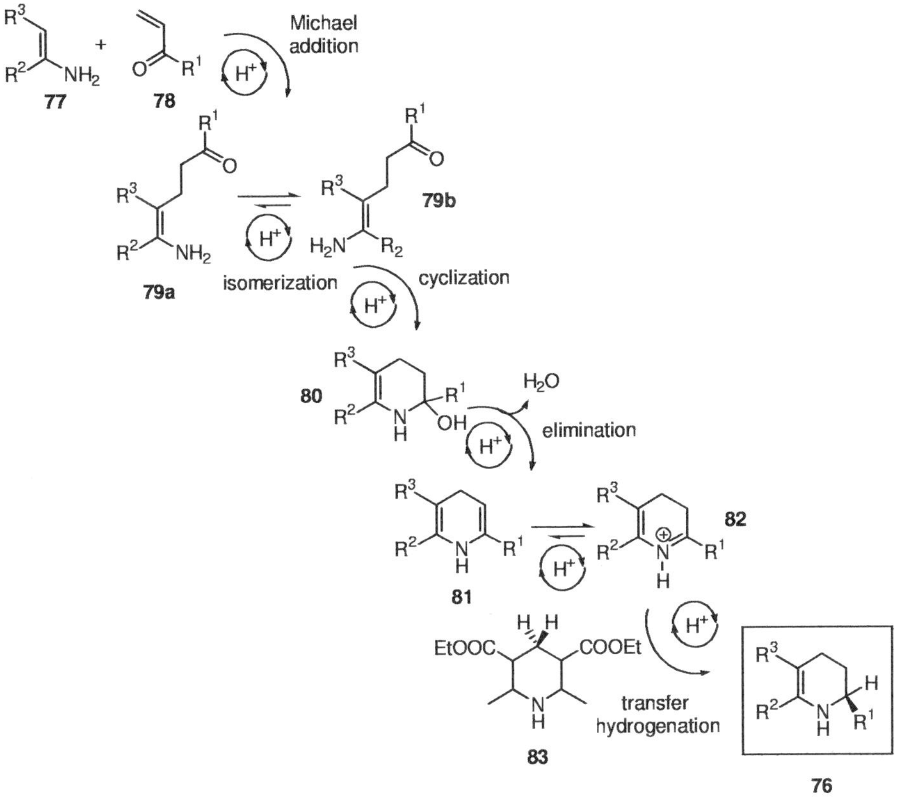

Scheme 24. Six step cascade reaction.

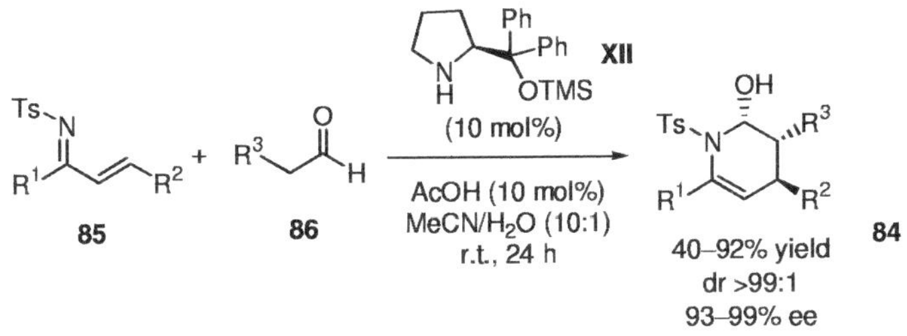

Scheme 25. Synthesis of piperidine derivatives **84** through an inverse-demand aza Diels-Alder.

Franzén *et al.*[22] published a very elegant synthesis of quinolozidine skeleton (**87** and **88**) through a one-pot, two-steps procedure (Scheme 26).

The organocatalytic conjugated addition of amide **89** to an α,β-unsaturated aldehyde **64** affords intermediate **90**, which cyclizes spontaneously under the reaction conditions to give hemiaminal **91**. The epimerization of the labile stereocentre at *C*3 establishes the most thermodynamically stable *trans* configuration. Treatment of **91** with catalytic amounts of acid results on the formation of acyliminium ion **92** which undergoes an electrophilic aromatic substitution with the aryl moiety to give the quinolizidine products **87** or **88** (Scheme 27).

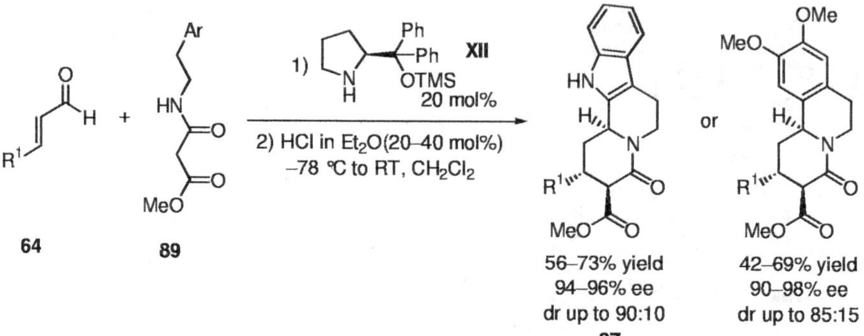

Scheme 26. Synthesis of quinolizidine skeleton (**87** and **88**) reported by Franzén and co-workers.

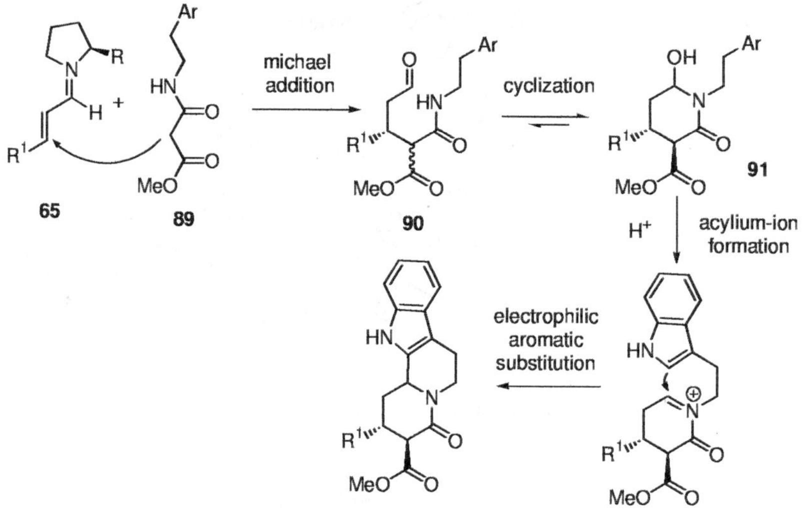

Scheme 27. Proposed mechanism for the one-pot synthesis of quinolizidine skeleton **87**.

The diastereoselectivity observed in the acidic-catalyzed cyclization of the acyliminium ion can be explained as follows: although the minor diastereomer detected is the more stable isomer which presents equatorial orientation of the indole moiety (thermodynamic product), the major isomer isolated is formed faster under kinetic conditions (–78 °C) owe to the less steric hindrance from the equatorial α proton in the transition state.

Finally, the formation of benzo[α]quinolizidine **88** requires stronger acidic conditions (40 mol%) than for indolo[2,3α]quinolizidine **87** (20 mol%), due the lower nucleophilicity of the phenyl ring compared to the 3-indolyl moiety.

Almost at the same time, Vesely, Moyano and Rios[23] reported an easy entry to the synthesis of pyperidines **93** based in the reaction of 3-amidoesters **94** and α,β-unsaturated aldehydes **64**. After the first malonate Michael addition over the enal **64**, the piperidine ring is formed through a hemiaminal cyclization, pushing the reaction toward the products with three new stereogenic centres (Scheme 28).

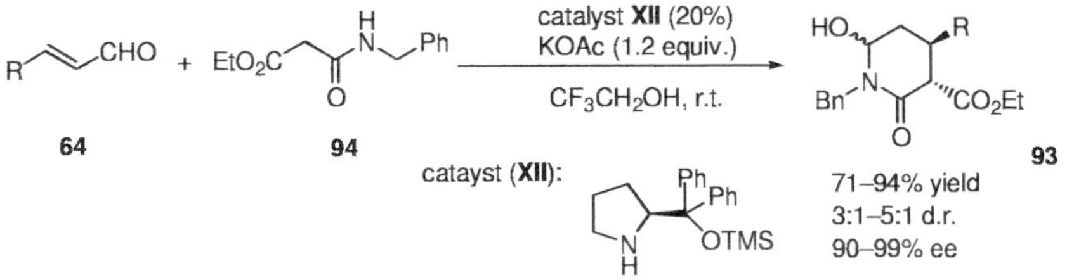

Scheme 28. Synthesis of pyperidines reported by Vesely, Moyano and Rios.

Among the eight possible isomers formed, only two are detected in the crude mixture. They correspond to the equatorial and axial position of the hemiaminalic hydroxyl piperidine. Their ratio can be controlled up to 5:1. The epimerization of the labile C3 stereocentre allows a total *trans* disposition between the other two stereocentres. Piperidines **93** are obtained in excellent yields, diastereo- and enantioselectivities.

3. Enantioselective synthesis of pyrrolidines

3.1. Metal catalyzed approaches

The enantioselective synthesis of pyrrolidines (**1**) *via* organometallic chemistry is one of the cornerstones of chemists nowadays. There are several highly efficient approaches, but, among all others, [3+2]-cycloadditions between azomethine ylides **96** and alkenes **97** are in our opinion the most easy and efficient in order to obtain this valuable scaffold.[24] Pyrrolidines (**1**) are important building blocks in the syntheses of many natural products and pharmaceuticals.[25]

1,3-Dipolar cycloaddition[26] reactions are fundamental processes in organic chemistry[27] and their asymmetric version offers a powerful and reliable synthetic methodology to access five-membered heterocyclic rings in a regio- and stereocontrolled fashion.[28]

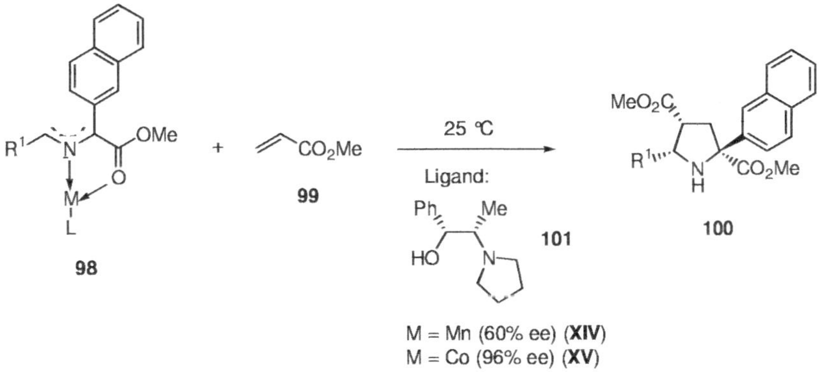

Scheme 29. Reaction reported by Grigg *et al.*.

Grigg *et al.*[29,30] were the first group to attempt to use chiral Mn(II) complex (**XIV**) of ephedrine derivative **101** in a stoichiometric amount as catalyst for the study of the enantioselective 1,3-dipolar cycloaddition reaction of the *N*-metalated AZY **98** with methyl acrylate **99**. Decreasing the amount of ligand lowered the ee, while an increased amount slowed the reaction dramatically. The use of a molar equivalent of the anhydrous $CoCl_2$ in presence of 2 mol of **101** provided the corresponding cycloadduct **100** in moderate yields and good ee (up to 80%). In order to increase the efficiency and stereoselectivity of the reaction, they decided to use methyl acrylate **99** as the solvent, enhancing the enantioselective excess up to 96% (Scheme 29).

Karlson *et al.*[31] described afterwards the use of a chiral Lewis acid (**XVI**) in order to catalyze the asymmetric cycloaddition between nonstabilized AZY **102** and a variety of α,β-unsaturated dipolarophiles **103**. This reaction proceeded with low enantioselectivity.

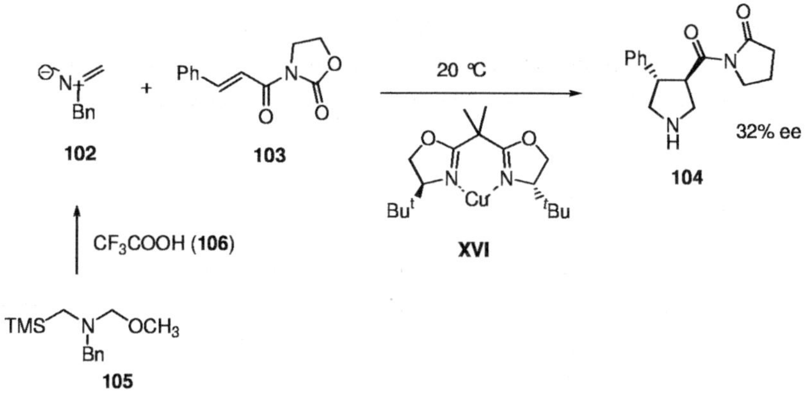

Scheme 30. Reaction developed by Karlson.

Zhang *et al.*[32] reported a highly enantioselective Ag(I)-catalyzed [3+2]-cycloaddition of **96** with dimethyl maleate **107** in the presence of various chiral phosphine ligands **109**. The reaction furnished only the *endo* diastereomer **108** in good yields and enantioselectivities. It should be noticed that authors used only 3% catalyst to perform the reaction, as shown in Scheme 31.

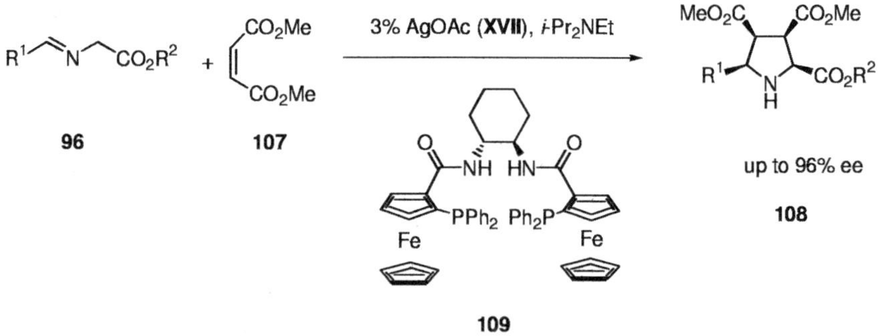

Scheme 31. Reaction performed by Zhang.

Jørgensen and coworkers described a catalytic asymmetric dipolar cycloaddition reaction of ylide **96** and methyl methacrylate **110** catalyzed by copper (**XVIII**) or zinc (**XIX**) in combination with *t*-Bu-BOX ligands. Both metals worked well in terms of yield and enantioselectivities, as shown in Scheme 32.

In 2003, Schreiber[33] reported a catalytic asymmetric dipolar cycloaddition of **96** with butyl acrylate **112** using silver as catalyst and different phosphines as ligands. The stereoselectivity of the reaction was high; concretely *P,N*-ligand QUINAP **112** showed excellent levels of both diastereo- (>20:1) and enantioselectivity (96–99%) even at a reduced catalyst loading of 1% (Scheme 33).

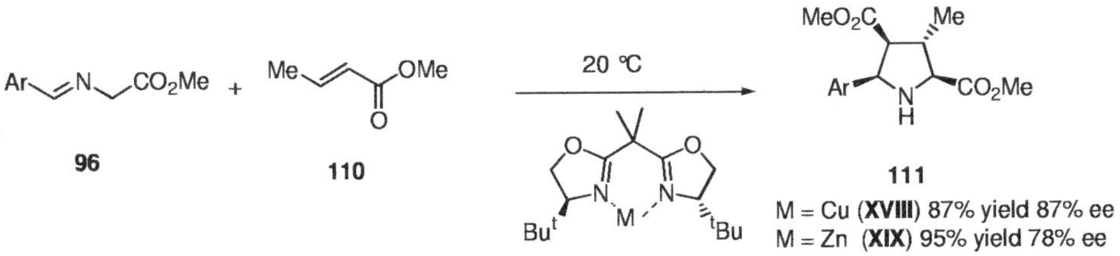

Scheme 32. Reaction reported by Jorgensen.

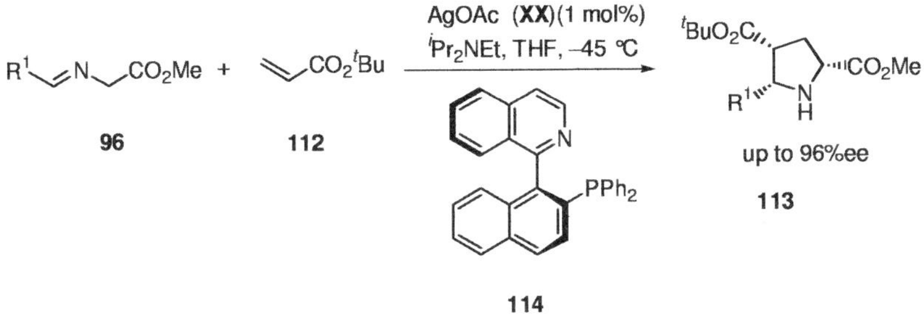

Scheme 33. [3+2]-Cycloaddition reported by Schreiber.

Soon after, intramolecular cycloaddition of azomethine ylides using Ag (I) complexes of QUINAP **114** was also reported,[34] with virtually complete diastereocontrol and enantiomeric excess of up to 99%.

Although most cycloadditions of azomethine ylides of AZY catalyzed by chiral metal complexes are *endo*-selective, Komatsu and co-workers[35] developed an *exo*-selective cycloaddition reaction between various *N*-metalated azomethine ylides **96** and *N*-phenylphtalimides **115** catalyzed by Cu(OTf)$_2$ (**XXI**) in combination with chiral phosphine ligands. Using 20 mol% of Cu(OTf)$_2$ (**XXI**) and 10 mol% of BINAP **117**, the best diastereo- and enantioselectivities (*exo/endo* 87:13; ee of *exo* adduct=34%) were obtained. Both diastereo- (*exo*) and enantioselectivities were improved at low temperature, employing BINAP **117** or BINAP derivatives and Cu(OTf)$_2$ (**XXI**).

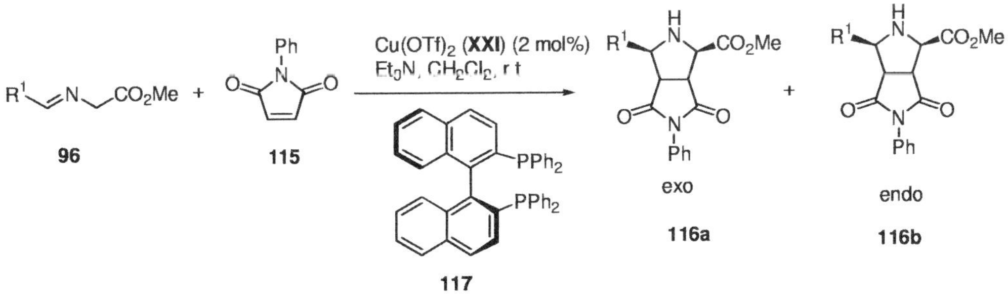

Scheme 34. *Exo*-cycloaddition reported by Komatsu.

In 2008, Najera and co-wokers reported a catalytic enantioselective 1,3-dipolar cycloaddition reaction of azomethine ylides **96** and alkenes **97** using phosphoramidite-silver(I) complexes.[36] This monodentate phosphoramidite-silver complex is a very efficient chiral catalyst for a wide range of 1,3-dipolar cycloaddition reactions between azomethine ylides **96** and different dipolarophiles. The advantage of this monodentate ligand **120** is the possibility to use sterically hindered components in the reaction.

For example, when *tert*-butyl acrylate **113** was used as dipolarophile with iminoglycinate **118**, the proline derivative **119** was afforded in 83% yield and in diastereo- (only *endo* adduct was observed) and enantiopure form (Scheme 35).

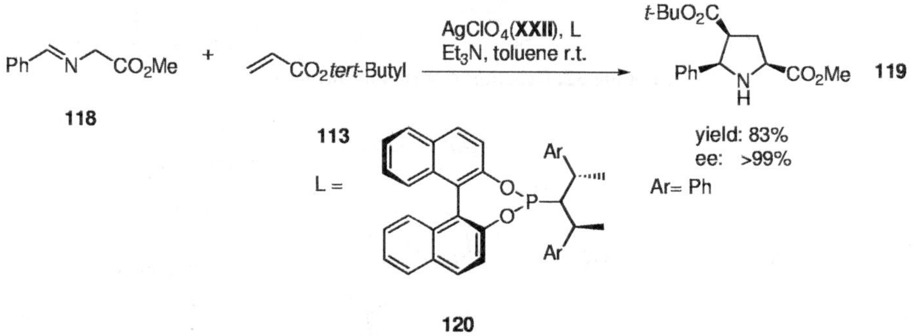

Scheme 35. Cycloaddition reported by Najera.

Another related reaction for the synthesis of pyrrolidines (**1**) by 1,3-dipolar cycloaddition of azomethine ylides **96** was reported by Carretero in 2009.[37] They described the use of phenylsulfonyl group **122** as regiochemical controller in the catalytic 1,3-dipolar cycloaddition. This strategy allowed them to obtain selectively the 2,3-dicarboxylic ester substituted pyrrolidines **121** with very high *exo* selectivity and enantioselectivity by using simple copper catalysts (**XXIII**). Moreover, desulfonylation of the adducts **121** makes this procedure very attractive due to the possibility of obtaining substituted pyrrolidines **121** with opposite regioselectivity to that obtained using typical acrylate dipolarophiles (Scheme 36).

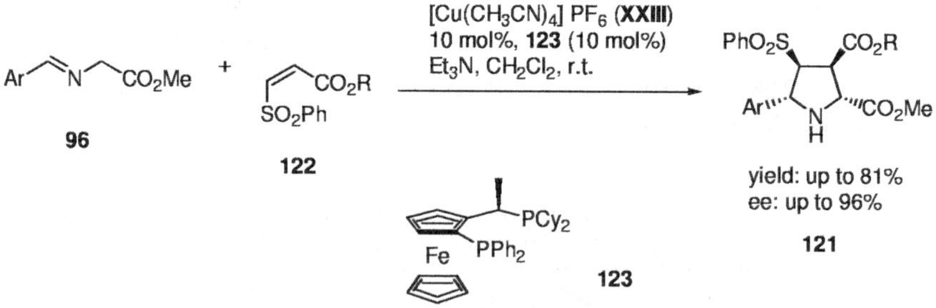

Scheme 36. 1,3 dipolar cycloaddition reported by Carretero.

Soon after, the same authors reported the 1,3 dipolar cycloaddition reaction of azomethine ylides and α,β-unsaturated ketones catalyzed by copper, affording *exo-trans* cyclic enone salts with excellent yields, diastereo- and enantiosectivities.[38]

In 2008, the same group reported the use of bis-sulfonylethylene as masked acetylene equivalent in the catalytic asymmetric [3+2]-cycloaddition of azomethine ylides, affording the desired pyrrolidine derivatives in excellent yields and enantioselectivities.[39]

Very recently, O'Brien and coworkers developed a new enantioselective approach to the synthesis of chiral pyrrolidines starting from simple achiral pyrrolidines.[40] In this approximation, authors use an enantioselective lithiation using (–)-sparteine as a chiral inductor. As shown in Scheme 37, N-Boc-pyrrolidine is treated with sec-BuLi and sparteine to promote the asymmetric deprotonation, then the anion was trapped with an aldehyde to form compound XX in good yields and enantioselectivities.

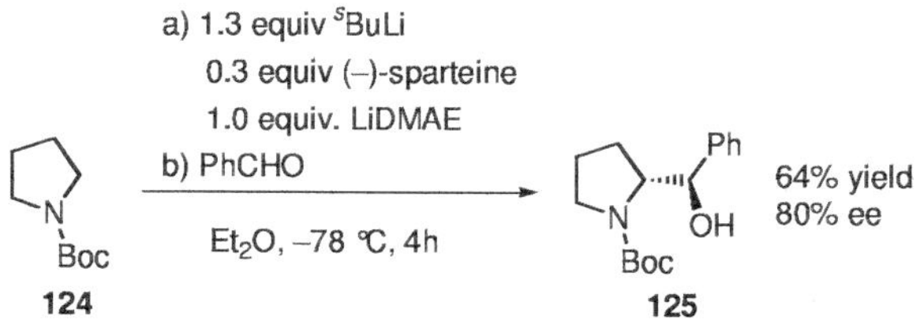

Scheme 37. Synthesis of pyrrolidines reported by O'Brien.

Other interesting approach to the enantioselective synthesis of pyrrolidines has recently been developed by Widenhoefer and coworkers.[41] They reported the use of gold-catalyzed dynamic kinetic enantioselective intermolecular hydroamination of allenes rendering the corresponding pyrrolidines in excellent yields, diastereo- and enantioselectivities using chiral diphosphines as ligands (Scheme 38).

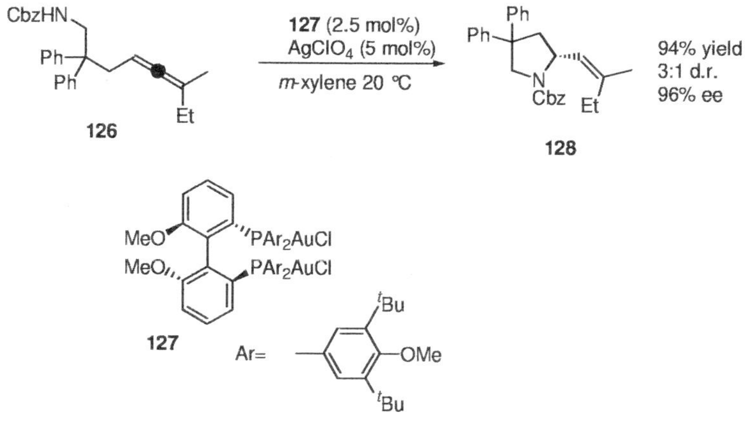

Scheme 38. Hydroamination reported by Widenhoefer.

3.2. Organocatalytic approaches

Vicario and coworkers described in 2007 a process that gave access to pyrrolidines **128** with three contiguous stereocentres.[42]

They designed the transformation depicted in Scheme 39. α-Aminoacid imines can undergo thermal 1,2-prototropy to produce azomethine ylides in a kinetically controlled process, being the acidity of the α-hydrogen atom a key parameter in terms of whether or not this process occurs. Imines **130** underwent this prototropy process very readily to afford a stabilized azomethine ylide (**131**), which reacted with an α,β-unsaturated aldehyde **64** under organocatalytic conditions upon the activation of the aldehyde as an iminium ion **132**.

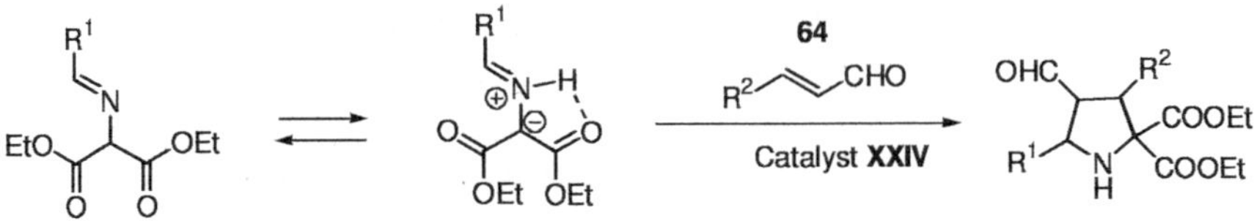

Scheme 39. Proposed enantioselective organocatalytic [3+2]-cycloaddition of azomethine ylides and α,β-unsaturated aldehydes.

The reaction was efficiently catalyzed by commercially available α,α-diphenyl prolinol (**XXIV**) and water as cocatalyst and could be carried out in THF. It worked with linear and branched aliphatic enals and with aldehydes having an aryl or heteroaryl moiety at β-position **64**, as well as different imine **130** substrates. In all the experiments, the cycloaddition product **129** was obtained as a single regioisomer, in excellent yields and with excellent enantio- and diastereoselectivities, as shown in Scheme 40.

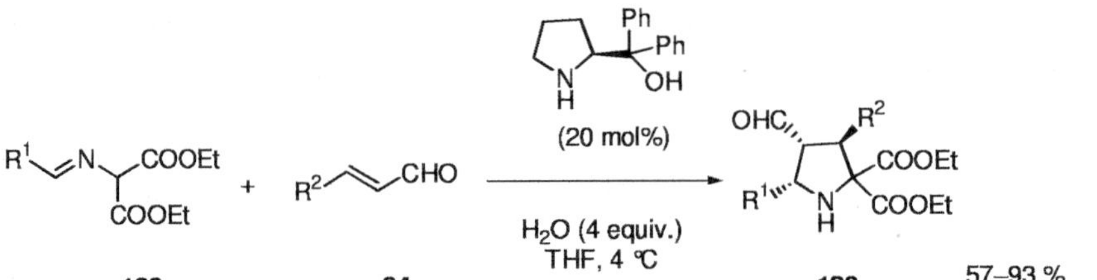

Scheme 40. Synthesis of pyrrolidines (**129**) developed by Vicario.

Their proposed mechanism involves efficient shielding of the *Si* face of the chiral iminium intermediate **132** by the bulky aryl groups of **XXIV**, which leads to a stereoselective *endo*-type approach of the *E* 1,3-dipole to the sterically less hindered *Re* face of the intermediate *E* iminium ion **133**, as illustrated in Scheme 41.

Soon after, Córdova and co-workers presented a highly chemo- and enantioselective [C+NC+CC] coupling process.[43] This multicomponent reaction was performed between aldehydes ("C") **86**, dialkyl 2-aminomalonates ("NC") **135** and α,β-unsaturated aldehydes ("CC") **64**. Diphenylprolinol TMS ether (**XII**) gave good results in terms of yield, medium to high diastereoselectivities and excellent enantioselectivities (Scheme 42).

One year later, Vicario, Badía and co-workers developed an improved procedure for the enantioselective Michael reaction between β-nitroacrolein dimethyl acetal **136** and a variety of aldehydes **86**, generating the corresponding highly functionalized nitroaldehydes **137** in high yields and stereoselectivities (Scheme 43).[44]

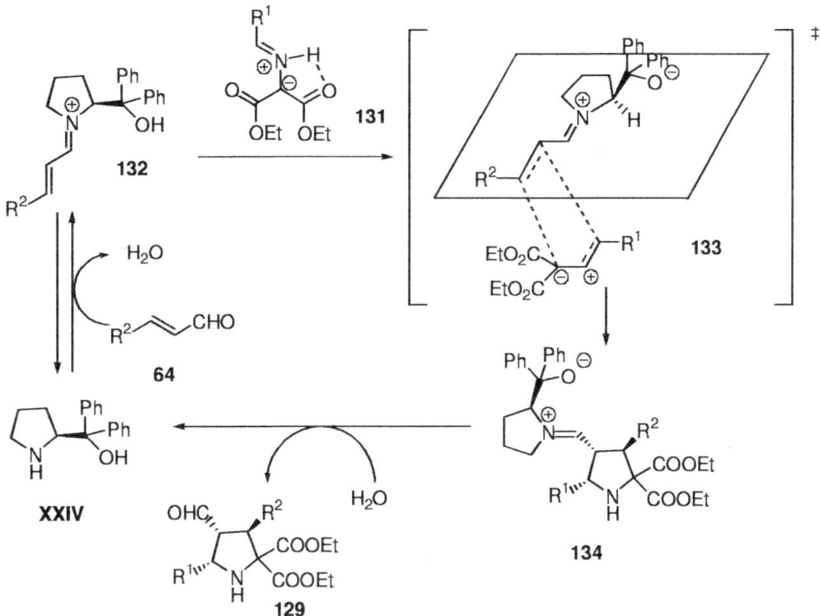

Scheme 41. A plausible reaction pathway for the enantioselective [3+2]-cycloaddition of azomethine ylides and α,β-unsaturated aldehydes with catalyst **XXIV**.

Scheme 42. Synthesis of highly functionalized pyrrolidine derivatives presented by Córdova.

Scheme 43. Conjugate addition of different aldehydes to β-nitroacrolein dimethyl acetal developed by Vicario and Badía.

The obtained adducts **137** were converted into enantiopure highly functionalized pyrrolidines **138** containing two or three contiguous stereocentres by means of two short and efficient protocols.

3,4-Disubstituted pyrrolidines **138** were obtained in a cascade process by Zn-mediated chemoselective reduction of the nitro group followed by intramolecular reductive amination, as shown in Scheme 44. Pyrrolidines **138** were isolated in excellent yields as single diastereoisomers. The reduction/reductive amination procedure proceeded without epimerization at the α-stereocentre of the starting material **137**, which was expected to be fairly prone to racemization during the reductive amination step through imine/enamine tautomerization.

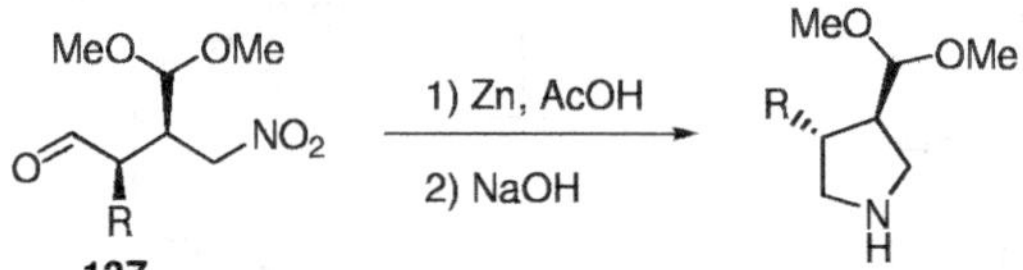

137 **138**

Scheme 44. Synthesis of 3,4-disubstituted pyrrolidines **138**.

On the other hand, trisubstituted homoproline derivatives **139** were prepared by means of an olefination reaction and a cascade process involving chemoselective reduction of the nitro group followed by a fully diastereoselective intramolecular aza-Michael reaction, cleanly furnishing the target pyrrolidines **138** as single diastereomers of high enantiomeric purity (Scheme 45).

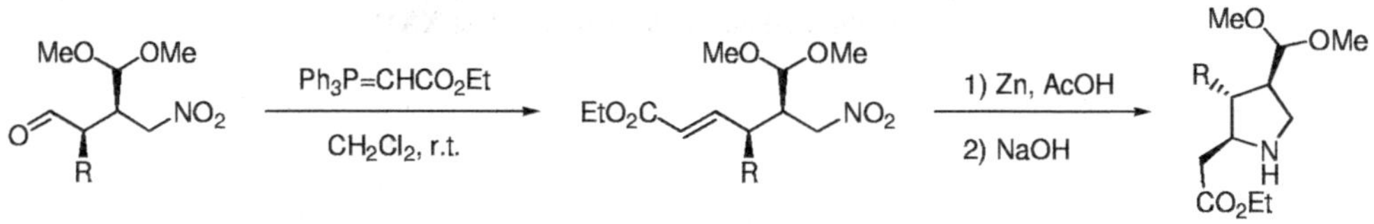

137 **140** **139**

Scheme 45. Synthesis of trisubstituted homoproline derivatives.

The same year, Wang and co-workers reported a direct method for the synthesis of trisubstituted chiral pyrrolidines **141**.[45] This unprecedented cascade aza-Michael-Michael reaction was performed between α,β-unsaturated aldehydes and *trans*-γ-Ts protected amino α,β-unsaturated ester **142**. The process, efficiently catalyzed by diphenylprolinol TMS ether (**XII**), was simple and practical, being its outcome excellent, achieving high levels of enantio and diastereoselectivity and, also, high yields.

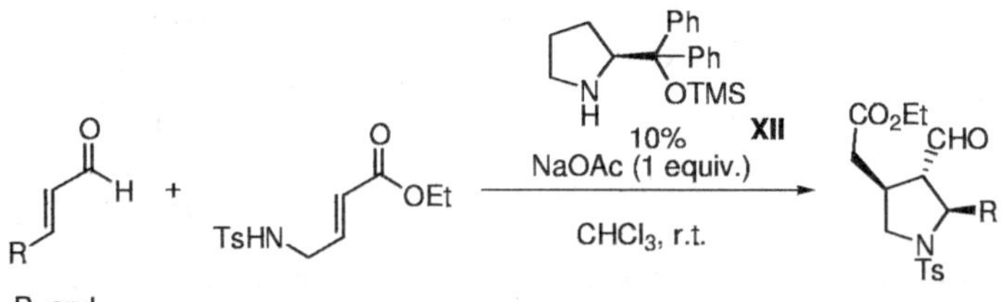

R. aryl

64 **142**

85–94% yield
96% to >99% ee
7:1 to >30:1 dr
141

Scheme 46. Synthesis of pyrrolidines described by Wang.

This one-pot transformation produced a complex molecular architecture formed with high stereocontrol of three new stereogenic centres and an array of exploitable orthogonal functionalities for further synthetic elaboration (Scheme 46).

Central to the implementation of this proposed cascade reaction was the recognition of reactivity and selectivity issues that needed to be addressed. Although *N*-centered nucleophiles had been employed for conjugate addition reactions, the more active specific species was required as a result of its weak nucleophilicity. The amine **143** should not function as catalyst for the formation of iminium intermediate (TS **A**), but must serve as nucleophile for conjugate addition to the catalyst activated iminium (TS **B**). The substrate amine ester **142** should be stable during the reaction without undergoing intramolecular lactamization. They designed a *trans*-γ-protected amino α,β-unsaturated ester. The protected amino-group inhibited forming an iminium with enal **145**, while the enhanced acidity rendered it to be readily deprotonated under basic conditions to produce a more nucleophilic nitrogen anion for the first Michael addition reaction. The *trans* geometry could significantly reduce intramolecular lactamization. With these molecules in hand, they found out that no reaction occurred with Ac, Boc and Cbz, but tosyl group gave good results. Once suitable conditions had been discovered, they tested the reaction with a set of aromatic enals, obtaining excellent results in all cases (Scheme 47).

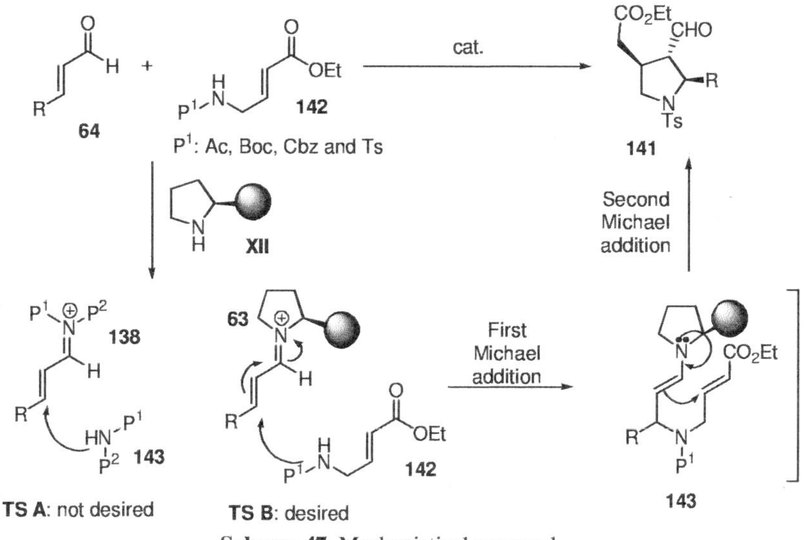

Scheme 47. Mechanistical proposal.

Until this point, we have discussed reactions based in iminium ion strategies that make use of chiral secondary amines. Organocatalytic syntheses of chiral pyrrolidine derivatives can be achieved, as well, by means of different activation modes. For instance, in 2008 Gong and co-workers[46] developed an asymmetric [3+2]-cycloaddition reaction of isocyanoesters **145** to nitroolefins **146** catalyzed by chiral cinchona alkaloids derivatives (**XXV**), which have been revealed as efficient organocatalysts for many asymmetric reactions. This reaction affords highly functionalized 2,3-dihydropyrroles **147** with excellent yields, diastereo- and enantioselectivities, as shown in Scheme 48.

169

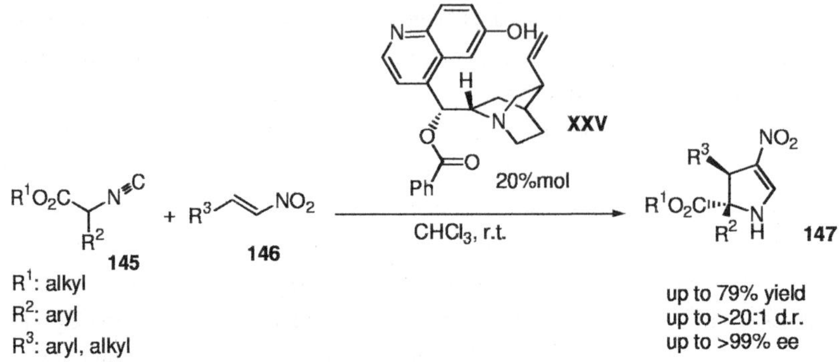

Scheme 48. Enantioselective synthesis of 2,3-dihydropyrroles developed by Gong.

Their mechanistic approach for the formal cycloaddition reaction catalyzed by a chiral base is shown in Scheme 49.

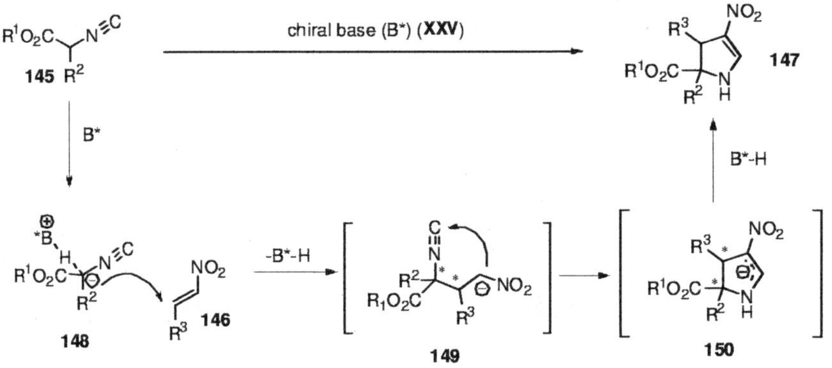

Scheme 49. Proposed mechanism for the asymmetric cycloaddition reaction of isocyanoesters to nitroolefins catalyzed by a chiral base.

In this approach, the chiral base (**XXV**) could promote an asymmetric Michael addition of isocyanoesters **145** to electron-deficient olefins, such as nitroolefins **146**, by activating the acidic α-carbon atom of **148** to generate intermediates **149**. Subsequent intramolecular cyclization reaction of intermediates **149** afforded precursor 1,2-dihydropyrroles **150**, which may be converted into dihydropyrrole **147** after protonation.

The same authors reported at the same time the asymmetric 1,3-dipolar cycloaddition of azomethine ylides **153** with nitroalkenes **146**.[47] The reaction was efficiently catalyzed by thiourea alkaloid derivatives (**XXVI**). In this approximation, benzophenone imines **151** reacted with different nitroalkenes, affording the corresponding highly substituted pyrrolidines **152** in high yields, high diastereoselectivities and moderate enantioselectivities, as shown in Scheme 50. It should be noticed that the relative configuration of the substituents in pyrrolidine ring is *trans*.

Azomethine ylides **153**, which can be formed by treatment of **151** with Lewis base, attack the reactive nitroalkenes **146** formed by hydrogen-bond activation between the nitro group and protons of Brønsted acids such as ureas and thioureas, **154** to principally undergo the 1,3-dipolar cycloaddition (it should be noticed that in the case of nitroalkenes, the 1,3-dipolar cycloaddition is a stepwise reaction and the intermediates can be easily isolated), as illustrated in Scheme 51.

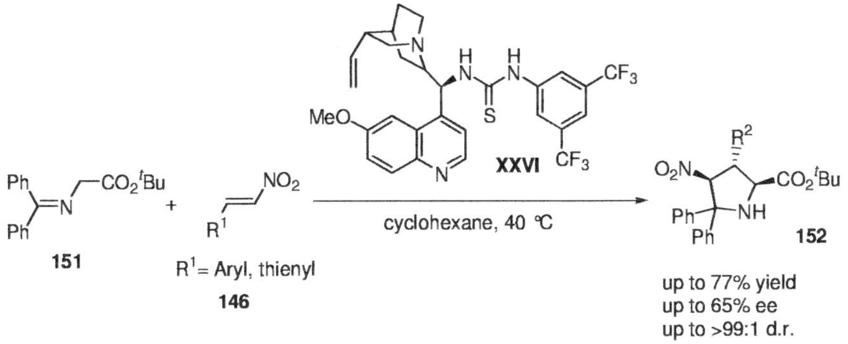

Scheme 50. 1,3-Dipolar cycloaddition of azomethine ylides with nitroalkenes.

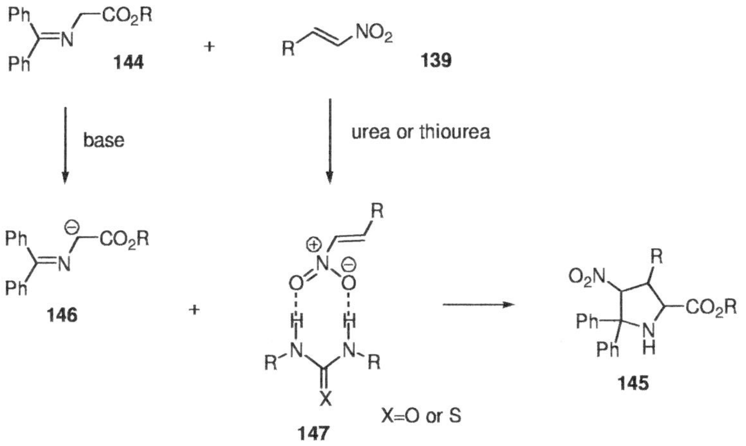

Scheme 51. The proposal pathway for bifunctional-organocatalyst promoted 1,3-dipolar cycloaddition of azomethine ylides with nitroalkenes.

Soon after, Chen[48] and Takemoto[49] reported, almost at the same time, the first asymmetric three-component 1,3-dipolar cycloaddition of aldehydes **86**, α-aminomalonates **135** and nitroalkenes **146**, catalyzed by chiral thioureas (**XVII**, **XXVIII**). The achiral diastereoselective version of this reaction had been published some months earlier by Rios and co-workers.[50]

In the case of Chen and co-workers, the reaction began with the formation of an imine **130** from α-aminomalonate **135** and aldehyde **86**, whereas Takemoto and co-workers started directly from α-amino-malonate imine **130**. This compound reacted with a nitrostyrene **146** *via* Michael addition and a subsequent aza-Henry reaction (formally, a [3+2]-cycloaddition) afforded highly substituted pyrrolidines **155**, as shown in Scheme 52. Both reactions were promoted by the thiourea catalyst.

The reaction worked well with aromatic aldehydes and an array of nitrostyrenes with electron-withdrawing or donating substituents, affording the corresponding pyrrolidine derivatives **155** in high yields, diastereo- and enantioselectivities. However, aliphatic nitroalkenes led to more modest ee values. Aliphatic aldehydes again could not be successfully used in the 1,3-dipolar cycloaddition (Scheme 53).

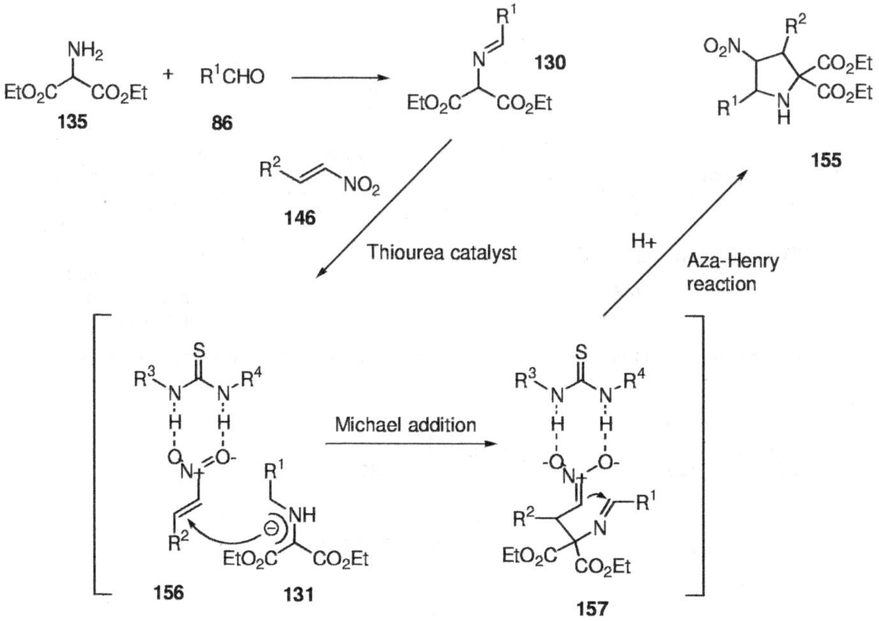

Scheme 52. Proposed mechanism for the formal [3+2]-cycloaddition.

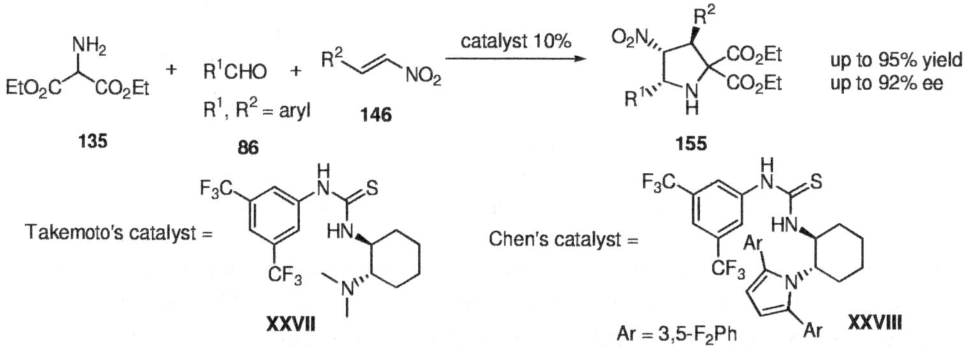

Scheme 53. Formal [3+2]-cycloaddition reported by Takemoto and Chen.

4. Conclusions

The synthesis of pyrrolidines and piperidines is nowadays one of the most important goals for synthetic chemists. As we have seen, there are several organocatalytic and organometallic methods that afford these privileged structures in exceptional yields and with high enantiomeric purities. In this review, we have summarized the most useful and selective procedures that allow us to perform this important transformations. Research in this field is continuing apace and, each year, we can find huge improvements in the literature. Organic chemists have been focusing their efforts to improve this reaction until achieving exceptional levels of stereoselectivity and atom economy. However, this field is in continuous expansion, the explosion of organocatalysis and the development of new organometallic methods suggest a brilliant future for the enantioselective synthesis of this privileged structures.

References
1. Nash, R. J.; Bell, E. A.; Fleet, G. W. J.; Jones, R. H.; Williams, J. M. *J. Chem. Soc., Chem. Commun.* **1985**, *11*, 738–740.
2. Colegate, S. M.; Dorling, P. R.; Huxtable, C. R. *Aus. J. Chem.* **1979**, *32*, 2257–2264.
3. Ikeda, K.; Takahashi, M.; Nishida, M.; Miyauchi, M.; Kizu, H.; Kameda, Y.; Arisawa, M.; Watson, A. A.; Nash, R. J.; Fleet, G. W. J.; Asano, N. *Carbohydr. Res.* **2000**, *323*, 73–80.
4. Strunz, G. M.; Findlay, J. A. In *The Alkaloids*; Brossi, A., Ed.; Academic Press: New York, 1985; Vol. 26, pp. 89–183.
5. For reviews on the synthesis of piperidines, see: (a) Lauchat, S.; Dickner, T. *Synthesis* **2000**, 1781–1813. (b) Weintraub, P. M.; Sabol, J. S.; Kane, J. M.; Borcherding, D. R. *Tetrahedron* **2003**, *59*, 2953–2989. (c) Buffat, M. P. G. *Tetrahedron* **2004**, *60*, 1701–1729.
6. (a) Williams, A. J.; Chakthong, S.; Gray, D.; Lawrence, R. M.; Gallagher, T. *Org. Lett.* **2003**, *5*, 811–814. (b) Bower, J. F.; Svenda, J.; Williams, A. J.; Charmant, J. P. H.; Lawrence, R. M.; Szeto, P.; Gallagher, T. *Org. Lett.* **2004**, *6*, 4727–4730. (c) Bower, J. F.; Szeto, P.; Gallagher, T. *Chem. Commun.* **2005**, 5793–5795. (d) Bower, J. F.; Chakthong, S.; Svenda, J.; Williams, A. J.; Lawrence, R. M.; Szeto, P.; Gallagher, T. *Org. Biomol. Chem.* **2006**, *4*, 1868–1877.
7. Bower, J. F.; Riis-Johannesen, T.; Szeto, P.; Whitehead, A. J.; Gallagher, T. *Chem. Commun.* **2007**, 728–730.
8. Cortez, G. A.; Schrock, R. R.; Hoveyda, A. H. *Angew. Chem. Int. Ed.* **2007**, *46*, 4534–4538.
9. Reddy, M. S.; Narender, M.; Rao, K. R. *Tetrahedron* **2007**, *63*, 331–336.
10. Krishna, P. R.; Dayaker, G. *Tetrahedron Lett.* **2007**, *48*, 7279–7282.
11. Gnamm, C.; Krauter, C. M.; Brödner, K.; Helmchen, G. *Chem. Eur. J.* **2009**, *15*, 2050–2054.
12. Esquivias, J.; Alonso, I.; Gomez Arrayas, R.; Carretero, J. C. *Synthesis* **2009**, 113–126.
13. Itoh, J.; Funchibe, K.; Akiyama, T. *Angew. Chem. Int. Ed.* **2006**, *45*, 4796–4788.
14. Akiyama, T.; Morita, H.; Fuchibe, K. *J. Am. Chem. Soc.* **2006**, *128*, 13070–13071.
15. He, M.; Struble, J. R.; Bode, J. W. *J. Am. Chem. Soc.* **2006**, *128*, 8418–8420.
16. Terada, M.; Machioka, K.; Sorimachi, K. *J. Am. Chem. Soc* **2007**, *129*, 10336–10337.
17. Sunden, H.; Rios, R.; Ibrahem, I.; Zhao, G.-L.; Eriksson, L.; Córdova, A. *Adv. Synth. Catal.* **2007**, *349*, 827–832.
18. Li, H.; Wang, J.; Xie, H.; Zu, L.; Jiang, W.; Duesier, E. N.; Wang, W. *Org. Lett.* **2007**, *9*, 965–968.
19. Hayashi, Y.; Gotoh, H.; Masui, R.; Ishikawa, H. *Angew. Chem. Int. Ed.* **2008**, *47*, 4012–4015.
20. Rueping, M.; Antonchick, A. P. *Angew. Chem. Int. Ed.* **2008**, *47*, 5836–5838.
21. Han, B.; Li, J.-L.; Ma, C.; Zhang, S.-J.; Chen, Y.-C. *Angew. Chem. Int. Ed.* **2008**, *47*, 9971–9974.
22. Franzén, J.; Fisher, A. *Angew. Chem. Int. Ed.* **2009**, *48*, 787–791.
23. Valero, G.; Schimer, J.; Cisarova, I.; Vesely, J.; Moyano, A.; Rios, R. *Tetrahedron Lett.* **2009**, *50*, 1943 1946.
24. For an authoritative review in copper-catalyzed [1,3]-dipolar cycloadditions, see: Stanley, L. M.; Sibi, M. P. *Chem. Rev.* **2008**, *108*, 2887–2902.
25. Obst, U.; Betschmann, P.; Lerner, C.; Seiler, P.; Diederich, F. *Helv. Chim. Acta* **2000**, *83*, 855–909. (b) Alvarez-Ibarra, C.; Csaky, A. G.; López, I.; Quiroga, M. L. *J. Org. Chem.* **1997**, *62*, 479–484. (c) Waid, P. P.; Flynn, G. A.; Huber, E. W.; Sabol, J. S. *Tetrahedron Lett.* **1996**, *37*, 4091–4094. (d) Bianco, A.; Maggini, M.; Scorrano, G.; Toniolo, C.; Marconi, G.; Villani, C.; Prato, M. *J. Am. Chem. Soc.* **1996**, *118*, 4072–4073. (e) Kolodziej, S. A.; Nikiforovich, G. V.; Skeean, R.; Lignon, M. F.; Martinez, J.; Marshall, G. R. *J. Med. Chem.* **1995**, *38*, 137–149. (f) Pearson, W. H. In *Studies in*

Natural Product Chemistry; Atta-Ur-Rahman, Ed.; Elsevier: New York, 1998; Vol. 1, p. 323. (g) Sebahar, P. R.; Williams, R. M. *J. Am. Chem. Soc.* **2000**, *122*, 5666–5667. (h) Denhart, D. J.; Griffith, D. A.; Heathcock, C. H. *J. Org. Chem.* **1998**, *63*, 9616–9617. (i) Overman, L. E.; Tellew, J. E. *J. Org. Chem.* **1996**, *61*, 8338–8340. (j) Sisko, J.; Henry, J. R.; Weinreb, S. M. *J. Org. Chem.* **1993**, *58*, 4945–4951.

26. For an authoritative review, see: Pandey, G.; Banerjee, P.; Gadre, S. R. *Chem. Rev.* **2006**, *106*, 4484–4517.

27. Padwa, A. In *1,3-Dipolar Cycloaddition Chemistry*; Padwa, A., Ed.; John Wiley & Sons: New York, 1984; Vol. 2, p. 277.

28. *Dipolar Cycloaddition Chemistry Toward Heterocycles and Natural Products*; Padwa, A.; Pearson, W. H., Eds.; John Wiley & Sons: New York, 2002; p. 169. (b) Gothelf, K. V. In *Cycloaddition Reactions in Organic Synthesis*; Kobayashi, S.; Jørgensen, K. A., Eds.; Wiley-VCH: Weinheim, 2002; Chapt. 6, p. 211. (c) Gothelf, K. V.; Jørgensen, K. A. *Chem. Rev.* **1998**, *98*, 863–909. (d) Karlsson, S.; Ho¨gberg, H. E. *Org. Prep. Proceed. Int.* **2001**, *33*, 103–172. (e) Najera, C.; Sansano, J. M. *Curr. Org. Chem.* **2003**, *7*, 1105–1150.

29. Allway, P.; Grigg, R. *Tetrahedron Lett.* **1991**, *32*, 5817–5820.

30. Grigg, R. *Tetrahedron: Asymmetry* **1995**, *6*, 2475–2486.

31. Karlsson, S.; Han, F.; Hogberg, H.-E.; Caldirola, P. *Tetrahedron:Asymmetry* **1999**, *10*, 2605–2616.

32. Longmire, J. M.; Wang, B.; Zhang, X. *J. Am. Chem. Soc.* **2002**, *124*, 13400–13401.

33. Chen, C.; Li, X.; Schreiber, S. L. *J. Am. Chem. Soc.* **2003**, *125*, 10174–10175.

34. Stohler, R.; Wahl, F.; Pfaltz, A. *Synthesis* **2005**, 1431–1435.

35. Oderaotoshi, Y.; Cheng, W.; Fujitomi, S.; Kasano, Y.; Minakata, S.; Komatsu, M. *Org. Lett.* **2003**, 5043–5046.

36. Najera, C.; Retamosa, M. G.; Sansano, J. M. *Angew. Chem. Int. Ed.* **2008**, *47*, 6055–6058.

37. Lopez-Perez, A.; Adrio, J.; Carretero, J. C. *Angew. Chem. Int. Ed.* **2009**, *48*, 340–343.

38. Hernandez-Toribio, J.; Gomez Arrayas, R.; Martin-Matute, B.; Carretero, J. C. *Org. Lett.* **2009**, *11*, 393–396.

39. Lopez-Perez, A.; Adrio, J.; Carretero, J. C. *J. Am. Chem. Soc.* **2008**, *130*, 10084–10085.

40. Bilke, J. L.; Moore, S. P.; O'Brien, P.; Gilday, J. *Org. Lett.* **2009**, *11*, 1935–1938.

41. Zhang, Z.; Bender, C. F.; Widenhoefer, R. A. *J. Am. Chem. Soc.* **2007**, *129*, 14148–14149.

42. Vicario, J. L.; Reboredo, S.; Badía, D.; Carrillo, L. *Angew. Chem. Int. Ed.* **2007**, *46*, 5168–5170.

43. Ibrahem, I.; Rios, R.; Vesely, J.; Córdova, A. *Tetrahedron Lett.* **2007**, *48*, 6252–6257.

44. Ruiz, N.; Reyes, E.; Vicario, J. L.; Badía, D.; Carrillo, L.; Uria, U. *Chem. Eur. J.* **2008**, *14*, 9357–9367.

45. Li, H.; Zu, L.; Xie, H.; Wang, J.; Wang, W. *Chem. Commun.* **2008**, 5636–5638.

46. Guo, C.; Xue, M.-X.; Zhu, M.-K.; Gong, L.-Z. *Angew. Chem. Int. Ed.* **2008**, *47*, 3414–3417.

47. Xue, M.-X.; Zhang, X.-M.; Gong, L.-Z. *Synlett* **2008**, *5*, 691–694.

48. Liu, Y.-K.; Liu, H.; Du, W.; Yue, L.; Chen, Y.-C. *Chem. Eur. J.* **2008**, *14*, 9873–9877.

49. Xie, J.; Yoshida, K.; Takasu, K.; Takemoto, Y. *Tetrahedron Lett.* **2008**, *49*, 6910–6913.

50. Crovetto, L.; Rios, R. *Synlett,* **2008**, *12*, 1840–1844.

SYNTHESIS OF PYRIDINES BY [2+2+2]-CYCLOTRIMERIZATION OF ALKYNES WITH NITRILES

Pavel Turek[a,b] and Martin Kotora*[a,b]

[a]Department of Organic and Nuclear Chemistry, Faculty of Science, Charles University in Prague, Hlavova 8, CZ-128 43 Praha 2, Czech Republic (e-mail: kotora@natur.cuni.cz)

[b]Institute of Organic Chemistry and Biochemistry, Academy of Sciences of the Czech Republic, Flemingovo 2, CZ-166 10 Praha 6, Czech Republic

Abstract. The rapid assembly of pyridine rings can be efficiently achieved by catalytic [2+2+2]-cyclo-trimerization of alkynes with nitriles in the presence of various transition metal compounds under mild reaction conditions. This review summarizes application of the aforementioned cyclotrimerization reaction for the synthesis of natural compounds and their analogs, biologically active compounds, pyridine based ligands for transition metal compounds and their chiral congeners.

Contents

1. General considerations
    1.1. History
    1.2. Reaction mechanism and regiochemistry
    1.3. Catalysts
    1.4. Regioselectivity
2. Synthesis of natural compounds and their congeners
3. Synthesis of potentially biologically active compounds
4. Synthesis of pyridines and bipyridines
5. Synthesis of pyridine based chiral ligands and catalysts
6. Enantioselective cyclotrimerization catalyzed by chiral catalysts
7. Other applications
8. Conclusion
Acknowledgment
References

1. General considerations

1.1. History

Substances with the pyridine framework constitute a large group of various compounds spanning from natural compounds to its heterocycle based ligands for transition metals. Despite the fact that there are numerous synthetic routes to the pyridine framework, the simplest and hence the attractive one, although not always easy to carry out, is cyclotrimerization of alkynes with nitriles. There are several points that make this approach highly desirable. Firstly, alkynes as well as nitriles are usually readily accessible starting material (also from the commercial point of view). Secondly, the cyclotrimerization of alkynes and nitriles has a high degree of synthetic efficiency. Last but not least, the cyclotrimerizations are usually carried out in the presence of the stoichiometric or catalytic amount of transition metals that ensure mild and neutral reaction conditions, under which a number of various functional groups is tolerated.

Although the first cyclotrimerization of alkynes with nitriles has been reported in the second half of 19th century[1] (synthesis of pyridine by passing a flux of ethyne with hydrogen cyanide through hot iron tubing), the first practical procedures started to appear in the 1970s and since then, this reaction has attracted a considerable amount of attention from chemists in both academic and industrial laboratories.

1.2. Reaction mechanism and regiochemistry

Regarding the course of the reaction mechanism of the cyclotrimerization of alkynes with nitriles catalyzed by late transition metal compounds, it has been discussed in detail in several reviews[2] and thus its subtle aspects will not be treated herein. Nonetheless, the generally accepted sequence of events can be outlined in the following terms (Scheme 1). It begins with the redox coupling of two alkynes on a coordinatively unsaturated transition metal compound (oxidative/reductive dimerization) forming metallacyclopentadiene **A**. Its formation is probably preceded by a stepwise formation of an alkyne-metal and a bis-alkyne-metal complexes. In the next step, a nitrile is coordinated to the central metal atom, followed by the insertion into the metal—C bond giving rise to azametallacycloheptatriene **B**. It should be emphasized that the nitrogen atom is always bound to the metal atom. Finally, the reductive elimination of pyridine completes the catalytic cycle by regenerating the coordinatively unsaturated transition metal compound, which can enter in the reaction again. This course of the reaction has been recently supported by theoretical calculations showing oxidative addition of two alkynes to give metallapentadienes to be the key step followed by the addition of the nitrile to these intermediates via a side-on fashion.[3] The rate of the pyridine formation is independent of the nitrile concentration, the amount of alkyne trimerization may be kept low by using nitrile in excess, if practical.

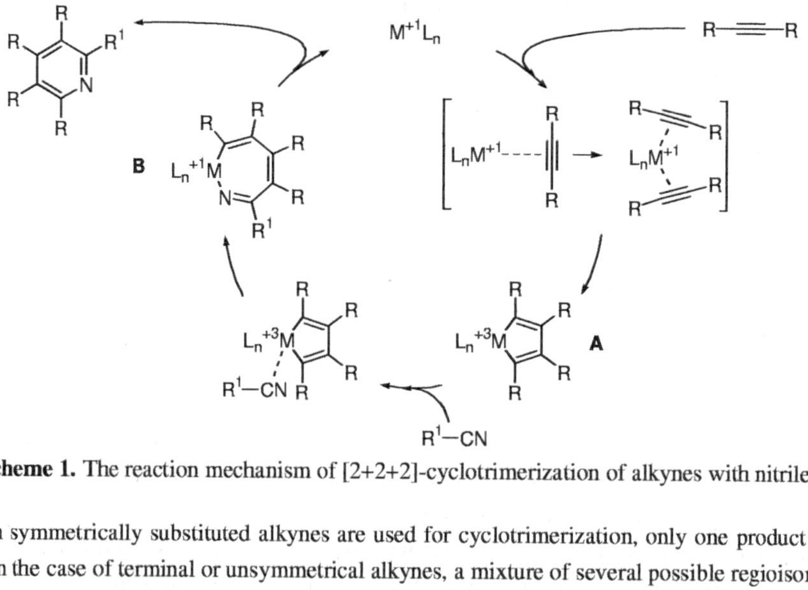

Scheme 1. The reaction mechanism of [2+2+2]-cyclotrimerization of alkynes with nitriles.

When symmetrically substituted alkynes are used for cyclotrimerization, only one product is formed. However, in the case of terminal or unsymmetrical alkynes, a mixture of several possible regioisomers might be formed as the result of the formation of variously substituted metallacyclopentadienes (Scheme 2).

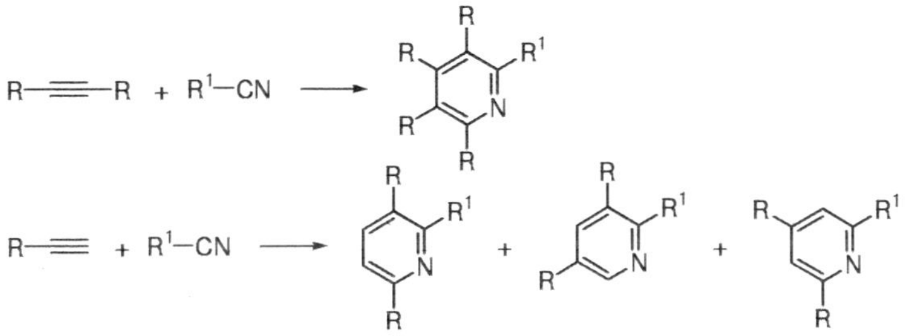

Scheme 2 is a chemical scheme showing cyclotrimerization reactions.

Scheme 2. The regioisomer formation in the cyclotrimerization of alkynes and nitriles.

1.3. Catalysts

A number of transition metal compounds are able to catalyze the selective cyclotrimerization of alkynes with nitriles. The use of various catalysts affecting the cyclotrimerization has been recently reviewed by Saá and Varela in 2003[2d] and 2008[2h] thus this area will be focused just on general aspects and the summary of catalysts or catalytic systems with the pertaining references is given in Table 1. Although a great deal of transition metal compounds has been used to affect the cyclotrimerization of alkynes and nitriles it is important to emphasize the pioneering works of Wakatsuki and Yamazaki[4] as well as that of Bergmann[5] and Bönnemann.[6]

Among the catalysts most often used there is the commercially available CpCo(CO)$_2$, known also as Vollhardt's catalyst and its derivatives such as CpCo(cod) and unstable, but highly reactive, CpCo(CH$_2$=CH$_2$)$_2$, known as Jonas' catalyst. The cobalt-catalyzed pyridine synthesis can be carried out also by *in situ* generated species from CoCl$_2$·6H$_2$O/NaBH$_4$. Popularity of the cobalt based-catalysts arises from their robustness and often the high effectiveness/price ratio and reasonably wide tolerance to various functional groups. Even today the use of CpCo(CO)$_2$ is widespread and the carrying out of the reactions under different reaction conditions (such as light irradiation at 0–25 °C, microwave irradiation, appropriate choice of the solvent) has enabled to expand horizons of its application. It is also worth mentioning that the *in situ* generation of catalytically active species from CoCl$_2$·6H$_2$O/dppe/Zn or Co(dppe)I$_2$/Zn brings considerable simplifications. The use of cobalt compounds is followed mainly by rhodium and recently by ruthenium complexes as well. In addition, it has been shown by Louie that cyclotrimerization of alkynes with nitriles can be catalyzed also by nickel-carbene complexes. A rare example of the Pd-complex catalysed cyclotrimerization of nitrile with diphenylacetylene was also reported (for pertaining references, see Table 1).

Last but not least, it should be mentioned that the pyridine ring could also be constructed by using stoichiometric processes. Typically those are based on the use of early transition metal compounds such as Zr and Ti. As for the former, it has been shown by Takahashi *et al.*[7] that the reaction of a preformed azazirconacyclopentadiene with an alkyne in the presence of the stoichiometric amount of NiCl$_2$(PPh$_3$)$_2$ resulted in the formation of highly substituted pyridines (Scheme 3). It is assumed that azanickelacyclopentadiene is the reactive intermediate able to undergo insertion of the second alkyne. Since this methodology is based on the sequential formation of azazirconacyclopentadiene from an alkyne and a nitrile, followed by the addition of a second alkyne, it enabled the regioselective synthesis of pyridines from two different alkynes (Figure 1). As for the titanium-based cyclotrimerization, the early examples of titanacyclopentadiene reactions with nitriles to give substituted pyridines were reported by Rothwell.[8]

Table 1. Catalytic systems for alkyne/nitrile cyclotrimerization.

Metal	Catalysts	Conditions	References
Co	CpCo(CO)$_2$	heating	5, 13, 14, 17, 25–27, 40, 44, 53, 55, 62, 64, 75
	Co-cyclopentadienes		4, 63
	CpCo(CO)$_2$	solvent effect	20
	CpCo(CO)$_2$	mw	16, 21–23, 41–43, 54
	CpCo(cod)		18, 19, 24, 31–36, 38–40, 66
	CpCo(cod)	hv, 0–25 °C	36, 45, 46
	CpCo(CH$_2$=CH$_2$)$_2$		15, 27, 40, 67, 75
	CpCo(CO)(fumarate)		77
	Cp$_2$Co		12
	CoCl$_2$·6H$_2$O/NaBH$_4$		6, 66
	CoCl$_2$·6H$_2$O/dppe/Zn		56
	Co(dppe)I$_2$/dppe/Zn		57
	CoCl$_2$/Mn or Zn		65
	Co-vapour		66, 69
	Co(0) complexes		70
Rh	[Rh(L)]$^+$X$^-$ [a]		47, 48, 49, 76
	Rh-complexes		68, 71, 73, 74
Fe	Fe(CH$_2$=CH$_2$)$_2$(toluene)		72
Ru	Cp*RuCl(cod)		3b, 28, 59
	Cp*Ru(MeCN)$_3$PF$_6$		58
Ni	Ni(cod)$_2$/SIPr [b]		60
Pd	[C$_3$H$_5$PdCl]$_2$/PPh$_3$		61
Ta	Ta(DIPP)$_2$Cl$_3$·(OEt$_2$) [c]		11a
	TaCl$_5$/Zn		11b

[a]L=bidentate chiral phosphines. [b]SIPr=1,3-bis-(2,6-diisopropylphenyl)-4,5-dihydroimidazol-2-ylidene. [c]DIPP=2,6-diisopropylphenoxy.

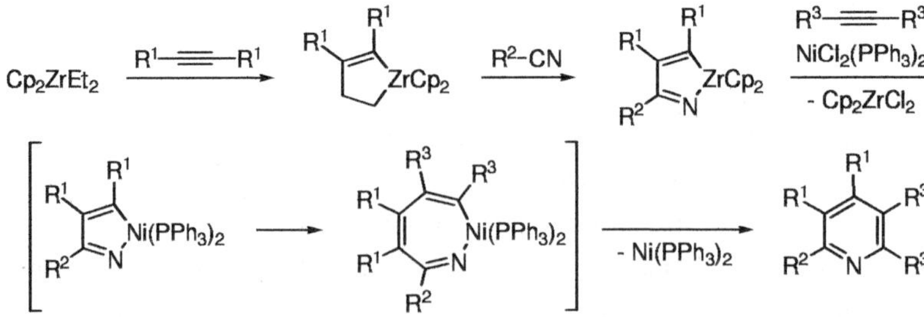

Scheme 3. The reaction mechanism of the Zr/Ni-mediated pyridine formation.

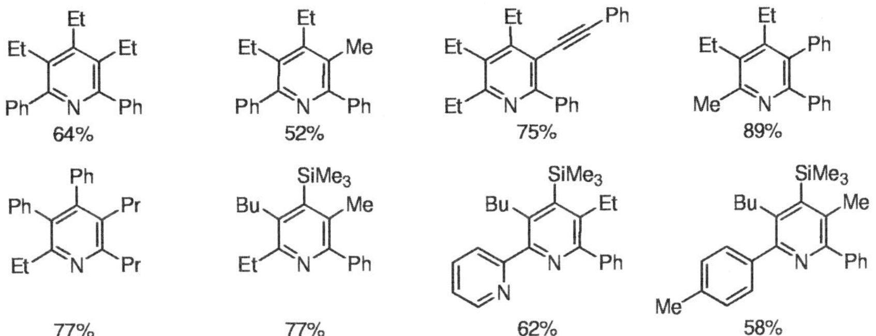

Figure 1. Typical examples of the Zr/Ni-mediated pyridine synthesis.

These results served as the basis for another selective formation of substituted pyridines from two different alkynes and a nitrile developed by Sato *et al.* (Scheme 4).[9] The advantage of the Ti-based methodology is in tolerance of ester groups present in reactants, which are otherwise incompatible with the Zr-based procedure. Fe-mediated reaction of two phenylethynes with acetonitrile was reported[10] as well. A tantalum complex[11a] or tantalum species[11b] formed by the reduction of $TaCl_5$ in the presence of Zn were also reported to mediate cyclotrimerization of alkynes with nitriles.

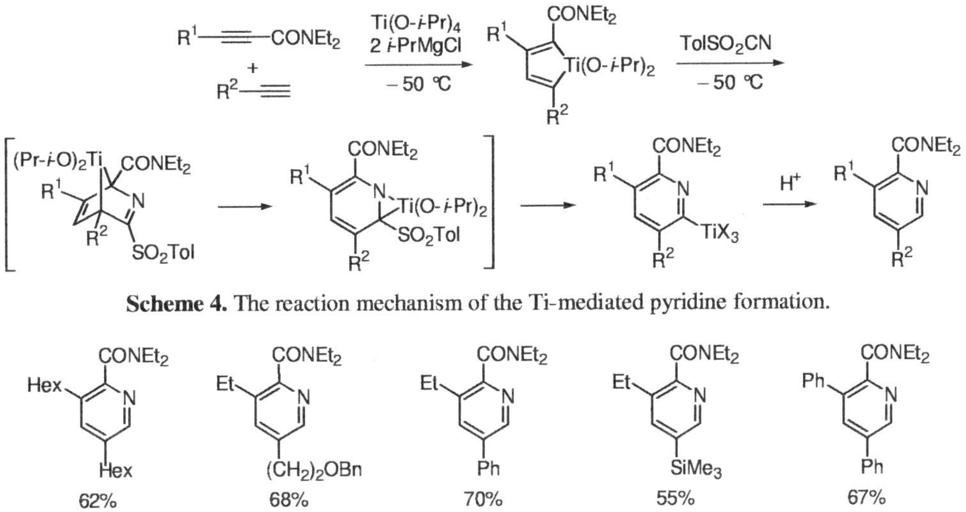

Scheme 4. The reaction mechanism of the Ti-mediated pyridine formation.

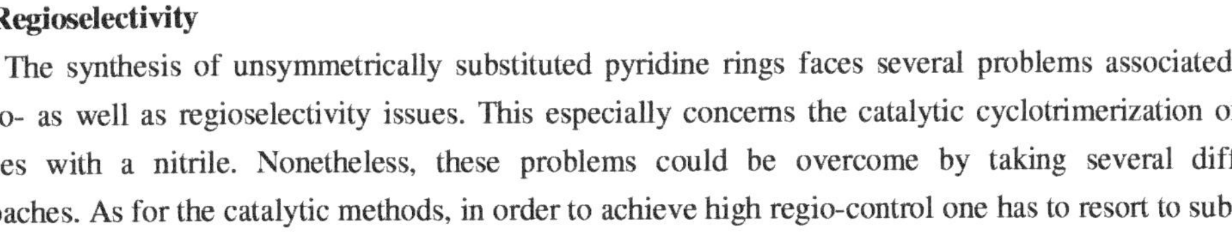

Figure 2. Some typical examples of the Ti-mediated pyridine synthesis.

1.4. Regioselectivity

The synthesis of unsymmetrically substituted pyridine rings faces several problems associated with chemo- as well as regioselectivity issues. This especially concerns the catalytic cyclotrimerization of two alkynes with a nitrile. Nonetheless, these problems could be overcome by taking several different approaches. As for the catalytic methods, in order to achieve high regio-control one has to resort to substrate limitations. High degree of chemoselectivity can be reached in cyclotrimerization of two identical alkynes with a nitrile as far as the formation of pyridine is concerned (Figure 3, **A**); however, the use of unsymmetrically substituted alkynes or two different alkynes (see Scheme 2) leads usually to a statistical mixture of all possible isomers. High degree of regioselectivity as well as chemoselectivity is achieved in cyclotrimerization of a diyne with a nitrile (Figure 3, **B**) and an ynenitrile with an alkyne (Figure 3, **C**). The methods **B** and **C** are complementary and give regioisomeric bicyclic pyridines. All these approaches have been utilized in syntheses of natural compounds.

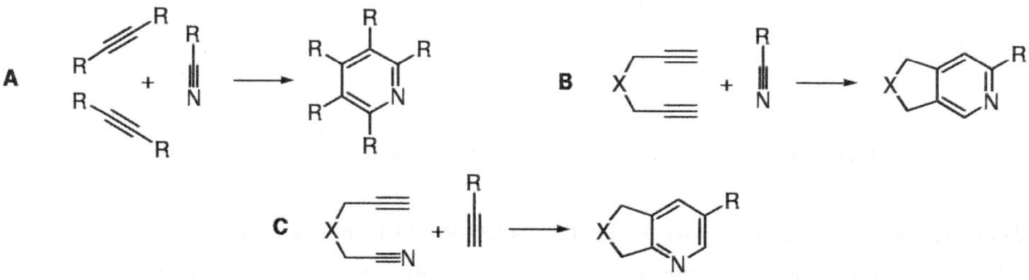

Figure 3. Different ways of the cyclotrimerization of alkynes with nitriles.

A stepwise construction of the pyridine framework based on the use of stoichiometric processes deserves special attention. This was already recognized in early papers on the cyclotrimerization of alkynes with nitriles: the stepwise formation of unsymmetrically substituted cobaltacyclopentadienes was followed by their reaction with nitriles. However, the necessity to purify intermediate cobalt complexes and also the formation of regioisomeric products in the last step make this approach impractical.[4] These problems were solved almost thirty years later thanks to advances in metallacyclopentadiene chemistry of zirconium and titanium.

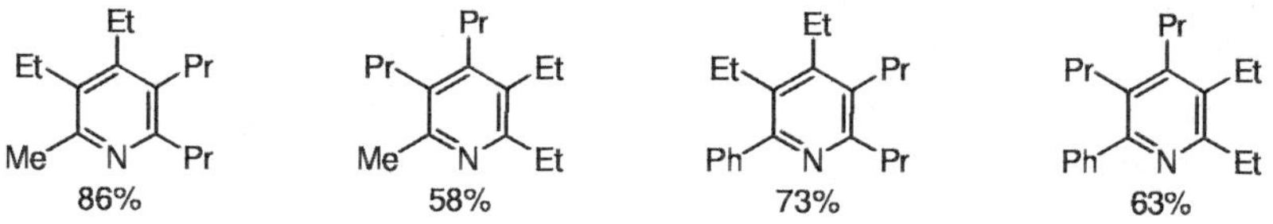

Figure 4. Selected examples of the regioselectively prepared pentasubstituted pyridines.

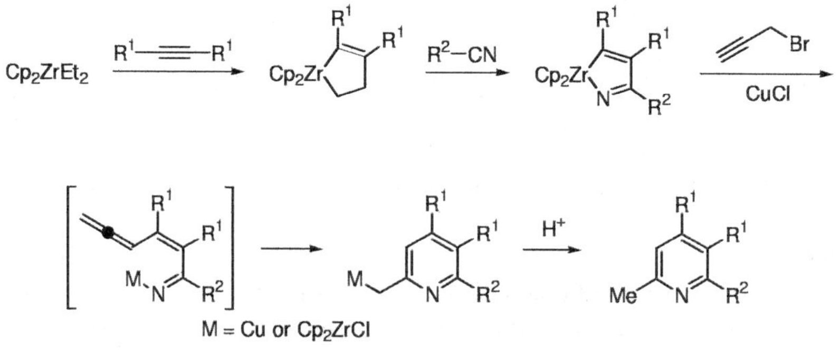

Scheme 5. The reaction mechanism of the Zr/Cu-mediated pyridine formation.

In the case of the polysubstituted pyridine synthesis, the stepwise cyclotrimerization approach based on the sequential reaction of alkynes and a nitrile through metallacycles could provide high degree of regiocontrol with respect to the substituent arrangement on the pyridine ring. Such is the example of the Zr-mediated synthesis of pentasubstituted pyridines (Scheme 3) where the switch in order to use two alkynes could lead to the synthesis of regioisomeric pyridines (Figure 4).[7] An alternative regioselective synthesis of tetrasubstituted pyridines is based on a similar approach; however, azazirconacyclopentadiene is allowed to react with propargyl bromide in the presence of the stoichiometric amount of CuCl (Scheme 5, Figure 5).[7b] Owing to the reaction mechanism regioselectively substituted pyridines can be also prepared by the titanium-mediated selective co-cyclotrimerization of two different alkynes.[9]

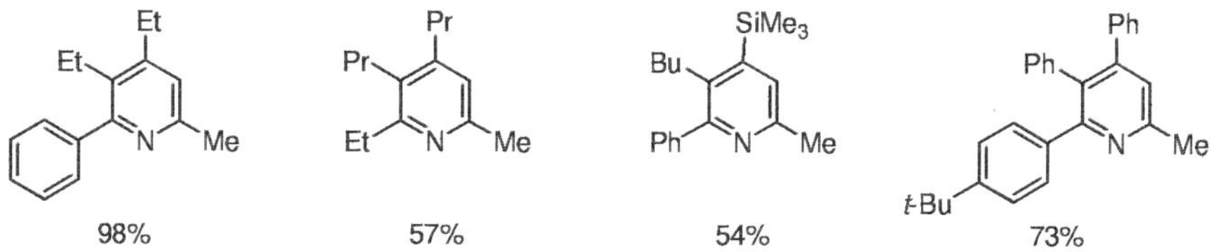

Figure 5. Selected examples of the regioselectively prepared 2,3,4,6-substituted pyridines.

2. Synthesis of natural compounds and their congeners

The pyridine ring is the essential structural feature in several groups of compounds found in Mother Nature. One of the most well known naturally occurring compounds possessing pyridine ring, albeit structurally simple, is probably (*S*)-nicotin, 3-[2(*S*)-1-methylpyrrolidin-2-yl]pyridine. Although the cyclotrimerization of alkynes with nitriles to pyridines constitutes an efficient synthetic method for the construction of six-membered heterocyclic nucleus, this synthetic strategy has been in last 30 years applied for synthesis of natural compounds, derivatives thereof and biologically active compounds in just a few cases so far.

One of the first cyclotrimerizations of alkynes with nitriles used in the synthesis of natural compounds or their derivatives appeared as early as in the mid 1980s, when there were almost simultaneously published two independent reports on the synthesis of pyridoxine (vitamin B6) by the group of Schleich[12] and that of Vollhardt.[13] The synthetic strategy of both the syntheses was similar, *i.e.* they were based on the co-cyclotrimerization of a diyne with a nitrile. The former synthesis (Schleich) started with the reaction of bis(trimethylsilyl)propargyl ether in the presence of an acetonitrile excess catalyzed by rarely used Cp$_2$Co. It provided the crucial intermediate bistrimethylsilylated cyclopenta[c]pyridine, which, after the additional elaboration, gave desired pyridoxine hydrochloride (vitamin B6) (Scheme 6).

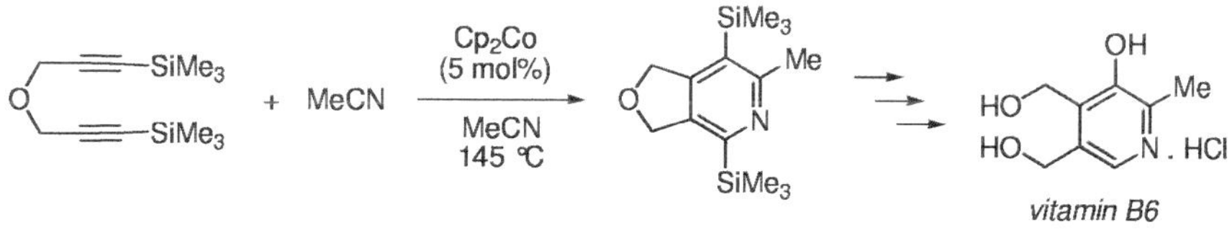

Scheme 6. The synthesis of vitamin B6 (Schleich).

The latter one (Vollhardt) applied the same strategy but used bis(trimethylstannyl)propargyl ether and CpCo(CO)$_2$ as the catalyst instead. The bistrimethylstannylated cyclopenta[c]pyridine obtained was subjected to additional functional group manipulation to give rise to vitamin B6 hydrochloride (Scheme 7).

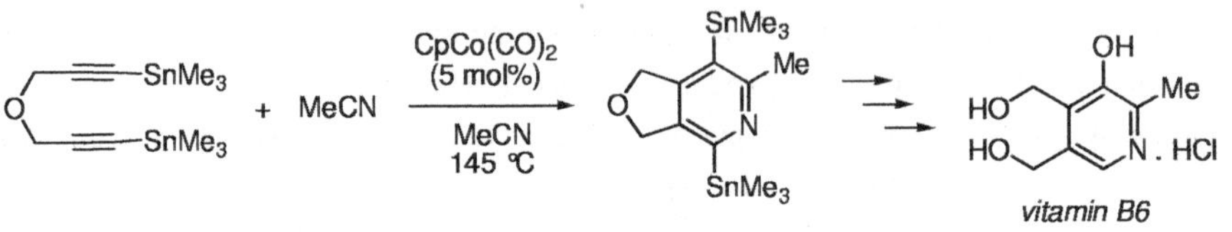

Scheme 7. The synthesis of vitamin B6 (Vollhardt).

The concept of the cyclotrimerization of an yne-nitrile with an alkyne was the underlying strategy for the synthesis of ergot alkaloid lysergene by Vollhardt.[14] The reaction of the yne-nitrile with the indol skeleton catalyzed by CpCo(CO)$_2$ yielded a tetracyclic compound with the ergot skeleton (Scheme 8) in regioselective manner. The target compound, lysergene, was obtained by further synthetic operations. An alternative approach to this type of alkaloids was outlined by Groth.[15] His strategy was based on the fully intramolecular cyclization of a diyne-nitrile having the alkyne moieties connected through a silicon tether (Scheme 9). Its cyclotrimerization was carried out under catalysis of CpCo(CH$_2$=CH$_2$)$_2$; interestingly, the use of CpCo(CO)$_2$ as the catalyst did not bring about the desired transformation. Moreover, the obtained tetracyclic intermediate showed considerable resistance to desired functional group transformations, namely the cleavage of the silicon tether. Thus its conversion to the desired ergot alkaloid derivative – lysergene – was not accomplished.

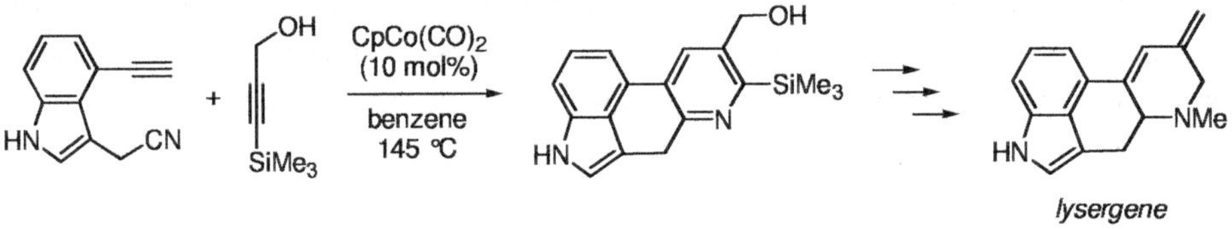

Scheme 8. The synthesis of lysergene based on intermolecular cyclotrimerization.

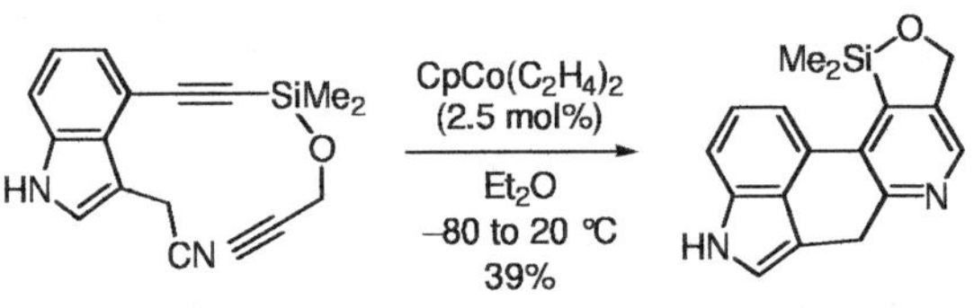

Scheme 9. The synthesis the lysergene framework based on intramolecular cyclotrimerization.

The reaction of a diyne with a nitrile bearing a suitable functional group in the side chain reported by Deiters *et al.* was the underlying strategy for the cyclotrimerization/intramolecular alkylation sequence utilized for the synthesis of phenanthroindolizidine class of alkaloids.[16] [2+2+2]-Cyclotrimerization of 2,2'-diethynylbiphenyl with 4-mesylbutanenitrile catalyzed by CpCo(CO)$_2$ was carried out under microwave irradiation (Scheme 10). In the course of the reaction, the non-nucleophilic *N*-centre of the nitrile was converted into the nucleophilic pyridine moiety that subsequently underwent an intramolecular nucleophilic substitution with tethered sulphonate leaving group. This one-pot reaction directly yielded dehydrotylophorine in 78% yield. Its reduction with NaBH$_4$ gave rise to tylophorine in quantitative yield.

The cyclotrimerization was also exploited in the synthesis of the heterocyclic moiety of antitumor alkaloid camptothecine by Vollhardt *et al.*[17] On the contrary to previous cases, a substrate with isocyanate moiety was used instead of the nitrile. The key synthetic operation was the cyclotrimerization of 1-isocyanato-1-pentyne with an unsymmetrically functionalized alkyne to give a bicyclic intermediate that was subsequently transformed into the target molecule (Scheme 11).

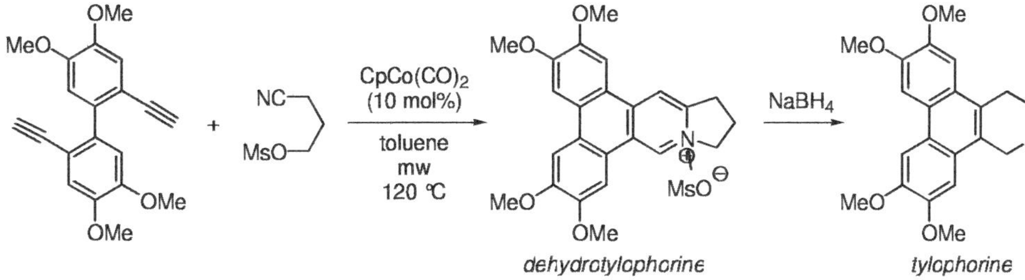

Scheme 10. The synthesis of tylophorine.

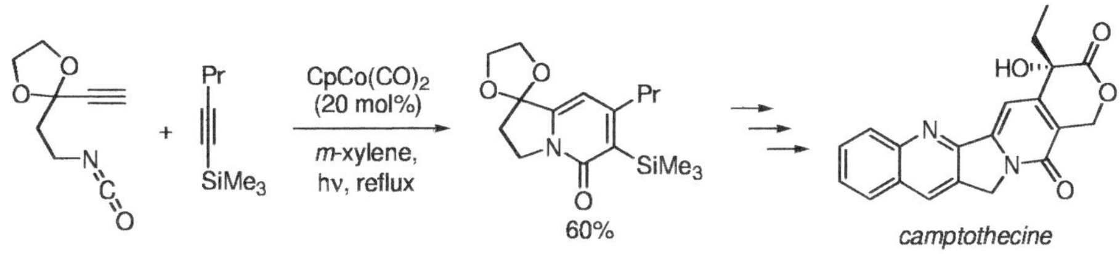

Scheme 11. The synthesis of camptothecine.

As for analogs of natural compounds, the synthesis of 2-nicotine (regioisomer of nicotine) by Chelucci *et al.* should be mentioned.[18] In this case, an intermolecular cyclotrimerization of a chiral nitrile with ethyne was applied. The pyridine ring was assembled by the reaction of ethyne with a boc-protected (2S)-2-cyanopyrrolidine, prepared from L-proline, catalyzed by CpCo(cod) under standard thermal conditions (Scheme 12). The corresponding chiral 2-(2-pyrrolidinyl)pyridine was obtained in very good 80% yield. Finally, the deprotection provided 2-nornicotine (R=H) and its methylation under usual conditions yielded (2S)-2-nicotine (R=Me).

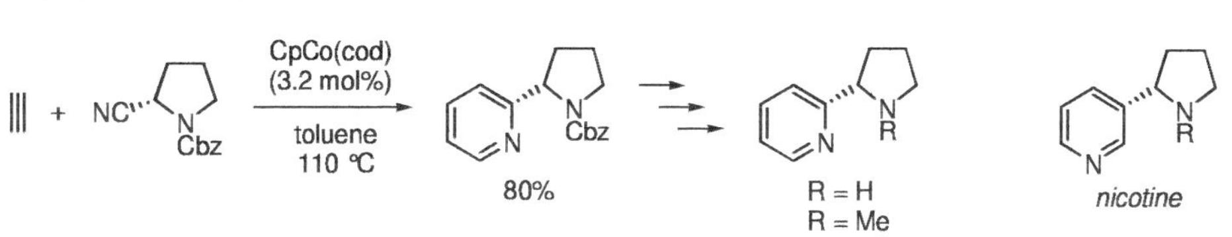

Scheme 12. The synthesis of 2-nicotine.

3. Synthesis of potentially biologically active compounds

Since the pyridine ring is found in many biologically active compounds (see above), the cyclotrimerization reaction offers an attractive approach for the synthesis of various compounds with potential medicinal application. One of the early examples was the synthesis of pheniramines that constitute an important class of antihistaminics agents by Botteghi *et al.* (Scheme 13).[19] Thus the cyclotrimerization of benzyl nitriles with variously substituted aryl moiety with ethyne catalyzed by CpCo(cod) gave the target compounds in good yields: 85% (X=H), 75% (X=Cl) and 70% (X=Br).

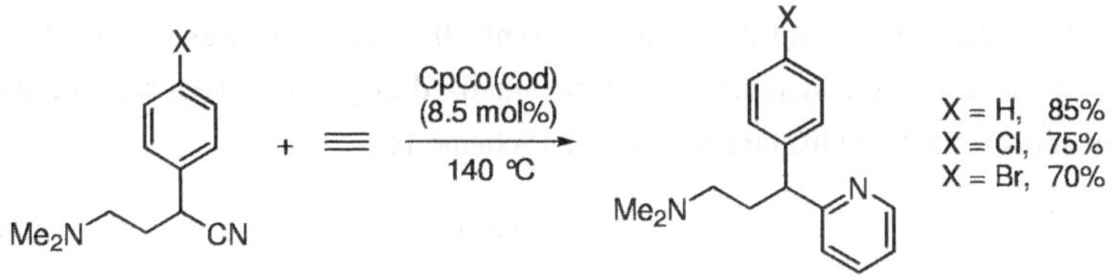

Scheme 13. The synthesis of pheniramines.

Recently the cyclotrimerization of nitriles with various silicon-tethered diynes was used by Schreiber *et al.* to generate a library of potential inhibitors of neuregulin-induced neurite outgrowth.[20] The reaction of silyloxaheptadiynes or silyoxaoctadiynes with nitriles catalyzed by CpCo(CO)$_2$ (25–30 mol%) at 140 °C in THF provided good yields of the desire bicyclic silylpyridines without the necessity of any external irradiation (Scheme 14).

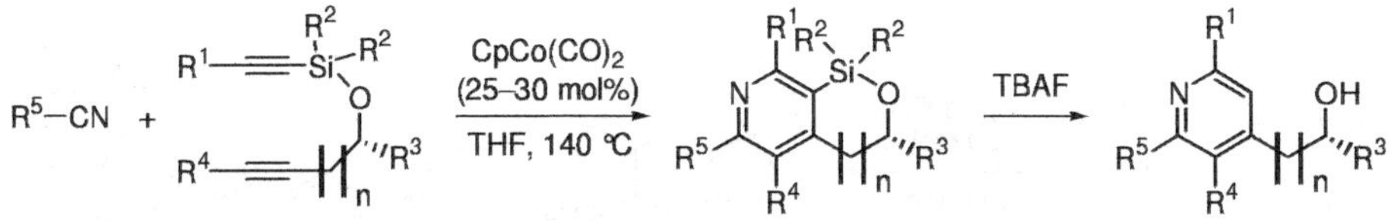

Scheme 14. The cyclotrimerization of nitriles with silyloxadiynes.

In some cases, better results were obtained by using the stoichiometric amount of the Co-complex with respect to the reactant (Figure 6).

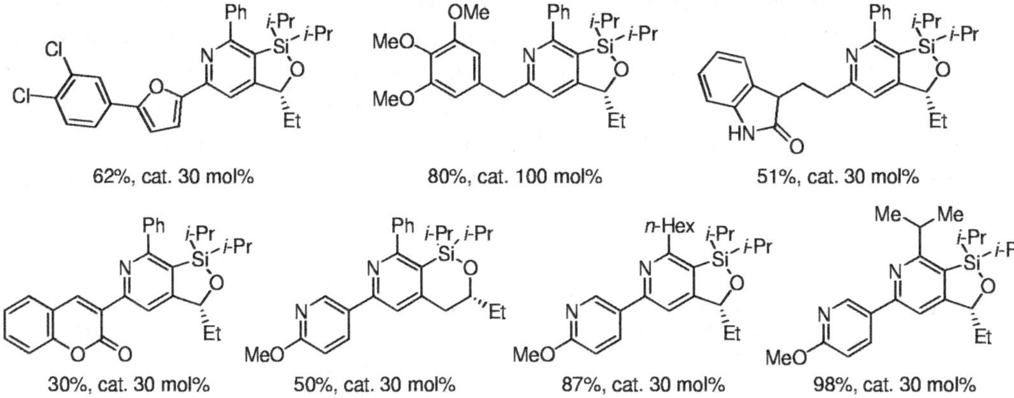

Figure 6. Some typical examples of the cyclotrimerization of silicon-tethered diynes with nitriles.

The bicyclic silylpyridines and monocyclic pyridines, prepared by treating the former with TBAF, were tested for the aforementioned biological activity. Using this assay, 1-[2-phenyl-6-(3,4,5-trimethoxy-benzyl)-pyridin-4-yl]-propan-1-ol (Figure 7) was discovered to be a potent inhibitor of the neuregulin/ErbB4 pathway, with an approximate EC$_{50}$ of 0.30 μM, while its bicyclic precursor was inactive at 20 μM.

A similar approach, *i.e.* the fully intramolecular cyclotrimerization of a diyne-nitrile, reported by Snyder *et al.*, was applied also for the synthesis of a library of unnatural heterocycles – tetrahydro-naphthyridines.[21] Suitably substituted diyne-nitriles were used as starting material in which the alkyne and nitrile functionalities were tethered by N and O atoms. The cyclotrimerization catalyzed by CpCo(CO)$_2$ under microwave irradiation led to the formation of four basic scaffolds that were further elaborated by functional group transformations (Scheme 15). The testing of the prepared library of compounds against *Mycobacterium tuberculosis* revealed three lead structures with promising MIC values (Figure 8).

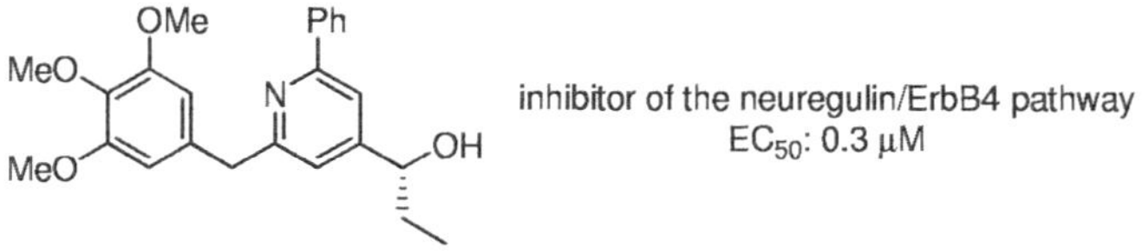

Figure 7. A potent inhibitor of the neuregulin/ErbB4 pathway.

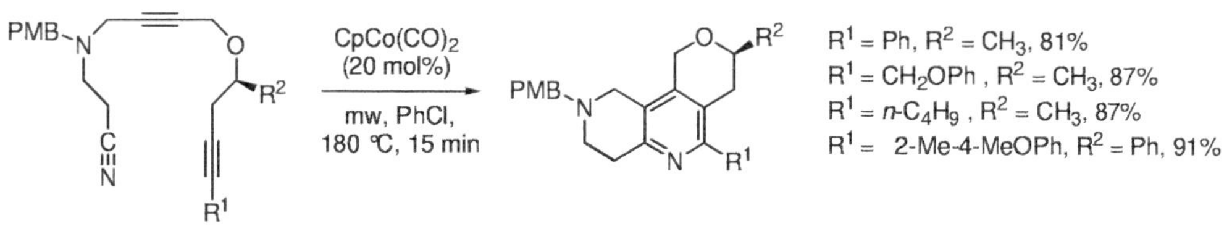

Scheme 15. The intramolecular cyclotrimerization of diyne-nitriles into tricyclic heterocycles.

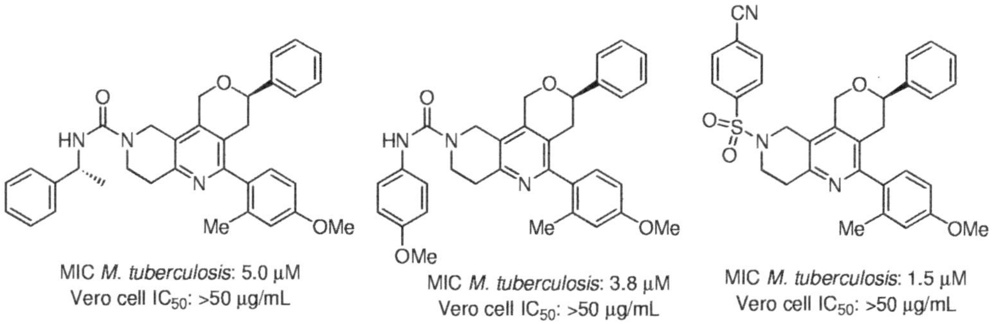

Figure 8. Three lead compounds showing high activity against *Mycobacterium tuberculosis*.

The cyclotrimerization of a diyne with a nitrile was also used for synthesis of new types of bicyclic 6-(tetrahydroisoquinolinyl)purines as potential purine based cytostatics (Kotora *et al.*).[22] In this instance, the successful reaction of purinyldiynes with nitriles was carried out in the presence of the stoichiometric or catalytic amount of CpCo(CO)$_2$ under microwave irradiation (Scheme 16). It should be emphasized that the application of microwave irradiation was essential for the successful course of the reaction (isolated yields were within the range of 29–89%), because under usual thermal conditions it gave the desired products either in low yields or it did not proceed at all. It should be noted that, in many cases, the yields obtained under catalytic conditions were superior to those obtained with the stoichiometric amount of the cobalt complex. Few typical examples of the prepared 6-heteroarylpurines are given in Figure 9. The cyclotrimerizations could also be expanded to the synthesis of 1,4-bis(purin-6-yl)tetrahydroisoquinolines (Scheme 17) under the identical conditions. The products were obtained in 28–42% isolated yields. Unfortunately, the further testing of the prepared compounds for cytostatic activity did not identify any promising leads among them.

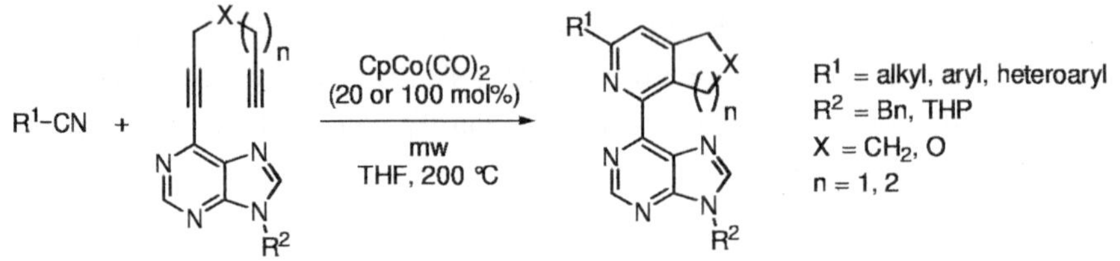

Scheme 16. The synthesis of 6-(pyridyl)purines.

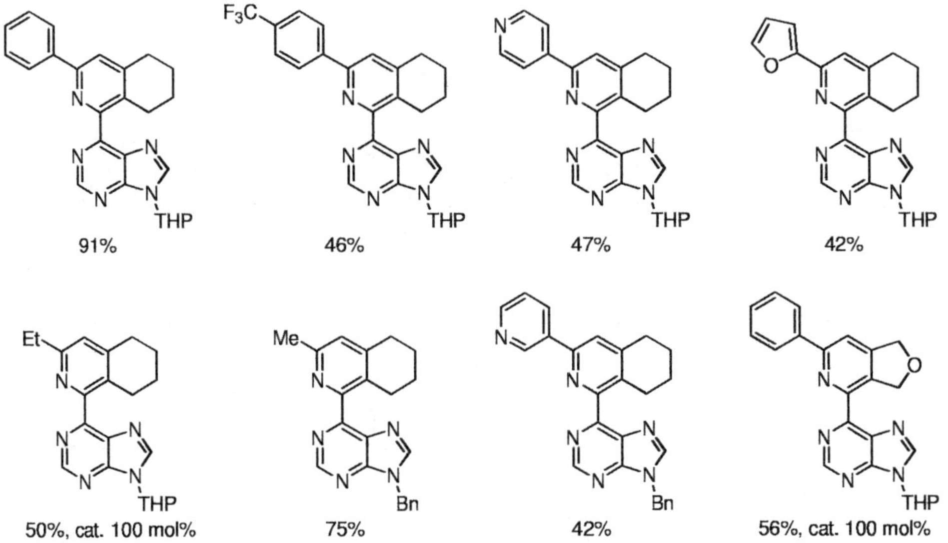

Figure 9. Some typical examples of the cyclotrimerization of 6-diynylpurines with nitriles (20 mol% of the catalyst unless otherwise noted).

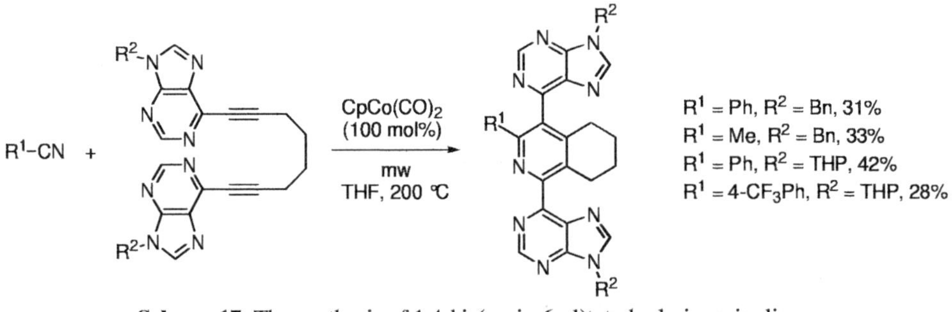

Scheme 17. The synthesis of 1,4-bis(purin-6-yl)tetrahydroisoquinolines.

The cyclotrimerization of 1,2-bis(propyn-3-yl)benzene with various nitriles was reported to be a convenient method for the rapid and high yielding synthesis of 2-azaanthracenes (Scheme 18).[23] The catalysis of the reaction by CpCo(CO)₂ under microwave irradiation led to almost quantitative yields of the corresponding products. The prepared 2-azaanthracenes exhibited generally higher fluorescence levels than the similar anthracenes.

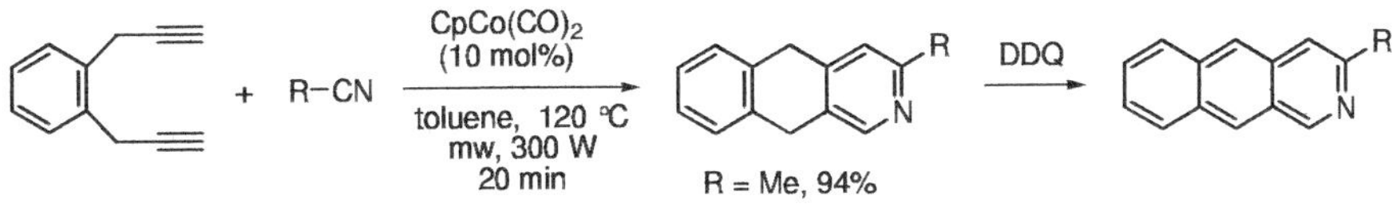

R = Me, 94%
R = Pr, 87%
R = $(CH_2)_2OH$, 83%
R = $CH=CH_2$, 90%
R = Ph, 87%
R = Pyr, 80%

Scheme 18. The synthesis of fluorescent 2-azaanthracenes.

4. Synthesis of pyridines and bipyridines

The synthesis of polypyridine compounds and namely bipyridines – potential ligands for transition metal compounds – has attracted considerable interest since the beginning of the application of the cyclotrimerization of alkynes with nitriles. One of the first examples was demonstrated by Bönnemann *et al.* who showed that the cyclotrimerization of cyanopyridines with ethyne catalyzed by CpCo(cod) led to the formation of the corresponding bipyridines (Scheme 19).[24] The reaction was carried out either with ethyne or terminal alkynes. Thus the reaction of 2-cyanopyridine with the former gave rise to unsubstituted 2,2'-bipyridines, whereas its reaction with the latter provided a mixture of 3,6- and 4,6-disubstituted 2,2'-bipyridines. Similar results were obtained with 3-cyano- and 4-cyanopyridines as well. This concept has later become the standard strategy for the synthesis of chiral pyridine base ligands (*vide infra*).

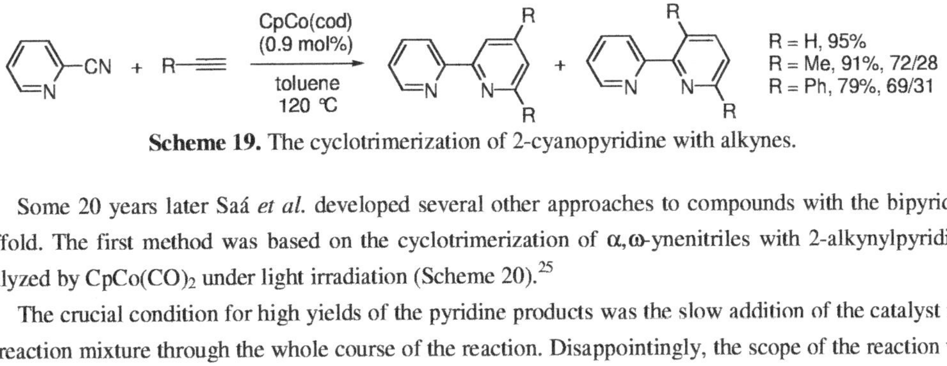

Scheme 19. The cyclotrimerization of 2-cyanopyridine with alkynes.

Some 20 years later Saá *et al.* developed several other approaches to compounds with the bipyridine scaffold. The first method was based on the cyclotrimerization of α,ω-ynenitriles with 2-alkynylpyridines catalyzed by $CpCo(CO)_2$ under light irradiation (Scheme 20).[25]

The crucial condition for high yields of the pyridine products was the slow addition of the catalyst into the reaction mixture through the whole course of the reaction. Disappointingly, the scope of the reaction was highly depending on the structure of substrates and also regioselectivity was rather poor: the products were always obtained as a mixture of 2,2'- and 2,3'-bipyridines in favour of the latter. This procedure was also used with some success for synthesis of terpyridines.

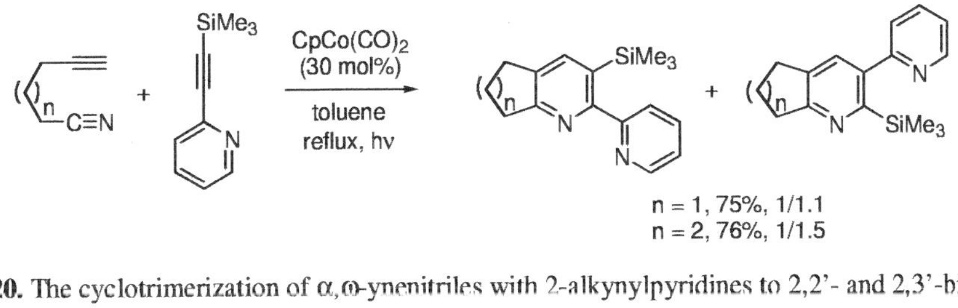

n = 1, 75%, 1/1.1
n = 2, 76%, 1/1.5

Scheme 20. The cyclotrimerization of α,ω-ynenitriles with 2-alkynylpyridines to 2,2'- and 2,3'-bipyridines.

The same authors later developed an alternative procedure that was based on the cyclotrimerization of 5-hexynenitrile with 1,4-substituted 1,3-butadiynes, allowing to synthesize compounds with the bipyridine scaffold in one step. As in the aforementioned case, the reaction was catalyzed by $CpCo(CO)_2$ under light irradiation (Scheme 21).[26] Unlike the previous cases, regioselectivity was shifted in favour of the formation of 2,2'-bipyridines.

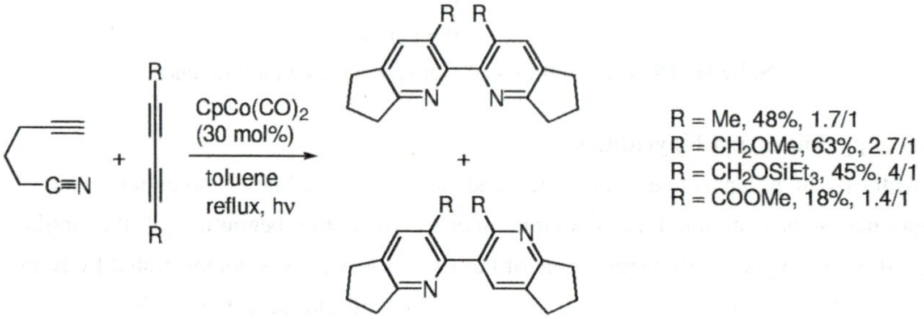

R
C≡N
R
CpCo(CO)2
(30 mol%)
toluene
reflux, hv
R R
N
N
R R
N
N
+
R = Me, 48%, 1.7/1
R = CH2OMe, 63%, 2.7/1
R = CH2OSiEt3, 45%, 4/1
R = COOMe, 18%, 1.4/1

Scheme 21. The cyclotrimerization of 5-ynenitrile with 1,3-butadiynes.

As for the synthesis of other compounds possessing several pyridine rings, the cyclotrimerization of bis-alkynylnitriles with alkynes is worth mentioning. The procedure was also developed by the group of Saá and opened a new approach to C_2-symmetric ligands.[27] The reaction was again catalyzed by $CpCo(CO)_2$ under light irradiation, but attempts to improve the yield by using CpCo(cod) and $CpCo(CH_2{=}CH_2)_2$ did not materialize (Scheme 22). In general, the cyclotrimerization provided the desired compounds – bis-pyridines – in rather low yields (7–33%) strongly depending on the substituents attached to the alkyne triple bond.

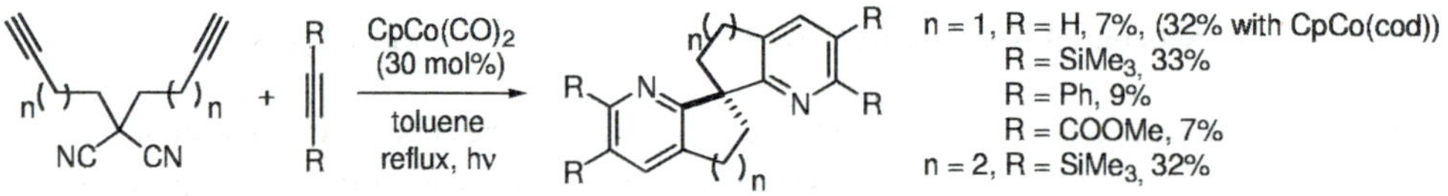

NC
CN
n
n
R
R
CpCo(CO)2
(30 mol%)
toluene
reflux, hv
R
R
N
N
R
R
n
n
n = 1, R = H, 7%, (32% with CpCo(cod))
R = SiMe3, 33%
R = Ph, 9%
R = COOMe, 7%
n = 2, R = SiMe3, 32%

Scheme 22. The cyclotrimerization of bis-ynenitriles with alkynes.

A similar method for the synthesis of bispyridinyl compounds was reported by Itoh *et al.* who utilized the cyclotrimerization of α,ω-diynes with nitriles.[28] Importantly, they demonstrated that Cp*RuCl(cod) is the alternative choice to the Co-based systems concerning the reaction conditions (lower reaction temperature, lower catalyst loads) and functional group variety attached to the diynes or nitriles. In addition they opened a new and efficient approach to bipyridines showing that they could be accessed in a single step by the cyclotrimerization of 1,6,8,13-tetradecatetraynes with nitriles (Scheme 23).

MeOOC
MeOOC
COOMe
COOMe
+ NC CN
Cp*RuCl(cod)
(10 mol%)
ClCH2CH2Cl
80 °C
95%
MeOOC
MeOOC
CN
N
N
COOMe
COOMe
NC

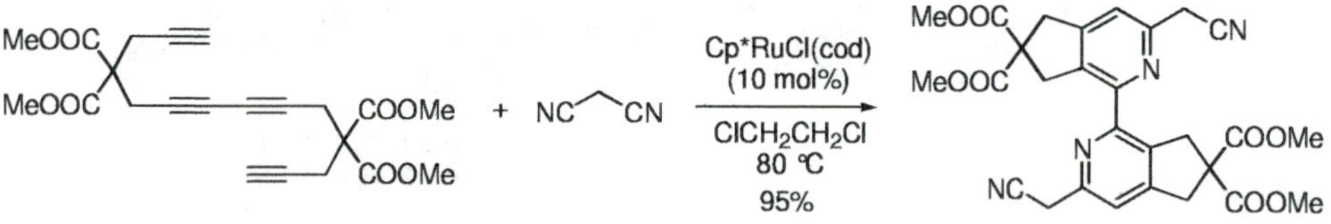

Scheme 23. The cyclotrimerization of a tetrayne with a nitrile.

This approach has later become the standard synthetic pathway to various bipyridines. Generally, it should be taken into the account that both catalytic systems are complementary and their choice depends on the substrates nature: Co-catalysts for electron-rich nitriles whereas Ru-catalysts for electron-poor nitriles.

5. Synthesis of pyridine based chiral ligands and catalysts

The discovery of chirality led to the development of new methods for obtaining enantioenriched compounds. The importance of chirality is underscored by the fact that nearly all natural compounds are chiral and their physiological and pharmacological properties depend upon their recognition by chiral receptor. Also many enantioselective syntheses are based on the use of transition metal complexes with chiral environment around the central metal atom. In this respect, a great deal of effort has been devoted to design and synthesise new chiral ligands. Moreover, the concept of organocatalysis and recent developments in enantioselective organocatalytic reactions indicate the importance of synthesizing of new pyridine-based Lewis basic catalysts. Since this area has been recently reviewed,[29] this part will be limited to the basic concepts of synthesizing of chiral pyridine ring containing compounds.

Generally, there are several strategies for the synthesis of chiral pyridines based on the use of the catalytic cyclotrimerization reaction: i) the cyclotrimerization with chiral nitriles, ii) the synthesis of racemic pyridine compounds followed by resolution, and finally, iii) the cyclotrimerization catalyzed by chiral catalysts (this will be dealt with in Section 6, because this approach has not yielded a useful ligand yet).

As for the first approach, *i.e.* the first cyclotrimerization with a chiral nitrile was reported already in 1975 by Botteghi *et al.*[30] who dealt with the cyclotrimerization of (+)-(*S*)-2-butyronitrile with ethyne catalyzed by CpCo(cod). It gave chiral pyridine in excellent 95% yield of high optical purity (Scheme 24).

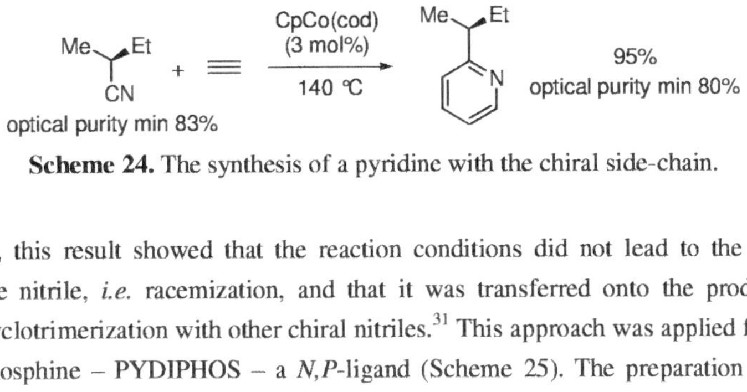

Scheme 24. The synthesis of a pyridine with the chiral side-chain.

Gratifyingly, this result showed that the reaction conditions did not lead to the loss of the chiral information of the nitrile, *i.e.* racemization, and that it was transferred onto the product. Soon, it was followed by the cyclotrimerization with other chiral nitriles.[31] This approach was applied for the synthesis of a chiral pyridylphosphine – PYDIPHOS – a *N,P*-ligand (Scheme 25). The preparation was based on the reaction of the chiral nitrile, prepared from diethyl L-(+)-tartrate, with ethyne (14 atm) and catalyzed by CpCo(cod) in toluene at 120 °C.[32] The pyridine intermediate was formed in excellent 91% yield.

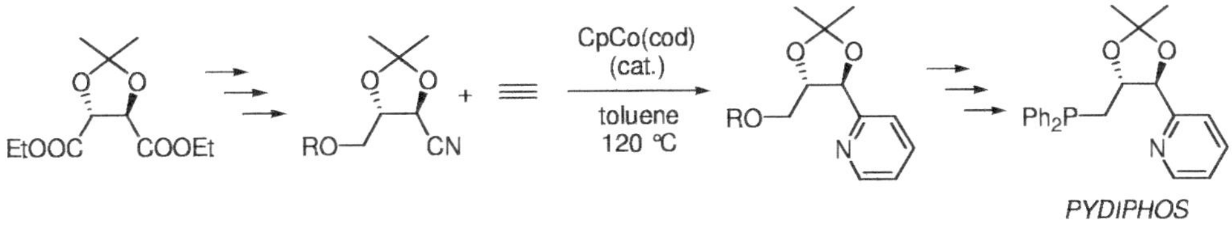

Scheme 25. The synthesis of pyridyl-phosphine *PYDIPHOS.*

Similarly, the reaction of a bisnitrile, also derived from the tartrate, with ethyne under the identical conditions yielded bispyridine in 70% yield (Scheme 26). The above mentioned approach was used to synthesize a number of various chiral hydroxyalkylpyridines,[33] aminoalkylpyridines[18,34] and also aminoalcohols.[35]

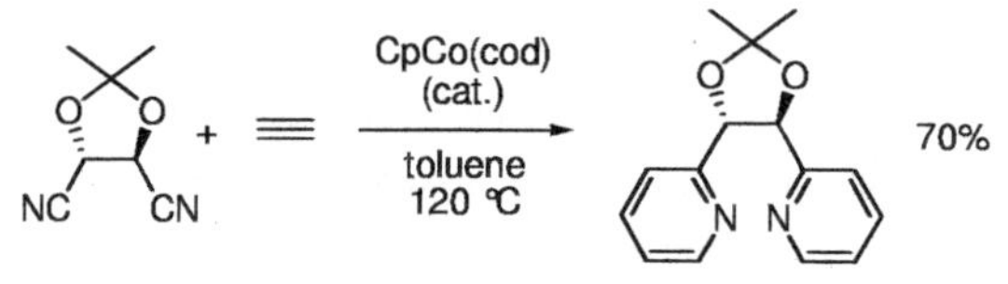

Scheme 26. The synthesis of chiral bispyridine.

Recently, it was shown by Heller *et al.* that the cyclotrimerization of alkynes with chiral nitriles could be carried out under much milder reaction conditions. They showed that catalysis of the reaction with CpCo(cod) (1 mol%) at 25 °C under light irradiation gave high yields of corresponding pyridines (Scheme 27).[36] Some typical examples are shown in Figure 10.

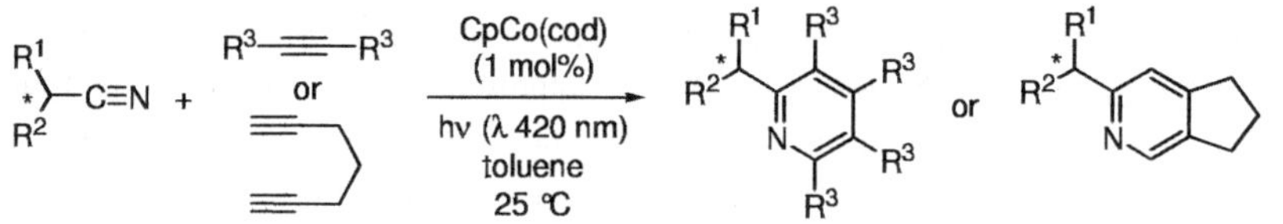

R^1 = alkyl; R^2 = alkyl, amido; R^3 = H, Me

Scheme 27. The synthesis of pyridines with a chiral side-chain.

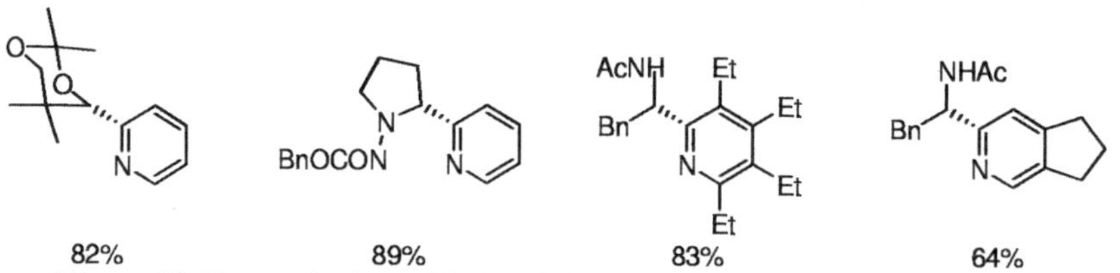

Figure 10. The synthesis of chiral pyridines by the reaction with chiral nitriles and chiral side-chain from nitriles.

The aforementioned approach was also successfully applied to the synthesis of chiral bipyridines. Chiral nitrogen containing ligands have distinct advantages over phosphines: i) they can often be employed in catalytic processes where the use of phosphines may be incompatible with the reaction conditions, ii) many nitrogen ligands are now available in the enantiomerically pure form and iii) ligands that bind through nitrogen are known to coordinate with a wide variety of metal ions and considerable progress has been made in understanding the role, which these ligands play in affecting catalytic processes. Out of variety of nitrogen ligands the most important are those with 2,2'-bipyridine scaffold.[37]

The first such an example was reported already in 1984[38] and was based on cyanation of (+)-(*S*)-2-*sec*-butylpyridine, which itself was prepared by the cyclotrimerization of (+)-(S)-2-*sec*-butyronitrile with ethyne (see Scheme 24), that provided two regioisomeric cyanopyridines (+)-2-cyano-6-((*S*)-*sec*-butyl)pyridine and (+)-4-cyano-(2-(*S*)-*sec*-butyl)pyridine. The subsequent cyclotrimerization of the obtained (+)-2-cyano-6-((*S*)-*sec*-butyl)pyridine provided the corresponding bipyridine in 80% yield (Scheme 28).

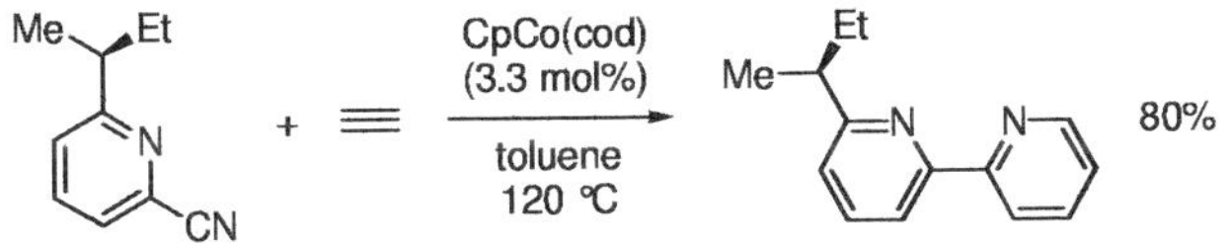

Scheme 28. The synthesis of a bipyridine with a chiral side-chain.

A similar approach was undertaken also in the synthesis of chiral bipyridine-phosphine (Scheme 29).[39] The bipyridine framework was constructed in a stepwise fashion: i) the cyclotrimerization of ethyne with a chiral nitrile (see Scheme 25), ii) the cyanation of the prepared chiral pyridine, and finally, iii) the cyclotrimerization of the prepared chiral cyanopyridine. The cyclotrimerization proceeded very well and provided the desired bipyridine in 86% yield. The cyclotrimerizations of other chiral cyanopyridines were carried out as well and in all instances the yields of the corresponding bipyridines were very high (up to 95%).[29a]

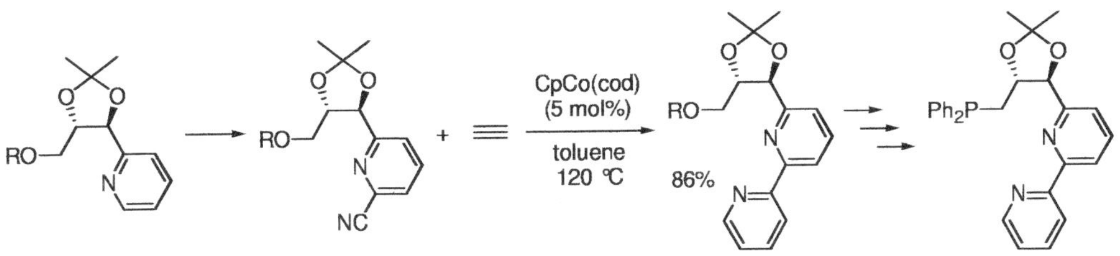

Scheme 29. The synthesis of a chiral bipyridine-phosphine.

As it has been mentioned above, the cyclotrimerization of suitably substituted alkynes (diynes) with nitriles can serve as a convenient pathway for the synthesis of various bipyridines. In this regard, a recent application of pyridine and bipyridine *N*-oxides as Lewis basic organocatalysts for activation of silicon compounds has provided an important impetus for renewed interest in this area. This approach has been taken by Kotora *et al.* who used cyclotrimerization of various diynes with nitriles for the synthesis of arylpyridines and bipyridines that were further used for the preparation of axially chiral *N*-oxides used in asymmetric allylation of aldehydes.[40,41] Thus the cyclotrimerization of 1-pyridyl-1,7-octadiynes with benzonitrile was carried out under two different conditions: A, classical thermal conditions and B, under microwave irradiation. The results obtained clearly show benefits of the conditions B, not only as far as yields are concerned but also in reaction time, which was shorten from 24 h to 20 minutes (Scheme 30).[41] The similar approach, *i.e.* the cyclotrimerization under microwave conditions of 1-isoquinolyl-1,7-octadiyne with various nitriles, was used by the same group for the synthesis of (tetrahydroisoquinolinyl)isoquinolines that were transformed into chiral (tetrahydroisoquinolinyl)isoquinoline *N,N'*-dioxides.[42]

A considerable improvement in the bipyridine synthesis was achieved later, when it was found that bipyridines derivatives – 1,1'-bis(tetrahydroisoquinolines) – could be prepared by the one-pot cyclotrimerization of hexadeca-1,7,9,15-tetrayne with nitriles under microwave irradiation (Scheme 31).[43] It should be emphasised that carrying out the reaction under conventional thermal conditions did not result in any reaction. When one takes into the account that the bipyridine formation consists in the formation of six new bonds and that the tetrayne is susceptible to oligomerization under the reaction conditions, then the isolated yields of the corresponding bipyridines (up to 50%) could be considered as very good. The selected bipyridines were oxidized to the corresponding N,N'-dioxides and, after separation into enantiomers, were used in asymmetric allylation of aldehydes. In addition, the cyclotrimerization was also carried out with chiral nitrile – 2-(R)-tetrahydrofuranecarbonitrile – yielding chiral bipyridine. Its oxidation provided two diasteroisomeric N,N'-dioxides (R,S,R) and (R,R,R) stereoisomers that could readily be separated by a simple column chromatography avoiding the often tedious and troublesome enantiomer separation. Few examples of chiral N,N'-dioxides are displayed in Figure 11.

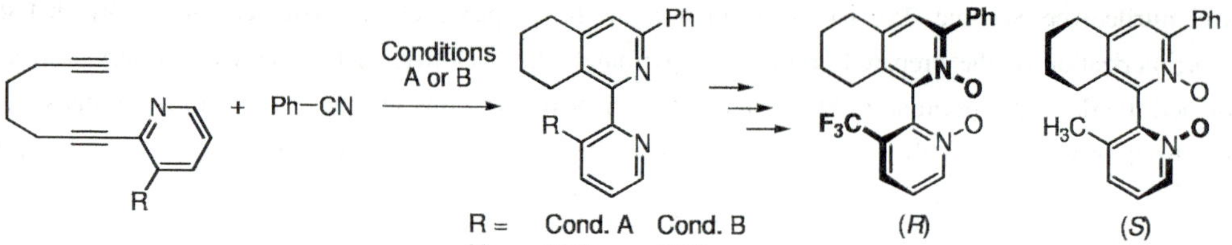

A: CpCo(CO)$_2$ (20 mol%), 140 °C, 24 h.
B: CpCo(CO)$_2$ (20 mol%), mw, 20 min.

R =	Cond. A	Cond. B
H	30%	43%
CH$_3$	27%	84%
CF$_3$	24%	35%
F	9%	65%
CN	18%	30%
OMe	0%	72%

Scheme 30. The synthesis of unsymmetrically substituted bipyridines.

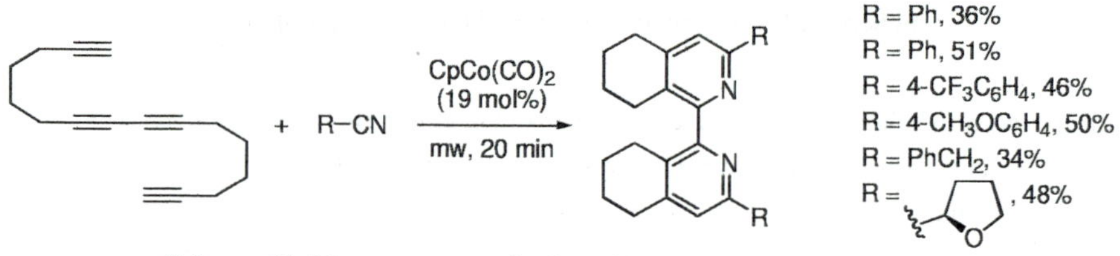

CpCo(CO)$_2$ (19 mol%), mw, 20 min

R = Ph, 36%
R = Ph, 51%
R = 4-CF$_3$C$_6$H$_4$, 46%
R = 4-CH$_3$OC$_6$H$_4$, 50%
R = PhCH$_2$, 34%
R = , 48%

Scheme 31. The one-pot synthesis of bis(tetrahydroisoquinolines).

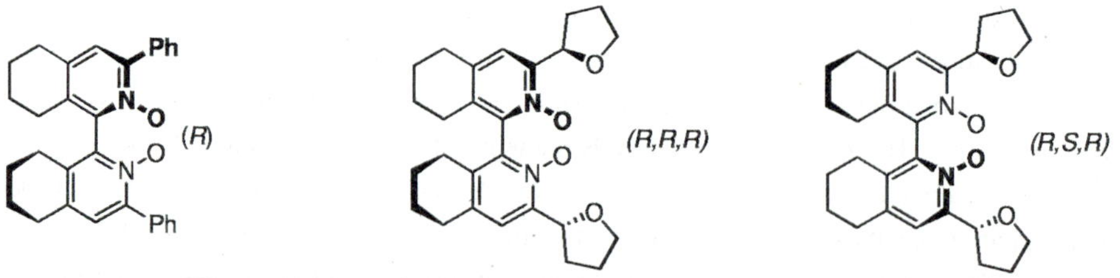

(R)   (R,R,R)   (R,S,R)

Figure 11. Typical examples of chiral bipyridine N,N'-dioxides.

The synthesis of chiral bipyridyl compounds with binapthyl scaffold is also worth mentioning. In this case the starting material, (R)-2,2'-dicyano-1,1'-binaphthyl, was cyclotrimerised with diynes yielding the corresponding (R)-bispyridyl-binaphthyls.[44] The reaction was catalyzed by CpCo(CO)$_2$ under standard thermal conditions (Scheme 32). Since racemization was not observed during the cyclotrimerization steps the expected products were obtained in ees over 99%.

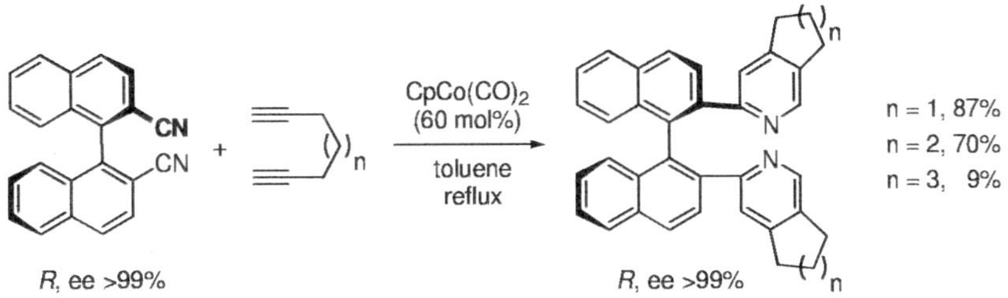

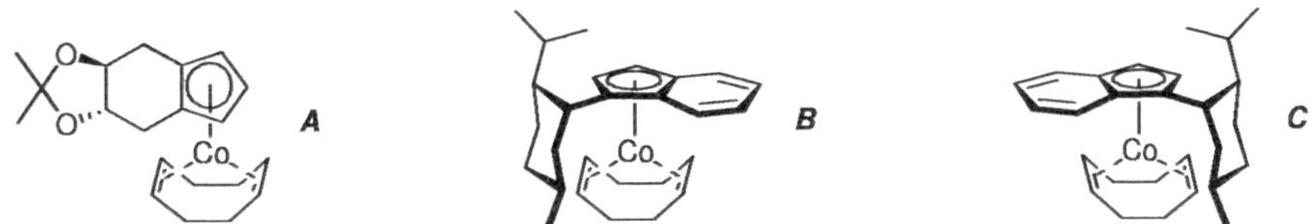

Scheme 32. The synthesis of bis-pyridines with binapthyl scaffold.

6. Enantioselective cyclotrimerizations catalysed by chiral catalysts

There is no doubt that carrying out the catalytic asymmetric cyclotrimerization of alkynes with nitriles providing chiral pyridine compounds (with axial chirality) would represent a state-of-art methodology as well as a useful synthetic tool for the synthesis of chiral ligands. However, the development of such a procedure has been hampered by issues associated with the reaction mechanism of the cyclotrimerization (sufficient rigidity of the transition state ensuring the transfer of the chiral information) as well as the availability of suitable catalysts or catalytic systems. Fortunately, there has been considerable progress in this area recently.

The first successful asymmetric cyclotrimerization was reported by Heller *et al.* who demonstrated that cobalt complexes bearing chiral cyclopentadienyl moiety (Figure 12) could efficiently catalyze the cyclotrimerization with nitriles providing atropoisomeric arylpyridines.[45] The crucial factor behind these outstanding results was a careful choice of the reaction conditions. Owing to the discovery of a photochemically induced cyclotrimerization at low temperatures (0–5 °C),[36,46] they were able to carry out the cyclotrimerization with high degree of enantioselectivity.

Figure 12. Typical examples of the used chiral cyclopentadienyl cobalt catalysts *A-C*.

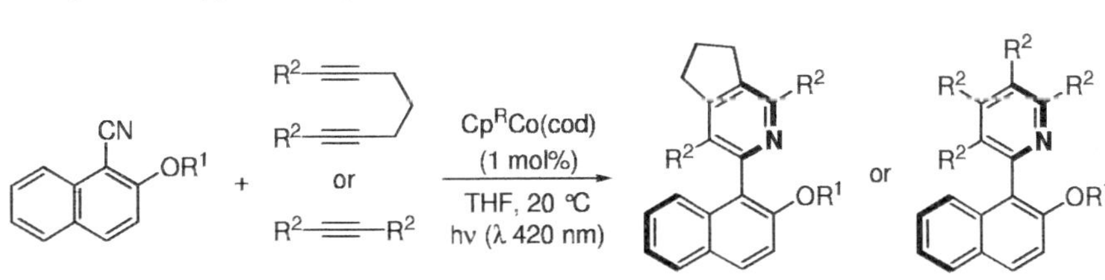

Scheme 33. The synthesis of atropoisomeric 2-arylpyridines.

The cyclotrimerization was carried out in two ways: either as the reaction of 2-methoxy-1-naphthyl-cyanide with alkynes or diynes (Scheme 33), or as the reaction of 2-methoxy-1-diynylnaphthalene with nitriles (Scheme 34). In the former the ees were up to 71% (for typical examples of the prepared arylpyridines, see Figure 13). On the other hand, the latter mode proceeded with substantially higher enantioselectivity yielding products with ees >98%.

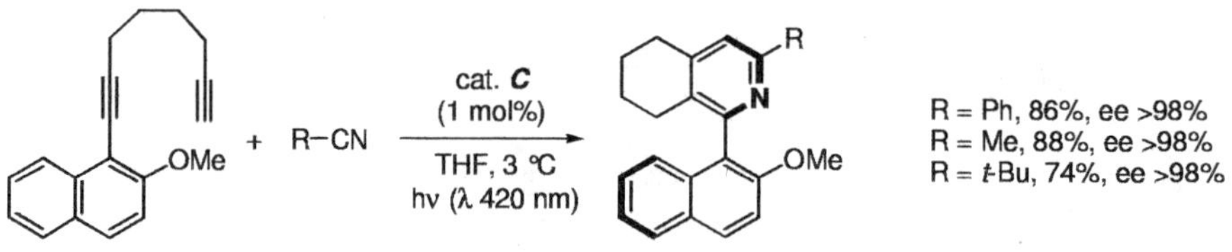

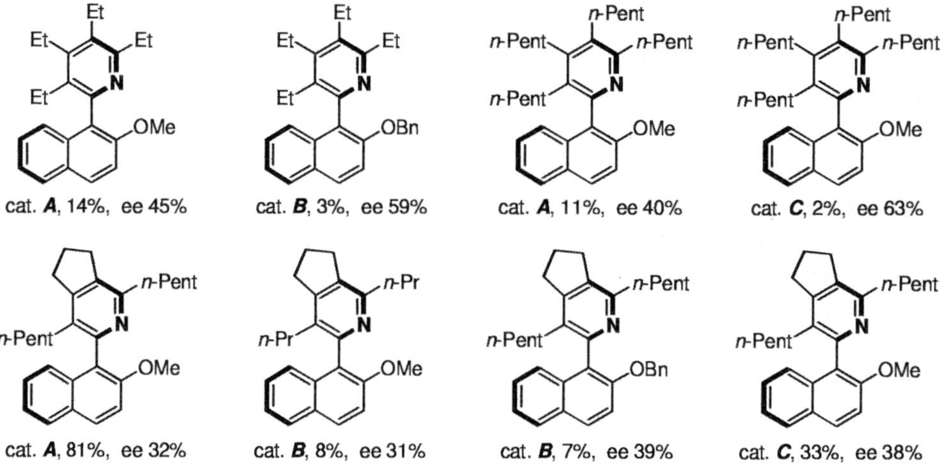

Scheme 34. The synthesis of atropoisomeric aryl-tetrahydroisoquinolines.

Figure 13. Typical examples of synthesized chiral arylpyridines.

Important advance in the cyclotrimerization of alkynes with nitriles has been achieved by the discovery of the possibility to catalyze this reaction by cationic Rh/phosphine complexes.[47] As for the phosphine a wide range of ligands such as BINAP, H_8-BINAP, Segphos, etc. were successfully tested. In addition, the potential of these catalytic systems was clearly demonstrated by the cyclotrimerization of a diyne with a dinitrile catalyzed by a combination of [Rh(cod)₂]BF₄/(R)-xyl-Solphos or (R)-BINAP that gave rise to the enantio-enriched pyridines with tertiary stereocentre in the side-chain (Scheme 35).

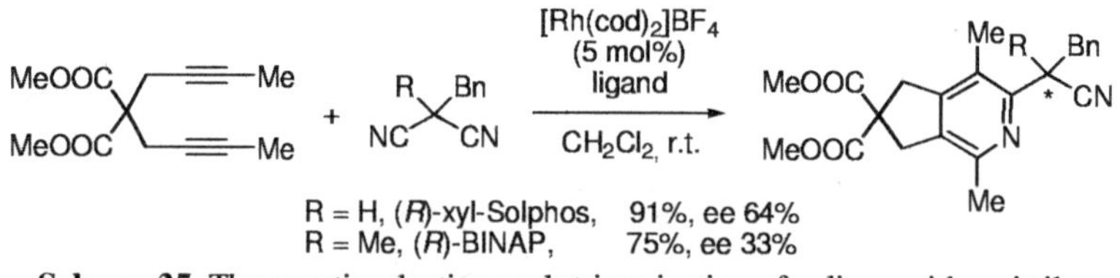

Scheme 35. The enantioselective cyclotrimerization of a diyne with a nitrile.

In a similar manner, the same catalytic system was used for the intramolecular cyclotrimerization of tetrayne-dinitriles (Scheme 36).[48] The reaction gave rise to C_2-symmetrical enantioenriched (up to 71% ee) bispyridyl compounds (Figure 14). The generality of the above mentioned approach was also demonstrated by asymmetric cyclotrimerization of a tetrayne with cyanoacetate to axially chiral bipyridine in 38% yield (ee 98%) along with achiral regioisomers (Scheme 37).[49] In view of the foregoing, it should be emphasized that the enantioselective cyclotrimerization methodology can be used also for the synthesis of chiral biaryls.[50]

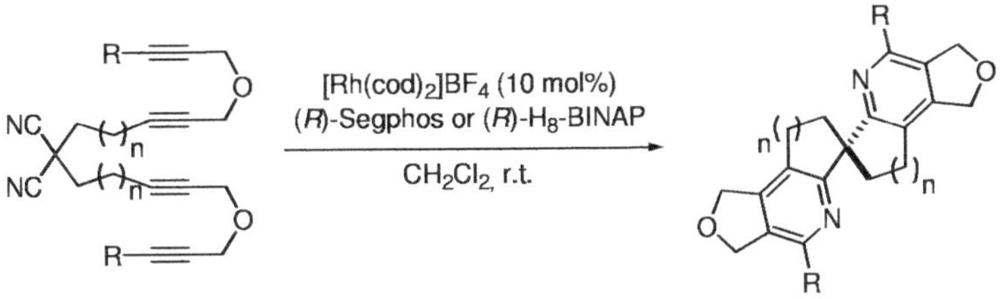
[Rh(cod)2]BF4 (10 mol%)
(R)-Segphos or (R)-H8-BINAP
CH2Cl2, r.t.

Scheme 36. The enantioselective cyclotrimerization of tetrayne-dinitriles.

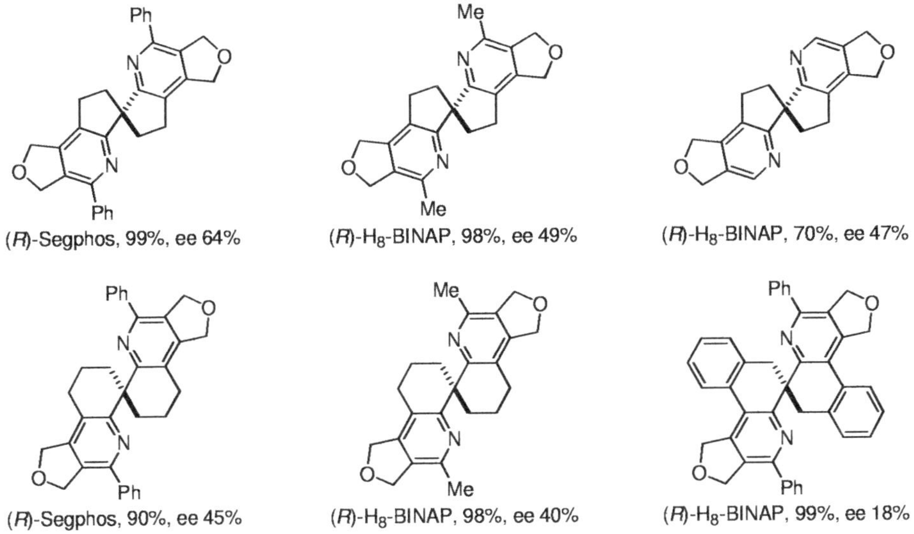
(R)-Segphos, 99%, ee 64%
(R)-H8-BINAP, 98%, ee 49%
(R)-H8-BINAP, 70%, ee 47%
(R)-Segphos, 90%, ee 45%
(R)-H8-BINAP, 98%, ee 40%
(R)-H8-BINAP, 99%, ee 18%

Figure 14. Typical examples of synthesized chiral bis-pyridines.

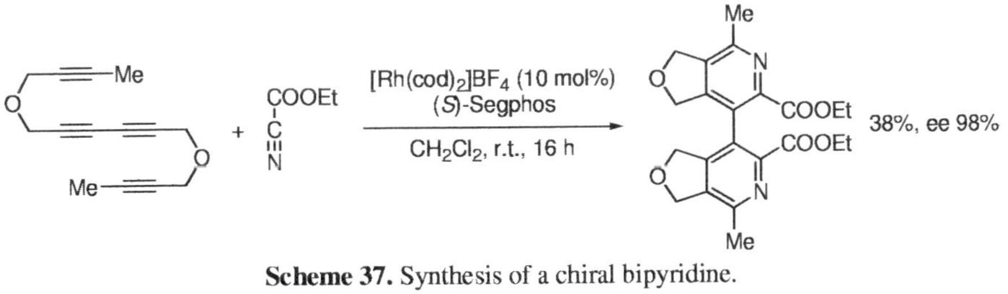
[Rh(cod)2]BF4 (10 mol%)
(S)-Segphos
CH2Cl2, r.t., 16 h
38%, ee 98%

Scheme 37. Synthesis of a chiral bipyridine.

7. Other applications

The cyclotrimerization of alkynes with nitriles has been used in many other applications. Early example of the cyclotrimerization carried out in water was reported by Heller as early as 1995.[46c] Later it was shown that the use of water as the solvent had a beneficial effect on the formation of pyridines.[46e,51] The use of a water soluble catalyst – a cyclopentadienyl cobalt complex with a hydrophilic side-chain – allowed to carry out the cyclotrimerization of various nitriles with propargyl alcohol or 1,4-butyndiol in a mixture water/MeOH (Scheme 38).[52]

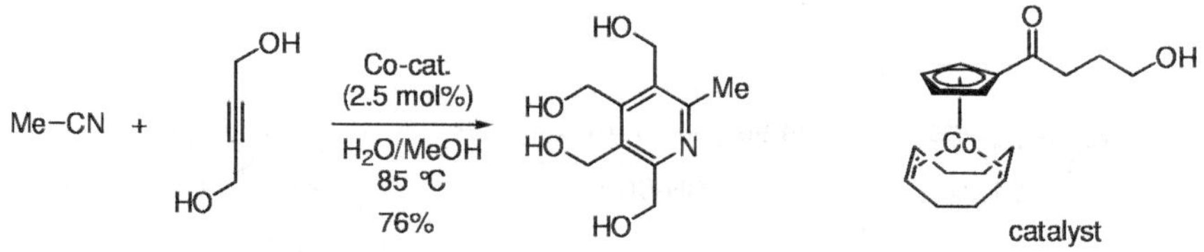

Scheme 38. The cyclotrimerization of alkynes with nitriles by a hydrophilic Co-complex.

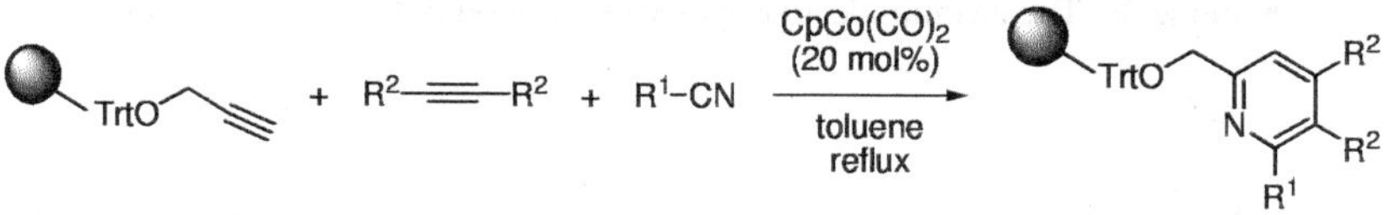

Scheme 39. The synthesis of pyridines on a solid-phase.

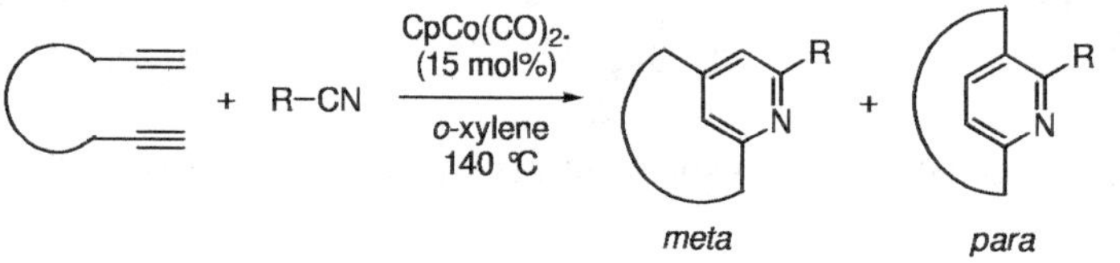

Scheme 40. The formation of *meta* and *para* regioisomers by the macrocyclization of a diyne with a nitrile.

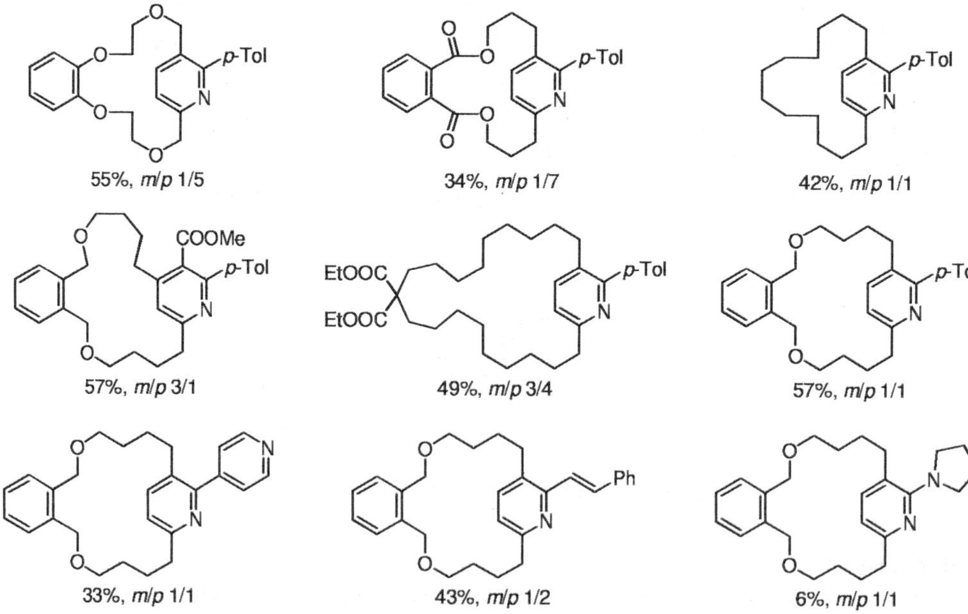

55%, *m/p* 1/5

34%, *m/p* 1/7

42%, *m/p* 1/1

57%, *m/p* 3/1

49%, *m/p* 3/4

57%, *m/p* 1/1

33%, *m/p* 1/1

43%, *m/p* 1/2

6%, *m/p* 1/1

Figure 15. Some typical examples of pyridine-cyclophanes.

196

The cyclotrimerization was also used in solid-phase synthesis of pyridines. This method was implemented in several modifications by immobilizing: i) one alkyne reaction partner on a resin,[53] ii) a diyne or an ynenitrile.[54] The former variant led to the formation of substituted pyridines (Scheme 39), whereas the latter provided bicyclic pyridines.

Against the odds, the cyclotrimerization of long chain bis-alkynes with nitriles was shown to be feasible and allowed the synthesis of fused pyridine macrocycles – pyridine-cyclophanes.[55] The reaction was catalyzed by CpCo(CO)$_2$ under light irradiation and to keep the low concentration of the bis-alkyne, in order to prevent its oligomerization, it had to be added by a syringe pump (Scheme 40). In such a fashion the corresponding pyridine-cyclophanes were obtained in reasonable yields (Figure 15). Because of the flexibility of the long-tether connecting triple bonds, the products were always obtained as a mixture of *para* and *meta* regioisomers with ring sizes ranging from 15 to 23. As for the nitriles, those bearing an aromatic, heteroaromatic, alkyl and alkenyl moieties were successfully used.

8. Conclusion

Transition metal catalyzed [2+2+2]-cyclotrimerization of alkynes with nitriles constitutes a convenient method for the preparation of various compounds possessing pyridine framework. The course of the reaction, *i.e.* yield of the products and the possible formation of regioisomeric products, often depends on the proper choice of the catalyst or catalytic system, on the electronic and steric nature of the substrates as well as on additional factors (heating, irradiation, solvent, etc.). However, by making the careful choice and variations of these factors, it is sometimes possible to control the overall outcome of the reaction. The broad scope and synthetic aspects of this methodology have demonstrated that it could be used for the preparation of not only natural compounds and their analogs but also of other classes of pyridine-ring containing compounds such as potentially biologically active compounds, ligands for transition metals, etc. It is obvious that the topic of the cyclotrimerization of alkynes with nitriles to pyridines has not been exhausted yet and there is ever-continuing demand for better, cheaper and more selective catalysts as well as applications in synthetic organic chemistry.

Acknowledgment

The authors gratefully acknowledge financial support from the Czech Science Foundation (Grant No. 203/08/0350) and Ministry of Education of the Czech Republic (Project No. LC06070, MSM0021620857).

References
1. (a) Ramsay, W. *Philos. Mag.* **1876**, *Ser. 5, 2*, 269–281. (b) Ramsay, W. **1877**, *Ser. 5, 3*, 241–253.
2. (a) Schore, N. E. *Chem. Rev.* **1988**, *88*, 1081–1119. (b) Bönnemann, H.; Brijoux, W. *Adv. Het. Chem.* **1990**, *48*, 177–222. (c) Dzehemilev, U. M.; Selimov, F. A.; Tolstikov, G. A. *Arkivoc* **2001**, *ix*, 85–116. (d) Varela, J. A.; Saa, C. *Chem. Rev.* **2003**, *103*, 3787–3802. (e) Chopade, P. R.; Louie, J. *Adv. Synth. Catal.* **2006**, *348*, 2307–2327. (f) Maryanoff, B. E.; Zhang, H.-C. *Arkivoc* **2007**, *xii*, 7–35. (g) Heller, B.; Hapke, M. *Chem. Soc. Rev.* **2007**, *36*, 1085–1094. (h) Varela, J. A.; Saa, C. *Synlett* **2008**, 2571–2578. (i) Agenet, N.; Buisine, O.; Slowinski, F.; Gandon, V.; Aubert, C.; Malacria, M. *Org. React.* **2007**, *68*, 1–302.
3. (a) Dazinger, G.; Torres-Rodrigues, M.; Kirchner, K.; Calhorda, M. J.; Costa, P. J. *J. Organomet. Chem.* **2006**, *691*, 4434–4445. (b) Yamamoto, Y.; Kinpara, K.; Saigoku, T.; Takagishi, H.; Okuda, S.; Nishiyama, H.; Itoh, K. *J. Am. Chem. Soc.* **2005**, *127*, 605–613. For additional study concerning a

nitrile coordination to rhodacyclopentadiene, see: (c) Orian L.; van Twist, W.-J.; Bickelhaupt, F. M. *Organometallics* **2008**, *27*, 4028–4030.

4. (a) Wakatsuki, Y.; Yamazaki, H. *J. Chem. Soc., Chem. Commun.* **1973**, 280–281. (b) Wakatsuki, Y.; Yamazaki, H. *Tetrahedron Lett.* **1973**, 3383–3384.

5. Vollhardt, K. P. C.; Bergman, R. C. *J. Am. Chem. Soc.* **1974**, *96*, 4496–4498.

6. Bönnemann, H.; Brinkmann, R.; Schenkluhn, H. *Synthesis* **1974**, 575–577.

7. (a) Takahashi, T.; Tsai, F.-Y.; Kotora, M. *J. Am. Chem. Soc.* **2000**, *122*, 4994–4995. (b) Takahashi, T.; Tsai, F.-Y.; Li, Y.; Wang, H.; Kondo, Y.; Yamanak, M.; Nakajima, K.; Kotora, M. *J. Am. Chem. Soc.* **2002**, *124*, 5059–5067.

8. Hill, J. E.; Balaich, G.; Fanwick, P. E.; Rothwell, I. P. *Organometallics* **1993**, *12*, 2911–2924.

9. (a) Suzuki, D.; Tanaka, R.; Urabe, H.; Sato, F. *J. Am. Chem. Soc.* **2002**, *124*, 3518–3519. (b) Tanaka, R.; Yuza, A.; Watai, Y.; Suzuki, D.; Takayama, Y.; Sato, F.; Urabe, H. *J. Am. Chem. Soc.* **2005**, *127*, 7774–7780.

10. Ferré, K.; Toupet, L.; Guerchais, V. *Organometallics* **2002**, *21*, 2578–2580.

11. (a) Smith, D. P.; Strickler, J. R.; Gray, S. D.; Bruck, M. A.; Holmes, R. S.; Wigley, D. E. *Organometallics* **1992**, *11*, 1275–1288. (b) Takai, K.; Yamada, M.; Utimoto, K. *Chem. Lett.* **1995**, 851–852.

12. Geiger, R. E.; Lalonde, M.; Stoller, H.; Schleich, K. *Helv. Chim. Acta* **1984**, *67*, 1274–1282.

13. Parnell, C. A.; Vollhardt, K. P. C. *Tetrahedron* **1985**, *41*, 5791–5796.

14. Saá, C.; Crotts, D. D.; Hsu, G.; Vollhardt, K. P. C. *Synlett* **1994**, 487–488.

15. Groth, U.; Huhn, T.; Kesenheimer, C.; Kalogerakis, A. *Synlett* **2005**, 1758–1760.

16. McIver, A.; Young, D. D.; Deiters, A. *Chem. Commun.* **2008**, 4750–4752.

17. (a) Earl, R. A.; Vollhardt, K. P. C. *J. Am. Chem. Soc.* **1983**, *105*, 6691–6693. (b) Earl, R. A.; Vollhardt, K. P. C. *J. Org. Chem.* **1984**, *105*, 4786–4790.

18. Chelucci, G.; Falorni, M.; Giacomelli, G. *Synthesis* **1990**, 1121–1122.

19. Botteghi, C.; Chelucci, G.; Del Ponte, G.; Marchetti, M.; Paganelli, S. *J. Org. Chem.* **1994**, *57*, 7125–7129.

20. Gray, B. L.; Wang, X.; Brown, W. C.; Kuai, L.; Schreiber, S. L. *Org. Lett.* **2008**, *10*, 2621–2624.

21. (a) Zhou, Y.; Beeler, A. B.; Cho, S.; Wang, Y.; Franzblau, S. G.; Snyder, J. K. *J. Comb. Chem.* **2008**, *10*, 534–540. (b) Zhou, Y.; Porco Jr, J. A.; Snyder, J. K. *Org. Lett.* **2007**, *9*, 393–395.

22. Turek, P.; Hocek, M.; Pohl, R.; Klepetářová, B.; Kotora, M. *Eur. J. Org. Chem.* **2008**, 3335–3343.

23. Zou, Y.; Young, D. D.; Cruz-Montanez, A.; Deiters, A. *Org. Lett.* **2008**, *10*, 4661–4664.

24. Bönnemann, H.; Brinkmann, R. *Synthesis* **1975**, 600–602.

25. Varela, J. A.;Castedo, L.; Saá, C. *J. Org. Chem.* **1997**, *62*, 4189–4192.

26. (a) Varela, J. A.; Castedo, L.; Saá, C. *J. Am. Chem. Soc.* **1998**, *120*, 12147–12148. (b) Varela, J. A.; Castedo, L.; Maestro, M.; Mahía, J.; Saá, C. *Chem. Eur. J.* **2001**, *7*, 5203–5213.

27. Varela, J. A.; Castedo, L.; Saá, C. *Org. Lett.* **1999**, *1*, 2141–2143.

28. Yamamoto, Y.; Ogawa, R.; Itoh, K. *J. Am. Chem. Soc.* **2001**, *123*, 6189–6190.

29. (a) Chelucci, G. *Tetrahedron: Asymmetry* **1995**, *6*, 811–826. (b) Chelucci, G. *Tetrahedron: Asymmetry* **2005**, *16*, 2353–2383.

30. Tatone, D.; Dich, T. C.; Nacco, R.; Botteghi, C. *J. Org. Chem.* **1975**, *40*, 2987–2990.

31. Botteghi, C.; Chelucci, G.; Marehetti, M. *Synth. Commun.* **1982**, *12*, 25–33.

32. Chelucci, G.; Cabras, M. A.; Botteghi, C.; Marchetti, M. *Tetrahedron: Asymmetry* **1994**, *5*, 299–302.

33. (a) Chelucci, G.; Cabras, M. A.; Saba, A. *Tetrahedron: Asymmetry* **1994**, *5*, 1973–1978. (b) Chelucci, G.; Falorni, M.; Giacomelli, G. *Gazz. Chim. Ital.* **1990**, *120*, 731–732.

34. Falorni, M.; Chelucci, G.; Conti, S.; Giacomelli, G. *Synthesis* **1992**, 972–976.

35. Cossu, S.; Conti, S.; Giacomelli, G.; Falorni, M. *Synthesis* **1994**, 1429–1432.

36. Heller, B.; Sundermann, B.; Fischer, C.; You, J.; Chen, W.; Drexler, H.-J.; Knochel, P.; Bonrath, W.; Gutnov, A. *J. Org. Chem.* **2003**, *68*, 9221–9225.

37. Chelucci, G.; Thunmmel, R. P. *Chem. Rev.* **2002**, *102*, 3129–3170.

38. Botteghi, C.; Caccia, G.; Chelucci, G.; Soccolini, F. *J. Org. Chem.* **1984**, *49*, 4290–4293.

39. Chelucci, G.; Cabras, M. A.; Botteghi, C.; Basoli, C.; Marchetti, M. *Tetrahedron: Asymmetry* **1996**, *7*, 885–895.

40. Hrdina, R.; Stará, I. G.; Dufková, L.; Mitchel, S.; Císařová, I.; Kotora, M. *Tetrahedron* **2006**, *62*, 968–976.

41. Hrdina, R.; Kadlčíková, A.; Valterová, I.; Hodačová, J.; Kotora, M. *Tetrahedron: Asymmetry* **2006**, *17*, 3185–3191.

42. Hrdina, R.; Valterová, I.; Hodačová, J.; Kotora, M. *Adv. Synth. Catal.* **2007**, *349*, 822–826.

43. Hrdina, R.; Dračínský, M.; Valterová, I.; Hodačová, J.; Císařová, I.; Kotora, M. *Adv. Synth. Catal.* **2008**, *350*, 1449–1456.

44. Hoshi, T.; Katano, M.; Nozawa, E.; Suzuki, T.; Hagiwara, H. *Tetrahedron Lett.* **2004**, *45*, 3489–3491.

45. Gutnov, A.; Heller, B.; Fischer, C.; Drexler, H.-J.; Spannenberg, A.; Sundermann, B.; Sundermann, C. *Angew. Chem. Int. Ed.* **2004**, *43*, 3795–3797.

46. (a) Heller, B.; Riehsig, J.; Schulz, W.; Oehme, G. *Appl. Organomet. Chem.* **1998**, *7*, 641–646. (b) Kabaret, F.; Heller, B.; Kortus, K.; Oehme, G. *Appl. Organomet. Chem.* **1998**, *7*, 641–646. (c) Heller, B.; Oehme, G. *J. Chem. Soc., Chem. Commun.* **1995**, 179–180. (d) Heller, B.; Heller, D.; Oehme, G. *J. Mol. Catal. A* **1996**, *110*, 211–219. (e) Heller, B.; Sundermann, B.; Buschmann, H.; Drexler, H.-J.; You, J.; Holzgrabe, U.; Heller, E.; Oehme, G. *J. Org. Chem.* **2002**, *67*, 4414–4422.

47. (a) Tanaka, K.; Suzuki, N.; Nishida, G. *Eur. J. Org. Chem.* **2006**, 3917–3922. (b) Tanaka, K. *Synlett* **2007**, 1977–1993.

48. Wada, A.; Noguchi, K.; Hirano, M.; Tanaka, K. *Org. Lett.* **2007**, *9*, 1295–1298.

49. Nishida, G.; Suzuki, N.; Noguchi, K.; Tanaka, K. *Org. Lett.* **2006**, 3489–3492.

50. Tanaka, K. *Chem. Asian J.* **2009**, 508–518.

51. Heller, B.; Heller, D.; Wagler, P.; Oehme, G. *J. Mol. Catal. A.* **1998**, 219–233.

52. Fatland, A. V.; Eaton, B. E. *Org. Lett.* **2000**, *2*, 3131–3133.

53. Senaiar, R. S.; Young, D. D.; Deiters, A. *Chem. Commun.* **2006**, *47*, 1313–1315.

54. Young, D. D.; Deiters, A. *Angew. Chem. Int. Ed.* **2007**, *47*, 5187–5190.

55. (a) Moretto, A.; Zhang, H.-C.; Maryanoff, B. E. *J. Am. Chem. Soc.* **2001**, *123*, 3157–3158. (b) Boñaga, L. V. R.; Zhang, H.-C.; Moretto, A.; Ye, H.; Gauthier, D. A.; Li, J.; Leo, G. C.; Maryanoff, B. E. *J. Am. Chem. Soc.* **2005**, *127*, 3473–3485.

56. (a) Kase, K.; Goswami, A.; Ohtaki, K.; Tanabe, E.; Saino, N.; Okamoto, S. *Org. Lett.* **2007**, *9*, 931–934. (b) Goswami, A.; Ohtaki, K.; Ito, T.; Okamoto, S. *Adv. Synth. Catal.* **2008**, *350*, 143–152.

57. Chang, H.-T.; Jeganmohan, M.; Cheby, C.-H. *Org. Lett.* **2007**, *9*, 505–508.

58. Varela, J. A.; Castedo, L.; Saá, C. *J. Org. Chem.* **2003**, *68*, 8595–8598.

59. (a) Yamamoto, Y.; Okuda, S.; Itoh, K. *Chem. Commun.* **2001**, 1102–1103. (c) Yamamoto, Y.; Kinpara, K.; Ogawa, R.; Nishiyama, H.; Itoh, K. *Chem. Eur. J.* **2006**, *12*, 5618–5631.

60. (a) McCormick, M. M.; Dugong, H. A.; Zuo, G.; Louie, J. *J. Am. Chem. Soc.* **2005**, *127*, 5030–5031. (b) Tekavec, T. N.; Zuo, G.; Simon, K.; Louie, J. *J. Org. Chem.* **2006**, *71*, 5834–5836.

61. Uhm, J.; An, H. W. *J. Kor. Chem. Soc.* **2001**, *45*, 268–272.

62. Naiman, A.; Vollhardt, K. P. C. *Angew. Chem. Int. Ed. Engl.* **1977**, *16*, 708–709.

63. Wakatsuki, Y.; Yamazaki, H. *J. Chem. Soc., Dalton Trans.* **1978**, 1278–1282.

64. Brien, D. J.; Naiman, A.; Vollhardt, K. P. C. *J. Chem. Soc., Chem. Commun.* **1982**, 133–134.

65. Chiusoli, G. P.; Pallini, L.; Terenghi, G. *Transition Met. Chem.* **1983**, *8*, 250–252.

66. Vitulli, G.; Bertozzi, S.; Lazzaroni, R.; Salvadori, P. *J. Organomet. Chem.* **1986**, *307*, C35–C37.

67. Diversi, P.; Ingrosso, G.; Lucherini, A.; Vanacore, D. *J. Mol. Catal.* **1987**, *41*, 261–270.

68. (a) Cioni, P.; Diversi, P.; Ingrosso, G.; Lucherini, A.; Ronca, P. *J. Mol. Catal.* **1987**, *40*, 337–357. (b) Diversi, P.; Ingrosso, G.; Minutillo, A. *J. Mol. Catal.* **1987**, *40*, 359–377.

69. (a) Vitulli, G.; Bertozzi, S.; Vignali, M.; Lazzaroni, R.; Salvadori, P. *J. Organomet. Chem.* **1987**, *326*, C33–C36. (b) Vitulli, G.; Bertozzi, S.; Pertici, P.; Lazzaroni, R.; Salvadori, P. *J. Organomet. Chem.* **1988**, *353*, C35–C38.

70. (a) Battaglia, L. P.; Delledonne, D.; Nardelli, M.; Predieri, G.; Chiusoli, G. P.; Costa, M.; Pelizzi, C. *J. Organomet. Chem.* **1989**, *363*, 209–211. (b) Zhou, Z.; Battaglia, L. P.; Chiusoli, G. P.; Costa, M.; Nardelli, M.; Pelizzi, C.; Predieri, G. *J. Organomet. Chem.* **1991**, *417*, 51–63.

71. Bianchini, C.; Meli, A.; Peruzzini, M.; Vacca, A.; Vizza, F. *Organometallics* **1991**, *50*, 645–651.

72. Schmidt, U.; Zenneck, U. *J. Organomet. Chem.* **1992**, *440*, 187–190.

73. Diversi, P.; Ermini, L.; Ingrosso, G.; Lucherini, A. *J. Organomet. Chem.* **1993**, *447*, 291–298.

74. (a) Costa, M.; Dias, F. S.; Chiusoli, G. P.; Gazzola, G. L. *J. Organomet. Chem.* **1995**, *488*, 47–53. (b) Costa, M.; Dalcanale, E.; Dias, F. S.; Graiff, C.; Tiripicchio, A.; Bigliardi, L. *J. Organomet. Chem.* **2001**, *619*, 179–193.

75. Garcia, P.; Moulin, S.; Miclo, Y.; Leboef, D.; Vandin, V.; Aubert, C.; Malacria, M. *Chem. Eur. J.* **2009**, *15*, 2129–2139.

76. Tanaka, K.; Hara, H.; Nishida, G., Hirano, M. *Org. Lett.* **2007**, 1907–1910.

77. Geny, A.; Agenet, N.; Iannazzo, L.; Malacria, M.; Aubert, C.; Gandon, V. *Angew. Chem. Int. Ed.* **2009**, *48*, 1810–1813.

RECENT DEVELOPMENTS IN THE SYNTHESIS OF DIHYDROPYRIDINES (DHPs) AND DIHYDROPYRIMIDINES (DHPMs)

Julie Moreau,[a] Jean-Pierre Hurvois,[a] Mbaye Diagne Mbaye[b,c] and Jean-Luc Renaud*[b]

[a]*UMR 6226: CNRS-Université de Rennes 1, Sciences Chimiques de Rennes, Catalyse et Organométalliques, Campus de Beaulieu, F-35042 Rennes Cedex, France*

[b]*Laboratoire de Chimie Moléculaire et Thioorganique, UMR 6507, INC3M, FR 3038, ENSICAEN-Université de Caen, F-14050 Caen, France (e-mail: jean-luc.renaud@ensicaen.fr)*

[c]*Université de Ziguinchor, BP: 523 Ziguinchor, Sénégal*

Abstract. *In this review, we will describe the latest achievements in the synthesis of two families of nitrogen containing heterocycles via (stereoselective) Brønsted or Lewis acid catalysis.*

Contents

1. Introduction

Nitrogen containing ring systems are key structural elements in a vast array of natural products as well as in a large class of biologically active natural products, being also often embedded within scaffolds recognized as privileged structures by medicinal chemists. Accordingly, new efficient and stereoselective (when possible) routes to these derivatives are of widespread interest. In this review, we will focus on the latest developments on the synthesis of two classes of these heterocyclic compounds: the non symmetrical dihydropyridines and the substituted dihydropyrimidones.

2. Brønsted and Lewis acids catalyzed Hantzsch type synthesis of 1,4-dihydropyridines (1,4-DHPs)

Current literature reveals that 1,4-DHPs exhibit interesting pharmacological and biological properties. Thus, they have been used as calcium channel modulators for the treatment of cardiovascular diseases.[1] They also present vasodilating activities[2] and are NADH mimics.[3] The best known procedure for the preparation of symmetrical 1,4-DHPs is the classical Hantzsch synthesis: a multicomponent condensation involving two molecules of β-ketoesters, one molecule of aldehyde and one molecule of ammonia.[4] Due to the importance of this class of heterocyclic system, many efforts have been devoted to the preparation of such compounds.[5]

The (*N*-substituted) non symmetrical 1,4-DHPs are also of particular interest for a systematic study of their biological activities and from a synthetic point of view (Figure 1). This first part of the review will be focussed on the recent developments of the synthesis of substituted 1,4-DHPs.

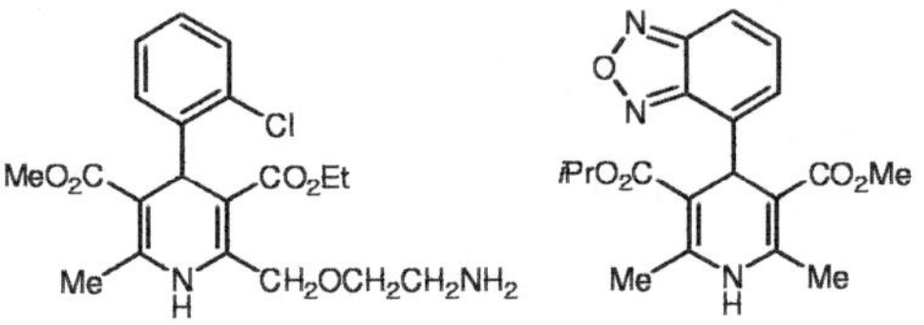

Figure 1

2.1. Recent synthesis of 1,4-DHPs

2.1.1. Brønsted acid catalysis

Catalytic strategies for the synthesis of 1,4-DHPs, involving different catalysts and conditions, have been developed since the last decade but they all suffer from one or more drawbacks including: low yields, costly reagents, and drastic reaction conditions. For the synthesis of 1,4-DHPs, catalysts can be either a Brønsted or a Lewis acid.

Electrophilic activation has emerged as an important tool for the organic catalysis, with new applications and developments appearing at a rapidly increasing pace.[6] Among the known carbonyl activators, Brønsted and Lewis acids have recently demonstrated their potential to serve as active catalysts for a variety of synthetically useful reactions in organic chemistry.[7,8]

In this chapter, we will report all the new and important methods which were described in the last decade and which lead to 1,4-DHPs *via* Brønsted or Lewis acids catalysis.

In 2001, J. L. Scott *et al.* reported studies of the condensation of an aromatic aldehyde with an appropriate acetoacetate ester in the presence of a catalytic amount of glacial acetic acid to form the corresponding benzylidene intermediate (Scheme 1).[9] The latter was reacted with 3-aminoacrylate to yield a set of unsymmetrical 1,4-DHPs. In all cases, the reactions were carried out under solvent-free conditions.

Yields were ranging between 5% and 99%, according to the reaction conditions and to the nature of the substituents.

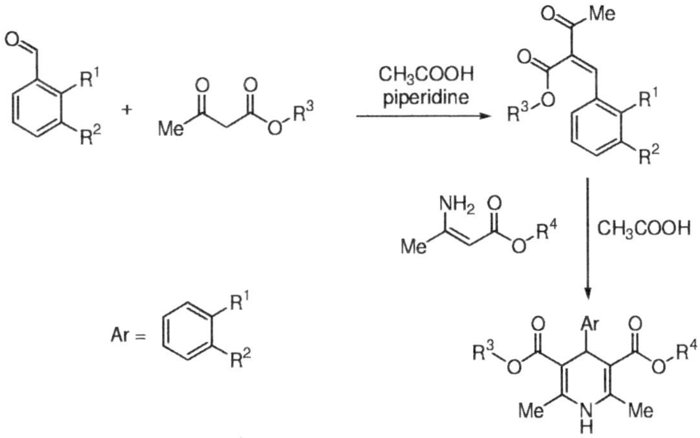

Scheme 1

In 2004, Tripathi *et al.* developed a novel method for the synthesis of several glycosyl 4-substituted DHPs at 80 °C in diethylene glycol using tetrabutylammonium hydrogen sulfate (TBAHS, Scheme 2).[10] Yields of this three-component reaction were ranging between 90% and 98%.

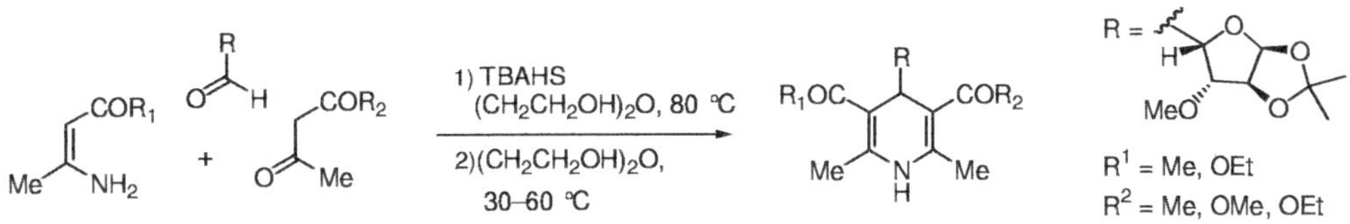

Scheme 2

The synthesis of potential anti-tubercular precursors, showed in Scheme 3, was carried out under these optimized reaction conditions utilizing an aldehyde which was obtained from the *L*-ascorbic acid (Scheme 2).[11]

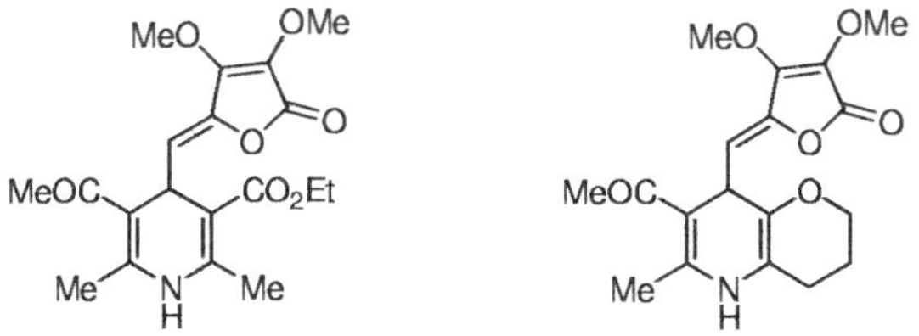

Scheme 3

H-donor organocatalysts such as amino acids or cinchona alkaloids have also been used as catalysts for the synthesis of unsymmetrical 1,4-DHPs such as polyhydroquinoline derivatives.[12] In all cases, *L*-proline proved to be the catalyst of choice. In the continuation of these studies, Kumar *et al.* reported the one-pot synthesis of 1,4-DHPs *via* the three-component coupling of cinnamaldehyde, anilines and β-ketoesters (Scheme 4). This reaction was carried out at room temperature under solvent-free conditions, in the presence of 10 mol% of *L*-proline.[13] No enantiomeric excess was reported by the authors for this process.

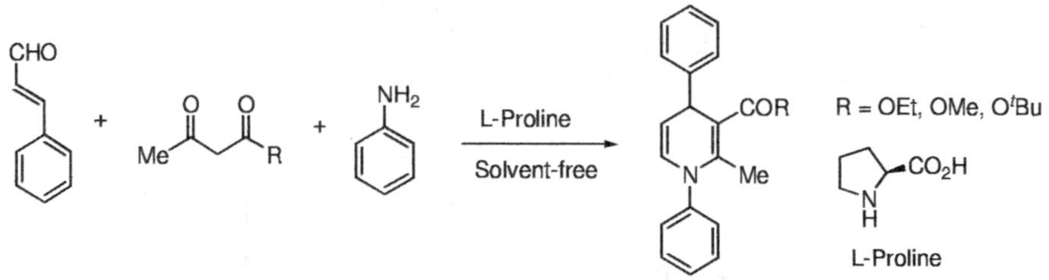

Scheme 4

Sulfonic acids proved to be efficient catalysts for the preparation of a variety of heterocyclic systems. The synthesis of polyhydroquinoline derivatives was performed in the presence of various sulfonic acids in aqueous micelles under ultrasonic irradiation.[14] Best results and reaction rates were obtained in the presence of *p*-toluene-sulfonic acid (Scheme 5).

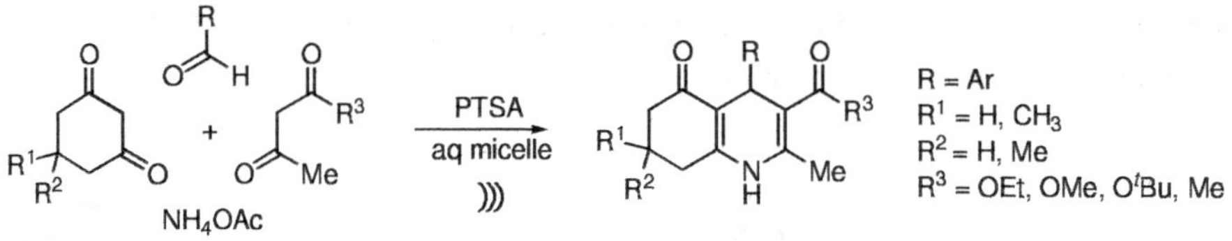

Scheme 5

Heterogeneous catalysts such as sulfonic acids functionalized silica have also been used to catalyze the condensation of cinnamaldehyde with aniline derivatives and β-ketoesters (Scheme 6).[15] The reaction proceeded at room temperature within 5 to 30 minutes under solvent free conditions. In that case, corresponding 1,4-DHPs were formed in high yields (80–89%) and the catalyst can be recycled after use.

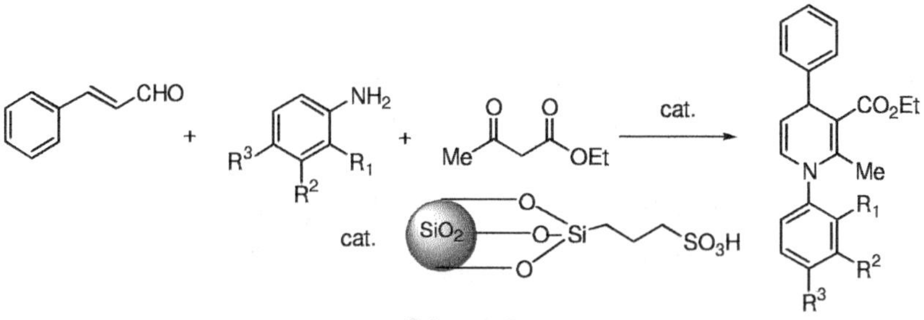

Scheme 6

Recently, other metal-free H-donor catalysts such as aryl phosphoric acids have been used for the synthesis of 4-alkyl and 4-aryl substituted 1,4-DHPs from α,β-unsaturated aldehydes and β-enamino-esters

(Scheme 7).[16] Under these mild reaction conditions, a substituted biaryl phosphoric acid was used as a catalyst (5 mol%) and yields ranging from 31% to 89% were obtained.

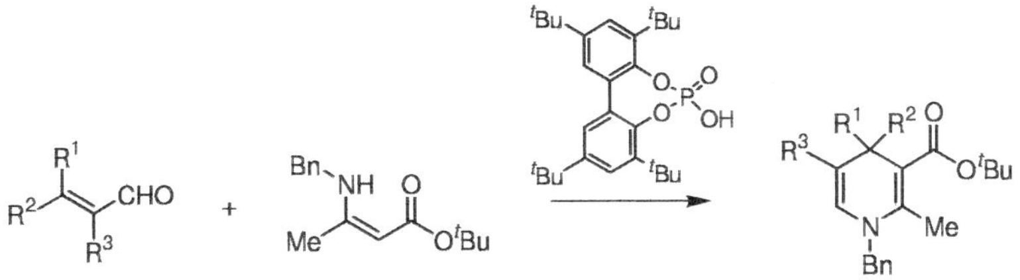

Scheme 7

The proposed mechanism for this Brønsted acid catalyzed cyclization reaction is outlined as follows (Figure 2). The Michael addition of the enaminoester on the α,β-unsaturated aldehyde is promoted by the interaction with the Brønsted acid (BH) catalyst (transition state **I**). After the catalyst release, the obtained enamine can react on the resulting aldehyde, according to the transition state **II**, which undergoes the cyclization. Finally, the intermediate **III** leads to the 1,4-DHPs after a dehydration reaction.

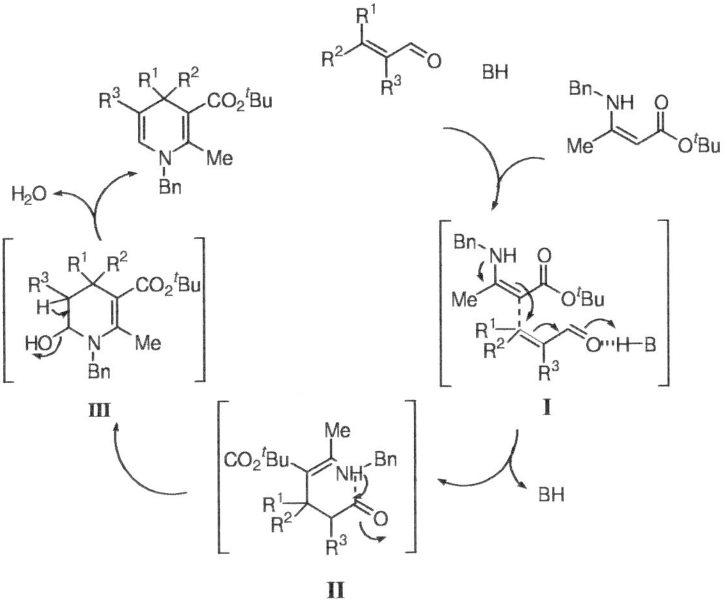

Figure 2

Chiral derivatives of this catalyst were also involved in this reaction and are described in the last chapter.

2.1.2. Lewis acid catalysis

An alternative to the Brønsted acid catalysis is the Lewis acid catalysis in the presence of metal salts. The Michael condensation of β-enaminoesters to α,β-unsaturated carbonyl compounds represents a powerful method for the synthesis of unsymmetrical 1,4-DHPs. This stepwise condensation is accelerated in the presence of Lewis acids catalysts. Renaud *et al.* reported on the iron(III) chloride and scandium(III) triflate catalyzed condensation of β-enaminoesters to α,β-unsaturated carbonyl compounds (Scheme 8).[17] The reaction was carried out at room temperature in dichloromethane in the presence of Na_2SO_4 as the dehydrating agent.

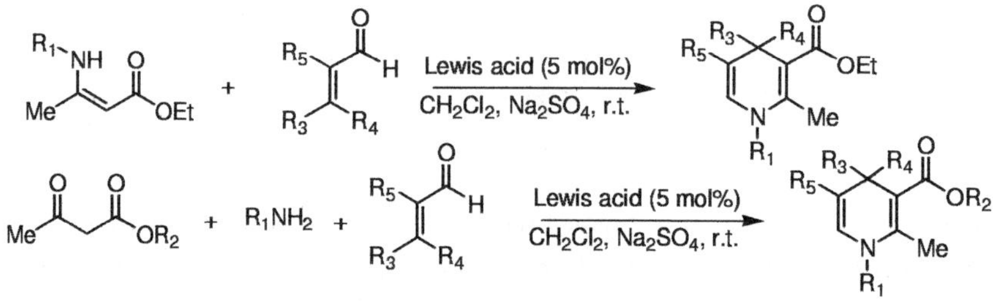

Scheme 8

In all cases, $FeCl_3·6H_2O$ was found to be the catalyst of choice. For example, the 1,4-DHP in which R_1=PhCH$_2$, R_2=Me, R_3=Ph, and R_4=R_5=H, was obtained in nearly quantitative yield from cinnamaldehyde and the corresponding β-enaminone. In the continuation of this study, the same authors carried out the synthesis of the 1,4-DHPs in a one-pot process. To this end, β-ketoesters or β-diketones were reacted first with the requisite primary amine in the presence of 5mol% of Lewis acid (Scheme 8). The addition of the α,β-unsaturated aldehyde afforded the 1,4-DHPs in yields comparable to those obtained from the isolated β-enaminoesters. Sambri *et al.* have shown that metal perchlorates behave also as cheap and efficient Lewis acids for the synthesis of unsymmetrical 1,4-DHPs (Scheme 9).[18] In preliminary experiments, β-enamino-ester in which R_1=Ot-Bu, R_2=Ph, R_3=i-Pr, and a slight excess (1.2 equiv.) of (E)-4-methyl-2-pentenal with R_3=i-Pr were stirred at ambient temperature in CH_2Cl_2 in the presence of various catalysts including: $Zn(ClO_4)_2·6H_2O$, $LiClO_4$ and $Mg(ClO_4)_2$. The conversion yields were high (up to 85%) and the expected 1,4-DHP was obtained in the presence of various amounts (8–25%) of the corresponding pyridinium salt whose formation could be attributed to the sensitivity of these adducts to work-up and probably to aerial oxidation.

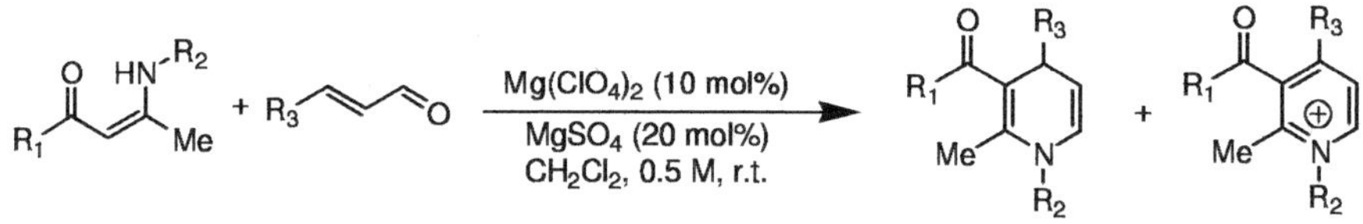

Scheme 9

In the continuation of their studies aimed at the discovery of multidrug resistance modifiers possessing a 1,4-DHP structure, Perumal *et al.* reported that boronic acids catalyzed the synthesis of various 4-[3-ethoxycarbonyl-1H-pyrazol-4-yl]-1,4-DHPs dicarboxylate in ionic liquid medium.[19] The authors found that, when pyrazole aldehyde (R_3=NO$_2$) reacted with ethyl acetoacetate and ethyl 3-amino-crotonate in [bmim]Cl, in the presence of 5 mol% of 3,4,5-trifluorobenzeneboronic acid, the unsymmetrical 1,4-DHP was obtained in 92% yield (Scheme 10).

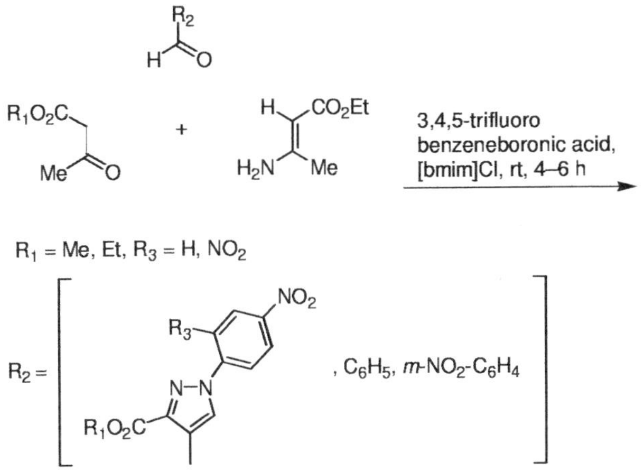

R$_1$ = Me, Et, R$_3$ = H, NO$_2$

R$_2$ = [...] , C$_6$H$_5$, m-NO$_2$-C$_6$H$_4$

Scheme 10

Debache *et al.* have shown that classical Hantzsch three-component reaction was efficiently catalyzed by the presence of 10 mol% of phenylboronic acid (Scheme 11).[20]

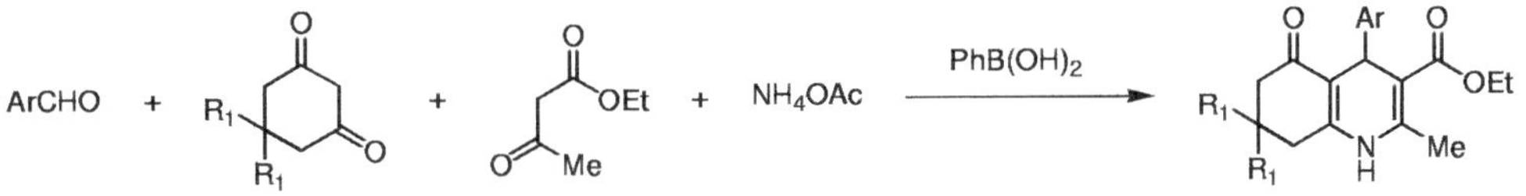

Scheme 11

Yao *et al.* have reported that molecular iodine was an efficient promoter for the condensation of benzaldehyde (Ar=C$_6$H$_5$), dimedone (R$_1$=Me) and ethyl acetoacetate.[21] The reaction was conducted using 15 mol% of iodine in ethanol at 40 °C for 30 minutes to afford a suitable entry into the polyhydroquinoline heterocyclic ring system.

The same authors have also investigated the catalytic activity of cerium ammonium nitrate (CAN).[22] As a typical experiment, one equivalent of each aldehyde (Ar=Ph, p-MeOC$_6$H$_4$, p-ClC$_6$H$_4$, p-HOC$_6$H$_4$, 2-thienyl), 1,3-cyclohexanedione (R$_1$=H), ethyl acetoacetate and ammonium acetate were stirred in ethanol in the presence of 5 mol% of CAN for 1.5 hours to yield the expected polyhydroquinoline in 98%. An increase in the quantity of CAN from 5 mol% to 10 mol% decreased both the reaction time (from 2 to 1 hours) and the yield (35%) of polyhydroquinoline. To address the role of CAN in this reaction, different Lewis acids such as CeF$_4$ and CeCl$_3$·7H$_2$O and Brønsted acids such as NH$_4$Cl were used. The Hantzsch-type condensation proceeded more rapidly and the yields were higher when CeF$_4$ and NH$_4$Cl were used instead of CeCl$_3$·7H$_2$O. This result indicates that Ce(IV) and ammonium ions which are present in CAN, act as Lewis and Brønsted acids, respectively.

Cerium ammonium catalyzed a three-component domino reaction between aromatic amines, α,β-unsaturated aldehydes and ethylacetoacetate, to afford a suitable entry into a new class of *N*-aryl-5,6-unsubstituted 1,4-DHPs (Scheme 12).[23] In a model reaction involving cinnamaldehyde (R$_1$=H), ethyl acetoacetate (Z=OEt) and m-toluidine (R$_2$=R$_3$=R$_5$=H, R$_4$=Me), the best results were obtained with 5 mol% of

CAN to afford the DHP derivative in a 71% yield after one hour stirring in ethanol at room temperature. These optimized reaction conditions were applied to a range of substrate including electron-donating (R_2=OCH$_3$, R_3=H) and electron-withdrawing groups (R_2=Cl, R_3=H) on the nitrogen aryl substituent. It is worth to note that the catalyst tolerates the use of *tert*-butyl β-ketothioesters as the dicarbonyl derivative. To clarify the role of CAN, a series of experiments were carried out. In one of these, the condensation between cinnamaldehyde, aniline and ethyl acetoacetate was carried out in the presence of a radical trap (1,1-diphenylethylene) and do not revealed a noticeable loss of yield. This result is similar to previous observations in which it was shown that CAN could behave as a Lewis acid rather than a single electron oxidant.

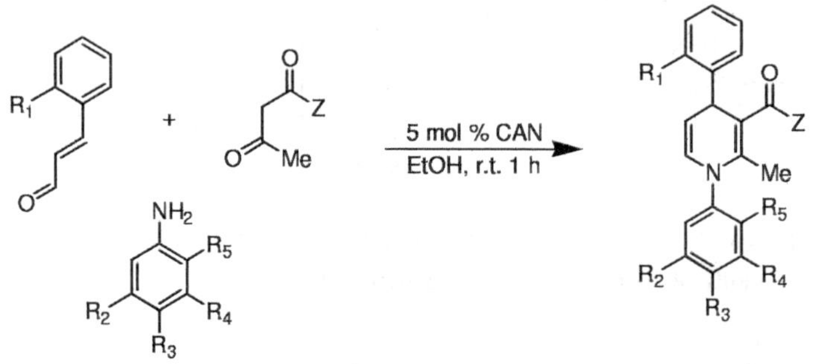

Scheme 12

Rare earth triflates such as Yb(OTf)$_3$[24] or Sc(OTf)$_3$[25] also catalyzed the four-component Hantzsch reaction of aldehydes, dimedone, ethyl acetoacetate and ammonium acetate. Polyhydroquinoline derivatives were obtained in good yields. Interestingly, the catalysts can be extracted by water and reused without significant loss of activity.

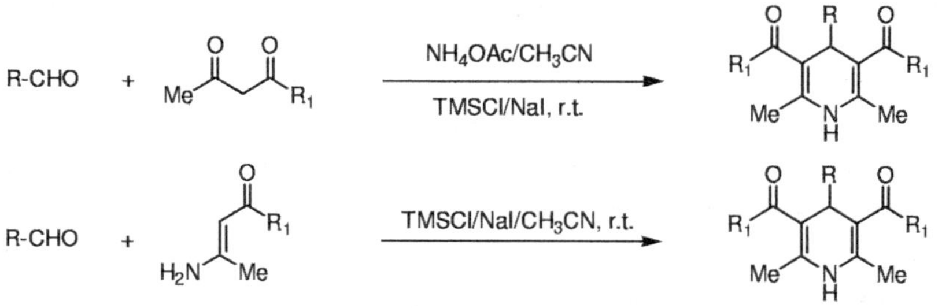

Scheme 13

Lewis acids have also been used to synthesize symmetrical Hantzsch 1,4-DHPs. Sabitha *et al.* reported the first condensation of an aldehyde with 2 equivalents of ethylacetoacetate and NH$_4$OAc in the presence of trimethylsilyl iodide (TMSI, generated *in situ* from trimethylsilyl chloride (TMSCl) and sodium iodide) (Scheme 13).[26] The reaction was carried out at room temperature and went to completion within 6 hours. Various substituted aliphatic and aromatic aldehydes carrying either electron-withdrawing or electron-donating substituents were converted into the expected 1,4-DHPs. It was also shown that TMSI also catalyzed the modified Hanzsch procedure employing ethyl aminocrotonate as nitrogen source. All the reactions proceeded smoothly to afford the expected 1,4-DHP in 2–2.5 hours.

More recently, convenient and cost-effective protocols involving $AlCl_3 \cdot 6H_2O$[27] and molecular iodine[28] as catalysts have been used for the synthesis of symmetrical Hantzsch 1,4-DHPs.

2.2. Synthesis of optically active 1,4-DHPs

The synthesis of unsymmetrical 1,4-DHPs involves the formation of a new stereogenic carbon at the C-4 position. It has been shown that the two enantiomers displayed different biological activities[29] and pharmacological effects.[30] Thus, efficient methods for the synthesis of optically pure 1,4-DHPs is not only of interest for drug discovery but also for organic synthesis.

The first optically active DHP derivative was purchased in 1992. Since then, several 1,4-DHPs derivatives have been patented for the treatment of neurodegenerative diseases.[31]

In 1991, Goldmann and Stoltefuss reviewed different methods for the formation of optically enriched 1,4-DHPs such as: the resolution of racemic mixture *via* the formation of diastereomeric esters, diastereoselective syntheses *via* chiral auxiliary groups, chemoenzymatic separations and diastereoselective reductions of pyridines. Pharmacological behaviours of 1,4-DHPs enantiomers were also discussed in this review.[32] Since the beginning of the 2000's, new advances in this field have been reported. This chapter covers the different methods leading to enantiomerically enriched 1,4-DHPs, namely the chemoenzymatic approach of optically active 1,4-DHPs, diastereoselective cyclizations and reductions of pyridines. Finally, the last advances in the organocatalytic enantioselective syntheses of 1,4-DHPs will be reported.

2.2.1. Enzyme-catalyzed enantioselective differentiation

Biotechnological approaches, based on enzyme-catalyzed enantiomeric differentiation, has been shown to be a promising technique for the synthesis of enantiopure 1,4-DHPs.[33] This method is based on the hydrolysis of simple prochiral 1,4-DHP-dicarboxylic diesters and showed a number of distinct advantages to other methods. Thus, enzymes are usually active under mild conditions and they often exhibit high stereoselectivities and possess a broad substrate tolerance.

Since 1991, chemoenzymatic syntheses of enantiopure 1,4-DHP derivatives have been largely reported by several research groups who well described the desymmetrization of prochiral 1,4-DHPs and the kinetic resolution of racemates.[34]

In 1993, Hirose *et al.* reported their results concerning the enzymatic hydrolysis of the bis(carbamoylmethyl)-1,4-dihydro-2,6-dimethyl-4-(3-nitrophenyl)-3,5-pyridine dicarboxylate using the protease Seaprose S (*Aspergillus muelleus*). This enzyme afforded the hydrolyzed 1,4-DHP in 83% yield and 99% e.e. (Scheme 14).[35]

The transesterification was then carried out in the presence of sodium methoxide, to lead to the (*R*)-methyl ester in a 54% overall yield (two steps), without any racemization. It is worth to note that the latter compound constitutes an advanced intermediate for the synthesis of the (*R*)-manidipine or the (*R*)-nitrendipine.

In 2001, Sobolev *et al.* described the enzyme catalyzed desymmetrization of racemic 3,5-bis(iso-butyryloxymethyl)-4-[2-(difluoromethoxy)phenyl]-2,6-dimethyl-1,4-dihydro-3,5-pyridinedicarboxylate by the *Candida rugosa* lipase (CRL) (Scheme 15). This latter was reported to be more efficient toward various substrates than the lipase AH previously described.[36] The enantioselectivity of the reaction was greater than 95%, and the *(R)* isomer was obtained in 55% yield.

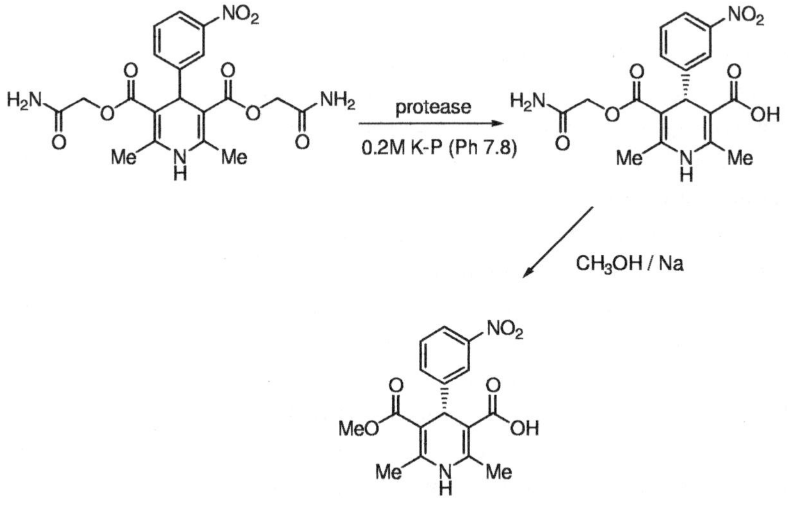

Scheme 14

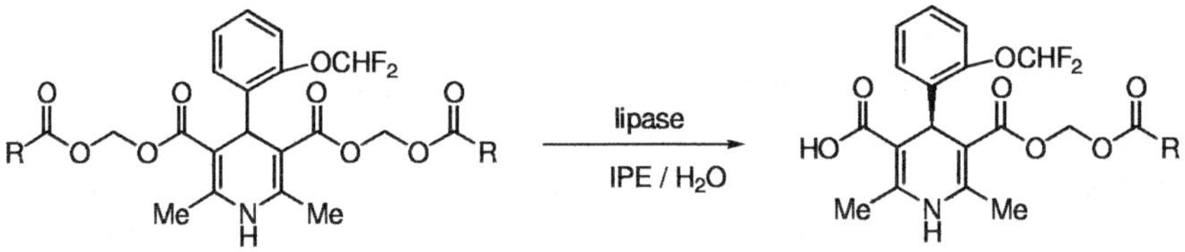

Scheme 15

2.2.2. Diastereoselective synthesis of 1,4-DHPs
2.2.2.1. Diastereoselective cyclizations

The major part of diastereoselective syntheses of 1,4-DHPs have been published before 2000. One of the first diastereoselective approach is reported in the review of Goldmann and Stoltefuss[32] and refers to the diastereoselective synthesis of a 1,4-DHP sulfone.[37] In that case, a chiral sulfoxide borne by an unsaturated ketone derivative was able to react with an enaminoester derivative to lead to a 1,4-DHP with 94% d.e. in which the C-5 ester was replaced by a sulfone group.

Diastereoselectivity can be brought as well by a chiral auxiliary group placed on the nitrogen. Chiral enaminoesters like hydrazones were part of the work of Enders *et al.* in 1988.[38] In this Hantzsch-type condensation, the chiral information is introduced *via* the nitrogen functionality. After the cleavage of the chiral auxiliary, the expected 1,4-DHP was isolated in high enantiomeric excess (98%) and good yield (72%).

210

Other methods involving amino acid derivatives as chiral auxiliary have also been described to synthesize a biological active 1,4-DHP with an enantiomeric excess greater than 94%.[39]

In 1992, a three component Hantzsch type synthesis involving a cyclic 1,3-diketone, an aromatic aldehyde, a chiral diketone and ammonia was reported by Rose and Dräger (Scheme 16).[40] This reaction led to the intermediary 1,4-DHP in 98% d.e. Cleavage of the chiral auxiliary by transesterification with sodium methoxide afforded the desired 1,4-DHP in 68% yield.

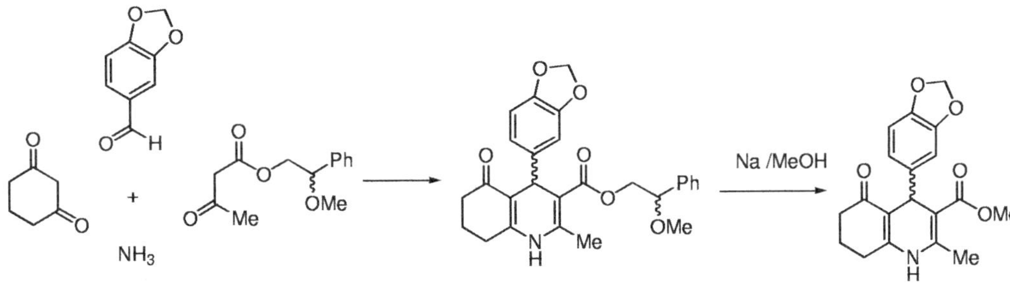

Scheme 16

In 2002, Patel *et al.* described a concise and convergent asymmetric synthesis of 4-substituted 1,4-DHPs *via* the Michael addition of an optically pure vinylogous amide to an α,β-unsaturated ketone (Scheme 17).[41] The intermediate enaminone was obtained in 70% d.e, and was directly treated with aqueous ammonia in the presence of ammonium chloride and finally dehydrated with concentrated HCl to afford the expected DHP in 95% e.e. and 29% overall yield.

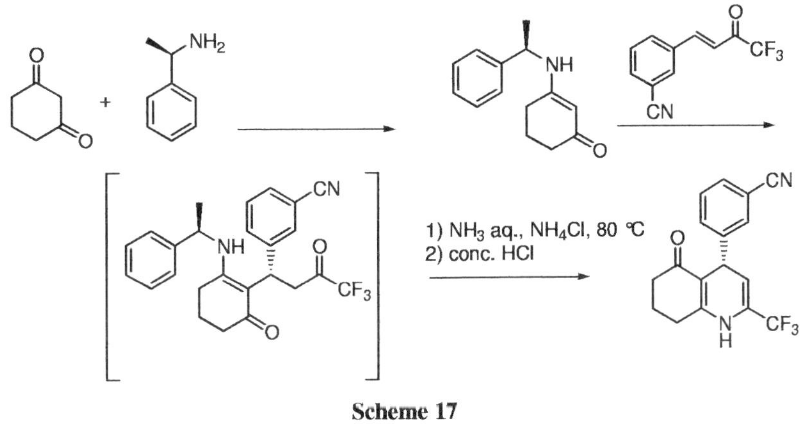

Scheme 17

Recently, Palacios *et al.* demonstrated the moderate activity of *N*-vinylic phosphazene as intermediates for the preparation of symmetrical and unsymmetrical 1,4-DHPs (Scheme 18).[42] Optically active *E*-vinylic phosphazene derived from β-aminoesters of (1*R*,2*S*,5*R*)-(−) menthol was prepared and condensed with methyl 2-(*p*-nitrophenyl)methyleneacetoacetate in CH₂Cl₂. The corresponding 1,4-DHP was obtained as a mixture of two diastereoisomers in a 63:34 ratio.

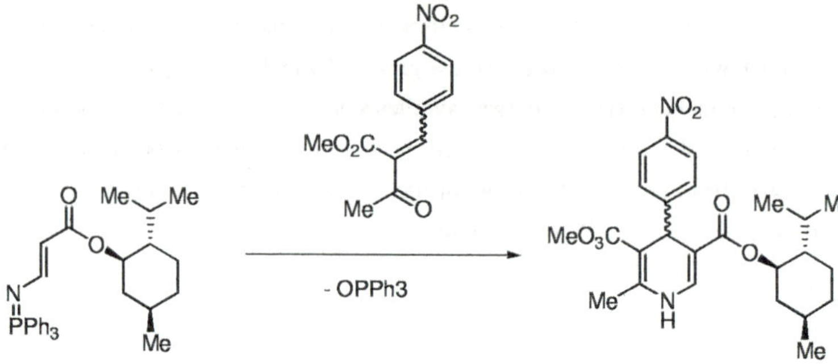

Scheme 18

2.2.2.2. Selective reduction of pyridines and pyridinium salts

As mentioned by Goldmann and Stoltefuss, the enantioselective reduction of pyridine and pyridinium salts is considered as an elegant route to optically active 1,4-DHPs.[32]

Thus, Meyers and Oppenlaender showed that the face selective nucleophilic addition of alkyl anions (such as methyllithium) to a pyridine nucleus can be successfully used for the synthesis of C-4 substituted non racemic 1,4-DHPs.[43] Schultz and Flood have shown that the addition of Grignard reagents to pyridinium salts was more selective.[44]

This reaction was extensively studied by Ohno *et al.* who pointed out that the stereochemical outcome of the reductive alkylation with Grignard reagents depended on the *syn* or *anti* conformation of the pyridinium salts derivatives.[45,46]

Likewise, Alexakis *et al.* showed the efficiency of organocopper derivatives for the 1,4-addition process (Scheme 19).[47] In that case, the chiral 1,4-DHP was obtained in 90% yield and 95% d.e. As depicted in Scheme 19, the metal chelated transition state can explain the observed stereoselectivity. According to X-ray studies, the *trans* relationship between the *N*-methyl substituent of the imidazolidine ring and the α-phenyl group creates a diastereofacial selectivity. It was hypothesized that the conformation shown in the transition state is locked by a chelation of the dimeric cuprate allowing for the *Re* face attack of the pyridine ring.

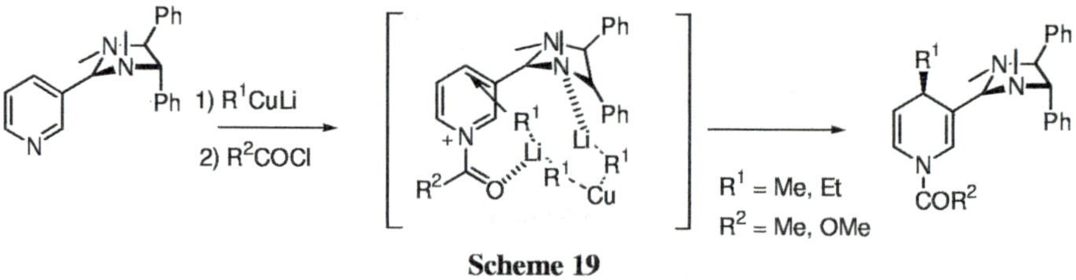

Scheme 19

This type of reduction has been extensively studied by Yamada's group, who prepared 1,4-DHPs by regio- and stereoselective addition of organometallic reagents[48] or ketene silyl acetals[49] on pyridinium salts (Scheme 20). They demonstrated that stereoselectivity could be relied to an intramolecular interaction between the 1,3-thiazolidine-2-thione moiety and the intermediary pyridinium salt derivative as shown in

Scheme 20.[50] After the addition of two equivalents of ketene silyl acetal at room temperature, the expected 1,4-DHP was obtained stereoselectively (95% d.e.).

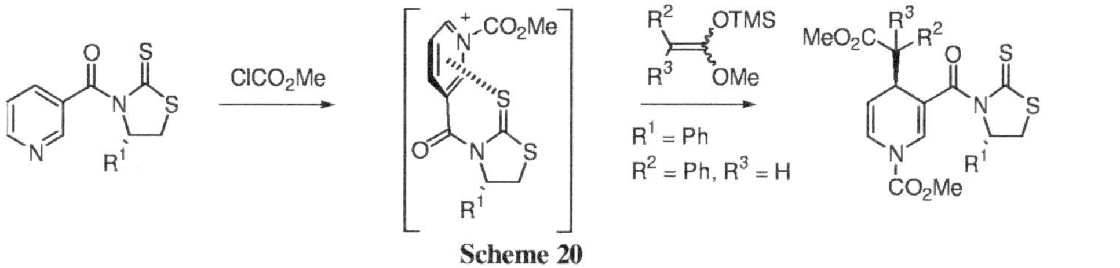

Scheme 20

Similarly, the same group also showed that intramolecular cation-π interactions can be used as a conformation-controlling tool for the face-selective addition of ketene silyl acetal onto pyridinium systems (Scheme 21).[51] High diastereoselectivities were then obtained (57% yield; 99% d.e.).

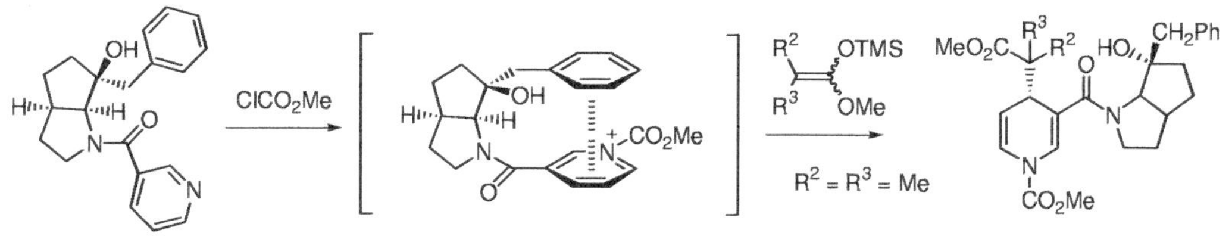

Scheme 21

2.3. Enantioselective catalytic syntheses of 1,4-DHPs

In 2007, Renaud *et al.* reported the first catalytic asymmetric synthesis of enantiomerically enriched 1,4-DHPs. 4-Aryl substituted 1,4-DHPs were obtained from the condensation of an α,β-unsaturated aldehyde with various enaminoester derivatives. This reaction was carried out in the presence of chiral phosphoric acids derived from different 3,3'-aryl substituted (S)-1,1'-bi-2-naphthol (Scheme 22).[16] As an example, the condensation of cinnamaldehyde and *tert*-butyl-*N*-benzylaminobut-2-enoate was carried out at $-7\ °C$ in dichloromethane to provide the optically enriched 1,4-DHP in 89% yield with 50% e.e.

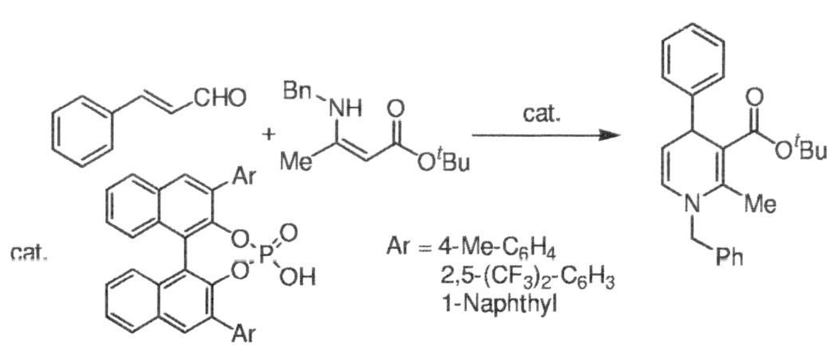

Scheme 22

Recently, Gong *et al.* reported the asymmetric three-component cyclization of an aryl α,β-unsaturated aldehyde, a primary aromatic amine and an acetoacetate in the presence of a chiral phosphoric acid.[52] Hydrogenated 1,1'-bi-2-naphthol derivatives proved to be the more efficient catalysts. From *p*-nitro-cinnamaldehyde, *p*-anisidine and ethyl acetoacetate, the corresponding 1,4-DHP was obtained in 97% e.e. and in 53% yield (Scheme 23). Lower enantiomeric excesses were obtained starting from unsaturated aliphatic aldehydes.

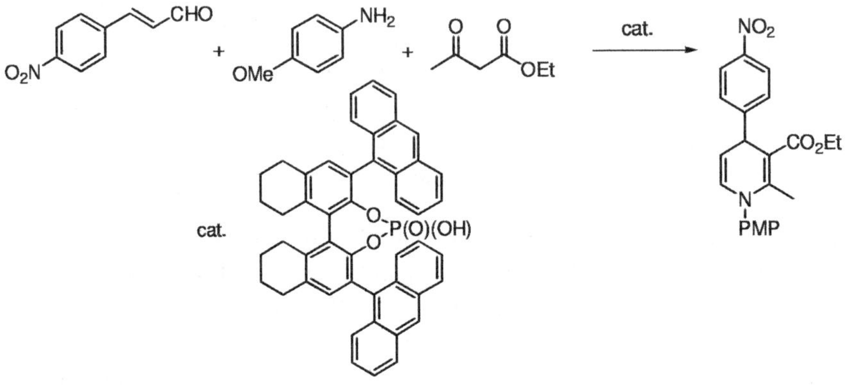

Scheme 23

Finally, Jørgensen *et al.* recently published their results[53] concerning the one-pot reaction of α,β-unsaturated aldehydes with β-diketones or β-ketoesters and primary amines. Optically active 2,3-disubstituted 1,4-DHPs were obtained in moderate yields and with good enantioselectivities (Scheme 24). The 1,4-DHP was synthesized in 55% yield and an enantiomeric excess of 90% when the reaction was performed in the presence of *L*-proline derivatives and benzoic acid as additive.

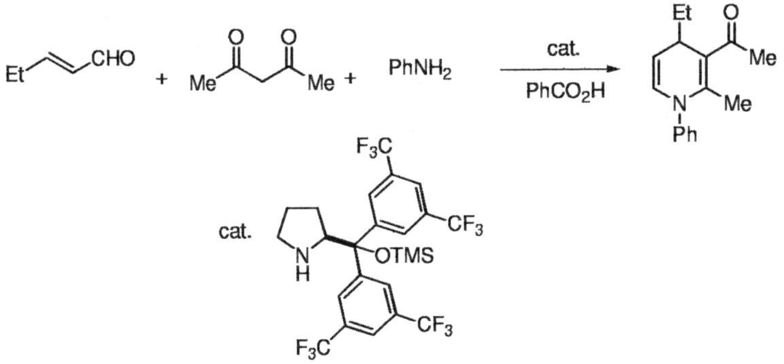

Scheme 24

3. Brønsted and Lewis acids catalyzed Biginelli reaction

The Biginelli reaction is a three-component condensation reaction between an aldehyde, an urea or thiourea and an easily enolizable carbonyl compound. It was originally described by the italian chemist Pietro Biginelli in 1893.[54] This reaction offers a straightforward approach to multifunctionalized 3,4-dihydropyrimidin-2-(1*H*)-ones and related heterocyclic compounds.[55] The first example of this reaction was carried out by simply heating the three components in refluxed ethanol in the presence of a catalytic amount of HCl. This multicomponent synthesis was largely unexplored until the late 1980's. Since then, the number of publications and patents on the subject increased tremendously because the dihydropyrimidines have a broad range of biological effects, including: antiviral, antitumor, antibacterial and antiinflammatory activities,[56] antihypertensive agents,[57] α_{1a} adrenoreceptor-selective antagonists,[58] mitotic kinesin Eg5 motor protein inhibitor and potential new lead for the development of anticancer drugs.[59] Moreover, several marine natural products with interesting biological activities contain the dihydro-pyrimidine-5-carboxylate core.[60] This second part of the review will report the recent advances in the Biginelli reaction catalyzed either by (chiral) Brønsted acids or (chiral) Lewis acids, *but it has to be mentioned that this chapter is not an exhaustive review of all the catalysts available and published in the literature during the last decade.*

3.1. Recent synthesis of dihydropyrimidines (DHPMs)

3.1.1. Lewis acid catalysis

The original Biginelli protocol for the preparation of dihydropyrimidines (DHPMs) consisted of heating a mixture of the three components included β-ketoester, aldehyde and urea in ethanol containing a catalytic amount of HCl. The major drawbacks associated with this protocol are the use of a strong acid as well as the low yields in the case of substituted aromatic and aliphatic aldehydes. As Biginelli reaction for the synthesis of DHPMs has received renewed attention, several improved procedures have been reported based on metal-catalyzed Biginelli reaction during the last decade and some of them are reported in Table 1.[61–73]

By comparison with the original reaction conditions, the yields are usually higher, even with substituted aromatic aldehydes. Among all the Lewis acids reported in Table 1, Fe(OTf)$_3$,[65] Fe(O$_2$CCF$_3$)$_3$,[65] Ce(NO$_3$)$_3$·6H$_2$O[82] and Yb(OTf)$_3$[79,84] seem to provide the best results in term of yield, reaction time and broad substrate compatibility. Yields of up to 95% can be reached within a few minutes with these catalysts.

An alternative method to improve the Biginelli condensation is the development of micro-wave heating associated to metal catalyzed reaction (Table 1). Copper(II)[67] and zirconium(IV)[72] led to encouraging results in this reaction as up to 99% yield could be reached in only 2 minutes.[66]

In the light of green chemistry, some works also focussed on:

i) the reuse of catalysts (Table 1): then, copper salts,[67,68,84] ytterbium triflate[79,83] or sulfonated zirconium[72] can be reused several times (3 to 5) without almost any decrease of the yields;

ii) solvent free reactions (Table 1): in neat conditions, iron salt at 70 °C,[65] copper salts at 100 °C,[67] and ruthenium chloride[73] provided good yields within few minutes;

iii) reactions in water: up to now, the most efficient catalysts in water were the metal triflimide (Cu, Ni, Yb).[66] By comparison with CeCl$_3$·7H$_2$O which was previously reported as efficient catalyst in water,[81] the catalyst loading is low (5 mol% of M(NTf$_2$)$_{2or3}$ and 25 mol% of CeCl$_3$·7H$_2$O) and the reaction occurred at lower temperature (room temperature), whatever the substituents on the aromatic ring, the yields remained quite high.

In the last ten years, many improvements have then been achieved and nowadays the Biginelli reaction might be one of the most studied multicomponent reactions. However, in these Biginelli reactions catalyzed by Lewis acids, some points have to be improved. For example, the role of metal halides such as FeCl$_3$, NiCl$_2$, ZrCl$_4$, or LaCl$_3$ is not well understood, and several authors suggest that these catalysts promote the *in situ* formation of HCl.[85]

$$R^1\text{CHO} \;+\; H_2N\text{-}C(=X)\text{-}NH_2 \;+\; R^2\text{-}C(=O)\text{-}CH_2\text{-}C(=O)\text{-}O\text{-}R^3 \quad \xrightarrow{\text{cat., solvent, temp.}} \quad \text{dihydropyrimidinone}$$

X = O, S

Table 1. Lewis acid catalyzed Biginelli reaction.

Catalyst	Yield (%, R^1=aryl)	Yield (%, R^1=alkyl)	Time	%cat.	solvent	Ref.
Mg(ClO$_4$)$_2$	60–92	/	4–8.5 h	10	EtOH, 80 °C	61
Al(H$_2$PO$_4$)$_3$	60–95 (thermal) 56–91 (M.W.)	/	20–60 min (Δ) 15–20 min (M.W.)	16 (Δ) or 31 (M.W.)	neat	62
SiCl$_4$	75–98	/	4 h	10	DMF/CH$_3$CN, t.a.	63
FeCl$_3$·6H$_2$O	53–96	53–72	4 h	25–60	EtOH, Δ	64
Fe(OTf)$_3$	86–99	90	15–40 min	5	Neat, 70 °C	65
Fe(O$_2$CCF$_3$)$_3$	86–99	90	15–40 min	5	Neat, 70 °C	65
NiCl$_2$.6H$_2$O/HCl	58–94	62–64	5 h	25	EtOH, Δ	64b
Ni(NTf$_2$)$_2$	59	/	24 h	5	Water, t.a.	66
CuCl$_2$·2H$_2$O CuSO$_4$·5H$_2$O	80–97 (thermal) 82–99 (M.W.)	80–85 (thermal) 80–88 (M.W.)	60–110 min (Δ) 1–2 min (M.W.)	10	neat, 100 °C or M.W.	67
Cu(OTf)$_2$	65–95	60		1	CH$_3$CN, t.a. to 70 °C	68
Cu(NTf$_2$)$_2$	65	/	24 h	5	Water, t.a.	66
ZnCl$_2$	56–84	60–68	6 h	10	neat	69
Sr(NO$_3$)$_2$	50–78	/	6 h	5	AcOH, Δ	70
ZrCl$_4$	40–99	/		10	neat	71
Zr(SO$_4$)$_2$	81–94	/	2 h (MeOH) 60–90 s in M.W	/(100mg).	MeOH, Δ neat, M.W.	72
RuCl$_3$	80–93	74–76	30–90 min	5	neat, 100 °C	73
InCl$_3$	84–95	75–85	6–9 h	10	THF, Δ	74
In(OTf)$_3$	75–95	/	5h	10	CH$_3$CN, 90 °C	75
SnCl$_2$·2H$_2$O	63–98	/	6 h	20	CH$_3$CN or EtOH, Δ	76
SbCl$_3$	51–89	/	18–24 h	20	CH$_3$CN, Δ	77
LaCl$_3$·7H$_2$O	68–96	56–60	5 h	50	EtOH, Δ	78
La(OTf)$_3$	89 (1 ex.)	/	20 min	5	neat, 100 °C	79
BiCl$_3$	50–92	85–95	5h	12	CH$_3$CN, Δ	80
CeCl$_3$·7H$_2$O	82–95 (EtOH) 80–93 (H$_2$O) 65–80 (neat)	80–83 (EtOH) 73–76 (H$_2$O) 65–73 (neat)	2.5–5 h	25	EtOH or H$_2$O or neat	81
Ce(NO$_3$)$_3$·6H$_2$O	84–98	84–87	10–40 min	5	neat, 80 °C	82
Yb(OTf)$_3$	81–98	83–87	20 min	5	neat, 100 °C	79,84
Yb(NTf$_2$)$_3$	81–98	83–87	24 h	5	water, r.t.	66

To the best of our knowledge, until now, there is also no really efficient catalytic system for the synthesis of a broad range of alkyl substituted DHMPs.

3.1.2. Brønsted acid catalysis

An alternative strategy for the synthesis of DHMPs is the use of Brønsted acids instead of metal salts as catalysts. Indeed, in the original procedure reported by P. Biginelli, the catalyst mediating the synthesis of DHPMs was hydrochloric acid. Because of some drawbacks (low yields, narrow scope, long reaction time...), these reaction conditions were revised during the last decade and numerous Brønsted type catalysts appeared in the literature. As for the Lewis acids, this chapter will not cover all the published catalysts and will not be an exhaustive review, but we will try to provide an overview of the different kinds of catalysts actually available (Table 2). By comparison with the original reaction conditions, some Brønsted acids such as acetic acids type[86] or sulfonic acid type[87–91] provide encouraging results when aromatic aldehydes and few aliphatic aldehydes were used in a such condensation. However, except for PTSA in water,[87] the catalyst loading remains usually high (from 10 to 300%).[87,88–91]

POMs (polyoxometallates) and zeolite seem to be more interesting in term of activity, recyclability (green chemistry) and scope.[92–98] In fact, in temperature ranging from 50 °C to 110 °C, DHPMs, substituted by an aromatic (bearing also a large variety of substituents: OH, Cl, NO_2, alkyl...) or an aliphatic moiety at the C4-position, can be isolated in very high yields within few minutes using low catalyst loading. Supported POMs were also effective for the synthesis of DHPMs and a slight decrease of the yield was noticed after the fourth cycle.[95] Even ammonium chloride, bromide or more functionalized ammonium chlorides (such as substituted hydrazine, proline) can act as efficient catalysts.[99–102] One might suggest that the ammonium cation is only a weak acid, but it appears (Table 2) that with such catalysts, results are better than those obtained with acetic acid or sulfonic acid.

Moreover, the most intriguing catalyst but also the most efficient one is the trialkyl ammonium bromide reported by Reddy in 2003, which has no acidic proton.[100] It is worth mentioning that most of the catalysts usually used for preparation of DHPMs are halogen containing Lewis acids. These results indicate that the halide present in the catalyst may be playing a crucial role in these transformations and may assume significance in the wake of the large number of publications which have appeared recently for the synthesis of DHPMs by modified Biginelli reactions using heavy metal halide catalysts. Maybe there is some doubt as to whether the reaction is catalyzed by Lewis acids *via* metal ion coordination or simply by the halide ions. However, whatever the activation type (H-bonding, Lewis acidity,...), the results presented in the Table 2 with trialkylammonium bromide indicate the scope and generality of the method, which is efficient, not only for urea or thiourea, but also for aliphatic as well as aromatic aldehydes (up to 93% yield and up to 99% yield, respectively). An important feature of this method is that electron releasing or withdrawing groups give excellent yields in high purity. It is pertinent to note that this trialkyl substituted ammonium salt gave consistently higher yields with aliphatic aldehydes.

3.1.3. Miscellaneous

Boron derivatives can be used as catalyst for the synthesis of DHPMs. Then boronic acid[104] (10 mol%) in refluxing acetonitrile or boric acid[105] in glacial acetic acid at 10 °C can lead to the heterocyclic compounds in 60–97% and 86–98% yield, respectively. It is worth to mention that only aromatic aldehydes have been engaged in the reaction.

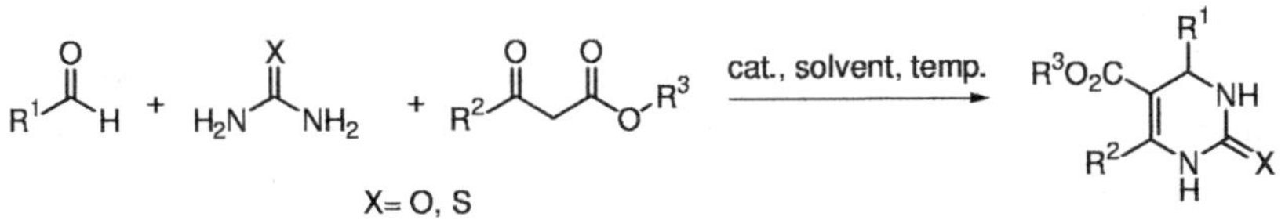

Table 2. Brønsted acid catalyzed Biginelli reaction.

Catalyst	Yield (%, R^1=aryl)	Yield (%, R^1=alkyl)	Time	%cat.	solvent	Ref.
ClCH$_2$CO$_2$H	79–98	47–75	3 h	10	neat, 90 °C	86
PTSA	81–94	/	15 min	2–3%	water, <50 °C	87
PSSA	86–92	/	20 min	3*weight of aldehyde	water, 80 °C under M.W.	88
1T3P	16–86	83	6 h	100	AcOEt, Δ	89
Al$_2$O$_3$-SO$_3$H	67–92	69–72	0.5–3 min	15	neat, M.W. 180W	90
KHSO$_4$	86–99	85–86	0.5–2 h	25	glycol, 100 °C	91
(NH$_4$)$_2$HPO$_4$	84–97	84–88	2–3.5 h	30	neat, 80 °C	92
H$_3$W$_{12}$PO$_{40}$	52–98	50–95	1–9 h	2–5	CH$_3$CN, Δ	93, 94
H$_3$Mo$_{12}$PO$_{40}$	57–97	47–89	1–9 h	2–5	CH$_3$CN, Δ	92, 94
H$_3$W$_{12}$SiO$_{40}$	60–96	52–65	1 h	5	CH$_3$CN, Δ	93
H$_3$W$_{12}$PO$_{40}$ on SiO$_2$	84–95	53–72	50–100 min	9%wt	CH$_3$CN, Δ	95
zeolite TS1	>93	>93	10–30 min	10%wt	neat, 50°C	96
zeolite HY	63–80	42–64	12 h	0.5 g	toluene, Δ	97
Dowex 50W	27–95	/	90–180 min	5mg/mmol	neat, 130 °C	98
NH$_4$Cl	77–90	42–78	3 h	40	neat, 100 °C	99
NH$_4$Br	77–93	64–86	20–60 min	35-50	neat, 100 °C	100
hydrazine·2HCl	62–97	93–95	2–24 h	5–10	i-PrOH or DMSO, r.t.	101
proline-OMe.HCl	10–99	63	18 h	10	MeOH, Δ	102
N,N,N-(Me$_2$, Et)-NCH(CH$_3$)Ph	87–99	86–93	20–60 min	35–50	neat, 100 °C	100
TMSCl	73–97	/	1–3 d	400	DMF, r.t.	103

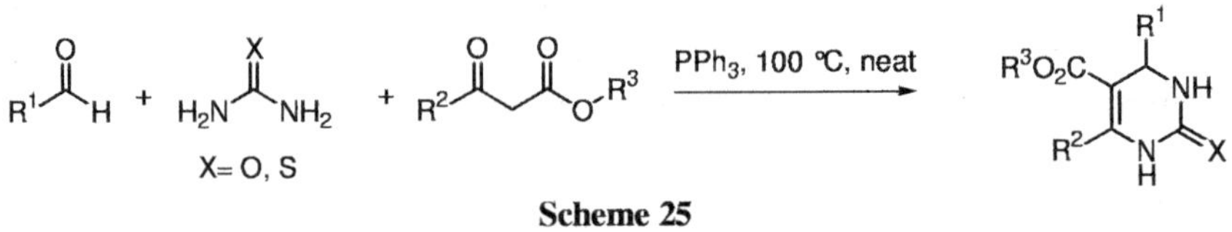

Scheme 25

Triphenylphosphine has been investigated by Debache and coworkers in 2008 as Lewis base catalyst for the Biginelli one-reaction (Scheme 25).[106] With only 10 mol% of PPh$_3$ at 100 °C, substituted dihydropyrimidones have been obtained within 10 hours with moderate to good yield (58–70%).

More interestingly, Khodaei *et al.* introduced Bi(NO$_3$)$_3$ salts in a Biginelli type reaction.[107] As these bismuth salts were able to oxidize benzyl bromide compounds into aryl aldehydes in the presence of TBAF (tetrabutyl ammonium fluoride), they prepared several DHPMs in high yields from different substituted benzyl bromide derivatives, urea and β-ketoesters (Scheme 26). The advantages of this method are not only the use of non toxic catalyst but also the introduction of benzylic halides instead of aromatic aldehydes which are prone to be readily oxidized into the corresponding acids.

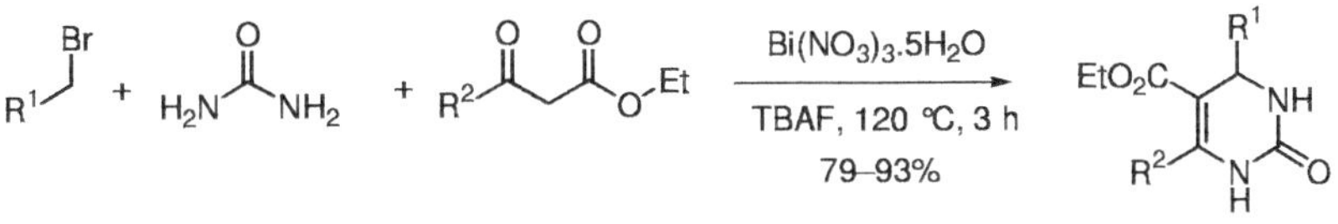

Scheme 26

3.2. Synthesis of chiral DHPMs

The considerable interest for DHPM-type products stems from their structural similarity to DHPs (Hantzsch products, DHPs), a class of compounds used as calcium channel antagonists. Therefore, they have found extensive use as therapeutics in the clinical treatment of cardiovascular diseases such as hypertension, cardiac arrhythmias or angina pectoris.[56]

Then, as biological relevant compounds, they have to be prepared in an enantiomerically pure form. For instance, (*S*)-L-771688 is a more potent and selective α$_{1a}$ receptor antagonist for the treatment of benign prostatic hyperplasia (BPH) than the corresponding *R* enantiomer (Figure 3).[58b]

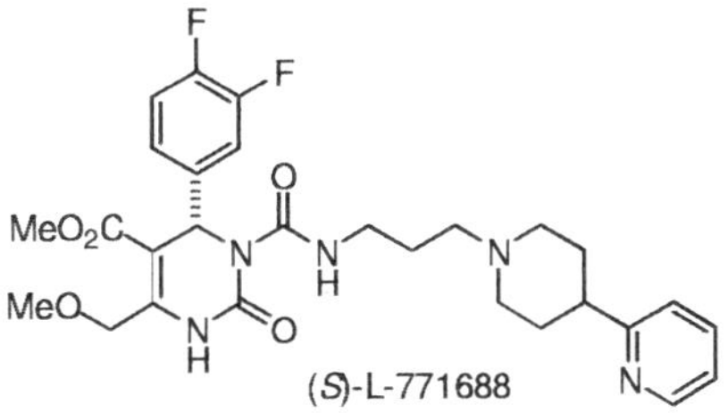

Figure 3

Despite the fact that the activity depends only on the absolute configuration of the C$_4$ stereocentre, the Biginelli reaction has been mostly carried out in an achiral version and methods to produce optically active DHPMs are scarce. Optically active DHPMs can be produced by three different strategies, *i.e.* (i) chemical or enzymatic resolution, (ii) diastereoselective synthesis and (iii) enantioselective synthesis.

3.2.1. Chemical and enzymatic resolution of racemic DHPMs

The easiest way to produce optically pure DHPMs is a chemical resolution of a racemic mixture, but this methodology is not broadly applicable. Such approach was described for the synthesis of SQ-32926 (*R* and *S*) *via* a transamidification reaction of the DHPM carbamate with the chiral (*R*)-α-methyl benzylamine as a key step. The two enantiomers could be isolated in an optically pure form after the release of the chiral auxiliary (Scheme 27).[108]

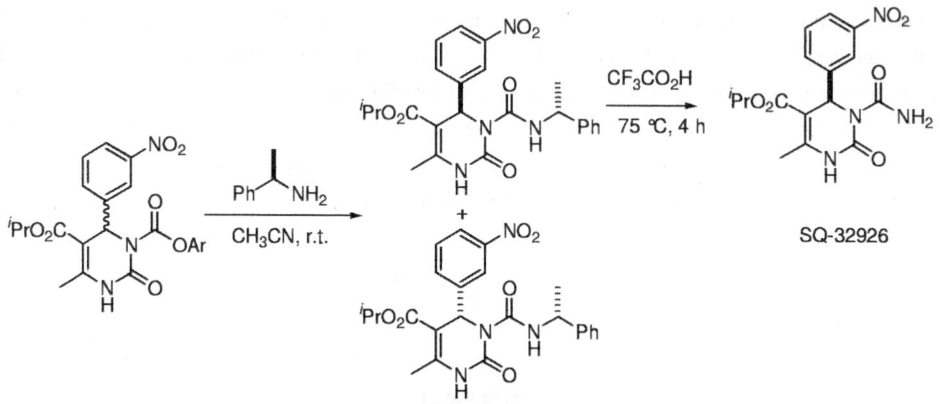

The use of C-glycosylcarbonyl chloride as a chiral auxiliary for the chemical resolution of monastrol was reported by Dondoni *et al.* After purification/separation of the diastereomers on silica gel and removal of the chiral auxiliary, both optically pure enantiomers were obtained (Scheme 28).[83b]

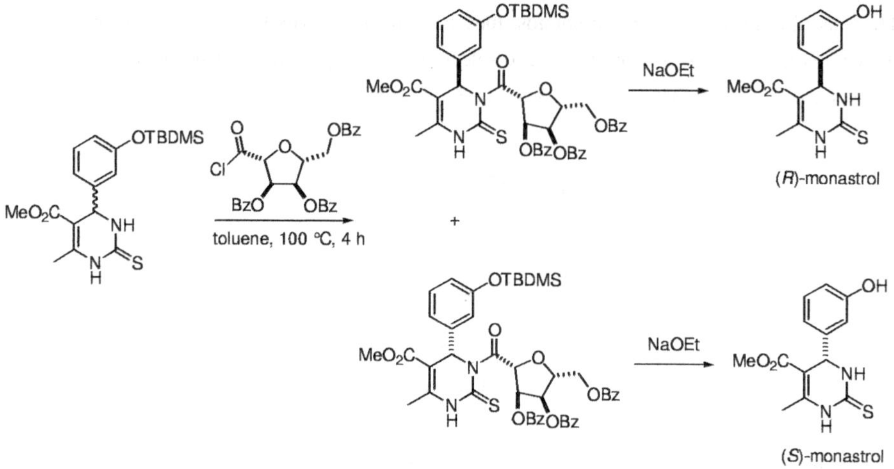

Scheme 28

The third route described in the literature to optically pure DHPMs using chemical resolution needs the preparation of chiral ammonium salts of racemic 5-dihydropyrimidinone carboxylic acid (Scheme 29). Separation of the diastereomers by simple crystallization and acidic treatment led to optically pure DHPMs, but this method cannot be extended to a broad range of DHPMs.[109]

The enzymatic resolution can also provide opportunities for the isolation of both enantiomers of DHMPs. Subtilisin was introduced for the first time by Ikemoto *et al.* in such kinetic resolution.[110] The kinetic hydrolysis of the ester function gave rise to the corresponding acid in 95% e.e. and the remaining ester was also isolated in similar optical purity (Scheme 30).

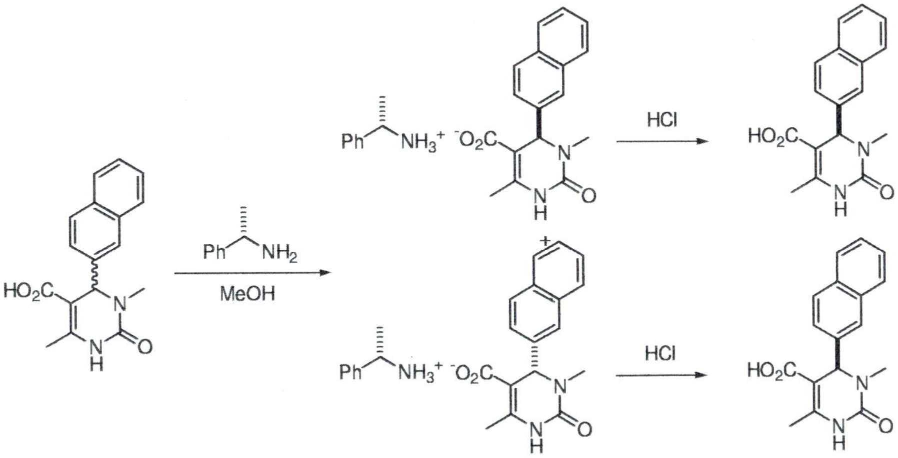

Scheme 29

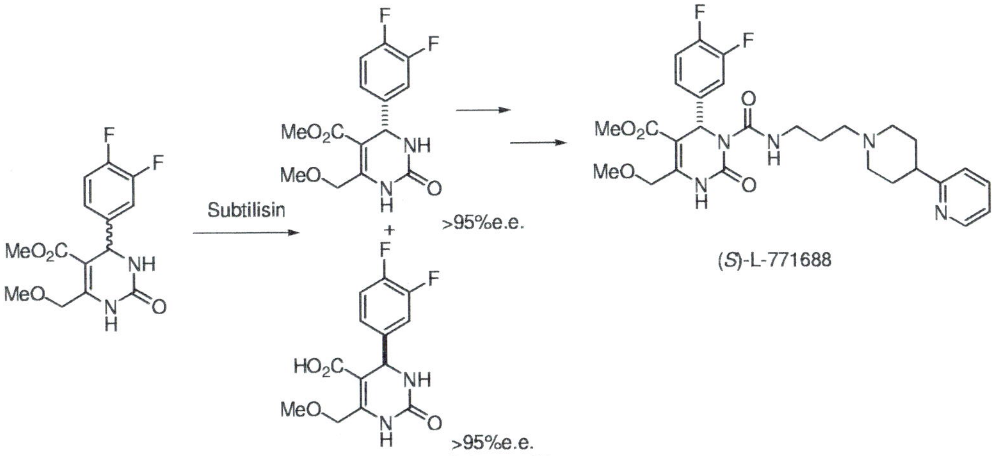

Scheme 30

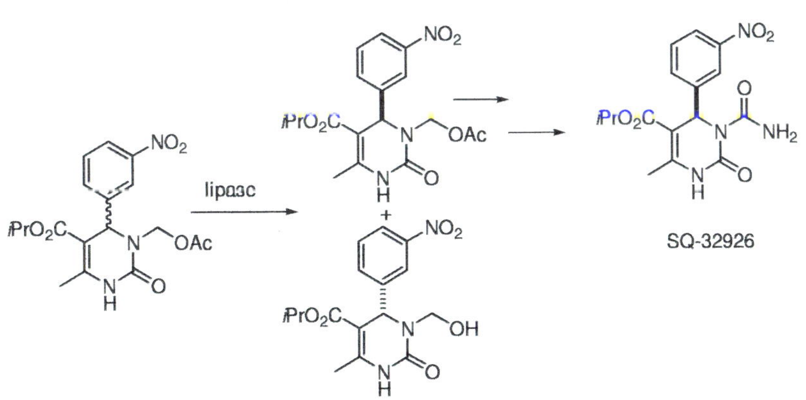

(S)-L-771688

Scheme 31

SQ-32926

221

Lipase can also be applied to synthesize optically pure intermediates of pharmaceutical derivatives (Scheme 31) *via* kinetic hydrolysis of an acetic ester moiety. After chemical arrangements, both enantiomers of SQ 32926, of which the (*R*)-isomer is a calcium channel blocker, can be obtained.[111]

3.2.2. Diastereoselective approaches

The diastereoselective synthesis of DHPMs was described independently by Kappe and Dondoni in 2002. The strategy developed by Kappe involved the use of the (−)-menthyl acetoacetate in the presence of urea, naphthaldehyde and a catalytic amount of H^+ in refluxing methanol. This reaction led to an inseparable mixture of diastereomers in a 1:1 ratio (Scheme 32).[112]

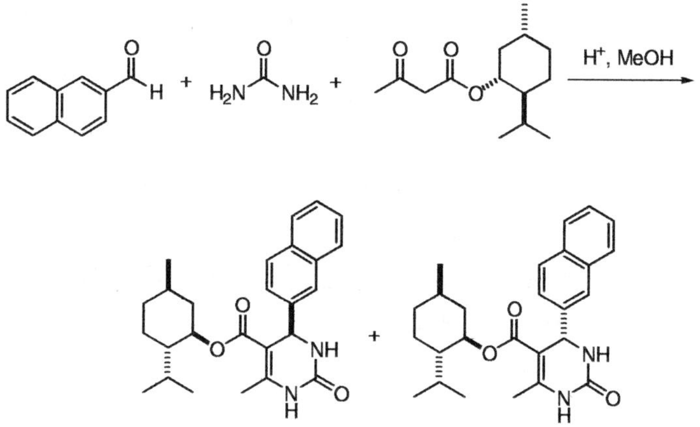

Scheme 32

Dondoni *et al.* introduced a C-glycosyl fragment on the DHPM nucleus. The purposes were to obtain and isolate all the DHPMs diastereomers (internal asymmetric induction), and to evaluate their pharmacokinetic or pharmacodynamic properties.[113] To this end, Dondoni *et al.* prepared several DHPMs containing C-glycosyl moieties either on the nitrogen, or at the C_2, and/or at the C_4 position, (Scheme 33). Diastereoselectivities were moderate when the C-glycosyl fragment was located either at the C_2, or the C_4 position, and no selectivity was observed when the nitrogen atom was linked to the C-glycosyl fragment.

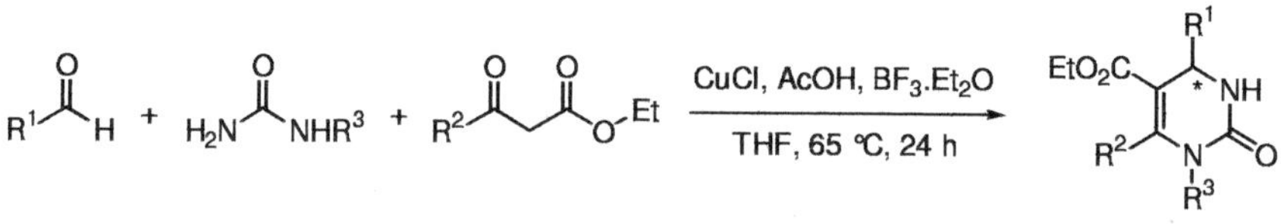

R^1 = C-glycosyl, R^2 = Me, R^3 = H; d.e. = 50–70%
R^2 = C-glycosyl, R^1 = Ph, R^3 = H; d.e. = 50–70%
R^3 = C-glycosyl, R^2 = Me, R^1 = Ph; d.e. = 0%
R^1 and R^2 = C-glycosyl, R^3 = H; d.e. = 76–82%
R^1, R^2, R^3 = C-glycosyl, no reaction

Scheme 33

3.2.3. Enantioselective synthesis

The most promising, economically and environmentally interesting approach to obtain enantiomerically pure DHPMs remains the catalytic asymmetric Biginelli reaction. The first report was disclosed by Juaristi in 2003 using Lewis acids [cerium(IV) and indium(III)] bearing chiral ligands (Scheme 34).[114] The best result was obtained by using a chiral cyclic urea and cerium(IV) chloride. However, the selectivities were quite low (up to only 28% e.e.) and the catalyst loading was high (20 mol%). The enantioselectivity could be improved to 40% if the intermediate imine of the urea was previously prepared.

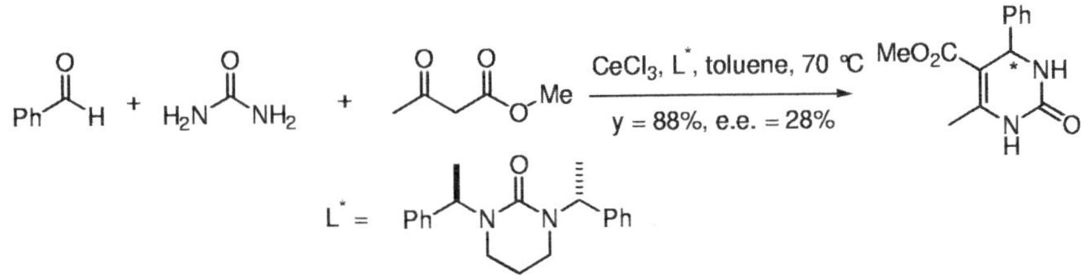

Scheme 34

The first breakthrough in the catalytic asymmetric Biginelli reaction was described by Zhu *et al.* in 2005. In the presence of 10 mol% of ytterbium triflate and chiral polydentate ligand, in THF at room temperature, DHPMs were isolated in high yields and excellent enantioselectivities (up to 99% e.e.) (Scheme 35).[115] These reaction conditions were also applied to the synthesis of SQ-32926 in 3 steps, 58% overall yield and 99% e.e. Zhu *et al.* demonstrated also that the catalyst could be reused several times without any loss of activity (yield and enantioselectivity).

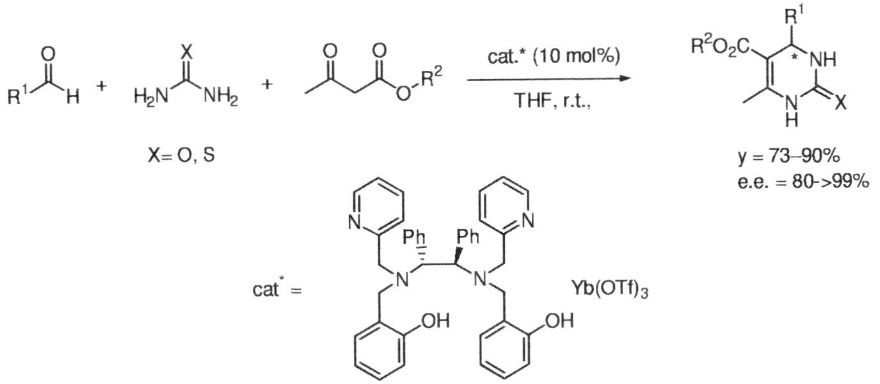

Scheme 35

As the first Biginelli reaction was carried out in refluxing ethanol with a catalytic amount of hydrochloric acid, the next challenge was to achieve this multicomponent reaction in the presence of chiral Brønsted acids. Their chemistry, and particularly that of chiral phosphoric acids, has known an explosive growth in the last few years.[6d,116] Given that phosphoric acids are known to activate imines (one of the intermediate of the Biginelli reaction) through the formation of an iminium ion, Gong *et al.* evaluate such catalysts in the reaction between (thio)urea, ethyl acetoacetate and several benzaldehydes.[117] Depending on the temperature and the substitution pattern at the 3,3' position of the phosphoric acid, the enantioselectivities ranged from 50 to 97% (Scheme 36). The reaction time, the nature of the substituent on the aromatic ring or on the ester group, have almost no influence on the stereochemical outcome of the reaction.

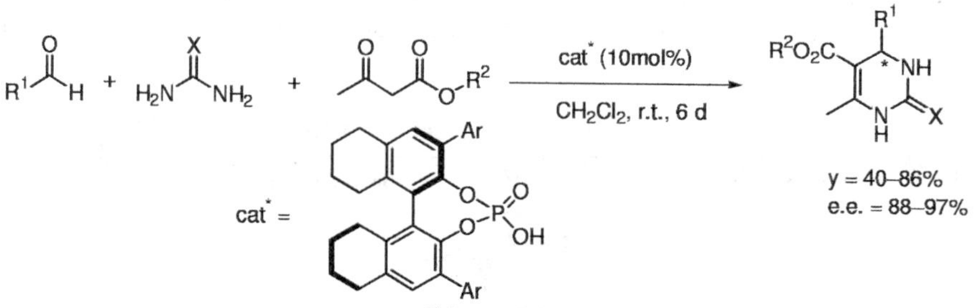

Scheme 36

Using the same methodology, Schaus *et al.* developed the synthesis of the key precursor of SNAP-7941 in high yeld (96%) and in 98% e.e. [118a] The target molecule was then obtained after functional arrangements (Scheme 37).

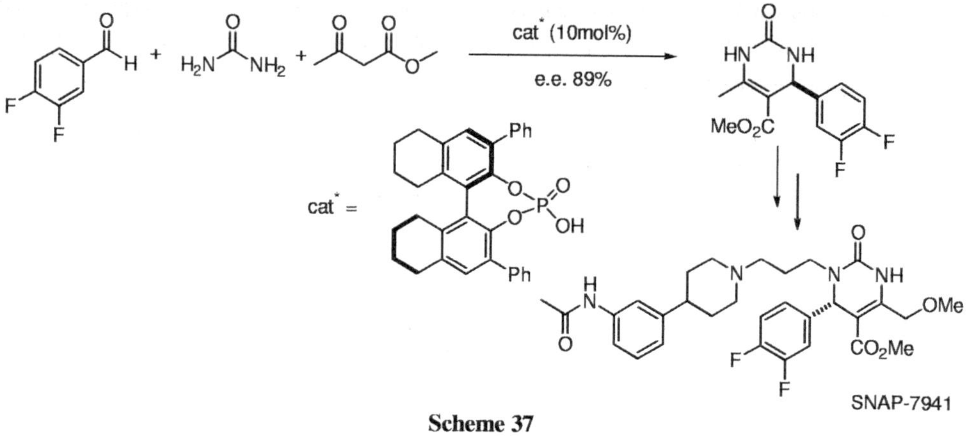

Scheme 37

Alternatively, the same group developed the synthesis of a variety of chiral DHPMs *via* an asymmetric Mannich reaction.[118b,c] Starting from β-ketoester, α-amidosulfones as acyl-imine precursors in the presence of 10 mol% of cinchonine as chiral inductor, the DHPMs were obtained in three steps in good overall yields and high selectivities (Scheme 38).

In the fertile area of organocatalysis, pyrrolidine derivatives belong to the most popular family of organocatalysts and they are now able to catalyze a wide range of reactions.[119] Zhao, Wang and co-workers introduced in 2009 chiral 5-(pyrrolidin-2-yl) tetrazoles as catalysts for the asymmetric Biginelli reaction.[120] Under the optimized reaction conditions (catalyst, solvent), high yields (63–88%) and enantioselectivities

(68–81%) can be reached (Scheme 39). To explain the stereochemical outcome, a double-activation by the catalyst was suggested. After addition of the pyrrolidine motive on the β-ketoester to produce the corresponding aminoacrylate, the tetrazole can activate the intermediate imine (formed from aldehyde and urea) *via* H-bonding interaction. The authors also showed that the final stereochemistry of the DHPM could be relied on the stereochemistry of the catalyst and more precisely on the stereochemistry of the tetrazole moiety.

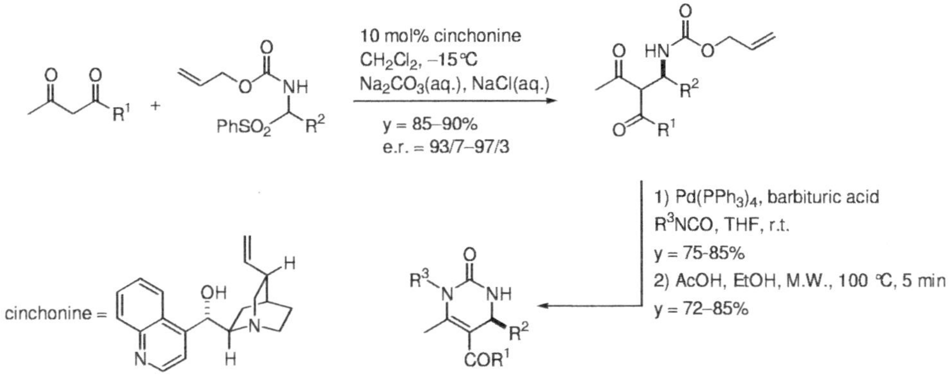

Scheme 38

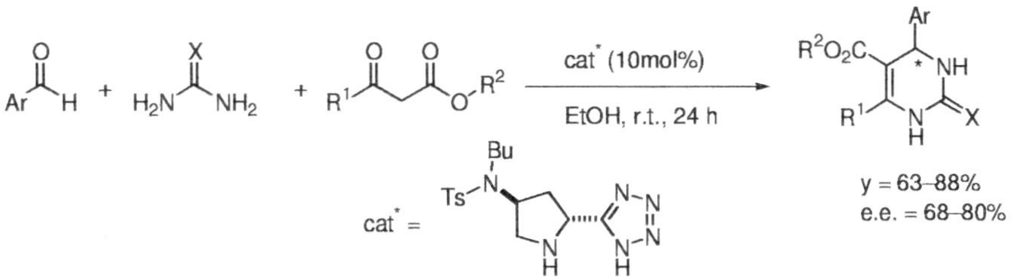

Scheme 39

A dual activation was also reported by Feng *et al.* Whereas, Zhao *et al.* and Wang *et al.* used only one catalytic species, Feng used a mixture of four components to catalyze the synthesis of DHMPs (Scheme 40).[121]

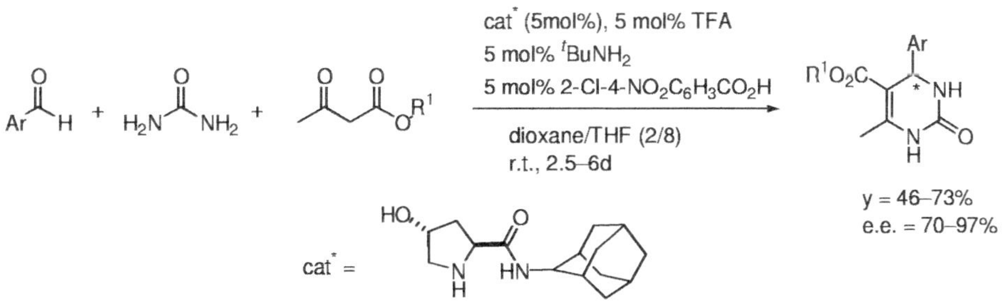

Scheme 40

Under the optimized conditions (yields, enantioselectivities), the reaction between aromatic aldehydes, urea and β-ketoesters has to be carried out in the presence of 5 mol% of chiral prolinamide, 5 mol% of 2-chloro-4-nitrobenzoic acid and 5 mol% of *tert*-butylamine and trifluoroacetic acid. Even if the role of the chiral proline and benzoic acid additive can be compared to the role of the chiral pyrrolidine tetrazole described by Zhao *et al.* and Wang *et al.*, the role of the two other additives remains unclear. However, yields up to 68% and enantiomeric excesses up to 97% were reached with this catalytic system.

4. Conclusion

During the last decade, development of efficient processes toward heterocyclic compounds, such as dihydropyridines and dihydropyrimidines, has been described. Lewis or Brønsted acids appeared to be good candidates for this purpose, and led to improvements compared to former procedures and allow new developments in green chemistry (recycling of the catalyst, reaction in aqueous medium, use of micro wave heating). Following these results, the first enantioselective catalysis of DHPs and DHPMs appeared, and excellent enantiomeric excesses were obtained using either organocatalysts (such as phosphoric acid derivatives) or metal salt catalysts (earth rare metal triflate, lanthanide). The enantioselective Hantzsch type and Biginelli reactions remain a hot topic in organic chemistry due to the high potential of the corresponding compounds in biology. The challenges lie in the next years on the development of a catalyst able to promote the synthesis of a broad range of alkyl substituted dihydropyrimidines and on the development of more efficient and enantioselective catalysts of the Biginelli and Hantzch type reactions.

Acknowledgments

The authors thank la Région Bretagne for a Ph. D. fellowship for J.M.. The University of Ziguinchor is warmly thank for a fellowship to M. D. M.. This work has also been performed within the «CRUNCH» interregional network. We gratefully acknowledge financial support from the "Ministère de la Recherche et des Nouvelles Technologies", CNRS (Centre National de la Recherche Scientifique), the "Région Basse-Normandie" and the European Union (FEDER funding).

References

1. (a) Suarez, M.; Verdecia, Y.; Illescas, B.; Martinez-Alvarez, R.; Alvarez, A.; Ochoa, E.; Seoane, C.; Kayali, N.; Martin, N. *Tetrahedron* **2003**, *59*, 9179. (b) Martin, N.; Seoane, C. *Quím. Ind.* **1990**, *36*, 115.
2. Boschi, D.; Caron, G.; Visentin, S.; Di Stilo, A.; Rolando, B.; Fruttero, R.; Gasco, A. *Pharm. Res.* **2001**, *18*, 987.
3. (a) Gaillard, S.; Papamicaël, C.; Marsais, F.; Dupas, G.; Levacher, V. *Synlett* **2005**, 441. (b) Xie, K.; Liu, Y-C.; Cui, Y.; Wang, J.-G.; Fu, Y.; Mak, T. C. W. *Molecules* **2007**, *12*, 415.
4. (a) Hantzsch, A. *Ann. Chem.* **1882**, *I*, 215. (b) Vanden Eynde, J. J.; Mayence, A. *Molecules* **2003**, *8*, 381.
5. (a) Lavilla, R. *J. Chem. Soc., Perkin Trans. 1* **2002**, 1141. (b) Stout, D. M.; Meyers, A. I. *Chem. Rev.* **1982**, *82*, 223.
6. (a) Pihko, P. M. *Angew. Chem. Int. Ed.* **2004**, *43*, 2062. (b) Bolm, C.; Rantanen, T.; Schiffers, I.; Zani, L. *Angew.Chem. Int. Ed.* **2005**, *44*, 1758. (c) Yamamoto, H.; Futatsugi, K. *Angew. Chem. Int. Ed.* **2005**, *44*, 1924. (d) Akiyama, T.; Itoh, J.; Fuchibe, K. *Adv. Synth. Catal.* **2006**, *348*, 999.
7. (a) Taylor, M. S.; Jacobsen, E. N. *Angew. Chem. Int. Ed.* **2006**, *45*, 1520. (b) Connon, S. J. *Chem. Eur. J.* **2006**, *12*, 5418.

8.   For the use of polyphosphates for the Biginelli reaction, see: Kappe, C. O.; Falsone, F. S. *Synlett* **1998**, 718.

9.   Correa, W. H.; Scott, J. L. *Green Chem.* **2001**, *3*, 296.

10. Tewari, N.; Dwivedi, N.; Tripathi, R. P. *Tetrahedron Lett.* **2004**, *45*, 9011.

11. Bisht, S. S.; Dwivedi, N.; Tripathi, R. P. *Tetrahedron Lett.* **2007**, *48*, 1187.

12. Kumar, A.; Maurya, R. A. *Tetrahedron* **2007**, *63*, 1946.

13. Kumar, A.; Maurya, R. A. *Tetrahedron* **2008**, *64*, 3477.

14. Kumar, A.; Maurya, R. A. *Synlett* **2008**, *6*, 883.

15. Dac, B.; Suneel, K.; Venkateswarlu, K.; Ravikanth, B. *Chem. Pharm. Bull.* **2008**, *56*, 366.

16. Moreau, J.; Duboc, A.; Hubert, C.; Hurvois, J.-P.; Renaud, J.-L. *Tetrahedron Lett.* **2007**, *48*, 8647.

17. Vohra, R. K.; Bruneau, C.; Renaud, J. L. *Adv. Synth. Catal.* **2006**, *348*, 2571.

18. Bartoli, G.; Babbiuch, K.; Bosco, M.; Carlone, A.; Galzerano, P.; Melchiore, P.; Sambri, L. *Synlett* **2007**, 2897.

19. Sridhar, R.; Perumal, P. T. *Tetrahedron* **2005**, *61*, 2465.

20. Debache, A.; Boulcina, R.; Belfaitah, A.; Rhouati, S.; Carboni, B. *Synlett* **2008**, *4*, 509.

21. Ko, S.; Sastry, M. N. V.; Lin, C.; Yao, C. F. *Tetrahedron Lett.* **2005**, *46*, 5771.

22. Ko, S.; Yao, C. F. *Tetrahedron* **2006**, *62*, 7293.

23. Sridharan, V.; Perumal, P. T.; Avendaño, C.; Menendez, J. C. *Tetrahedron* **2007**, *63*, 4407.

24. Wang, L.-M.; Sheng, J.; Zhang, L.; Han, J.-W.; Fan, Z.-Y; Tian, H.; Qian, C.-T. *Tetrahedron* **2005**, *61*, 1539.

25. Donelson, J. L.; Gibbs, R. A.; De, S. K. *J. Mol. Catal. A: Chemical* **2006**, 309.

26. Sabitha, G.; Reddy, G. S. K. K.; Reddy, C. S.; Yadav, J. S. *Tetrahedron Lett.* **2003**, *44*, 4129.

27. Sharma, S. D.; Hazarika, P.; Konvar, D. *Catal. Commun.* **2008**, *9*, 709.

28. Akbari, J. D.; Tala, S. D.; Dhaduk, M. F.; Joshi, H. S. *Arkivoc*, **2008**, *12*, 126.

29. Vo, D.; Matowe, W. C.; Ramesh, M.; Iqbal, N.; Wolowyk, M. W.; Howlett, S. E.; Knauss, E. E. *J. Med. Chem.* **1995**, *38*, 2851.

30. (a) Franckowiak, G.; Bechem, M.; Schramm, M.; Thomas, G. *Eur. J. Pharamacol.* **1985**, *114*, 223. (b) Hof, R. P.; Rüegg, R. T.; Hof, A.; Vogel, A. *J. Cardiovasc. Pharamacol.* **1985**, *7*, 689.

31. Caccamese, S.; Chillemi, R.; Principato, G. *Chirality* **1996**, *8*, 281.

32. Goldmann, S.; Stoltefuss, J. *Angew. Chem. Int. Ed. Engl.* **1991**, *30*, 1559.

33. Holdgrün, X. K.; Sih, C. J. *Tetrahedron Lett.* **1991**, *32*, 3465.

34. Sobolev, A.; Franssen, M. C. R.; Duburs, G.; de Groot, A. *Biocatalysis and Biotransformation* **2004**, *22*, 231.

35. Hirose, Y.; Kariya, K.; Sasaki, I.; Kurono, Y.; Achiwa, K. *Tetrahedron Lett.* **1993**, *34*, 3441.

36. Sobolev, A.; Franssen, M. C. R.; Vigante, B.; Cekavicus, B.; Makarova, N.; Duburs, G.; de Groot, A. *Tetrahedron: Asymmetry* **2001**, *12*, 3251.

37. Satzinger, G.; Hartenstein, J.; Mannhardt, K.; Kleinschroth, J.; Fritschi, E.; Osswald, H.; Weinheimer, G. *DE 3431 303* **1986**, Goedecke AG.

38. Enders, D.; Müller, S.; Demir, A. S. *Tetrahedron Lett.* **1988**, *29*, 6437.

39. (a) Kakuiri, A.; Ikawa, H.; Kobayashi, N.; Isowa, Y. Jpn Patent, **1988** JP 63208573; *Chem. Abstr.* **1989**, *110*, 95014. (b) Kataoka, K.; Sugano, H.; Yamaguchi, H. Jpn Patent, **1990** JP 2212474; *Chem. Abstr.* **1991**, *114*, 143156.

40. Rose, U.; Dräger, M. *J. Med. Chem.* **1992**, *35*, 2238.

41. Ashworth, I.; Hopes, P.; Levin, D.; Patel, I.; Salloo, R. *Tetrahedron Lett.* **2002**, *43*, 4931.

42. Palacios, F.; Herrán, E.; Rubiales, G.; Alonso, C. *Tetrahedron* **2007**, *63*, 5669.

43. (a) Meyers, A. I.; Oppenlaender, T. *Chem. Commun.* **1986**, 920. (b) Meyers, A. I.; Oppenlaender, T. *J. Am. Chem. Soc.* **1986**, *108*, 1989.

44. Schultz, A. G.; Flood, L. *J. Org. Chem.* **1986**, *51*, 838.

45. Ohno, A.; Oda, S.; Yamazaki, N. *Tetrahedron Lett.* **2001**, *42*, 399.

46. Rezgui, F.; Mangeney, P.; Alexakis, A. *Tetrahedron Lett.* **1999**, *40*, 6241.

47. (a) Gosmini, R.; Mangeney, P.; Alexakis, A.; Commerçon, M.; Normant, J.-F. *Synlett*, **1991**, 111. (b) Mangeney, P.; Gosmini, R.; Rassou, S.; Commerçon, M.; Alexakis, A. *J. Org. Chem* **1994**, *59*, 1877.

48. (a) Yamada, S.; Ichikawa, M. *Tetrahedron Lett.* **1999**, *40*, 4231. (b) Yamada, S.; Jahan, I. *Tetrahedron Lett.* **2005**, *46*, 8673.

49. Yamada, S.; Misono, T.; Ichikawa, M.; Morita, C. *Tetrahedron* **2001**, *57*, 8939.

50. Yamada, S.; Morita, C. *Chem. Lett.* **2001**, 1034.

51. (a) Yamada, S.; Saitoh, M.; Misono, T. *Tetrahedron Lett.* **2002**, *43*, 5853. (b) Yamada, S.; Morita, C. *J. Am. Chem. Soc.* **2002**, *124*, 8184.

52. Jiang, J.; Yu, J.; Sun, X.-X.; Rao, Q.-Q.; Gong, L.-Z. *Angew. Chem. Int. Ed.* **2008**, *46*, 1.

53. Franke, P. T.; Johansen, R. L.; Bertelsen, S.; Jørgensen, K. A. *Chem. Asian J.* **2008**, *3*, 216.

54. Biginelli, P. *Gazz. Chim. Ital.* **1893**, *23*, 360.

55. (a) Dondoni, A.; Massi, A. *Acc. Chem. Res.* **2006**, *39*, 451. (b) Kappe, C. O. In *Multicomponent Reactions*; Zhu, J.; Bienaymé, H., Eds.; Wiley-VCH: Weinheim, 2005; pp. 95. (c) Kappe, C. O. *Acc. Chem. Res.* **2000**, *33*, 879. (d) Simon, C.; Constantieux, T.; Rodriguez, J. *Eur. J. Org. Chem.* **2004**, 4957.

56. Kappe, C. O. *Eur. J. Med. Chem.* **2000**, *35*, 1043.

57. (a) Atwal, K. S.; Swanson, B. N.; Unger, S. E.; Floyd, D. M.; Moreland, S.; Hedberg, A.; O'Reilly, B. C. *J. Med. Chem.* **1991**, *34*, 806. (b) Rovnyak, G. C.; Atwal, K. S.; Hedberg, A.; Kimball, S. D.; Moreland, S.; Gougoutas, J. Z.; O'Reilly, B. C.; Schwartz, J.; Malley, M. F. *J. Med. Chem.* **1992**, *35*, 3254. (c) Grover, G. J.; Dzwonczyk, S.; McMullen, D. M.; Normandin, D. E.; Parham, C. S.; Sleph, P. G.; Moreland, S. *J. Cardiovasc. Pharm.* **1995**, *26*, 289.

58. (a) Nagarathnam, D.; Miao, S. W.; Lagu, B.; Chiu, G.; Fang, J.; Dhar, T. G. M.; Zhang, J.; Tyagarajan, S.; Marzabadi, M. R.; Zhang, F. Q.; Wong, W. C.; Sun, W. Y.; Tian, D.; Wetzel, J. M.; Forray, C.; Chang, R. S. L.; Broten, T. P.; Ransom, R. W.; Schorn, T. W.; Chen, T. B.; O'Malley, S.; Kling, P.; Schneck, K.; Benedesky, R.; Harrell, C. M.; Vyas, K. P.; Gluchowski, C. *J. Med. Chem.* **1999**, *42*, 4764, and subsequent papers in this issue. (b) Barrow, J. C.; Nantermet, P. G.; Selnick, H. G.; Glass, H. G.; Rittle, K. E.; Gilbert, K. F.; Steele, T. G.; Homnick, C. F.; Freidinger, R. M.; Ransom, R. W.; Kling, P.; Reiss, D.; Broten, T. P.; Schorn, T. W.; Chang, R. S. L.; O'Malley, S. S.; Olah, T. V.; Ellis, J. D.; Barrish, A.; Kassahun, K.; Leppert, P.; Nagarathnam, D.; Forray, C. *J. Med. Chem.* **2000**, *43*, 2703.

59. (a) Maliga, Z.; Kapoor, T. M.; Mitchison, T. J. *Chem. Biol.* **2002**, *9*, 989. (b) Debonis, S.; Simorre, J. P.; Crevel, I.; Lebeau, L.; Skoufias, D. A.; Blangy, A.; Ebel, C.; Gans, P.; Cross, R.; Hackney, D. D.; Wade, R. H.; Kozielski, F. *Biochemistry* **2003**, *42*, 338.

60. (a) Heys, L.; Moore, C. G.; Murphy, P. J. *Chem. Soc. Rev.* **2000**, *29*, 57. (b) Patil, A. D.; Kumar, N. V.; Kokke, W. C.; Bean, M. F.; Freyer, A. J.; De Brosse, C.; Mai, S.; Truneh, A.; Faulkner, D. J.; Carte, B.; Breen, A. L.; Hertzberg, R. P.; Johnson, R. K.; Westley, J. W.; Potts, B. C. M. *J. Org. Chem.* **1995**, *60*, 1182.

61. Zhang, X.; Li, Y.; Liu, C.; Wang, J. *J. Mol. Catal. A: Chemical* **2006**, *253*, 207.

62. (a) Shaterian, H. R.; Hosseinian, A.; Ghashang, M. *Phosphorus, Sulfur, and Silicon* **2009**, *184*, 126. For other phosphate salts, see: (b) Niknam, K.; Zolfigol, M. A.; Hossieninejad, Z.; Daneshvar, N. *Chinese Journal of Catalysis* **2007**, *28*, 591.

63. Ramalingan, C.; Kwak, Y.-W. *Tetrahedron* **2008**, *64*, 5023.

64. (a) Lu, J.; Ma, H. *Synlett* **2000**, 63. (b) Lu, J.; Bai, Y. *Synthesis* **2002**, 466. (c) Zorkun, I. S.; Saraç, S.; Çelebib, S.; Erol, K. *Bioorg. Med. Chem.* **2006**, *14*, 8582. (d) Cepanec, I.; Litvić, M.; Bartolinčić, A.; Lovrić, M. *Tetrahedron* **2005**, *61*, 4275.

65. Adibi, H.; Samimi, H. A.; Beygzadeh, M. *Catal. Commun.* **2007**, *8*, 2119.

66. Suzuki, I.; Iwata, Y.; Takeda, K. *Tetrahedron Lett.* **2008**, *49*, 3238.

67. Gohain, M.; Prajapati, D.; Sandhu, J. S. *Synlett* **2004**, 235.

68. Paraskar, A. S.; Dewkar, G. K.; Sudalai, A. *Tetrahedron Lett.* **2003**, *44*, 3305.

69. Lee, K.-Y.; Ko, K.-Y. *Bull. Korean Chem. Soc.* **2004**, *25*, 1929.

70. Liu, C.; Wang, J.; Li, Y. *J. Mol. Catal. A: Chemical* **2006**, *258*, 367.

71. Rodríguez-Domínguez, J. C.; Bernardi, D.; Kirsch, G. *Tetrahedron Lett.* **2007**, *48*, 5777.

72. Kumar, D.; Sundaree, M. S.; Mishra, B. G. *Chem. Lett.* **2006**, *35*, 1074.

73. De, S. K.; Gibbs, R. A. *Synthesis* **2005**, 1748.

74. Ranu, B. C.; Hajra, A.; Jana, U. *J. Org. Chem.* **2000**, *65*, 6270.

75. Godoi, M. N.; Costenaro, H. S.; Kramer, E.; Machado, P. S.; Montes D'Oca, M. G.; Russowsky, D. *Quim. Nova* **2005**, *28*, 1010.

76. Russowsky, D.; Lopes, F. A.; da Silva, V. S. S.; Canto, K. F. S.; D'Oca, M. G. M.; Godoi, M. N. *J. Braz. Chem. Soc.* **2004**, *15*, 165.

77. Cepanec, I.; Litvić, M.; Filipan-Litvić, M.; Grüngold, I. *Tetrahedron* **2007**, *63*, 11822.

78. Lu, J.; Bai, Y.; Wang, Z.; Yang, B.; Ma, H. *Tetrahedron Lett.* **2000**, *41*, 9075.

79. Ma, Y.; Qian, C.; Wang, L.; Yang, M. *J. Org. Chem.* **2000**, *65*, 3864.

80. Ramalinga, K.; Vijayalakshmi, P.; Kaimal, T. N. B. *Synlett* **2001**, 863.

81. Bose, D. S.; Fatima, L.; Mereyala, H. B. *J. Org. Chem.* **2003**, *68*, 587.

82. Adib, M.; Ghanbary, K.; Mostofi, M.; Ganjali, M. R. *Molecules* **2006**, *11*, 649.

83. (a) Dondoni, A.; Massi, A. *Tetrahedron Lett.* **2001**, *42*, 7975. (b) Dondoni, A.; Massi, A.; Sabbatini, S. *Tetrahedron Lett.* **2002**, *43*, 5913.

84. For the use of other copper salts, see: (a) Kamal, A.; Krishnaji, T.; Azhar, M. A. *Catal. Commun.* **2007**, *8*, 1929. (b) Kalita, H. R.; Phukan, P. *Catal. Commun.* **2007**, *8*, 179.

85. For a similar discussion in the Lewis acid mediated hetero-Michael addition reactions, see: Wabnitz, T. C.; Yu, J.-Q.; Spencer, J. B. *Chem. Eur. J.* **2004**, *10*, 484.

86. Yu, Y.; Liu, D.; Liu, C.; Luo, G. *Bioorg. Med. Chem. Lett.* **2007**, *17*, 3508.

87. (a) Bose, A. K.; Manhas, M. S.; Pednekar, S.; Ganguly, S. N.; Dang, H.; He, W.; Mandadi, A. *Tetrahedron Lett.* **2005**, *46*, 1901. (b) Manhas, M. S.; Ganguly, S. N.; Mukherjee, S.; Jain, A. K.; Bose A. K. *Tetrahedron Lett.* **2006**, *47*, 2423.

88. Polshettiwar, V.; Varma, R. S. *Tetrahedron Lett.* **2007**, *48*, 7343.

89. Zumpe, F. L.; Flüß, M.; Schmitz, K.; Lender, A. *Tetrahedron Lett.* **2007**, *48*, 1421.

90. Shaterian, H. R.; Hosseinian, A.; Ghashang, M. *Phosphorus, Sulfur, and Silicon* **2009**, *184*, 197.

91. Tu, S.; Fang, F.; Zhu, S.; Li, T.; Zhang, X.; Zhuang, Q. *Synlett* **2004**, 537.

92. Salehia, P.; Dabiri, M.; Khosropour, A. R.; Roozbehniya, P. *J. Iran. Chem. Soc.* **2006**, *3*, 98.

93. Fazaeli, R.; Tangestaninejad, S.; Aliyan, H.; Moghadam, M. *Appl. Catal. A* **2006**, *309*, 44.

94. Rafiee, E.; Jafari, H. *Bioorg. Med. Chem. Lett.* **2006**, *16*, 2463.

95. Rafiee, E.; Shahbazi, F. *J. Mol. Catal. A: Chemical* **2006**, *250*, 57.

96. Kulkarni, M. G.; Chavhan, S. W.; Shinde, M. P.; Gaikwad, D. D.; Borhade, A. S.; Dhondge, A. P.; Shaikh, Y. B.; Ningdale, V. B.; Desai, M. P.; Birhade, D. R. *Beilstein Journal of Organic Chemistry* **2009**, *5*, No. 4 (doi:10.3762/bjoc.5.4).

97. Rani, V. R.; Srinivas, N.; Kishan, M. R.; Kulkarni, S. J.; Raghavan, K. V. *Green Chem.* **2001**, 305.

98. Singh, K.; Arora, D.; Singh, S. *Tetrahedron Lett.* **2006**, *47*, 4205.

99. Shaabani, A.; Bazgir, A.; Teimouri, F. *Tetrahedron Lett.* **2003**, *44*, 857.

100. Reddy, K. R.; Reddy, Ch. V.; Mahesh, M.; Raju, P. V. K.; Reddy, V. V. N. *Tetrahedron Lett.* **2003**, *44*, 8173.

101. Suzuki, I.; Iwata, Y.; Takeda, K. *Tetrahedron Lett.* **2008**, *49*, 3238.

102. (a) Yadav, J. S. ; Kumar, S. P.; Kondaji, G.; Rao, R. S.; Nagaiah, K. *Chem. Lett.* **2004**, *33*, 1168. (b) Mabry, J.; Ganem, B. *Tetrahedron Lett.* **2006**, *47*, 55.

103. Ryabukhin, S. V.; Plaskon, A. S.; Ostapchuk, E. N.; Volochnyuk, D. M.; Tolmachev, A. A. *Synthesis* **2007**, 417.

104. Debache, A.; Boumoud, B.; Amimour, M.; Belfaitah, A.; Rhouati, S.; Carboni, B. *Tetrahedron Lett.* **2006**, *47*, 5697.

105. Tu, S.; Fang, F.; Miao, C.; Jiang, H.; Feng, Y.; Shi, D.; Wang, X. *Tetrahedron Lett.* **2003**, *44*, 6153.

106. Debache, A.; Amimour, M.; Belfaitah, A.; Rhouati, S.; Carboni, B. *Tetrahedron Lett.* **2008**, *49*, 6119.

107. Khodaei, M. M.; Khosropour, A. R.; Jowkar, M. *Synthesis* **2005**, 1301.

108. (a) Atwal, K. S.; Rovnyak, G. C.; Kimball, S. D.; Floyd, D. M.; Moreland, S.; Swanson, B. N.; Gougoutas, J. Z.; Schwartz, J.; Smillie, K. M.; Malley, M. F. *J. Med. Chem.* **1990**, *33*, 2629. (b) Atwal, K. S.; Swanson, B. N.; Unger, S. E.; Floyd, D. M.; Moreland, S.; Hedberg, A.; O'Reilly, B. C. *J. Med. Chem.* **1991**, *34*, 806.

109. Kappe, C. O.; Uray, G.; Roschger, P.; Lindner, W.; Kratky, C.; Keller, W. *Tetrahedron* **1992**, *48*, 5473.

110. (a) Roberge, C. M.; Li, W.; Bills, G. F.; Larsen, R. D.; Sidler, D. R.; Chartrain, M.; Ikemoto, N.; Taylor, C. S. *PCT Int. Appl.* WO9907695, **1999**. (b) Sidler, D. R.; Barta, N.; Li, W.; Hu, E.; Matty, L.; Ikemoto, N.; Campbell, J. S.; Chartrain, M.; Gbewonyo, K.; Boyd, R.; Corley, E. G.; Ball, R. G.; Larsen, R. D.; Reider, P. J. *Can. J. Chem.* **2002**, *80*, 646.

111. Schnell, B.; Strauss, U. T.; Verdino, P.; Faber, K.; Kappe, C. O. *Tetrahedron: Asymmetry* **2000**, *11*, 1449.

112. Kappe, C. O.; Uray, G.; Roschger, P.; Lindner, W.; Kratky, C.; Keller, W. *Tetrahedron* **1992**, *48*, 5473.

113. (a) Dondoni, A.; Massi, A.; Sabbatini, S.; Bertolasi, V. *J. Org. Chem.* **2002**, *67*, 6979. (b) Dondoni, A.; Massi, A. *Mol. Diversity* **2003**, *6*, 261.

114. Muñoz-Muñiz, O.; Juaristi, E. *Arkivoc* **2003**, *xi*, 16.

115. Huang, Y.; Yang, F.; Zhu, C. *J. Am Chem . Soc.* **2005**, *127*, 16386.

116. (a) Terada, M. *Chem. Commun.* **2008**, 4097. (b) Connon, S. J. *Angew. Chem. Int. Ed.* **2006**, *45*, 3909. (c) Akiyama, T. *Chem. Rev.* **2007**, *107*, 5744.

117. (a) Chen, X.-H.; Xu, X.-Y.; Liu, H.; Cun, L.-F.; Gong, L.-Z. *J. Am. Chem. Soc.* **2006**, *128*, 14802. (b) Gong, L. Z.; Xu, X. Y. *Chem. Eur. J.* **2007**, *13*, 8920.

118. (a) Goss, J. M.; Schaus, S. E. *J. Org. Chem.*, **2008**, *73*, 7651. (b) Lou, S.; Taoka, B. M.; Ting, A.; Schaus, S. E. *J. Am. Chem. Soc.*, **2005**, *127*, 11256. (c) Lou, S.; Dai, P.; Schaus, S. E. *J. Org. Chem.* **2007**, *72*, 9998.

119. (a) Lattanzi, A. *Chem. Commun.* **2009**, 1452. (b) Mukherjee, S.; Yang, J. W.; Hoffmann, S.; List, B. *Chem. Rev.* **2007**, *107*, 5471. (c) Xu, L.-W.; Luo, J.; Lu, Y. *Chem. Commun.* **2009**, 1807. (d) Xu, L.-W.; Lu, Y. *Org. Biomol. Chem.* **2008**, *6*, 2047. (e) Melchiorre, P.; Marigo, M.; Carlone, A.; Bartoli, G. *Angew. Chem. Int. Ed.* **2008**, *47*, 6138. (f) Mielgo, A.; Palomo, C. *Chem. Asian J.* **2008**, *3*, 922.

120. Wu, Y. Y.; Chai, Z.; Liu, X. Y.; Zhao, G.; Wang, S. W. *Eur. J. Org. Chem.* **2009**, 904.

121. Xin, J.; Chang, L.; Hou, Z.; Shang, D.; Liu, X.; Feng, X. *Chem. Eur. J.* **2008**, *14*, 3177.

C2-FUNCTIONALIZED FURANS AS DIENES
IN [4+3] CYCLOADDITION REACTIONS

Ángel M. Montaña,* Pedro M. Grima and Consuelo Batalla

Industrial and Applied Organic Chemistry Research Unit, Department of Organic Chemistry,
University of Barcelona, c/Martí i Franquès 1-11, E-08028 Barcelona, Spain
(e-mail: angel.montana@ub.edu)

Abstract. *[4+3] Cycloaddition reactions of C2-functionalized furans with 1,3-dimethyl-oxyallyl cation afford versatile polyfunctionalized cycloheptane synthons, which are building blocks of a wide variety of natural and synthetic products with biological and/or structural interest. A study of the influence of stereo-electronic properties of the substituent at C-2 of furan on the yield and diastereoselectivity has been carried out. Also a general method of assignment of the relative stereochemistry in cis-endo, cis-exo and trans diastereoisomeric pairs of cycloadducts has been established. On the other hand, an improved methodology to carry out [4+3] cycloaddition reactions of dienes and oxyallyl cations, under mild thermal or sonochemical conditions at low temperatures (from 0 °C to –44 °C) and for short reaction times (<15 minutes) has been developed. In this particular study, the analysis of the influence of the nature of reducing metal in the cycloaddition outcome has been very useful.*

Contents

1. Introduction

1.1. The [4+3] cycloaddition reaction

The reaction of an allyl cation with a diene to generate a seven-membered ring represents a highly convergent and very useful route to such ring systems (Scheme 1).[1] Essentially this reaction is a

[4C(4π)+3C(2π)] cycloaddition that may have inter- and intramolecular variants. In this review it will be discussed, in particular, the intermolecular [4+3] cycloaddition of C2-substituted furans and 1,3-dimethyl-2-oxyallyl cation, generated *in situ* from 2,4-dihalo-3-pentanone and a reducing metal (Scheme 2).[2,3]

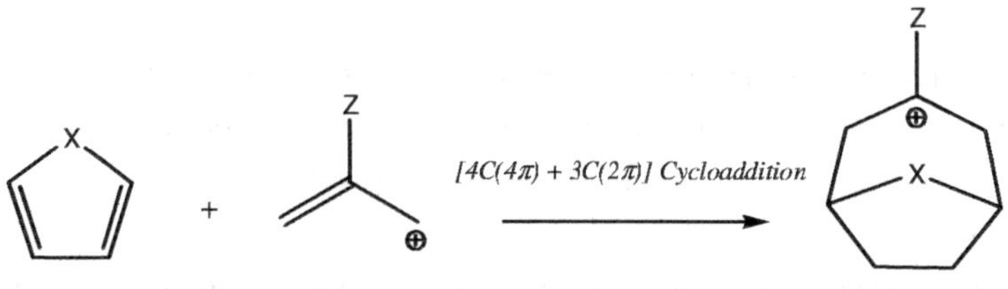

Scheme 1. [4+3] Cycloaddition reaction generating seven-membered ring systems.

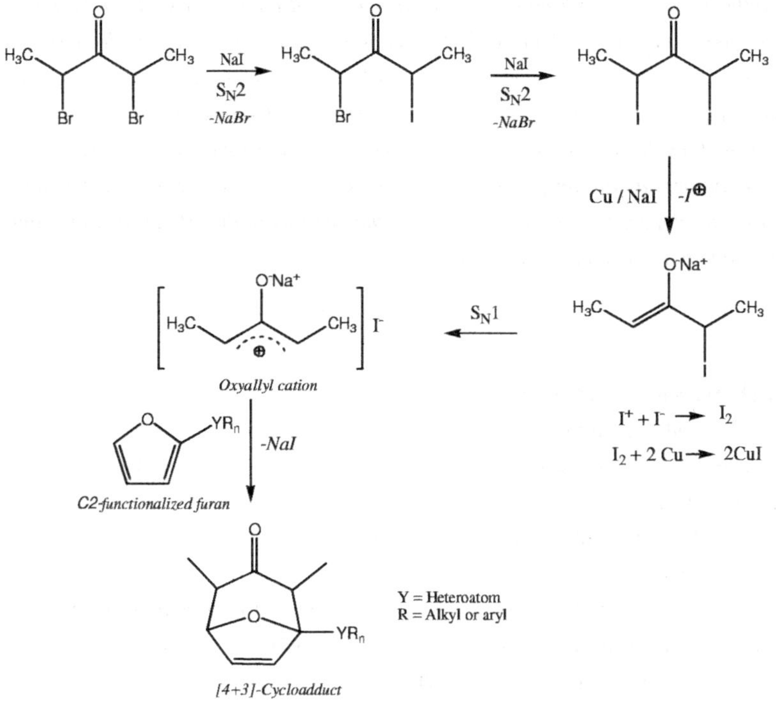

Scheme 2. Generation of the oxyallyl cation from α,α'-dihaloketones.

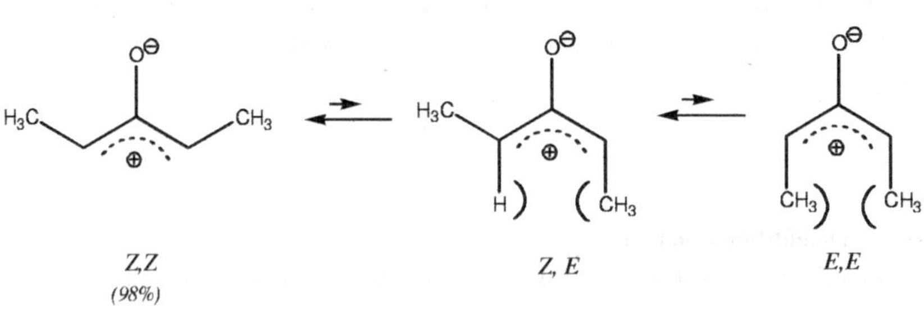

Scheme 3. Configuration of the oxyallyl cation.

232

The configuration of the oxyallyl cation used in these studies, *i.e.* 1,3-dimethyl-2-oxyallyl cation, is mainly (Z,Z) or "W-configuration" (Scheme 3) in a 98% extent, according to trapping studies, under kinetic control.[3] The other configurations (Z,E and E,E) are quite minor as demonstrated by the distribution of diastereomeric cycloadducts formed in the reaction with a cyclic diene (see below).

Why to use C2-functionalized furans as dienes in this cycloaddition process? There are several reasons: in first place, because in the resulting cycloadducts the bridge-head of the oxabicyclic system (in case of using furan as diene) is an acetalic carbon and this functionalization allows the easy opening of the oxygen bridge in order to convert the cycloadducts in other molecules and synthons of higher added value and complexity (Scheme 4). In second place, because it is possible to convert both bridge-heads into different organic functions, which may be modified and transformed, in a more easy way, than in the case of conventional cyclic ethers. In a third place, because the C2-functionalization may be designed with the adequate stereo-electronic nature in order to condition the kinetics, the diastereoselectivity and the enantioselectivity of the cycloaddition reaction, in this last case by using a chiral auxiliary as a C2-substituent of the furan diene.[4e]

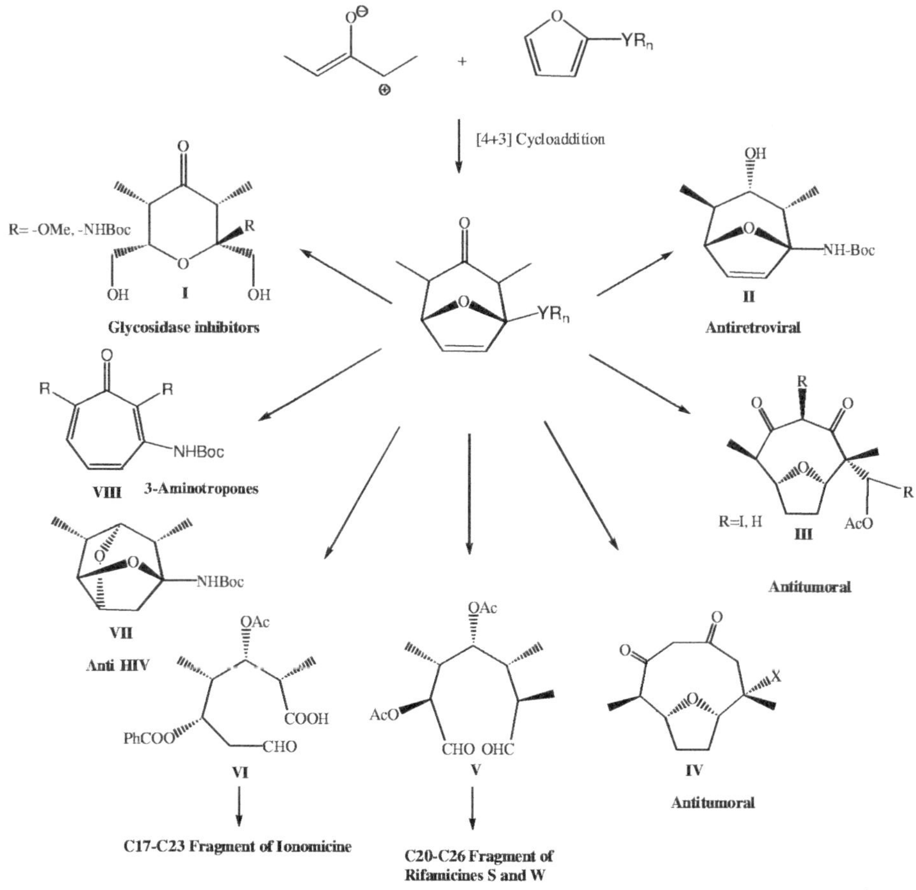

Scheme 4. Added value synthons prepared from [4+3] cycloadducts derived from C2-functionalized furans.

1.2. Concerted and stepwise mechanisms

Intermolecular [4+3] cycloaddition reactions have been classified into two types based on mechanistic considerations (Scheme 5):[1,3] concerted cycloadditions are categorized as type A and stepwise cycloadditions are designed as type B. Moreover, there is another mechanistic alternative in the reaction of allyl cations and dienes: it is the type C mechanism, which is a stepwise process that does not lead to cycloadducts but to products derived from an initial electrophilic addition of the allyl cation on the diene. This last type of reaction is common in the case of using electron rich dienes as substrates. In the case of aromatic electron rich heterocyclic dienes like pyrrole the resulting products are formally S_EAr products.

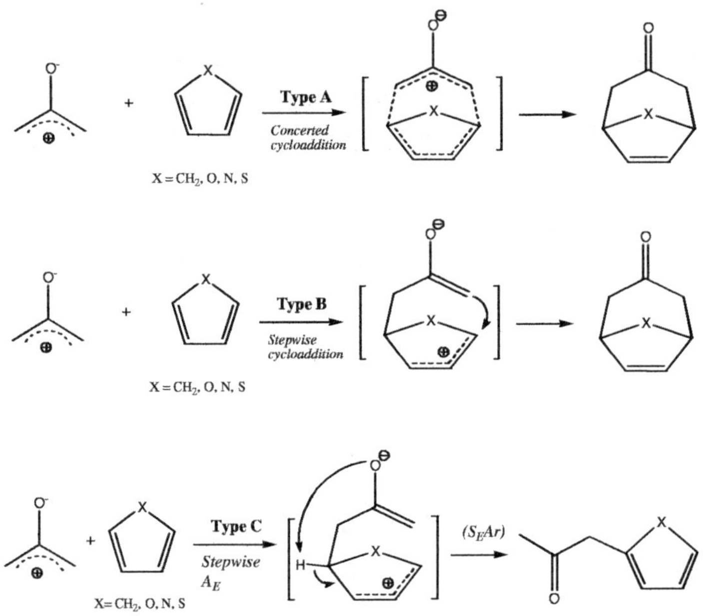

Scheme 5. Mechanistic alternatives in the reaction of allyl cation and dienes.

Experimental data demonstrate that the reaction proceeds mainly *via* a concerted process (Type A) but, when the diene is electron rich or when the electrophilicity of the oxyallyl cation increases, the reaction proceeds also *via* Type B and Type C mechanisms in a certain extent. It has been established that the electrophilic character of oxyallyl cations ranges from the very electrophilic and poorly nucleophilic hydroxyallyl through metal oxyallyls to the free oxyallyl species. The electron density of oxygen atom increases along the series from left to right and the electrophilicity decreases in the same order (Scheme 6).[1a,3]

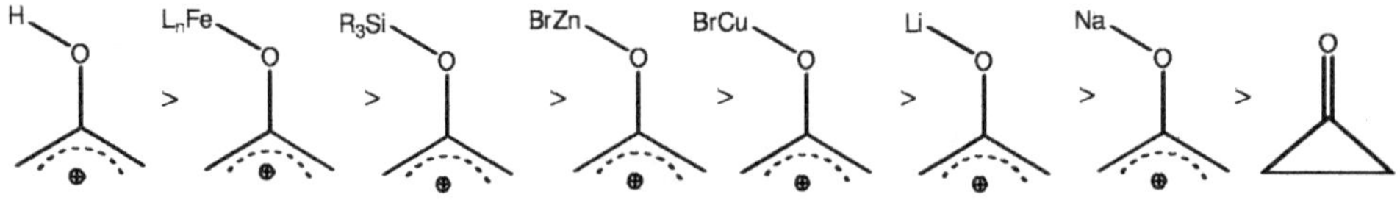

Scheme 6. Electrophilicity of oxyallyl cations.

The concerted [4+3] cycloaddition (Type A) of an allyl cation to cyclic conjugated dienes can proceed *via* both a compact transition state (**I**) and an extended transition state (**II**) (Scheme 7).

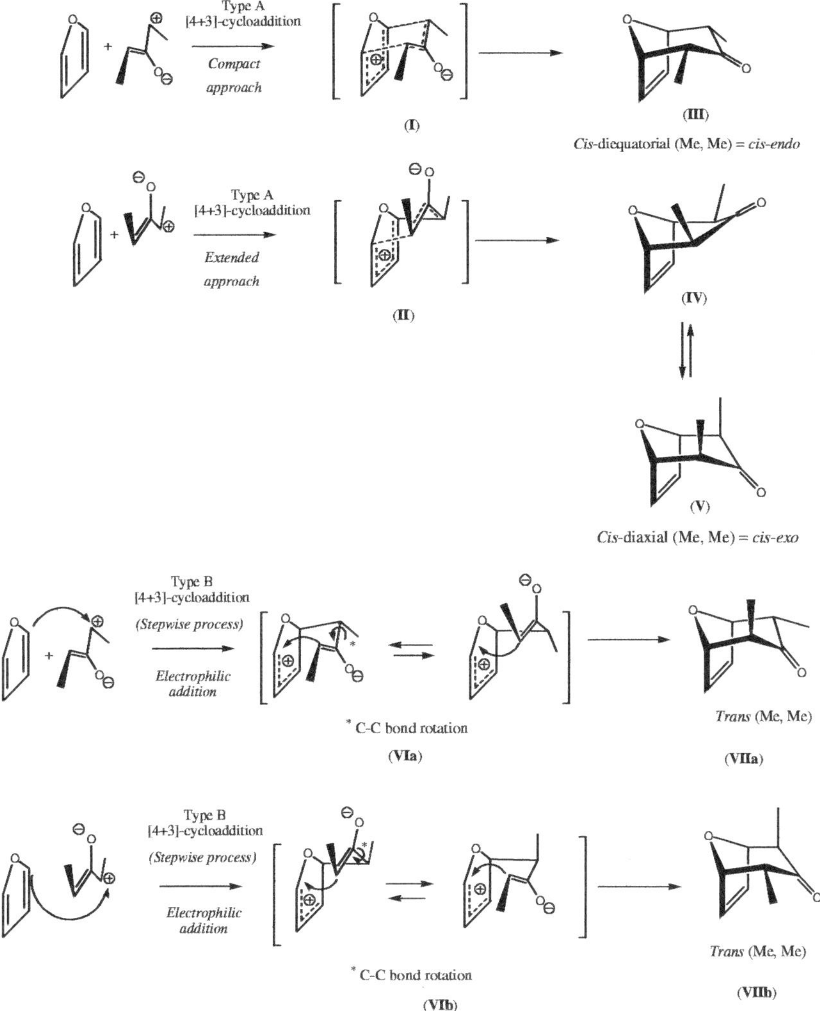

Scheme 7. Formation of diastereomeric *cis*-diequatorial, *cis*-diaxial and *trans* [4+3] cycloadducts.

In the case of working with 1,3-dimethyl-2-oxyallyl cation (derived from 2,4-dibromo-3-pentanone) and assuming that it predominantly adopts a W configuration, the resulting [4+3] cycloadduct (**III**) derived from a compact transition state will have a chair-like conformation for the newly formed oxane-4-one ring and the methyl groups on C-2 and C-4 will adopt a *cis*-diequatorial disposition. However, the product (**IV**) derived from an extended transition state will adopt a boat-like conformation for the oxane-4-one ring. This conformation is unfavourable and the ring flipping occurs to give the more stable conformer (**V**); simultaneously the methyl groups on C-2 and C-4 will adopt a *cis*-diaxial disposition. The *cis*-diequatorial/

cis-diaxial ratio indicates the proportion of compact *versus* extended attack. Experimentally, it could be observed that the proportion of the extended attack increases when increasing electrophilicity of the allyl cation. On the other hand, formation, in a measurable extent, of the *trans* (Me, Me) cycloadducts, under kinetic control, from a (Z,Z)-allyl cation will require a configurational loss and suggests a stepwise process (mechanism Type B) *via* an intermediate, which must be sufficiently long lived to allow rotation about the newly formed σ bonds. The likelihood of rotation in the case of using C-2 functionalized furans as substrates will depend, *inter alia*, on the nature of atom or group Y (Scheme 2).

2. C2-Functionalized furans as dienes in [4+3] cycloadditions: a study of the stereo-electronic factors controlling the diastereoselectivity

In this section, the results found in the literature on the [4+3] cycloaddition reactions of 2-functionalized furans with oxyallyl cations (Scheme 8) are presented.[4] This work is related with several research efforts oriented to the synthesis of polyfunctionalized cycloheptanes as versatile synthons and precursors of biologically active natural products.[5] The attention has been focused on furans as dienes, having a function attached at C-2 through an heteroatom (in particular atoms from groups IVA and VIA of the periodic table).

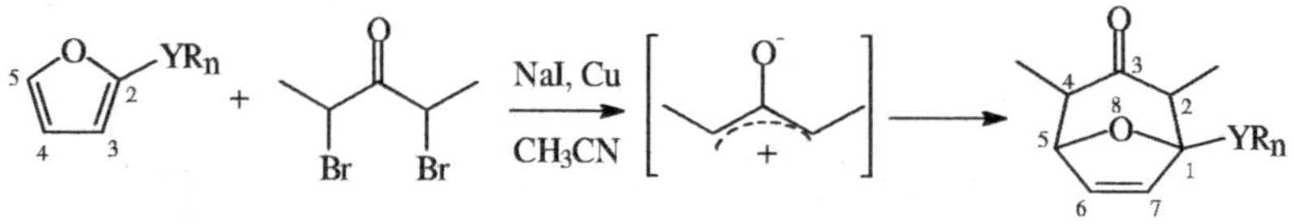

Y= N, C, Si, Ge, Sn, O, S, Se

Scheme 8. [4+3] Cycloaddition reaction.

The resulting functionalization on C-1 of cycloadducts, together with the wide variety of heteroatom linkers (Y=N, O, S, Se, Si, Ge, Sn) and their different reactivity, give to these products high versatility regarding the modification of their structures: interconversion of functional groups, cleavage of the oxygen bridge, etc. Starting from these bicyclic cycloadducts, it has been possible to prepare polyfunctionalized cycloheptanes or linear heptane synthons having five different organic functions and up to five stereocentres.[5f,g]

Apart from the afore-referred works, we have not found in the literature any reference regarding a similar systematic study. Only, there are a few examples[1b,i] where furans substituted at C-2 (through C-C bond) by aryl or alkyl groups are used as dienes in such reactions, but none with a function linked to C-2 through an heteroatom which is of major interest here.

From C-2 functionalized furans, it is possible to obtain bicyclic cycloadducts having on C-1 an organic function that facilitates opening of the oxygen bridge. This methodology has some advantages with respect to other synthetic approaches because at least five stereocentres and five different organic functions could be introduced in the cycloadduct, since the beginning, maintaining at the same time the relative stereochemistry of substituents in the cycloheptane system. These features make the aforementioned cycloadducts very versatile and useful synthons (Schemes 4 and 9).[5]

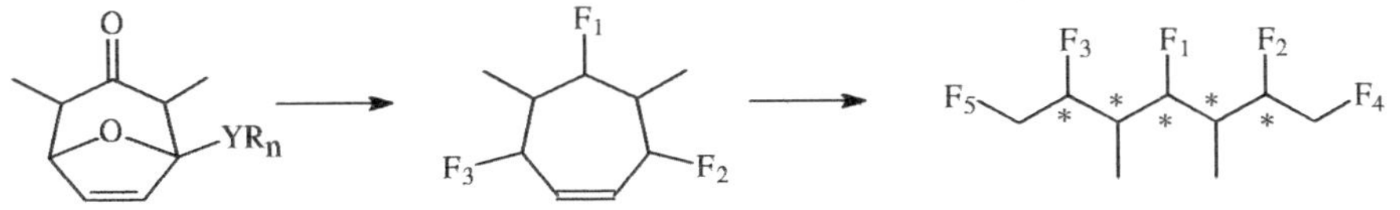

Scheme 9. Generation of polyfunctionalized synthons with up to five stereocentres from [4+3] cycloadducts.

Table 1. Results obtained from the [4+3] cycloaddition of C2 substituted furans.

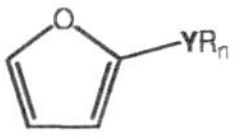

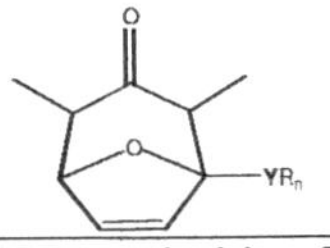

Entry	Furan diene YR$_n$ =	Reference of furan preparation method N°	Cycloaddition method[b]	Conversion (%)[c]	Product N°	Yield (%)	Diastereoselectivity cis:trans (%)	Diastereoselectivity endo:exo (%)
1	H	1	-[a]	H	100	**28**	63	100:0	80:20
2	NO$_2$	2	8[a]	N	82	**29**	13	100:0	96:4
3	NHBoc	3	9	N	100	**30**	76	45:55	89:11
4	CH$_3$	4	-[a]	H	92	**31**	77	100:0	92:8
5	CH$_3$	4	-[a]	H^2	100	**31**	95	100:0	95:5
6	C(CH$_3$)$_3$	5	-[a]	H	60	**32**	9	100:0	67:33
7	Ph	6	10	H	66	**33**	60	100:0	96:4
8	CO$_2$Et	7	-[a]	H	97	**34**	27	100:0	97:3
9	CO$_2$Ch	8	11	H	63	**35**	22	100:0	100:0
10	CHO	9	-[a]	H	10	**36**	25	68:32	78:22
11	COMe	10	-[a]	H^1	59	**37**	85	100:0	100:0
12	CH(OCH$_2$CH$_2$O)	11	12	H	100	**38**	54	100:0	100:0
13	SiMe$_3$	12	13	H	100	**39**	32	100:0	100:0
14	GeMe$_3$	13	14	H	100	**40**	30	100:0	100:0
15	SnMe$_3$	14	15	H	96	**41**	67	100:0	94:6
16	SnBu$_3$	15	14[a]	H	31	**42**	30	100:0	100:0
17	OCO$_2$Et	16	16	H	37	**43**	100	100:0	100:0
18	OCO^{t}Bu	17	16,17	H	100	**44**	93	100:0	95:5
19	OCOPh	18	17	H	86	**45**	16	100:0	83:17
20	OPO(NMe$_2$)$_2$	19	-[a]	H	98	**46**	21	100:0	93:7
21	OMe	20	-[a]	H	84	**47**	96	100:0	67:33
22	OMe	20	-[a]	H^2	100	**47**	95	100:0	70:30
23	OMe	20	-[a]	H^3	100	**47**	95	100:0	52:48
24	OMe	20	-[a]	N	82	**47**	80	88:12	92:8
25	OSiMe$_3$	21	-[a]	H	100	**48**	86	100:0	95:5
26	OSiMe$_2{}^t$Bu	22	18	H	100	**49**	29	100:0	92:8
27	SMe	23	15	H	37	**50**	41	100:0	93:7
28	SCh	24	15	H	63	**51**	27	100:0	100:0
29	SPh	25	19	H	60	**52**	79	100:0	93:7
30	SeMe	26	20	H	62	**53**	31	100:0	97:3
31	SeCH$_2$Ph	27	20	H	57	**54**	25	100:0	100:0

[a]Commercially available. [b]H=Hoffmann's cycloaddition method (Cu/NaI, stirring, and heating at 60 °C); H^1=(Zn/NaI, stirring at r.t.); H^2=(Cu/NaI, ultrasound sonication at 0 °C); H^3=(Zn/NaI, ultrasound sonication at −44 °C); N=Noyori's cycloaddition method (Fe$_2$(CO)$_9$, ACN, stirring at −10 °C then r.t.). [c]Conversions of furan substrates and yields of cycloadducts are not optimized. Conversion was evaluated by GC in front of a standard. Diastereoselectivities *cis-trans* and *endo-exo* were determined by NMR and GC.

In these studies, twenty seven different C2-substituted furan substrates were submitted to [4+3] cycloaddition.[4] In these reactions, a 1,3-dimethyloxyallyl cation model was used as dienophile, due to its symmetry, to avoid regiochemistry problems which would complicate the interpretation of results of this study. This cation was generated *in situ* by reaction of 2,4-dibromo-3-pentanone with a reducing metal by several alternative procedures: Hoffmann's method[2] (NaI, Cu, MeCN, 60 °C) and Noyori's method[1b,6] (Fe$_2$(CO)$_9$, C$_6$H$_6$, 80 °C) and modifications of both procedures (see footnotes of Table 1). The different electrophilic character of oxyallyl cations, generated by both methodologies, is quite important because it conditions the mechanism of cycloaddition reactions and their stereochemical results.[1f,2,7]

In this review, we have analyzed both the reactivity of furans in front of the oxyallyl cation (conversion and yield) and the stereoselectivity of cycloaddition: *cis-trans* and *endo-exo* diastereoselectivity. The results obtained from the cycloaddition reactions of these twenty seven furan models are quoted in Table 1. Furans were commercially available or synthesized according to shown references.

Peering at experimental data from Table 1, it is possible to establish a relationship between conversion, yield and diastereoselectivity *versus* stereoelectronic properties of groups (YR_n) attached at C-2 of furans or of their heteroatom linkers (Y).

First of all, it is worth noting the high *cis* stereoselectivity obtained in practically all the studied models. This fact could indicate that, under the used reaction conditions, the [4+3] cycloadditions take place *via* a concerted mechanism where the oxyallyl cation preferably adopts a "W" or (Z,Z) configuration.[2] However, in the case of using the Noyori's reaction conditions (entries 3 and 24), which generate a very electrophilic oxyallyl cation (Scheme 6), it is possible to observe *trans* product, which becomes the major product in the case of YR_n=NHBoc (entry 3). In entry 10, the presence of *trans* isomer may be due to the inductive deactivation of the furan ring by the CHO group, which leads the reaction though a stepwise mechanism.

For substituents of type YMe_3 (Y=C, Si, Ge, Sn), (see entries 6, 13, 14, 15), it is possible to appreciate in all cases, apart from a *cis* stereospecificity, a moderated yield of cycloaddition (due to the large volume of YMe_3 group and its close proximity to the reactive C-2 carbon of furan) and a very high *endo* diastereoselectivity.

A similar effect could be noted in the model system YMe (Y=O, S, Se): when descending group VIA, an improvement of *endo-exo* diastereoselectivity and a decrease of yield is observed (see entries 21–24, 27, 30). This last fact could be interpreted taking into account that, when the volume of atom Y in YMe increases, it is more difficult for the cation to approximate the furan diene system, making cycloaddition less feasible (Scheme 10). According to these experimental data, the steric effect of atom Y of groups YMe_3 and YMe should be of larger magnitude than the concomitant electronic effect, when conditioning the cycloaddition outcome. This could explain how, when descending group IVA or VIA (even though decreases electronegativity of atom Y and consequently the attached diene would be richer in electron density), the tendency to undergo cycloaddition decreases, affording lower yields.

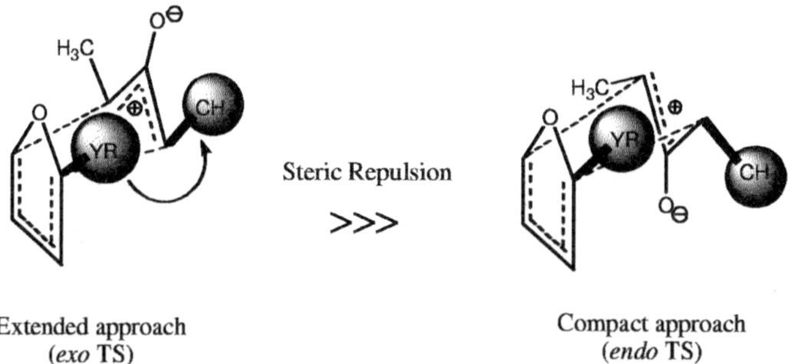

Scheme 10. Influence of the steric hindrance in the approach diene-dienophile.

On the other hand, comparing organic functions with the same heteroatom linker (Y), it is possible to observe the steric effects of anchored R groups on the yield and/or diastereoselectivity *endo/exo* of cycloadducts. So, looking at entries 27 and 28 (Y=S), entries 15 and 16 (Y=Sn) or entries 30 and 31 (Y=Se), we can appreciate how, when increasing the size of the group R, the yield decreases and the *endo/exo* selectivity increases. In all cases, we appreciated a higher influence of the steric effect than the inductive effect on the cycloaddition results. Entries 28 and 29 (cyclohexyl and phenyl groups) show how the steric effect exerted by cyclohexyl overwhelms the negative inductive effect of phenyl group.

The presence of electron-withdrawing groups at C-2 (see entries 2, 8–10, 13–16, 20) weakens substituted furans as dienes affording low to moderate yields of cycloadducts. If alkyl groups, with slight electron-donating character, are anchored in position C-2 of the furan ring, cycloadduct yields turn from moderate to good (compare entry 1 *versus* 4 and 5). Inserting groups with marked electron donating properties (YR$_n$=OR, OCOR, OSiRRR'), it is possible to obtain the best yields and conversions.

Even though it is not possible to observe neat steric and/or electronic effects, because both act simultaneously (stereoelectronic effects),[21] we have selected appropriate furan substrates in order to rationalize how the type of substituents at C-2 of furans can affect conversion, yield and stereoselectivity. Thus, after this systematic study, it is possible to establish structure-reactivity and structure-diastereoselectivity relationships in the [4+3] cycloaddition reactions of C2-functionalized furans with 1,3-dimethyloxyallyl cation. First of all, it is possible to conclude that *cis-trans* diastereoselectivity does not depend on the nature of the function on C-2 of furan substrate but on the reaction conditions (mechanism). Secondly, it is possible to remark that both yield and *endo-exo* diastereoselectivity of [4+3] cycloaddition reactions of furans with 1,3-dimethyl-oxyallyl cation are highly affected by the pattern of substitution at C-2 of the furan diene. Bulky and/or electron-withdrawing functions at C-2 afford moderated to low yields. Small and/or electron-donating groups improve considerably conversion of reagents and yield of cycloadducts.

Regarding *endo-exo* diastereoselectivity, it is quite clear, from these results quoted in Table 1, that, increasing the size of the function attached at C-2 of furan, increases *endo* diastereoselectivity of cycloaddition, becoming stereospecific when the substituent is bulky enough. In third place, it is possible to design the adequate model of function at C-2 of furan in order to get good yields and very high *endo* diastereoselectivities. For this purpose, it is necessary to insert a "spacer", like for example an atom of oxygen (Y=O), which activates furan diene by an electron-donating effect and put the steric demanding R groups (YR$_n$), apart from the reactive centres of furan, without losing its steric discriminating effect when approaching to the oxyallyl cation.

3. General method of assignment of relative stereochemistry in C-1 substituted cycloadducts by ^{1}H- and ^{13}C-NMR correlations

In this type of reactions, the stereochemical outcome is of capital importance. Thus, in every cycloaddition experiment, the diastereoisomeric cycloadducts were isolated from the reaction mixture and purified by column chromatography and spectroscopically characterized. The stereochemical assignment of diastereoisomeric cycloadducts was carried out, in an unequivocal way, by careful correlation of the spectroscopic properties of *cis/trans* and *endo/exo* stereoisomers, on the basis of 1D and 2D ^{1}H- and ^{13}C-NMR experiments: DEPT, COSY-45, COSY-90, DQFCOSY, HETCOR, HMBC, HMQC and NOESY.

Due to the stereochemical complexity of the reaction, it has been necessary to establish a model of assignment of stereochemistry based on NMR correlations. Here this NMR based model is exemplified for stereoisomers **20a** (*endo*) and **20b** (*exo*), whose relative stereochemistry was confirmed by X-ray diffraction analysis of single crystals.[4a]

3.1. Complete assignment of the ^{1}H and ^{13}C-NMR spectra of the diastereomeric pairs

The structural problem of the complete assignment of the ^{1}H- and ^{13}C-NMR spectra of the diastereoisomeric pairs **a/b** (Scheme 11) is discussed here. Correlations of clearly identified ^{1}H- and ^{13}C-NMR data were made and the effects responsible for the upfield or downfield shifts were analyzed in signals, within each pair of diastereoisomers and among the pairs of the studied series of cycloadducts. The observed effects depend on both the kind of function attached to C-1 of the bicyclic skeleton and on the relative position of methyl groups H_3-C-9 and H_3-C-10 with respect to the bridging oxygen, following certain trends which allow establishing a general method of assignment of relative stereochemistry in this type of oxabicyclic structures.[22]

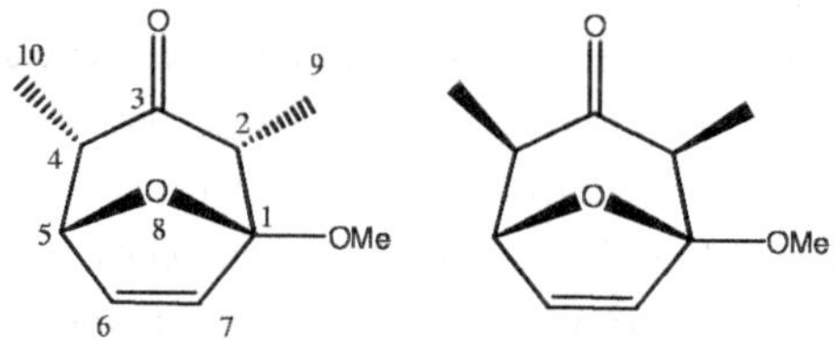

20a   **20b**

Scheme 11. Models for the assignment of relative stereochemistry of C-1 substituted diastereomeric cycloadducts.

The observed phenomena have been exemplified for the pair of diastereoisomers (**20a**, **20b**) where the function attached at C-1 is YR_n=OMe. The structural assignment was confirmed by X-ray diffraction analyses on single crystals of these particular molecules.[4a] Afterwards, the observed phenomena were extended to the remaining pairs of cycloadducts of the studied series.[23] In this way, it was possible to establish a general method of assignment of stereochemistry for this type of cycloadducts.

For the aforementioned objectives, a complete assignment of ^{1}H- and ^{13}C-NMR spectral data was required. In some cases, the signals of H-2 and H-4 appeared very close to each other or overlapped, so to distinguish them we followed one of next approaches: a) to run the spectra at 500 MHz to increase resolution; b) to carry out selective 1D double-irradiation experiments on methyl groups H_3-C-9 and H_3-C-10 to observe the chemical shifts and multiplicities of H-2 and H-4; c) to run the spectra in C_6D_6 in order to resolve the overlapped signals by aromatic solvent-induced shifts (ASIS).[24] The final assignment of signals to the corresponding hydrogen atoms was done by COSY-90 experiments.

The ^{13}C signals from the 1D totally decoupled spectra were analyzed by DEPT experiments (in order to know multiplicities) and 2D hetero-correlated spectroscopy (HETCOR). In the minor diastereoisomer **b**, the assignment of ^{1}H and ^{13}C signals corresponding to positions C-2, C-4, C-9 and C-10 was initially not clear due to H-4 appearing as a quartet without coupling to H-5. To obviate this difficulty, HMQC and HMBC experiments were carried out, modifying the parameter τ_{MB} from 0.06 to 0.11 in order to appreciate

240

the H-C long-distance couplings. From these experiments, it was possible to observed next couplings: H_3-C-10 with C-3, C-4 and C-5 and on the other hand, H_3-C-9 with C-1, C-2 and C-3. This information allowed to clearly assign all abovementioned signals and to corroborate previous assignments carried out for the remaining carbons and hydrogen atoms of the molecule, and also to confirm the C-C connectivities of the carbon skeleton. So, as C-1 and C-5 are clearly identified by the HETCOR experiment and, based on the previous observations, it was possible to assign the signals corresponding to H-2, H-4, H-9, H-10, C-2, C-4, C-9 and C-10, which have interesting stereochemical diagnostic value, as will be discussed below.

Running NOESY experiments for both diastereoisomers (**a,b**), it was possible to establish the relative stereochemistry for both of them. The stereochemistry of the molecule in both cases was confirmed by a careful correlation and comparative study of ^{1}H and ^{13}C spectra (δ, multiplicity and J), as will be analyzed later on. From NOESY spectra, we observed the NOE effects[25] shown on Scheme 12 (exemplified for **20a** and **20b**). It is worth noting that the optimum mixing-time parameter for the NOESY pulse sequence was 750 ms, lower values did not give NOE signals and higher values gave COSY signals as artefacts.

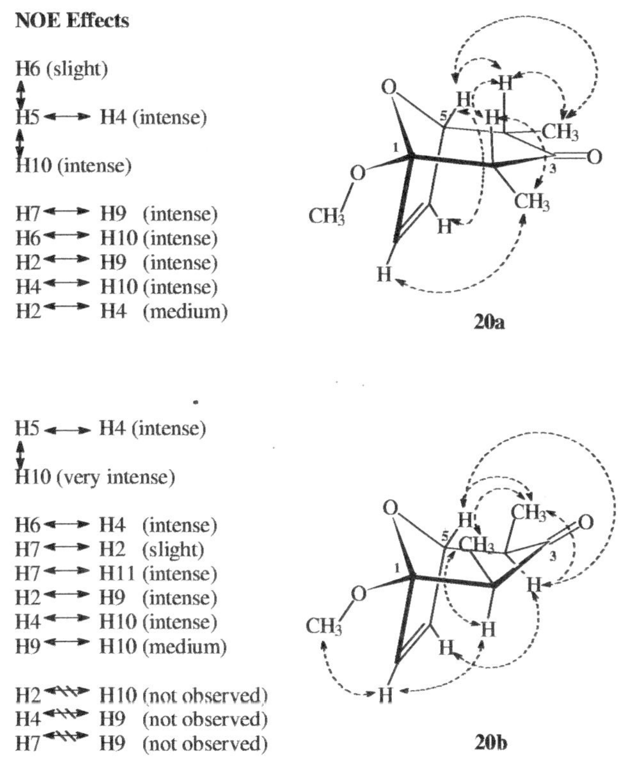

Scheme 12. NOE effects observed in the NOESY spectra of **20a** and **20b**.

The NOE observations allowed establishing a working model of relative stereochemistry for both diastereoisomers. So, it was assigned to **a**, the major isomer, a 1,3-*cis*-diequatorial disposition of methyl groups H_3-C-9 and H_3-C-10, attached respectively to C-2 and C-4, and a chair-like conformation to the

1-oxane-4-one ring. For the minor diastereoisomer **b**, in the light of NOE effects a *quasi*-1,3-*cis*-diequatorial disposition for methyl groups and a boat-like conformation for the 1-oxane-4-one ring were considered. Additionally, it was deduced that the configuration of C-2 and C-4 stereocentres in **a** were simultaneously opposite to those of **b**.

3.2. Comparative analysis of ^{1}H-NMR spectra for 20a and 20b

The chemical shifts (δ, ppm), multiplicities and coupling constants (J, Hz) of signals corresponding to ^{1}H spectra of **20a** and **20b** have been quoted for comparative purposes in Table 2. The main differences observed in both spectra are commented below; some of them give considerable structural information.

The multiplicity of H-4 varies from **20a** to **20b**, observing in **20a** a $J_{4,5}$=4.8 Hz and in **20b** a $J_{4,5}$=0 Hz. Also it is appreciated in **20b**, $J_{4,10}$=$J_{2,9}$=7.5 Hz while in **20a**, $J_{4,10}$=$J_{2,9}$=7.0 Hz. This phenomenon is observed in all pairs of diastereoisomers of the studied series, independently of the function attached to C-1 in the cycloadduct. From Table 2, we can observe that the most significant differences in δ (ppm), multiplicity and J (Hz) between both diastereoisomers are at H-2, H-4, H$_3$-C-9 and H$_3$-C-10. The fact that in **20b** $J_{4,5}$=0 Hz, is only possible for a dihedral angle (H-4)-(C-4)-(C-5)-(H-5) with an approximate value of 90°, which is consistent with a major population of conformers having a 1,3-*cis-quasi*-diequatorial disposition for methyl groups H$_3$-C-9 and H$_3$-C-10, as previously observed by NOESY experiments. Also, according to this J value, the 1-oxane-4-one ring would adopt necessarily a halfboat-like conformation, which is logical from the thermodynamic point of view in order to decrease the destabilizing steric repulsions existing in a typical 1,3-*cis*-diaxial disposition for methyl groups attached to C-2 and C-4. This phenomenon is known as inverse reflex effect (Scheme 13).[26]

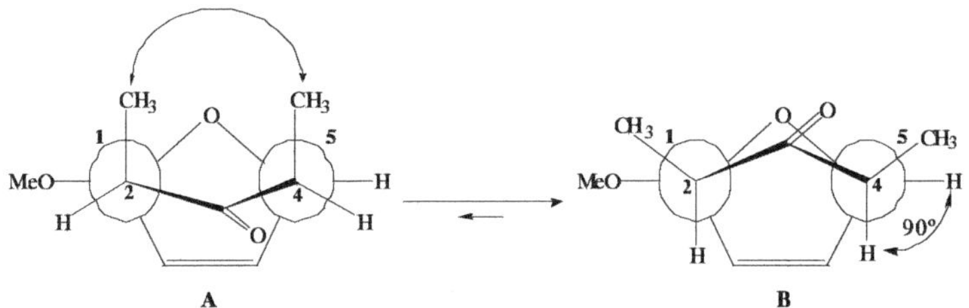

Scheme 13. Inverse Reflex Effect in **20b**: A) destabilization by steric compression for 1,3-*cis*-diaxial orientation of H$_3$-C-9 and H$_3$-C-10. B) 1,3-*Cis-quasi*-equatorial orientation for methyl groups on C-2 and C-4.

A phenomenon with high stereochemical diagnostic value is that H$_3$-C-9 and H$_3$-C-10 in **20b** isomer are deshielded with respect to their homologous methyl groups in **20a** (Table 2). This behaviour is due to a deshielding 1,3-dipolar interaction (electric field effect)[27] of methyl groups in **20b** with the bridging oxygen, interaction which is not possible in **20a**. At the same time, H-2 and H-4 are deshielded in **20a** compared to **20b**, because of the same aforementioned interaction with the bridging oxygen (Scheme 14 and Table 2). These observations are consistent with a simultaneous opposite configuration at C-2 and C-4 stereocentres of **20a** and **20b** diastereoisomers.

Table 2. ^{1}H-NMR data of **20a** and **20b** in CDCl$_3$.

Hydrogen	20a			20b			Δδ$_{20a-20b}$	ΔJ
	δ (ppm)	m	J (Hz)	δ (ppm)	m	J (Hz)		
H$_3$-10	0.84	d	$J_{10,4}$=7.0	1.36	d	$J_{10,4}$=7.5	−0.52	$\Delta J_{10,4}$=−0.5
H$_3$-9	0.91	d	$J_{9,2}$=7.0	1.27	d	$J_{9,2}$=7.5	−0.36	$\Delta J_{9,2}$=−0.5
H-2	2.62	q	$J_{2,9}$=7.0	2.54	q	$J_{2,9}$=7.5	0.08	$\Delta J_{2,9}$=−0.5
H-4	2.60	dq	$J_{4,10}$=7.0	2.23	q	$J_{4,10}$=7.5	0.37	$\Delta J_{4,10}$=−0.5
			$J_{4,5}$=4.8			$J_{4,5}$=0		$\Delta J_{4,5}$=4.8
H-11	3.28	s	0	3.42	s	0	−0.14	
H-5	4.73	dd	$J_{5,6}$=1.9	4.67	d	$J_{5,6}$=1.9	0.06	$\Delta J_{5,6}$=0
			$J_{5,4}$=4.8			$J_{5,4}$=0		$\Delta J_{5,4}$=4.8
H-7	6.06	d	$J_{7,6}$=6.1	6.09	d	$J_{7,6}$=6.0	−0.03	$\Delta J_{7,6}$=0.1
H-6	6.28	dd	$J_{6,5}$=1.9	6.36	dd	$J_{6,5}$=1.9	−0.08	$\Delta J_{6,5}$=0
			$J_{6,7}$=6.1			$J_{6,7}$=6.0		$\Delta J_{6,7}$=0.1

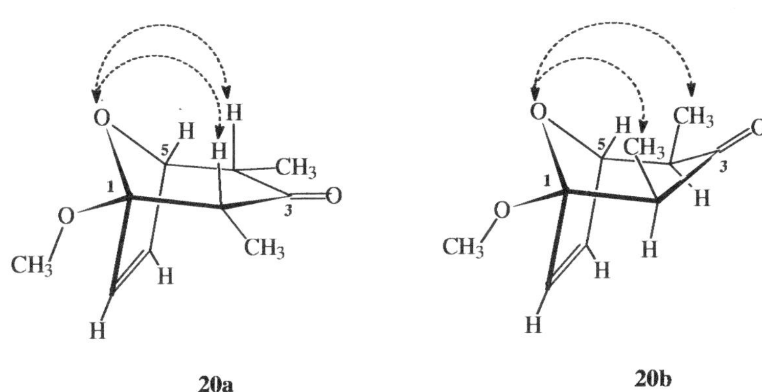

Scheme 14. Electric field effect exerted by the bridging oxygen.

Other effect contributing to the upfield shift of H$_3$-C-9 and H$_3$-C-10 in **20a** *versus* **20b** is the shielding influence of the anisotropy cone of carbonyl group[28] of C-3, which could affect in a certain extent to these methyl groups in **20a**.

The molecules under study, due to their bicyclic structure, have a low degree of conformational freedom, as can be judged from molecular models. There is a major conformation with an energy minimum, as calculated by molecular modelling (MOPAC); so, the weighed contribution of this preferred conformation to the chemical shifts and coupling constants should be very important. On the basis of this consideration, for the value of $J_{4,5}=4.8$ Hz in **20a**, it was made an estimation of the dihedral angle (H-4)-(C-4)-(C-5)-(H-5) by the Karplus equation[25c, 29] obtaining an approximate value of 44° for that angle. This indicated that the 1-oxane-4-one ring has a halfboat-like conformation in which carbons C-1, C-2, C-3, C-4 and C-5 are almost on the same plane (Scheme 15).

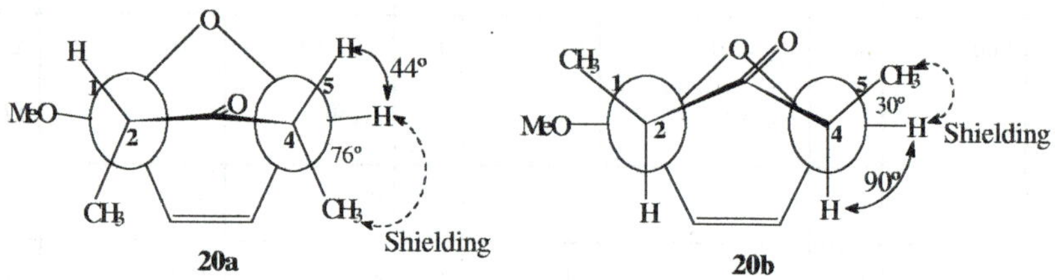

Scheme 15. Different intensity of shielding effects H₃-C-9↔H-5 in **20a** and **20b**: approximate values of dihedral angles obtained from $J_{4,5}$ coupling constants.

Other observation, consistent with the previous stereochemical assignment, is that in **20b** hydrogen H-5 appears at higher field than its homologous hydrogen **20a**. The origin of this phenomenon may be in the shielding 1,2-interaction between methyl group H₃-C-10 and H-5, which is more intense in **20b** than in **20a** because of the close proximity of both kind of interacting hydrogen atoms in the former stereoisomer. This could be appreciated in Scheme 14, where the estimated values for the dihedral angle (H-5)-(C-5)-(C-4)-(C-10) are shown: 30° for **20b** and 76° for **20a**.

The protons H-6 and H-7 have not stereochemical diagnostic value because δ(H-6)≥ or ≤δ(H-7) and/or δ(H-6, H-7)[20a]≥ or ≤δ(H-6, H-7)[20b], depending on the stereoelectronic effects exerted by the organic function attached to C1 of cycloadduct (Table 2).

3.3. Correlation of ^{13}C spectra of 20a and 20b

The correlation of ^{13}C-NMR spectra of diastereoisomers **a** and **b** afford interesting information for the establishment of configuration at C-2 and C-4 stereocentres. The main differences between the spectra of both stereoisomers are observed on δ values of C-1, C-3, C-9 and C-10 carbons (Table 3).

In first place, it might be observed an upfield shift of C-6, C-7, C-9 and C-10 in compound **a** with respect to the homologous carbons in its diastereoisomer **b** (Table 3). This shielding effect could be due to a γ-*gauche* interaction[30] between C-7 and C-9 and also between C-6 and C-10 in compound **a**. However, this interaction is not possible in the stereoisomer **b** (Scheme 16).

The important Δδ(C-3) observed between **a** and **b** stereoisomers could be interpreted on the basis of the different intensity of destabilizing repulsive interactions between the non-shared electron pairs of the bridging oxygen and the π-electrons of the carbonyl group. In structure **b**, due to the close proximity of orbitals, the electron density of the π orbital in the carbonyl group is dissymmetrically pushed towards its

244

oxygen atom, leaving the carbonyl carbon deshielded. This effect would not be observed or it would be produced in a lesser extent in isomer **a**, where the 1-oxane-4-one ring adopts a half-chair like conformation (Scheme 17).

Table 3. ^{13}C NMR data of **20a** and **20b** in CDCl$_3$.

Carbon	δ_{20a} (ppm)	δ_{20b} (ppm)	$\Delta\delta_{20a-20b}$
C-1	112.16	110.27	1.89
C-2	54.68	54.05	0.63
C-3	208.11	213.74	−5.63
C-4	48.01	47.82	0.19
C-5	79.00	79.72	−0.72
C-6	136.10	137.01	−0.91
C-7	132.42	133.39	−0.97
C-9	10.18	13.14	−2.96
C-10	8.64	17.83	−9.19
C-11	51.14	51.39	−0.25

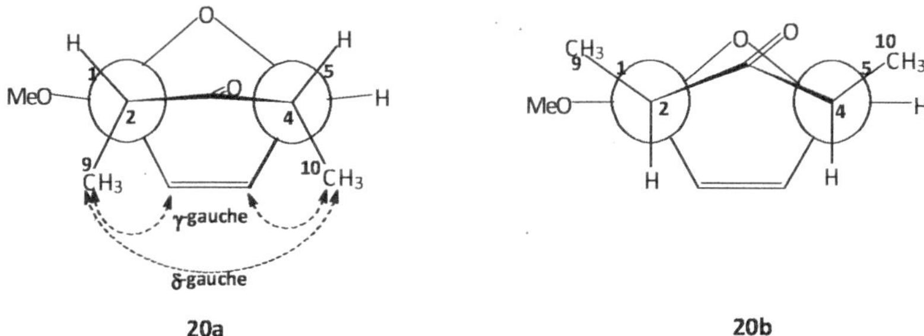

Scheme 16. γ-*Gauche* and δ-*gauche* interactions in **20a**.

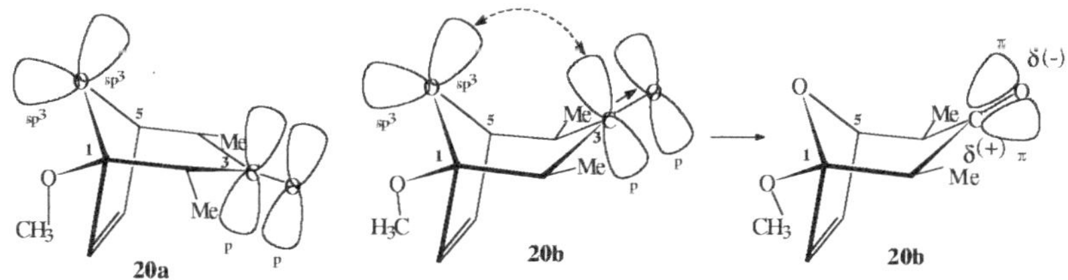

Scheme 17. Downfield shift of C-3 in **20b**.

In summary, the ^{1}H and ^{13}C chemical shifts of the 25 pairs of studied diastereomeric cycloadducts have been carefully assigned.[22,23,31] The identification of every individual signal was carried out unequivocally by the general methodology commented here. Peering at these data, it was possible to draw the same conclusions and observations than those appreciated in the **20a/20b** system. This is an indication of the wide scope and applicability of this method of stereochemical assignment for the C-1-functionalized oxabicyclic structures under study.

3.4. General method of relative stereochemistry assignment based on NMR correlation

The general trends in the correlation of ^{1}H- and ^{13}C-NMR spectra of diastereomeric **a/b** pairs of cycloadducts are summarized as follows:

1.- The most significant differences in chemical shifts (δ), multiplicity and coupling constants between both diastereoisomers are at the level of H-2, H-4, H-9 and H-10.

2.- δ(H-9) and δ(H-10) of major diastereoisomer **a** are smaller than the homologous signals in the minor stereoisomer **b**. This is caused by the different interaction of methyl groups with the bridging oxygen.

3.- $J_{2,9}$ (**a**)$<J_{2,9}$ (**b**)

   $J_{4,10}$ (**a**)$<J_{4,10}$ (**b**)

   $J_{4,5}$ (**a**)$\neq$0, normally 4$<J_{4,5}<$5 Hz; $J_{4,5}$ (**b**)$\approx$0

4.- δ(H-2) and δ(H-4) of **a** are higher than δ(H-2) and δ(H-4) of **b**, because of the different interaction of those hydrogen atoms with the bridging oxygen.

5.- δ(H-5)(**a**)$>\delta$(H-5)(**b**) by 1,2 shielding effects with Me-(C-4).

6.- The main differences in δ (^{13}C) are observed for carbons: C-1, C-3, C-9 and C-10.

7.- δ(C-9)(**a**)$<\delta$(C-9)(**b**) and δ(C-10)(**a**)$<\delta$(C-10)(**b**), due to the γ-*gauche* interactions in the **a** diastereoisomer.

8.- δ(C-3)(**a**) $< \delta$(C-3)(**b**), caused by electron compression between a sp^3 orbital of the bridging oxygen and a p_z orbital of the C3 atom.

4. Improvement and optimization of the [4+3] cycloaddition reaction: generation of oxyallyl cations by reduction of α,α'-diiodoketones under very mild sonochemical or thermal conditions

The optimization work here presented deals with the development of an improved methodology for the [4+3] cycloaddition reactions based on the *in situ* generation of oxyallyl intermediates from diiodoketones in the presence of reducing metals under very mild conditions and for short reaction times. This optimization study was performed in order to get high conversion of substrates and also high yield (chemical selectivity) and diastereoselectivity of cycloadducts under kinetic control, *i.e.* under very mild reaction conditions and in short reaction times.[32] For this purpose, the reaction was carried out under such conditions that the generation of the oxyallyl cation should be performed by using selective reagents (compatible with a wide variety of functional groups), which also would improve the general applicability and versatility of this important reaction, and at low temperatures to improve the diastereoselectivity of the process.

The substrate models used in this study are illustrated in Scheme 18. These molecules are important precursors of both versatile linear and cyclic synthons for the preparation of biologically active natural products.[5]

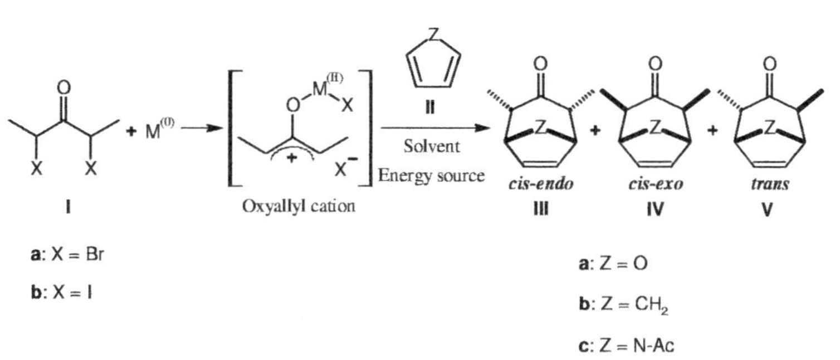

a: X = Br
b: X = I

a: Z = O
b: Z = CH$_2$
c: Z = N-Ac

Scheme 18. Diastereoisomeric products resulting from the [4+3] cycloaddition reaction of dienes **II** and 1,3-dimethyl-2-oxy-allyl cation.

Classical [4+3] cycloaddition methodologies[1] frequently use, to generate allyl cations, unstable and lachrymator α,α'-dibromoketones which in the reaction medium are reduced by Fe$_2$(CO)$_9$ (in benzene at 80 °C for 18 hours)[1b,6] or by Cu/NaI (in acetonitrile at 60 °C for 4 hours).[2]

Precedents could be found in the literature to prepare allyl cations at low temperatures. In most cases their application is oriented to certain structural models, for example: the generation of allyl cations by reaction of α-chloro-silyl-enol ethers,[33] α-chloro-enamines,[34] α-chloro-imines[35] or α-chloroalkyl-enol ethers[36] with Ag$^+$ salts. Also allyl cations could be prepared starting from α,α'-dialkoxy-silyl-enol ethers,[37] α-bromoalkyl-enol ethers,[38] α-bromo-silyl-enol ethers,[39] α-sulfono-alkyl-enol ethers[40] or α-carbonyl-silyl-enol ethers[41] in the presence of Lewis acids. On the other hand, there are methodologies that use strong reaction conditions (*i.e.*, the generation of allyl cations by reaction of chloroketones in an alkaline medium),[42] which make these procedures non-compatible with a wide variety of organic functions attached to the diene. On the basis of these precedents, a methodology was developed starting from non-volatile and non-lachrymator α,α'-diiodoketones that afforded cycloadducts in high yield (up to 90%), at low temperatures (from 0 °C to −44 °C) and in very short reaction times (<15 minutes).

With the purpose of analyzing the factors that condition the reduction of dihaloketones in the [4+3] cycloaddition reaction, an exhaustive study was carried out on all experimental parameters affecting the reaction: (a) type of dihaloketone (iodo and bromo-derivatives); (b) type and morphology of reducing agent (Fe$_2$(CO)$_9$, Cu powder, Cu bronze, Cu submicron, Zn, Zn-Cu pair); (c) reaction activators (NaI, LiI, TMSCl); (d) temperature (reflux of solvent, r.t., 0 °C, −20 °C, −44 °C, −78 °C); (e) source of energy (heating, ultrasound sonication); (f) type of solvent (polar, non-polar, protic, aprotic); (g) concentration of reagents; (h) stoichiometry. On the other hand, the following parameters were evaluated: reaction time (up to total or constant conversion), conversion of starting materials, yield, *cis-trans* diastereoselectivity and *endo-exo* diastereoselectivity.

From this study, it was possible to draw the following conclusions (Table 4):

1) The use of α,α'-diiodoketones as precursors of the oxyallyl cation instead of α,α'-dibromoketones (maintaining constant the remaining reaction factors) considerably improve the reaction rate, in such a way that it it possible to get complete conversion and high yield, at reaction temperatures down to −44 °C, much lower than the usual reaction temperatures found in the literature for these proccesses, and affording cycloadducts in a reaction time also much shorter (10 to 30 minutes instead of 4 to 18 hours).

2) From the operational and safety points of view, to work with diiodoketones, which are non-volatile (in same cases they are solids) and non-lachrymator, is easier and safer than to work with the corresponding dibromoderivatives that are relatively volatile and very lacrimator. Even though dibromoketones are usually stored in the freezer, after a while they decompose and every time you need to use them, it is necessary to filter them throught a pad of neutral alumina, in order to remove Br_2 and HBr formed, or even it is necessary to redistil them. However, diiodoketones, when stored in the freezer and protected from light, crystallize and they are stable for long periods of time.

3) The use of reducing agents based on Zn (Zn powder and Zn-Cu pairs)[43] improves the reactivity of the system, affording cycloadducts at lower temperatures and shorter reaction times than in the case of using copper powder as a reducing agent. At the same time, the use of Zn based reductors allows doing the reaction without activators such as NaI or TMSCl, which, on the other hand, facilitate the formation of polymers and /or electrophilic addition products to the detriment of [4+3] cycloadducts.

4) The use of ultrasound (generated by a high-frequency immersion sonication probe) considerably accelerates the kinetics of the reaction. Energy is released in sonication *via* a cavitation process that involves the formation and collapse of microbubbles with liberation of sufficient heat to reach temperatures up to 5000 °K and pressures up to 1000 atm within the microbubble in microseconds.[44] It is worth noting the use of ultrasound in heterogeneous reactions involving metals, where sonication produces four main effects:[45] (a) activation of the surface of metal particles (removal of passivating metallic oxides); (b) removal of reaction products from the surface of solid particles, facilitating the access by new molecules of reactants; (c) vigorous dispersion of the metallic or solid particles in the reaction medium; and (d) generation of the activation energy necessary for the reaction process to take place, increasing the number of efficient collisions among metallic particles.[46] This raising of the reaction rate allows to carry out the cycloaddition reaction, with similar yields, at lower temperature (−44 °C) than the necessary one (0 °C) to carry out reactions under stirring (without sonication).

A selection of significant results from this study is summarized in Table 4, which comported almost 300 experiments! In all cases a 2,4-dihalo-3-pentanone was used as precursor of the oxyallyl cation. As dienes three different cyclic dienes were used (Scheme 18).

In entries 1 and 2 from Table 4 it is possible to observe the results obtained by using the standard conditions found in the literature.[1b,2,6] Under these conditions, good yields and diastereoselectivities are obtained but the reaction conditions require reflux of solvent (60–80 °C) for long periods of time.

When this new cycloaddition methodology was applied to furan as a diene (entries 5, 7 and 8 from Table 4), it was possible to appreciate that the reaction went faster (10–30 minutes instead of 240–960 minutes) and at a much lower reaction temperature (−44 °C to 0 °C instead of 60–80 °C) than in the case of using the Hoffmann's or the Noyori's methodologies (entries 1 and 2, respectively).[1,2,6] On the other hand, in entry 1 the yield is lower (with similar *cis-trans* and *endo-exo* diastereoselectivities) than in the present study case. Also, in entry 2 the *cis-trans* diastereoselectivity is low (46:54) meanwhile the *endo-exo* diastereoselectivity is complete (100:0). It is worth noting (compare entries 7 and 8) the use of sonication as an energy source, which allows carrying out the reaction in one third of the usual reaction time, maintaining invariable the rest of the reaction conditions. The use of NaI as reaction activator decreases the reaction time down to 7 minutes, but considerably decreases the yield of cycloadducts (favouring the formation of electrophilic addition products on C-2 of dienes), even though the *endo-exo* diastereoselectivity is slightly increased (see entries 5 and 6). When this new cycloaddition methodology was applied to dienes other than furan (entries 8–10), the reaction afforded diastereomeric cycloadducts **III**, **IV** and **V** with similar yields and diastereoselectivities than those observed in the literature[1] (under more drastic reaction conditions), but in the present case, with much shorter reaction times and at lower reaction temperature.

The role of the metal is to transform *in situ* the starting α,α'-dihaloketones into the oxyallyl cation. Cleaning effectively the surface of these metals prior to react and activating their surfaces by ultrasound during the reaction considerably accelerates the reaction rate and it allows carrying it out at temperatures as low as −44 °C. This fact represents a great improvement of this synthetic methodology.

Table 4. Comparative results from [4+3] cycloaddition reactions of several dienes, between standard and newly designed reaction conditions.

Entry	1[a]	2[b]	Reducing agent	Additional reagent	Energy source[c]	T (°C)	t (min)	Conv.[d] (%)	Yield[e] (%)	DS[f] (cis: trans) (%)	DS[f] (endo: exo) (%)
1	**Ia**	**IIa**	Cu	NaI	Therm.	60	240	100	63	100:0	90:10
2	**Ia**	**IIa**	Fe$_2$(CO)$_9$	-	Therm.	80	960	100	90	46:54	100:0
3	**Ia**	**IIa**	Zn(Cu)	-	Sonic	−44	150	100	83	100:0	80:20
4	**Ia**	**IIa**	Zn(Cu)	NaI	Sonic	−44	90	100	78	100:0	90:10
5	**Ib**	**IIa**	Zn(Cu)	-	Sonic.	−44	15	100	90	100:0	91:9
6	**Ib**	**IIa**	Zn(Cu)	NaI	Sonic	−44	7	100	58	100:0	94:6
7	**Ib**	**IIa**	Zn(Cu)	-	Sonic.	0	10	100	70	100:0	90:10
8	**Ib**	**IIa**	Zn(Cu)	-	Stirr.	0	30	100	75	100:0	90:10
9	**Ib**	**IIb**	Zn(Cu)	-	Stirr.	0	60	100	71	100:0	66:34
10	**Ib**	**IIc**	Zn(Cu)	-	Stirr.	0	30	100	42	100:0	51:49

[a]I=Dihaloketone. [b]II=Diene. [c]Energy source: Therm.=Thermal (heating and magnetic stirring). Sonic.=Ultrasounds sonication. Stirr.=Magnetic Stirring at r.t.. [d]Conv.=conversion of raw materials. [e]Yield=% of raw material reacted that afforded cycloadducts **III** to **V**. [f]DS=Diastereoselectivity: *cis/trans*=**III**+**IV**:**V**; *endo/exo*=**III**:**IV**.

From the operational point of view, stirring was carried out by a magnetic stirrer and sonication with ultrasound was performed with an ultrasound horn sonicator (with higher releasing energy: >100 W.cm^{-2}). For this last system an ultrasonic reactor was specifically designed[32b] to work on small scale (2 to 50 mL volume, adequate to react a few mg of starting materials). For the experiments carried out under ultrasound, a three-necked sonication reactor, fitted with septa and partially immersed in a cooling bath (0 °C, −44 °C, −78 °C), was used. A 1/8" sonication probe (from Branson EDP ref. n° 101-148-062) was immersed in the reaction mixture and adapted to obtain a conveniently closed system. The reaction was carried out under inert argon atmosphere and the inner temperature was monitored by a flexible thermo-couple (PT-100).

Acknowledgments

Financial support for our research program from the Spanish Ministry of Education and Science, the University of Barcelona and the Bosch i Gimpera Foundation is gratefully acknowledged.

References

1.   (a) Hoffmann, H. M. R. *Angew. Chem.* **1973**, *20*, 877. (b) Noyori, R.; Hayakawa, Y. *Org. React.* **1983**, *29*, 163. (c) Joshi, N. N.; Hoffmann, H. M. R. *Angew. Chem., Int. Ed. Eng.* **1984**, *23*, 1. (d) Mann, J.

Tetrahedron **1986**, *42*, 4611. (e) Pavzda, A.; Schoffstall, A. *Adv. Cycloaddit.* **1990**, *2*, 1. (f) Hosomi, A.; Tominaga, Y. *Comprehen. Org. Chem.* **1995**, *5*, 593. (g) West, F. G. *Adv. Cycloaddit.* **1995**, *4*, 1. (h) Harmata, M. *Tetrahedron* **1997**, *53*, 6235. (i) Rigby, J. H.; Pigge, F. C. *Org. React.* **1997**, *51*, 351. (j) Harmata, M. *Recent Res. Dev. Org. Chem.* **1997**, *1*, 523. (k) Harmata, M. *Adv. Cycloaddit.* **1997**, *4*, 41. (l) Cha, J. K.; Oh, J. *Curr. Org. Chem.* **1998**, *2*, 217. (m) El-Wareth, A.; Sarhan, A. A. O. *Curr. Org. Chem.* **2001**, *5*, 827. (n) Harmata, M. *Acc. Chem. Res.* **2001**, *34*, 595. (o) Schall, A.; Reiser, O. *Chemtracts* **2004**, *17*, 436. (p) Huang, J.; Hsung, R. P. *Chemtracts* **2005**, *18*, 207. (q) Harmata, M. *Adv. Synth. Catal.* **2006**, *348(16, 17)*, 2297.

2.   (a) Ashcroft, M. R.; Hoffmann, H. M. R. *Org. Synth.* **1978**, *58*, 17. (b) Vinter, J. G.; Hoffmann, H. M. R. *J. Am. Chem. Soc.* **1973**, *95*, 3051. (c) Hoffmann, H. M. R. *Angew. Chem., Int. Ed. Eng.* **1984**, *23*, 1–88. (d) Hoffmann, H. M. R.; Wagner, D.; Wartchow, R. *Chem. Ber.* **1990**, *123*, 2131. (e) Schottelius, T.; Hoffmann, H. M. R. *Chem. Ber.* **1991**, *124*, 1673.

3.   Rawson, D. I.; Carpenter, B. K.; Hoffmann, H. M. R. *J. Am. Chem. Soc.* **1979**, *101*, 1786.

4.   (a) Montaña, A. M.; Ribes, S.; Grima, P. M.; García, F.; Solans, X.; Font-Bardia, M. *Tetrahedron* **1997**, *53*, 11669. (b) Montaña, A. M.; Ribes, S.; Grima, P. M.; García, F. *Chem. Lett.* **1997**, *9*, 847. (c) Montaña, A. M.; Ribes, S.; Grima, P. M.; García, F. *Acta Chem. Scand.* **1998**, *52*, 453. (d) Montaña, A. M.; García, F.; Grima, P. M. *Tetrahedron* **1999**, *55*, 5483. (e) Montaña, A. M.; Grima, P. M. *Tetrahedron* **2002**, *58*, 4769.

5.   (a) Montaña, A. M.; García, F.; Grima, P. M. *Tetrahedron Lett.* **1999**, *40*, 1375. (b) Montaña, A. M.; Fernández, D. *Tetrahedron Lett.* **1999**, *40*, 6499. (c) Montaña, A. M.; Fernández, D.; Pagès, R.; Filippou, A. C.; Kociok-Köhn, G. *Tetrahedron* **2000**, *56*, 425. (d) Montaña, A. M.; Grima, P. M. *Tetrahedron Lett.* **2001**, *42*, 7809. (e) Montaña, A. M.; Grima, P. M. *Tetrahedron Lett.* **2002**, *43*, 2017. (f) Montaña, A. M.; García, F.; Batalla, C. *Lett. Org. Chem.* **2005**, *2*, 480. (g) Montaña, A. M.; García, F.; Batalla, C. *Lett. Org. Chem.* **2005**, *2*, 475. (h) Montaña, A. M.; Ponzano, S. *Tetrahedron Lett.* **2006**, *47*, 8299. (i) Montaña, A. M.; Ponzano, S.; Kociok-Köhn, G.; Font-Bardia, M.; Solans, X. *Eur. J. Org. Chem.* **2007**, *26*, 4383. (j) Montaña, A. M.; Barcia, J. A.; Kociok-Kohn, G.; Font-Bardia, M. *Tetrahedron* **2009**, *65*, 5308.

6.   Noyori, R.; Hayakawa, Y. *Tetrahedron* **1985**, *41*, 5879.

7.   (a) Shimizu, N.; Tanaka, M.; Tsuno, Y. *J. Am. Chem. Soc.* **1982**, *104*, 1330. (b) Jin, S. J.; Choi, J. R.; Oh, J.; Lee, D.; Cha, J. K. *J. Am. Chem. Soc.* **1995**, *117*, 10914, and references cited therein.

8.   2-Nitro-furan is an electron-poor diene and to react it needs a more electrophilic oxyallyl cation, afforded by the Noyori's reaction conditions. This is just a pull-push electronic effect: electron-poor dienes need strong electrophilic cations and electron-rich dienes can react smoothly with "Hoffmann's oxyallyl cations".

9.   Padwa, A.; Brodney, M. A.; Lynch, S. M. *Org. Synth.* **2000**, *78*, 202.

10.   Pelter, S.; Rowlands, M.; Clements, G. *Synthesis* **1987**, 51.

11.   Prepared from the commercially available furoyl chloride and cyclohexanol.

12.   Fieser, L. F.; Stevenson, R. *J. Am. Chem. Soc.* **1954**, *76*, 1731.

13.   Burkenser, K. A.; Currie, R. B. *J. Am. Chem. Soc.* **1947**, *70*, 1780.

14.   (a) Liebeskind, L. S.; Wang, J. *J. Org. Chem.* **1993**, *58*, 3550. (b) Bailey, J. R. *Tetrahedron Lett.* **1986**, *27*, 4407. (c) Yamamoto, M.; Munakata, H.; Kishikawa, K.; Kohmoto, S.; Yamada, K. *Bull. Chem. Soc. Japan* **1992**, *65*, 2366.

15.   (a) Stutz, P.; Stadler, P. A. *Org. Synth.* **1988**, *56*, 109. (b) Lumbroso, H.; Bertin, D. M.; Fringuelli, F.; Taticchi, A. *J. Chem. Soc., Perkin Trans. 2* **1977**, 776.

16.   Hormi, O. E. O.; Näsman, J. H. *Synth. Commun.* **1986**, *16*, 69.

17.   Näsman, J. H. *Synthesis* **1985**, 788.

18.   (a) Bednarski, M.; Danishefsky, S. *J. Am. Chem. Soc.* **1986**, *105*, 3716. (b) Jefford, C. W. *Gazz. Chim. Ital.* **1993**, *123*, 317–320.

19.   (a) Adams, R.; Reifchneider, W.; Nair, M. D. *Croatica Chem. Acta* **1957**, *29*, 279. (b) Adams, R.; Ferretti, A. *J. Org. Chem.* **1959**, *81*, 4927.

20.   Niwa, L.; Aoki, H.; Tanaka, H.; Munakata, K.; Namiki, M. *Chem. Ber.* **1966**, *99*, 3215.

21.   Deslongchamps, P. In *Stereoelectronic Effects in Organic Chemistry*; Pergamon Press: New York, 1983; pp. 2–53.

22. Montaña, A. M.; Ribes, S.; Grima, P. M.; García, F. *Magn. Reson. Chem.* **1998**, *36*, 174.

23. Montaña, A. M.; Grima, P. M.; García, F. *Magn. Reson. Chem.* **1999**, *37*, 507.

24. (a) Blunt, J. W.; Burrit, A.; Coxon, J. M.; Steel, P. J. *Magn. Reson. Chem.* **1996**, *34*, 131. (b) Jackman, L. M.; Sternhell, S. In *Nuclear Magnetic Resonance Spectroscopy in Organic Chemistry*; Pergamon Press: New York, 1969; pp. 52 and 104. (c) Laszlo, P. *Progr. Nucl. Magn. Reson. Spectrosc.* **1967**, *3*, 231.

25. (a) Neuhans, D; Williamson, M. In *The Nuclear Overhauser Effect in Structural and Conformational Analysis*; VCH Publishers: New York, 1989; pp. 211, 253, 353 and 421. (b) Noggle, J. H.; Schirmer, R. E. *The Nuclear Overhauser Effect: Chemical Applications*; Academic Press: New York, 1972. (c) Davies, H. M. L.; Huby, N. J. S.; Cantell, W. R.; Olive, J. L. *J. Am. Chem. Soc.* **1993**, *115*, 9470.

26. (a) Hoffmann, H. M. R.; Clemens, K. E.; Smithers, R. H. *J. Am. Chem. Soc.* **1972**, *94*, 3940. (b) Waegell, B.; Ourisson, G. *Bull. Soc. Chim. Fr.* **1963**, 495. (c) Waegell, B.; Jefford, C. W. *Bull. Soc. Chim. Fr.* **1964**, 844. (d) Jefford, C. W.; Baretta, A.; Fournier, J.; Waegell, B. *Helv. Chim. Acta* **1970**, *53*, 1180. (e) Jefford, C. W.; Burger, U. *Chimia* **1970**, *24*, 385.

27. (a) Schweizer, M. P.; Chan, S. I.; Helmkamp, G. K.; Tso, P. O. P. *J. Am. Chem. Soc.* **1964**, *86*, 696. (b) Hoffmann, H. M. R. *Angew. Chem., Int. Ed. Engl.* **1973**, *12*, 819. (c) Jackman, L. M.; Sternhell, S. In *Nuclear Magnetic Resonance Spectroscopy in Organic Chemistry*; Pergamon Press: New York, 1969; pp. 67.

28. Jackman, L. M.; Sternhell, S. In *Nuclear Magnetic Resonance Spectroscopy in Organic Chemistry*; Pergamon Press: New York, 1969; pp. 88.

29. (a) Montaña, A. M.; Cano, M. *Magn. Reson. Chem.* **2002**, *40*, 261. (b) Montaña, A. M.; Nicholas, K. M. *Magn. Reson. Chem.* **1990**, *28*, 486.

30. Wehrli, F. W.; Wirthling, T. In *Interpretation of Carbon 13 NMR Spectra*; John Wiley: New York, 1976; pp. 37.

31. (a) Montaña, A. M.; Barcia, J. A. *Tetrahedron Lett.* **2005**, *46*, 8475. (b) Montaña, A. M.; Barcia J. A.; Kociok-Kohn, G.; Font-Bardía, M.; Solans, X. *Helv. Chim. Acta* **2008**, *91*, 187.

32. (a) Montaña, A. M.; Grima, P. M. *Tetrahedron Lett.* **2001**, *42* 7809. (b) Montaña, A. M.; Grima, P. M. *J. Chem. Educ.* **2000**, *77*, 754. (c) Montaña, A. M.; Grima, P. M. *Synth. Commun.* **2003**, *33*, 265.

33. Shimizu, N.; Tanaka, M.; Tsuno, Y. *J. Am. Chem. Soc.* **1982**, *104*, 1330.

34. (a) Schmid, R.; Schmid, H. *Helv. Chim. Acta* **1974**, *57*, 1883. (b) Oh, J.; Ziani-Cherif, C.; Choi, J. R.; Cha, J. K. *Org. Synth.* **2000**, *78*, 212.

35. De Kimpe, N.; Palamareva, M.; Verhe, R.; De Buyck, L.; Schamp, N. *J. Chem. Res. (S)* **1986**, 190.

36. Hill, A. E.; Hoffmann, H. M. R. *J. Am. Chem. Soc.* **1974**, *96*, 4597.

37. Murray, D. H.; Albizati, K. F. *Tetrahedron Lett.* **1990**, *31*, 4109.

38. Hoffmann, H. M. R. *Chem. Ber.* **1980**, *113*, 3837.

39. Sakurai, H.; Shirahata, A.; Hosomi, A. *Angew. Chem., Int. Ed. Engl.* **1979**, *18*, 163.

40. Harmata, M.; Elahmad, S.; Barnes, C. L. *J. Org. Chem.* **1994**, *59*, 1241.

41. Sasaki, T.; Ishibashi, Y.; Ohno, M. *Tetrahedron Lett.* **1982**, *23*, 1693.

42. (a) Mann, J.; Wilde, P. D.; Finch, M. W. *J. Chem. Soc., Chem. Commun.* **1985**, 1543. (b) Föhlisch, B.; Gehrlach, E.; Herter, R. *Angew. Chem., Int. Ed. Engl.* **1982**, 2,137. (c) Herter, R.; Föhlisch, B. *Synthesis* **1982**, 976.

43. Joshi, N. N.; Hoffmann, H. M. R. *Tetrahedron Lett.* **1986**, *27*, 124.

44. (a) Pestman, J. M.; Engberts, J. B. F. N.; de Jong, F. *Recl. Trav. Chim. Pays-Bas* **1994**, *113*, 533. (b) Suslick, K. S. *Science* **1990**, *247*, 1439. (c) Mason, T. J.; Lorimer, P. J. In *Sonochemistry: Theory, Applications and Uses of Ultrasound in Chemistry*, 1st. Edit.; John Wiley: New York, 1988; pp. 1–245. (d) Mason, T. J. In *Chemistry with Ultrasound*, 1st. Edit.; Elsevier: London, 1990; pp. 1–186.

45. (a) Einhorn, C.; Einhorn J.; Luche, J. L. *Synthesis* **1989**, 787. (b) Addulla, R. F. *Aldrichimica Acta* **1988**, *21*, 31. (c) Lindley, J.; Mason, T. J. *Chem. Soc. Rev.* **1987**, *16*, 275. (d) Boudjouk, Ph. *J. Chem Educ.* **1986**, *63*, 427. (e) Suslick, K. S.; Doktycz, S. J. *J. Am. Chem. Soc.* **1989**, *111*, 2342. (f) Doktycz, S. J.; Suslick, K. S. *Science* **1990**, *247*, 1067.

46. Montaña, A. M.; Grima, P. M. *Microscopy and Analysis* **2000**, *1*, 9.

METHODS FOR THE SYNTHESIS OF RHAZINILAM AND ITS ANALOGUES

Inga Kholod and Reinhard Neier*

Department of Chemistry, University of Neuchâtel, Av. Bellevaux 51, CH-2000 Neuchâtel, Switzerland
(e-mail: reinhard.neier@unine.ch)

Abstract. *Pyrrole containing natural products have been relatively rare. (R)-(−)-Rhazinilam is an unusual monopyrrolic product isolated from nature but its formation is probably an artefact of the isolation procedure. The discovery and the identification of rhazinilam and analogues of rhazinilam are reviewed. The seven total syntheses of rhazinilam are described and critically evaluated. Some general conclusions concerning the use of protecting groups during the synthesis are presented. The comparison of the first synthesis with the more recent syntheses shows that the number of steps needed starting from commercial materials has stayed almost constant despite the spectacular development of organic synthesis in recent years. Only one of the published syntheses gives a considerable improved overall yield. Most other syntheses give the natural product in comparable overall yield.*

Contents

Acknowledgments

References

1. Pyrrole containing natural products

Alkaloids are the archetypical nitrogen containing natural products. Many important heterocyclic systems used by nature contain one or several nitrogen atoms. Until the sixties of last century, an impressive structural variety of nitrogen containing natural products has been isolated and their structure determined. Before 1954, the number of monopyrrolic natural products isolated was however rather small (Figure 1).[1]

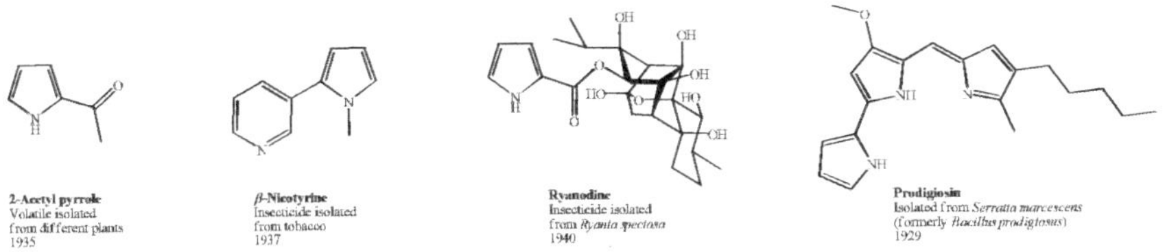

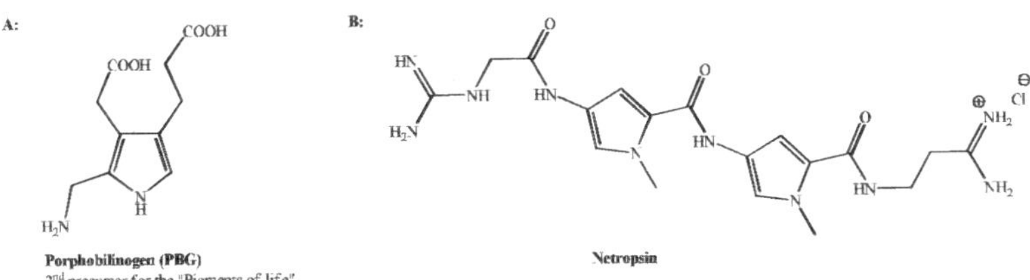

Figure 1. Monopyrrolic natural products known before 1954.

This is all the more surprising because the more complex tetrapyrrolic macrocyclic natural products are synthesized by nature in large quantities. The tetrapyrrolic pigments fulfill essential tasks as inevitable cofactors for important processes. Porphobilinogen or PBG is the dedicated pyrrolic intermediate for the biosynthesis of uroporphyrinogen III, the precursor of all tetrapyrrolic macrocycles (Figure 2A). The structure of PBG was determined by Cookson and Rimington in 1954.[2] Only three years later, the initial proposals for the structure of netropsin an oligopeptidic antibiotic isolated from *Streptomyces netropsis* were published (Figure 2B).[3,4]

Figure 2. A) Structure of porphobilinogen and B) structure of netropsin.

Since these two important discoveries, an increasing number of monopyrrolic natural products have been isolated and their structure determined. The beautiful review article by Gossauer presents a systematic overview of all monopyrrolic natural products known before 2003. The multitude of structures is impressive. The monopyrrolic natural products are incorporated in an astonishing variety of different skeletons. One structural factor is present with many of the monopyrrolic natural products: most natural products contain a functional group directly linked to the pyrrolic aromatic ring stabilizing the electron rich heterocycle against oxidation and degradation. Very often esters, amides or lactam rings are directly bound to the heteroaromatic ring. In some cases, an aromatic ring directly connected to the heterocycle plays the role of the stabilizing group. The number of pyrrole containing natural products lacking a stabilizing group is relatively small.

Pyrroles isolated from ants are a notable exception (Figure 3).[5] However, the myrmicarins 215 B and 215 C contain additional conjugated double bonds which will increase the stability of the heterocycle.

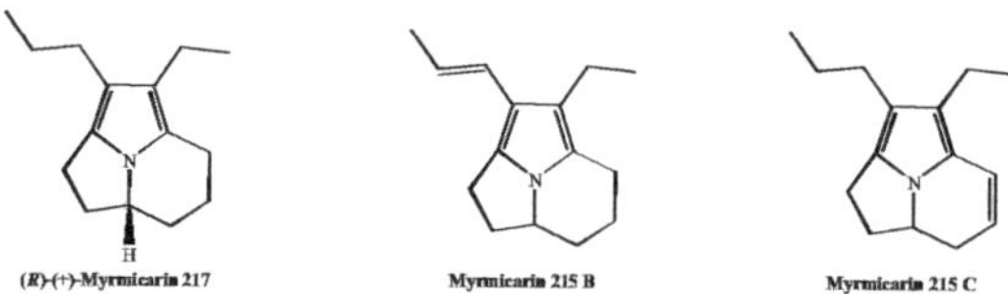

Figure 3. Alkaloids isolated from African ants *Mirmicaria opaciventris*.

The most remarkable exception is porphobilinogen, the second dedicated precursor of the tetrapyrrolic natural macrocycles. The absence of stabilizing groups in porphobilinogen bestows a high reactivity and a low stability to this compound. Nature uses this reactivity for the synthesis of uroporphyrinogen III, the reduced precursor of the natural tetrapyrroles. From a chemical standpoint nature's choice, to introduce electron attracting and/or conjugating groups onto the pyrrole ring is very reasonable. At least in one case, where nature has chosen not to introduce stabilizing groups on the pyrrole ring, this choice makes (chemical) sense, because the increased reactivity is used for the next step of the biosynthesis. One can therefore conclude that electron withdrawing groups juxtaposed on pyrrolic natural products or on precursor of natural products fulfill a purpose. These groups are probably necessary to make these compounds sufficiently stable to be used by nature and based on the same logic that these compounds can be isolated and manipulated by humans.

2. Isolation and structure determination of rhazinilam and its analogues
2.1. Isolation of rhazinilam

(*R*)-(−)-Rhazinilam (**1**) was first isolated by Linde from *Melodinus australis* in 1965.[6] Subsequently, it has been found in other South-East Asian members of the family Apocynaceae such as *Rhazya stricta* (1970),[7,8] *Aspidosperma quebracho-blanco* (1972),[9–11] *Aspidosperma marcgravianum* (1983),[12] *Leuconotis eugenifolia* (1986),[13] *Kopsia singapurensis* (1987),[14] *Leuconotis griffithii* (1989),[15] *Kopsia teoi* (1993)[16] and *Vallesia glabra* (1995).[17] The alkaloid **1** has also been obtained from somatic hybrid intergenic cell cultures of two Apocynaceae, namely *Rauvolfia serpentina* and *Rhazia stricta* (2000).[18] More recently, (*R*)-(−)-rhazinilam (**1**) was isolated from the plant species *Kopsia arborea* (2007).[19,20]

2.2. Structure determination of rhazinilam

The structure of (*R*)-(−)-rhazinilam (**1**) was established in 1972 concomitantly through spectroscopic analysis (¹H-NMR, fluorescence studies) and chemical degradation studies[8] and confirmed by X-ray crystallographic techniques (Figure 7).[10] Later, this alkaloid was identified by infrared, mass spectra and ¹H- and ¹³C-NMR analysis.[21] Compound **1** possesses four rings: a phenyl A-ring, a nine-membered lactam B-ring, a pyrrole C-ring and a piperidine D-ring. The molecule incorporates two stereogenic elements, a quaternary carbon centre at C20 and a phenyl-pyrrole chirality axis. Through X-ray analysis, it was determined that the A-C dihedral angle is almost 90°, the amide bond possesses s-*cisoid* conformation and the nine membered lactam ring adopts a boat-chair conformation. However, it was not possible to determine the absolute configuration (*R*, a*R*) by X-ray analysis. The absolute configuration was deduced *via* semi-

254

synthesis from an aspidosperma alkaloid, namely (+)-1,2-didehydroaspidospermidine (**3**).[22] The systematic name for rhazinilam used by Chemical Abstracts is (8a*R*, 14a*R*)-8a-ethyl-7,8,8a,9,10,11-hexahydro-indolizino[8,1-ef][1]benzazonin-6(5*H*)-one. The numbering system used is indicated in Figure 4 and it is based on the supposed biosynthesis.[23,24]

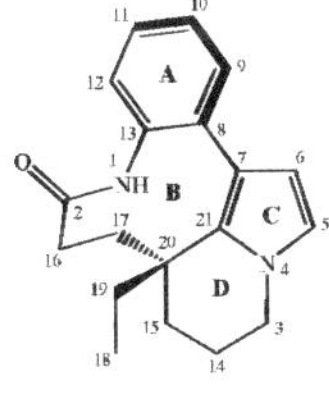

Figure 4. (*R*)-(−)-Rhazinilam (**1**).

2.3. Putative formation of (*R*)-(−)-rhazinilam from other natural products

(*R*)-(−)-Rhazinilam (**1**) is now considered to be an artifact of the extraction procedure from the plants. Its immediate natural precursor is believed to be the unstable 1,2-dihydrorhazinilam (**5**),[8] which aromatizes spontaneously on exposure to air.[13,14] (*R*)-(−)-Rhazinilam (**1**) can also be synthesized starting from (+)-vincadifformine (**2**) (Scheme 1).

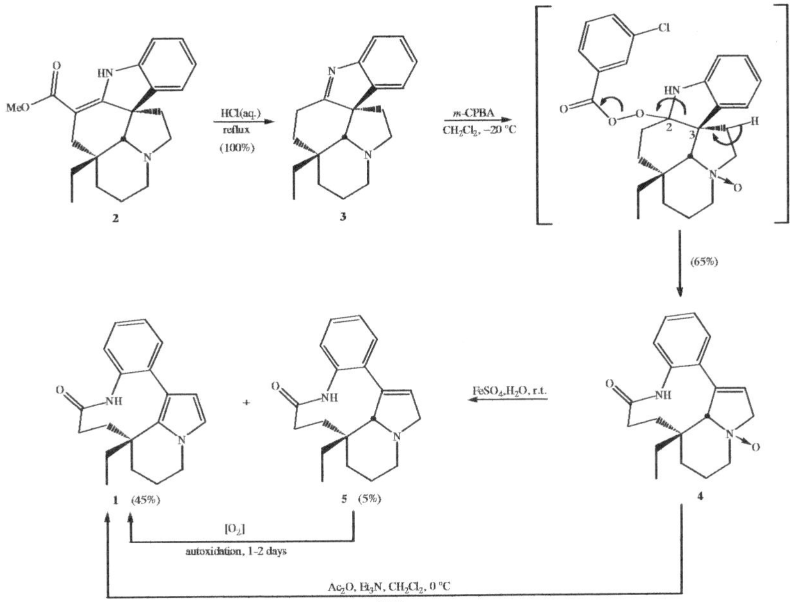

Scheme 1. Semi-synthesis of (*R*)-rhazinilam (**1**) from (+)-vincadifformine (**2**).

The mechanism of the stepwise conversion was proposed by Smith[22] and later experimentally confirmed by Guenard.[25,26] The sequence started with the acid-catalyzed removal of the ester group in (+)-vincadifformine (**2**) resulting in the formation of (+)-1,2-didehydroaspidospermidine (**3**) in quantitative

255

yield. The MCPBA-promoted oxidative cleavage of the C2-C3 indoline bond in (+)-1,2-didehydro-aspidospermidine (**3**) produced the 5,21-dihydrorhazinilam *N*-oxide (**4**) in 65% yield *via* a non-isolable *bis*-oxidized intermediate. The configuration at carbon C2 is not known. However, the isomer with the absolute configuration (*S*) at C2 is sterically much less hindered than its diastereomer with the inverted configuration. The *N*-oxide **4** was reduced with Fe(II) to give a 9:1 mixture of (*R*)-(−)-rhazinilam (**1**) and 5,21-dihydrorhazinilam (**5**). Exposing 5,21-dihydrorhazinilam (**5**) to air for several days gave rhazinilam. This slow conversion (*R*)-5,21-dihydrorhazinilam (**5**)→(*R*)-(−)-rhazinilam (**1**) suggested that the formation of rhazinilam from the 5,21-dihydrorhazinilam *N*-oxide (**4**) occurred *via* a Polonovsky-type reaction.[27] Subjecting the *N*-oxide **4** to classical Polonovski conditions using acetic anhydride and triethyl amine afforded (*R*)-(−)-rhazinilam (**1**) in 81% yield.[26] Moreover, starting from (+)-1,2-didehydroaspidospermidine (**3**), the one-pot reaction (without isolation of the intermediate) gave (*R*)-(−)-rhazinilam (**1**) in reproducible 50% yield. Therefore the air-sensitive 5,21-dihydrorhazinilam (**5**) is almost certainly the direct precursor of (*R*)-(−)-rhazinilam (**1**) in plants. 5,21-Dihydrorhazinilam (**5**) and (*R*)-(−)-rhazinilam (**1**) were co-isolated from the Apocynaceae *Leuconotis eugenifolia*,[13] *Leuconotis griffithii*,[15] *Kopsia singapurensis*[14,28] and *K. Arborea*.[14,19] This is further circumstantial evidence for this hypothesis. (+)-1,2-Didehydroaspido-spermidine (**3**) has never been detected *in vivo* together with 5,21-dihydrorhazinilam (**5**) or (*R*)-(−)-rhazinilam (**1**). If (+)-1,2-didehydroaspidospermidine (**3**) is the precursor of (*R*)-(−)-rhazinilam (**1**) remains an open question.

2.4. Isolation of analogues of rhazinilam

Other alkaloids with the same tetracyclic system as (*R*)-(−)-rhazinilam (**1**) have been isolated from different members of the family Apocynaceae: rhazinicine (**6**) [*Kopsia dasyrachis* (1998),[29] intergeneric somatic hybrid cell of *Rauvolfia serpantina* and *Rhazia stricta* (2000),[30] *Kopsia arborea* (2006),[31] *Kopsia singapurensis* (2007)[32]], 3-oxo-14,15-dehydrorhazinilam (**7**) [*Aspidosperma quebracho-blanco* (1991)[33]], (−)-leuconolam (**8**) [*Leuconotis griffithii* (1984),[15,34] *Leuconotis eugenifolia* (1986),[13,35] *Kopsia singapurensis* (2007)[32]], (+)-*epi*-leuconolam (**9**) [*Leuconotis eugenifolia* (1986),[13,15] *Leuconotis griffithii* (1989)[15]], *N*-methylleuconolam (**10**) [*Rhazia stricta* (1995)[36]], (*R*)-(−)-rhazinal (**11**) [*Kopsia teoi* (1998),[37] *Kopsia singapurensis* (2007)[32]], Kopsiyunnanines C1 (**12**), C2 (**13**), C3 (**14**) [*Kopsia arborea* (2009)[20]] (Figure 5). Such alkaloids are also believed to be derived from 5,21-dihydrorhazinilam (**5**) *via* oxidative pathways.

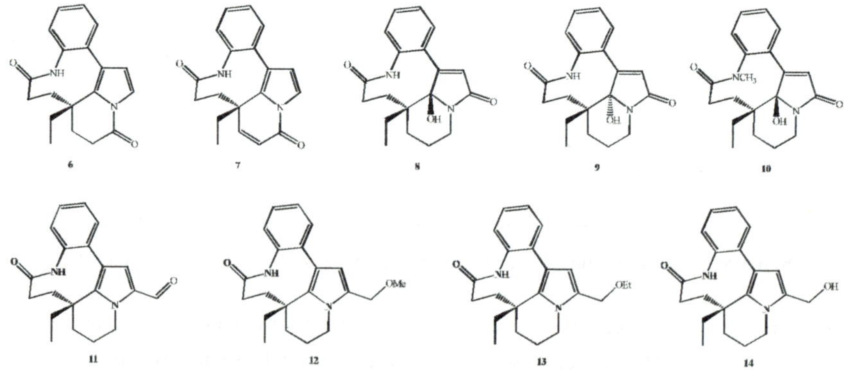

Figure 5. Naturally occurring analogues of (*R*)-rhazinilam (**1**).

3. Overview of the published total syntheses of *rac*-rhazinilam and of (*R*)-(−)-rhazinilam

3.1. Published total syntheses of rhazinilam

Seven total syntheses of rhazinilam have been reported, four of racemic (±)-rhazinilam and three of the naturally occurring (−)-enantiomer. The main objective of these studies were to overcome the synthetic challenges of these unusual structures: the A-C biaryl axis of chirality, the fused pyrrole-piperidine C-D rings, the nine-membered lactam B ring and the quaternary-carbon at C-20.

3.1.1. First total synthesis of rhazinilam by Smith

The first total synthesis of the racemic (±)-rhazinilam (**1**) was reported by Smith and co-workers in 1973[22] together with the semisynthesis of (*R*)-(−)-rhazinilam from (+)-1,2-dehydroaspidospermidine. The two intermediates **21** and **26** had to be prepared first. The reaction of ethyl magnesium bromide with diethyl 4-ketopimelate **15** was the first step for the synthesis of the lactone **16** (Scheme 2). The resulting γ-lactone of 4-ethyl-4-hydroxyheptane-1,7-dioic acid ethyl ester **16** was hydrolyzed to give the corresponding acid **17** (40% yield over two steps). The acid **17** was converted into the corresponding acid chloride **18** which on Rosenmund reduction[38] gave the aldehyde **19**. Reduction of the aldehyde **19** by NaBH$_4$ led to the alcohol **20** (70% yield from **17**). This compound was treated with tosyl chloride in pyridine to provide the desired tosyl derivative of γ-lactone **21** in 78% yield (22% over six steps from commercially available **15**).

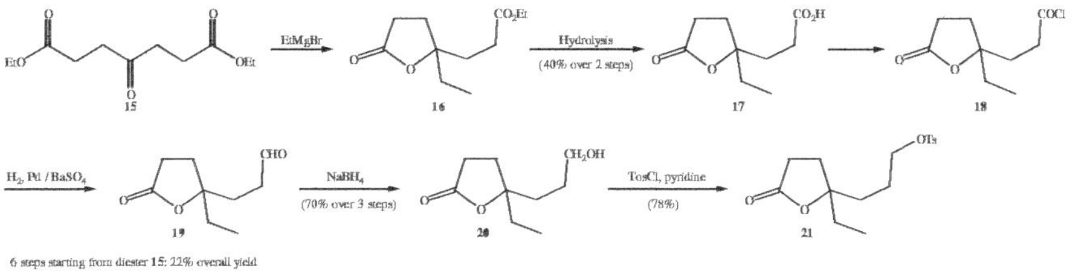

Scheme 2. Synthesis of intermediate **21**.

The second intermediate, pyrrole **26**, was prepared *via* Knorr-type condensation (Scheme 3). The reaction of 3-nitro-α-oxo-benzenepropanoic acid **22** with 2,2-dimethoxyethylamine gave the pyrrole **23** in 20% yield. Subsequent Vilsmeier formylation of compound **23** produced formyl pyrrole **24** which was oxidized using silver oxide in aqueous methanol to give the corresponding carboxylic acid **25**. Subsequent esterification with diazomethane generated the pyrrole **26** in 56% yield from compound **23** (11% yield over five steps from commercially available **22**).

Combining the two advanced intermediates was a crucial step in this total synthesis. The *N*-alkylation of the sodium salt of pyrrole **26** with the tosyl derivative of the lactone **21** provided the pyrrole **27** in 90% yield. The intramolecular Friedel-Crafts cyclization induced by aluminium trichloride in nitromethane gave the tetrahydroindolizine derivative **28** in 50% yield. The intermediate **28** contained all the carbons present in the target molecule. The additional methyl ester group at the pyrrolic C-5 position was needed to make the pyrrole ring sufficiently resistant to the following transformations. Catalytic reduction of the nitro group using Adams' catalyst gave compound **29** in 86% yield. Lactamization of compound **29** was carried out with DCC in THF at room temperature to afford compound **30** in 95% yield. The synthesis was completed in two steps. Saponification of the ester group with methanolic sodium hydroxide gave, after acidic work-up, the corresponding acid **31**. Decarboxylation of the acid **31** occurred at 240 °C and at 0.05 mmHg to produce *rac*-rhazinilam (**1**) in 88% yield over the last two steps. The synthesis proceeds through ten steps starting from commercially available acid **22** in 3.6% overall yield.

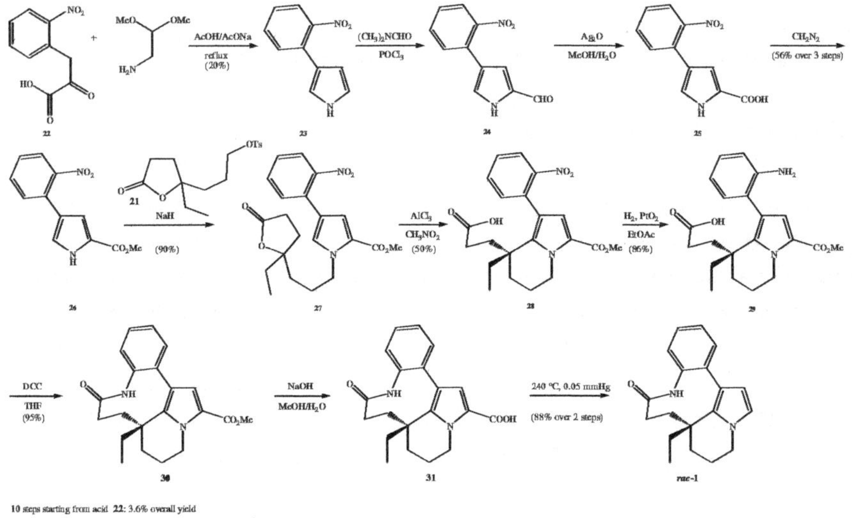

Scheme 3. Smith's total synthesis of *rac*-rhazinilam.

3.1.2. The two total syntheses of rhazinilam of Sames: *rac*-rhazinilam and (*R*)-rhazinilam

In 2000 and 2002, two very elegant total syntheses of rhazinilam were reported by Sames and co-workers. The initial work was directed towards the preparation of *rac*-(±)-rhazinilam (*rac*-**1**).[39] Two years later, Sames reported the first synthesis of the natural (−)-enantiomer.[40] Both syntheses proceeded through the achiral intermediate **41** containing the A-C-D ring system of rhazinilam. This compound was synthesized in the efficient sequence depicted in Scheme 4. Iminium salt **37** was obtained in 90% yield *via* the reaction of *o*-nitrocinnamyl bromide **36**[41] and imine **34**. The imine **34** had been generated by cyclization of the nitrile **33**.[42] The allyl iminium salt **37** in the presence of silver carbonate underwent a Grigg-type 1,5-electro-cyclization reaction[43] to produce the achiral intermediate **38** in 70% yield. The sensitive pyrrole ring was then protected with a methyl carboxylate group to give the ester **40** *via* intermediate **39**. Selective reduction of **40** provided the aniline **41** in 88% yield over three steps (25% over seven steps from commercially available nitrile aldehyde **32**).

3.1.2.1. The first total synthesis of racemic rhazinilam of Sames

The completion of the synthesis of racemic rhazinilam (*rac*-**1**) involved the formation of the Schiff base by condensation of aniline **41** with the commercially available achiral ketone **42** to give the intermediate **43** in 83% yield. Treatment of this compound with the dimethyl platinum reagent [Me₂Pt(*μ*-SMe₂)]₂[44] provided the pivotal platinum complex **44** in 88% yield. Intermediate **44** was then treated with triflic acid, resulting in liberation of methane and formation of complex **45**. The thermolysis of **45** could be induced in trifluroethanol at 70 °C and afforded the alkene-hydride **46**. Decomplexation of the platinum with potassium cyanide followed by hydrolysis of the resulting Schiff base **47**, in the presence of

258

hydroxylamine, provided the racemic alkene **48** in 60% yield over four steps. The total synthesis was then completed *via* a formal one-carbon extension of the vinyl group and the subsequent closure of the medium sized lactam ring. *N-tert*-Butyloxycarbonylation of the anilino group gave the Boc-protected derivative **49** in 76% yield. This intermediate was subjected to a Lemieux-Johnson oxidation with osmium tetroxide and sodium periodate to provide aldehyde **50** which underwent Horner-Emmons olefination followed by selective reduction of alkene **51** to afford the racemic intermediate **52** in 70% over three steps. Boc-deprotection of this compound with trifluoroacetic acid gave the indolizine intermediate **53** in 75% yield.

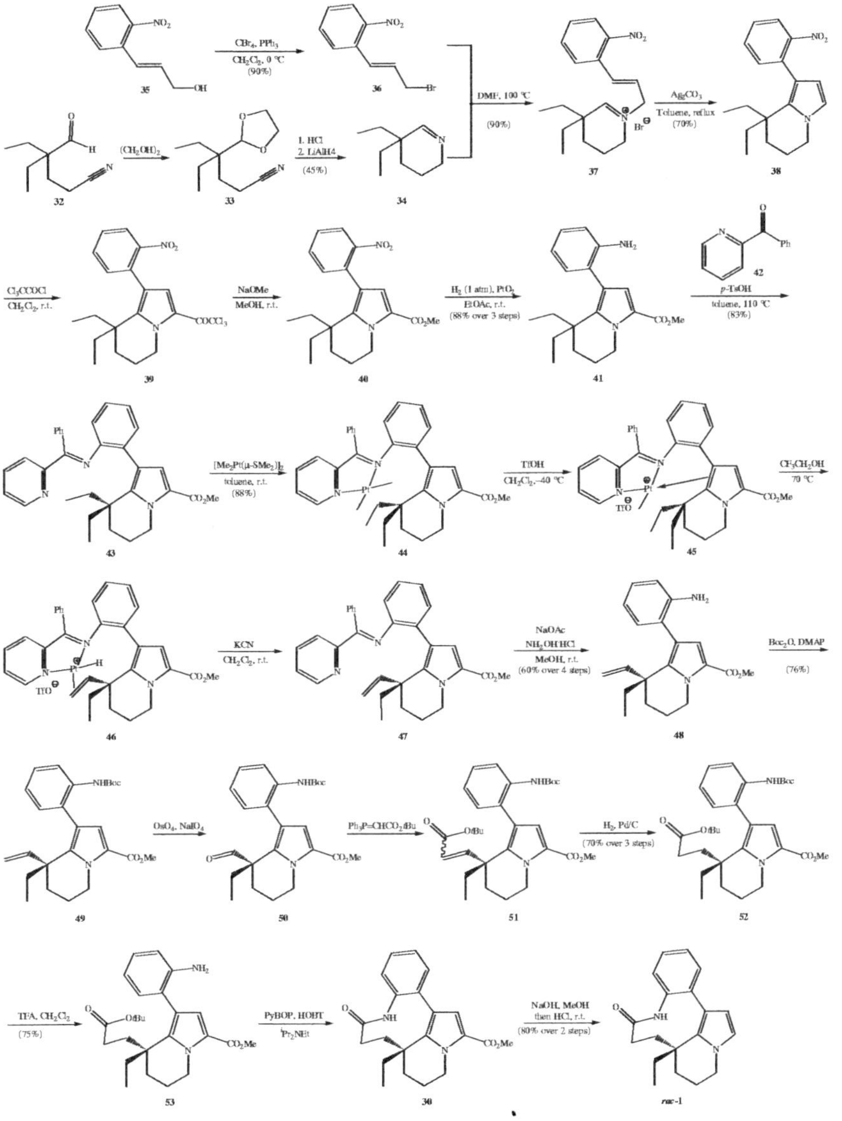

20 steps starting from nitrile aldehyde **32**: 3.5% overall yield

Scheme 4. Sames' total synthesis of *rac*-rhazinilam.

259

Lactamization of compound **53** was carried out in the presence of coupling reagents such as benzotriazol-1-yl-oxytripyrrolidinophosphonium hexafluorophosphate and hydroxybenzotriazole to provide the previously reported ester **30**.[22] Removal of the methyl ester protection group of the pyrrole under conditions previously described by Smith afforded *rac*-**1** in 80% yield over two steps. The synthesis proceeds through twenty steps starting from nitrile aldehyde **32** in 3.5% overall yield.

3.1.2.2. Sames' second enantioselective total synthesis of (*R*)-(−)-rhazinilam

The enantioselective synthesis diverged from the synthesis of the racemate on the step forming the Schiff base. The enantiopure ketone phenyl-(5*R*)-cyclohexyl-2-oxazolinone **58** was used as chiral auxiliary for the Schiff base formation. This chiral ligand was prepared in three steps from commercially available mandelonitrile **54**. Treating **54** with methanol/acetyl chloride followed by treatment of formed intermediate **55** with the amino alcohol **56** and finally Dess-Martin oxidation of the resulting alcohol **57** gave the chiral auxiliary **58** in 32% overall yield (Scheme 5). Thus, the desymmetrization of the two enantiotopic ethyl groups in compound **41** was achieved by metal-induced C(sp^3)-H activation *via* the five steps sequence. Schiff base condensation of **41** with the chiral ligand **58** afforded compound **59** in 65% yield (Scheme 6).

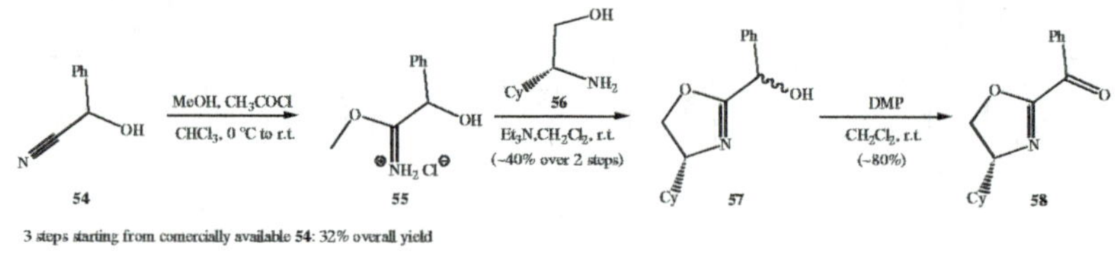

Scheme 5. Synthesis of chiral oxazolyl ligand **58**.

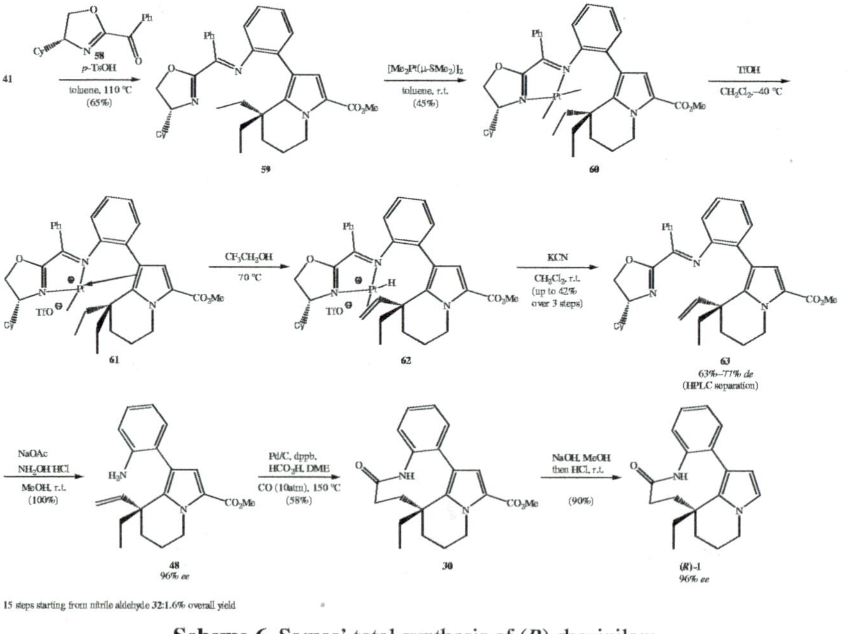

Scheme 6. Sames' total synthesis of (*R*)-rhazinilam.

Complexation of the imine **59** with the dimethyl platinum reagent [Me$_2$Pt(μ-SMe$_2$)]$_2$ provided the chiral platinum complex **60** in 45% yield. Subsequent addition of triflic acid resulted in liberation of methane and formation of complex **61**. The thermolysis of **61** produced the alkene-hydride complex **62**. Decomplexation of the platinum with potassium cyanide afforded a mixture of the diastereomers of the Schiff base **63** (63–77% *de*) in 42% overall yield. After separation of the diastereomers on preparative HPLC, the ligand was removed to give quantitatively the alkene **48** (96% *ee*). The palladium-catalyzed carbonylation of the alkene **48** gave, in 58% yield, the required nine-membered ring lactam **30** which had been reported[22,39] previously in its racemic form. The enantiomerically pure (*R*)-(−)-rhazinilam (**1**) was finally obtained in 90% yield and 96% *ee* following the procedure described by Smith.[22] The synthesis proceeds through fifteen steps starting from nitrile aldehyde **32** in 1.6% overall yield.

3.1.3. Total synthesis of *rac*-rhazinilam of Magnus

In 2001, a straightforward and elegant synthesis of *rac*-(±)-rhazinilam (*rac*-**1**) was reported by Magnus and co-workers (Scheme 7).[41] The starting material was commercially available 2-piperidone **64**. Treatment of **64** with 2 equivalents of *n*-BuLi followed by 1.5 equivalents of ethyl iodide provided α-mono-alkylated 2-piperidone **65** in 90% yield. For the second alkylation, the piperidone **65** was treated with *n*-BuLi and trimethylsilyl chloride first followed by lithium diisopropylamide and a large excess of allyl bromide. Under these conditions, the racemic *gem*-dialkyl piperidone **66** was obtained in 61% yield.

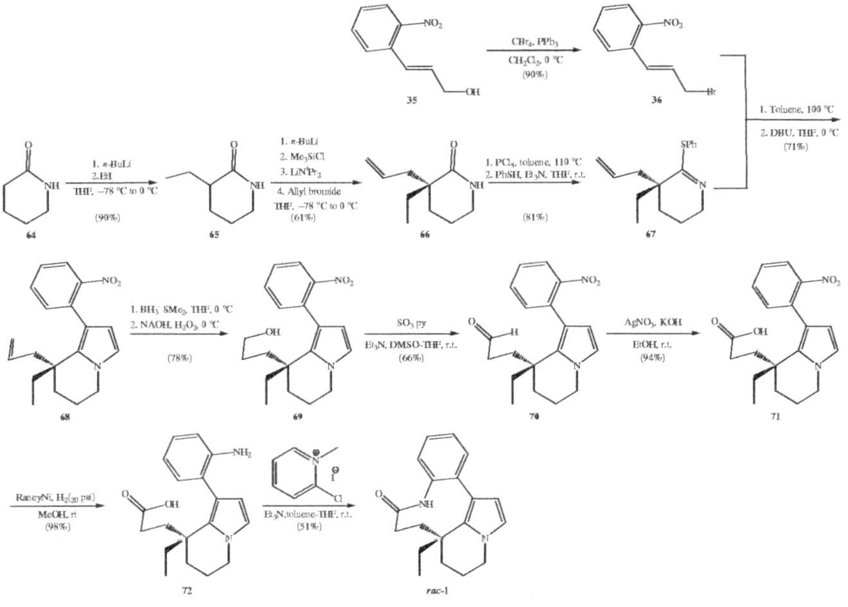

Scheme 7. Magnus' total synthesis of *rac*-rhazinilam.

The key step of the synthesis was the transformation of **66** into the annulated pyrrole derivative **68**. The thioimino ether intermediate **67** was obtained in 81% yield by treatment of the lactam **66** in sequence with phosphorous pentachloride followed by thiophenol and triethylamine. Subsequent *N*-alkylation of this

261

thiophenyl imino ether with 2-nitrocinnamyl bromide **36** gave the corresponding iminium intermediate. This intermediate underwent a Grigg-type 1,5-electrocyclization/thiophenol elimination reaction[43] to provide compound **68** in 71% yield. The compound **68** contains all the carbons required for the synthesis of the final target. Hydroboration of **68** followed by oxidative work-up gave the alcohol **69** in 78% yield. Oxidation of **69** to the aldehyde **70** was accomplished by a Swern-type oxidation with pyridine/sulphur trioxide/dimethyl-sulfoxide[45] in 66% yield. The aldehyde **70** was converted into the corresponding carboxylic acid **71** (94% yield) using silver nitrate under alkaline conditions. Raney nickel reduction of nitro-group and subsequent lactamization of compound **72** afforded *rac*-**1** in 50% yield over two steps. Thus, the synthesis proceeds through nine steps in 7.6% overall yield.

3.1.4. Total synthesis of *rac*-rhazinilam of Trauner

In 2005, Trauner and co-workers published a concise synthesis of *rac*-(±)-rhazinilam (*rac*-**1**) using an intramolecular Heck-type reaction as the key step (Scheme 8).[46]

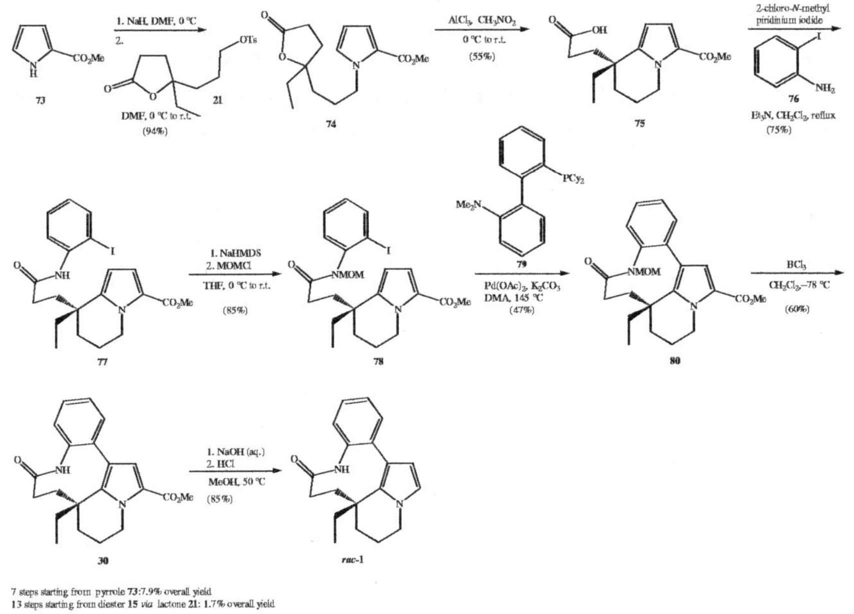

Scheme 8. Trauner's total synthesis of *rac*-rhazinilam.

The synthesis started with the *N*-alkylation of sodium salt of 2-carbomethoxy pyrrole **73** with the tosyl derivative of γ-lactone **21** prepared according to the procedure described by Smith.[22] The *N*-alkylated pyrrole **74** was obtained in 94% yield. Treatment of **74** with aluminium trichloride induced the intramolecular Friedel-Crafts cyclization and produced the racemic intermediate **75** in 55% yield. Coupling of **75** with the commercially available 2-iodoaniline **76** provided the amide **77** in 75% yield. Protection of the amide nitrogen with methoxymethyl (MOM) protecting group afforded the key intermediate **78** in 85% yield. Heating of compound **78** with 10 mol% of Buchwald's "DavePhos" ligand **79**[47] and Pd(OAc)$_2$ in the presence of potassium carbonate as base gave the strained, nine-membered lactam **80** in 47% yield. The

MOM protecting group was subsequently removed with a large excess of boron trichloride at −78 °C to afford the previously reported[22,39,40] ester **30** in 60% yield. The total synthesis was completed by saponification of the ester in compound **30** followed by acid-catalysed decarboxylation to produce *rac*-**1** in 85% yield. Thus, the synthesis proceeds through seven steps starting from commercially available pyrrole **73** in 7.9% overall yield or through thirteen steps starting from commercially available diester **15** in 1.7% overall yield.

3.1.5. Total synthesis of (*R*)-(−)-rhazinilam of Nelson

In 2006, an elegant and original enantioselective total synthesis of (*R*)-(−)-rhazinilam (**1**) was reported by Nelson and co-workers (Scheme 9).[48] The key step of this synthesis was an intramolecular asymmetric Au(I)-catalyzed pyrrole addition to the enantioenriched allene **85**. This key transformation installed the quaternary carbon stereocentre with good control of the absolute configuration.

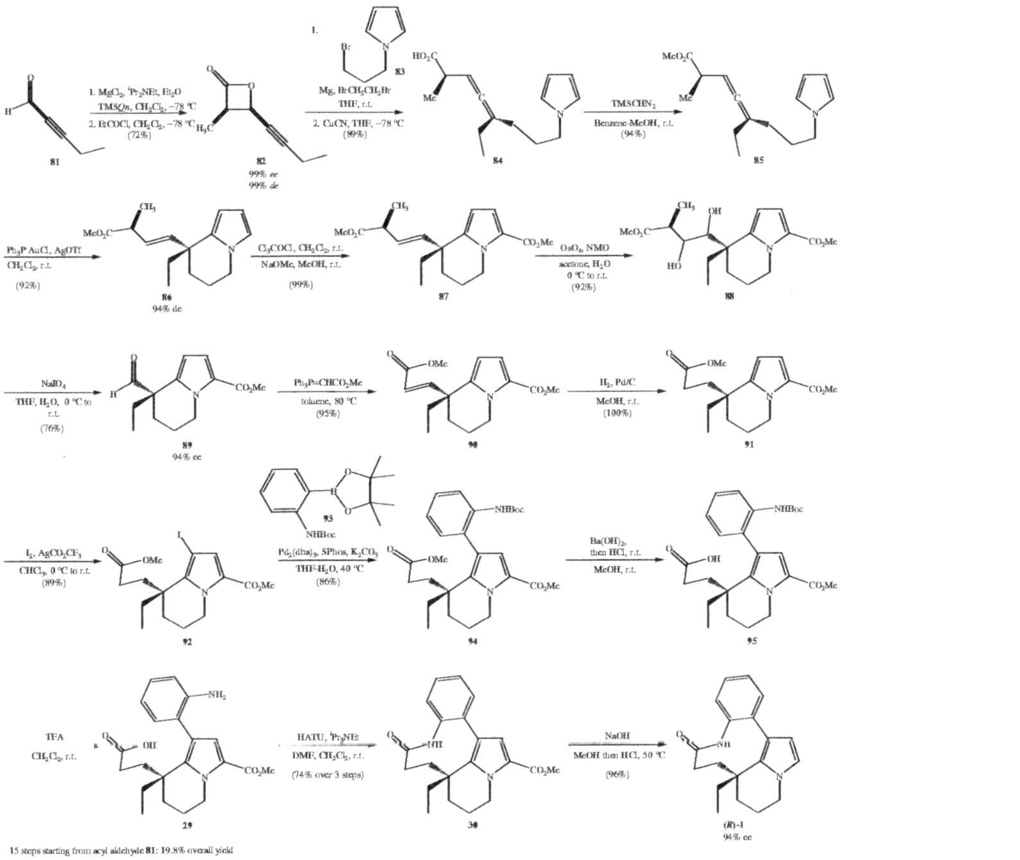

Scheme 9. Nelson's total synthesis of (*R*)-rhazinilam.

The synthesis started with the Mg-catalyzed acyl halide-aldehyde cyclocondensation[49–51] of 2-pentynal **81** and propionyl chloride to give (*R*,*R*)-β-lactone **82** in 72% yield[52] using TMSQn (*o*-trimethylsilylquinine) as a chiral catalyst. The β-lactone ring opening *via* an S$_N$2'-type reaction with (3-(1*H*-pyrrole-1-yl)propyl)magnesium bromide **83** afforded the allene derivative **84** in 89% yield as a single diastereoisomer. Methylation of compound **84** with trimethylsilyldiazomethane produced the methyl ester **85** in 94% yield. Intermediate **85** was then treated with silver triflate and Ph$_3$P·AuCl (*in situ* formation of Ph$_3$P·AuOTf), resulting in the key annulation step producing the pivotal tetrahydroindolizine **86** in 92% yield with nearly complete translation of the allene chirality (94% *de*). Subsequent protection of the pyrrole C-5 position with a methyl ester group gave **87** in 99% yield. An Upjohn-type dihydroxylation gave the *syn*-diol **88** in 92% yield from the alkene **87** using OsO$_4$ as a catalyst and *N*-methylmorpholine-*N*-oxide as an oxidant. Subsequent treatment of compound **88** with sodium periodate led to the formation of the aldehyde **89** in 76% yield. Horner-Wittig homologation of this compound afforded the alkene **90** in 95% yield. The Pd-catalyzed hydrogenation of **90** gave quantitatively the indolizine **91**. The completion of the synthesis of the enantiopure (*R*)-**1** involved the installation of the aniline moiety and subsequent lactamization. The intermediate **91** was subjected to electrophilic iodination[53] to provide iodo-pyrrole **92** in 89% yield. Suzuki-Miyaura cross-coupling[54] of compound **92** with the commercially available 2-(*N*-Boc-amino)phenylboronic acid pinacol ester **93** using Buchwald's SPhos ligand[55] afforded the 3-aryl pyrrole **94** in 86% yield. Subsequent chemoselective ester saponification and TFA-mediated aniline *N*-deprotection resulted in the formation of the previously reported[22] amino acid **29**. Lactamization of this compound was carried out with the HATU coupling reagent to produce the previously described[22,39,40,46] ester **30** in 74% yield over three steps. Removal of the methyl ester group in pyrrole **30**, as previously reported by Smith,[22] furnished (*R*)-(−)-rhazinilam (**1**) in 96% yield and 94% *ee*. Thus, the synthesis proceeds through fifteen steps starting from commercially available acyl aldehyde **81** in 19.8% overall yield.

3.1.6. Total synthesis of (*R*)-(−)-rhazinilam of Banwell

In the same year, another elegant enantioselective total synthesis of (*R*)-(−)-rhazinilam (*rac*-**1**) was reported by Banwell and co-workers.[56] The key step of this synthesis is an intramolecular enantioselective Michael addition reaction promoted by MacMillan's chiral organocatalyst.[57] The synthetic route to the pivotal intermediate **114** followed chemistry developed earlier by Smith and Banwell's own group in the context of the synthesis of *rac*-(±)-rhazinal (*rac*-**11**)[53] (Scheme 10). Thus, the *N*-alkylated pyrrole carboxylic acid **98** was obtained, after acidic work-up, in 60–90% yield by the reaction of potassium salt of pyrrole **96** generated *in situ* with γ-butyrolactone **97** under conditions defined by Li and Snyder.[58] The acid **98** was converted in 87% yield, into the corresponding Weinreb amide **99** using modified Mukaiyama conditions.[59] Subsequent treatment of amide **99** with ethyl magnesium bromide furnished the ketone **100** in 95% yield. Wadsworth-Horner-Emmons olefination of compound **100** with commercially available methyl diethylphosphonoacetate provided the β,β-disubstituted methyl acrylate **101** in 77% yield as a *ca.* 1:1 mixture of *E*- and *Z*-isomers. The initial synthesis of the racemic (±)-rhazinal (*rac*-**11**) involved, as a key step, the Lewis-acid mediated intramolecular Michael addition of the C2 of pyrrole to an *N*-tethered acrylate. The compound **101** was treated with aluminium trichloride in diethyl ether to produce the tetrahydroindolizine derivative **102** in 83% yield. The reduction of **102** to the corresponding alcohol **103** could be achieved in 75% yield. For the racemic synthesis, the final product was *rac*-(±)-rhazinal. The completion of the synthesis of *rac*-**11** involved the one-carbon homologation and the installation of the aniline moiety. Thus, alcohol **103** was converted into the corresponding mesylate **104** in 95% yield.

Treatment of this compound with sodium cyanide provided the nitrile **105** in 91% yield. Hydrolysis of the nitrile **105** and protection of the resulting acid **106** produced the methyl ester **107** in 63% yield over two steps. Compound **107** was subjected to Vilsmeier-Haack formylation to give the aldehyde **108** in 78% yield. Subsequent regioselective electrophilic iodination of the formylated pyrrole **108** using iodine in the presence of silver(I) trifluoroacetate afforded iodo-pyrrole **109** in quantitative yield. Biaryl intermediate **111** was obtained in 64% yield by Suzuki-Miyaura cross-coupling[54] of **109** with the commercially available pinacolate ester of *o*-aminophenylboronic acid **110**. Saponification of the ester group and lactamization of the ensuing carboxylic acid **112** using EDCI and DMAP delivered *rac*-(±)-rhazinal (**11**) in 68% over 2 steps. This synthesis is at the same time a formal synthesis of racemic rhazinilam as the decarboxylation step has been developed by the Banwell group.[56]

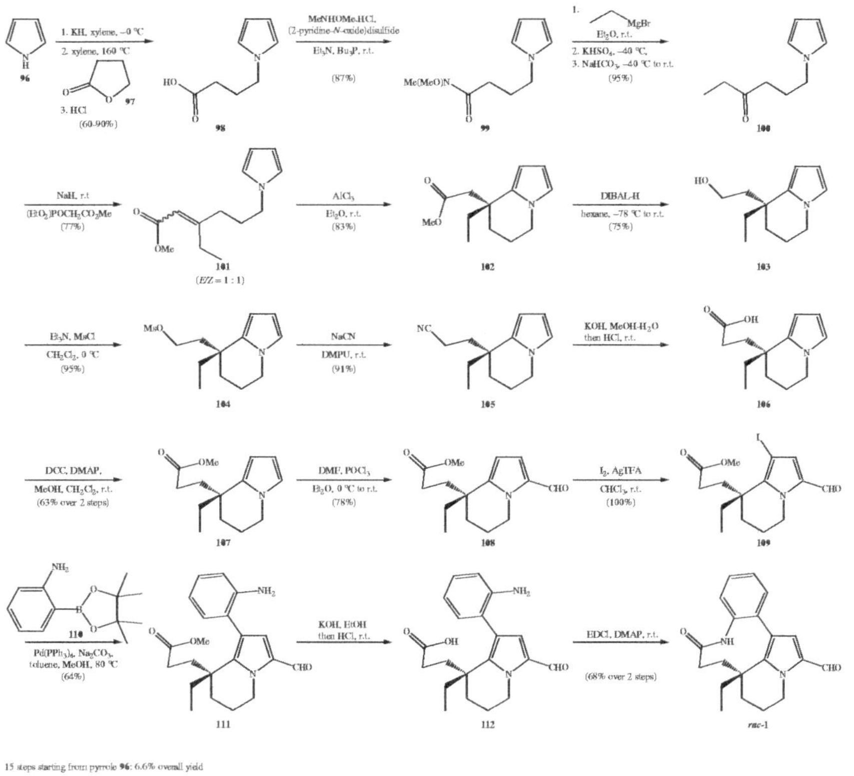

15 steps starting from pyrrole **96**: 6.6% overall yield

Scheme 10. Banwell's total synthesis of *rac*-rhazinal.

The enantioselectivity in the asymmetric version of the synthesis was provided by a chiral MacMillan's organocatalyst (Scheme 11).[57] The synthetic route to the substrate **101** was identical to the synthesis of this compound described above. Thus, reduction of the mixture of acrylates with excess DIBAL-H gave the corresponding alcohol **113** in 91% yield. The alcohol **113** was oxidized with barium manganate[60] to the aldehyde **114** obtained in 76% as a roughly 1:1 mixture of *E*- and *Z*- isomers. The pivotal intramolecular Michael addition reaction involved exposure of this mixture of aldehydes to (5*S*)-2,2,3-tri-methyl-5-phenylmethyl-4-imidazolidinone monotrifluoroacetate **115** (MacMillan's first generation

265

organocatalyst) resulting in formation of the unstable indolizine aldehyde **116** in 81% yield. Subsequent reduction of compound **116** using sodium borohydride afforded the stable alcohol **103** in 84% yield and in 74% *ee*. The completion of the synthesis of (*R*)-(−)-rhazinilam (**1**) was analogous to the synthesis of *rac*-(±)-rhazinal (**11**) and involved the one-carbon homologation. Finally, saponification of the ester group and lactamization of the ensuing carboxylic acid **112** using EDCI and DMAP delivered synthetic (*R*)-(−)-rhazinal (**11**) in 68% yield and 74% *ee*. Conversion of [(*R*)-**11**] into (*R*)-(−)-rhazinilam (**1**) was achieved by heating of [(*R*)-**11**] with stoichiometric quantities of Wilkinsons "catalyst" in refluxing 1,4-dioxane. Under these conditions, (*R*)-(−)-rhazinilam (**1**) was obtained in 89% yield and 74% *ee*. Thus, the synthesis proceeds through eighteen steps starting from pyrrole **96** in 4.4% overall yield. Treatment of (*R*)-(−)-rhazinilam (**1**) with an excess of pyridinium chlorochromate (PCC) in the presence of 4 Å molecular sieve resulted in the conversion of this substrate into a chromatographically separable mixture of (−)-leuconolam (**8**) (28%) and (+)-*epi*-leuconolam (**9**) (46%).

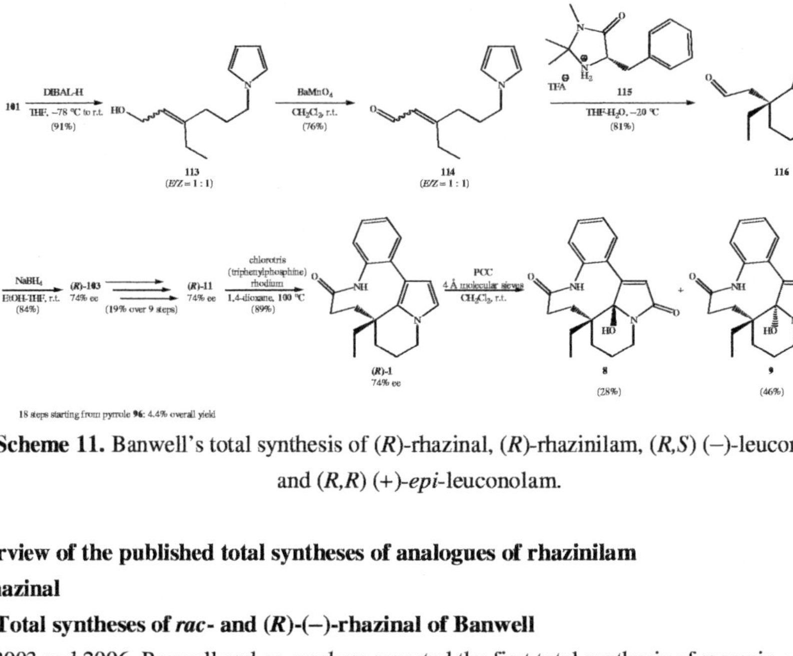

Scheme 11. Banwell's total synthesis of (*R*)-rhazinal, (*R*)-rhazinilam, (*R,S*) (−)-leuconolam and (*R,R*) (+)-*epi*-leuconolam.

3.2. Overview of the published total syntheses of analogues of rhazinilam

3.2.1. Rhazinal

3.2.1.1. Total syntheses of *rac*- and (*R*)-(−)-rhazinal of Banwell

In 2003 and 2006, Banwell and co-workers reported the first total synthesis of racemic *rac*-(±)-rhazinal (*rac*-**11**)[53] achieved in fifteen steps with 6.6% overall yield and then of the natural (*R*)-(−)-rhazinal [(*R*)-**11**][56] achieved in seventeen steps with 5.0% overall yield. Both syntheses involved as a key step an intramolecular Michael addition reaction. They were described in previous chapter (Schemes 10 and 11).

3.2.1.2. Total synthesis of *rac*-rhazinal of Trauner

In 2009, Trauner and co-workers reported[61] a concise synthesis of *rac*-(±)-rhazinal (**11**) using a Heck-type Pd-catalyzed oxidative direct cyclization (involving **127**) and a direct coupling (involving **124** and **132**) (Scheme 12). The synthesis proceeded through the *E*-trisubstituted ester **122** which was constructed from known[62] aldehyde **117**. The α-methylenation of the aldehyde **117** provided the acrolein **118** in 98% yield.

266

Treatment of **118** with methyllithium afforded the allylic alcohol **119** in 50% yield, which underwent a Claisen rearrangement to produce the ester **120** in 94% yield. Deprotection of the ester **120** with tetra-*n*-butylammonium fluoride gave the alcohol **121** in 80% yield. Subsequent tosylation of **121** afforded the desired intermediate **122** in 75% yield. The tosyloxy group of **122** was then substituted with the potassium salt of pyrrole **126** to give the *N*-alkylated pyrrole **127** in 67% yield. The key oxidative direct cyclization of compound **127** using Pd(OAc)₂ in the presence of *t*-BuOOH resulted in the formation of tetrahydroindolizine **128** in 69% yield. Subsequent Vilsmeier-Haack formylation of compound **128** afforded the aldehyde **125** in 88% yield (41% over three steps from compound **122**).

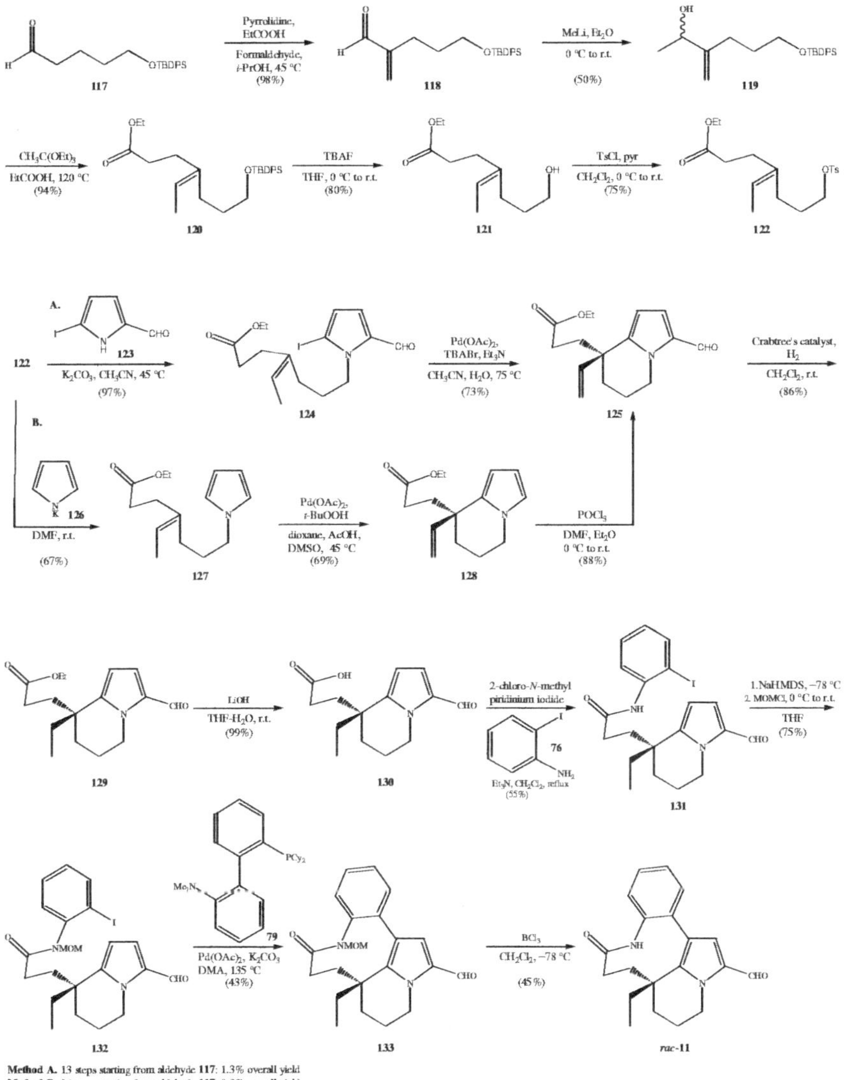

Method A. 13 steps starting from aldehyde 117: 1.3% overall yield
Method B. 14 steps starting from aldehyde 117: 0.8% overall yield

Scheme 12. Trauner's total synthesis of *rac*-rhazinal.

In parallel, the intermediate **125** was prepared by a direct Heck coupling reaction involving the iodopyrrole **123**. The compound **123** was obtained in 65% yield by iodination of azafulvene **134** and subsequent hydrolysis of resulting iodo azafulvene **135** (Scheme 13). Nucleophilic substitution of the tosylate **122** by potassium salt of pyrrole **123** gave the *N*-alkylated pyrrole **124** in 97% yield. The intramolecular Heck-type direct coupling of compound **124** under modified Jeffery conditions provided the tetrahydroindolizine **125** in 75% yield (71% over two steps from compound **122**).

The selective reduction of compound **125** using Crabtree's catalyst afforded the saturated ester **129** in 86% yield. Hydrolysis of the ester **129** yielded the corresponding acid **130** in 99% yield. Following previous procedure,[46] coupling of the acid **130** with 2-iodoaniline **76** under Mukaiyama's conditions provided the amide **131** in 55% yield. Protection of the amide as a methoxymethyl (MOM) derivative gave the key intermediate **132** in 75% yield. A Heck macrocyclization of this compound was carried out with 10 mol% of Buchwald's "DavePhos" ligand **79**[47] and Pd(OAc)$_2$ in the presence of potassium carbonate to produce the *N*-MOM rhazinal **133** in 43% yield. The total synthesis was completed by removal of the MOM protecting group with a large excess of boron trichloride to afford the racemic *rac*-(±)-rhazinal (**11**) in 45% yield. The synthesis proceeds through thirteen steps in 1.3% overall yield (Method A) or through fourteen steps in 0.8% overall yield (Method B). We have calculated the yield starting from aldehyde, which is not commercially available. In the original paper, no reference is given for the preparation of this aldehyde.

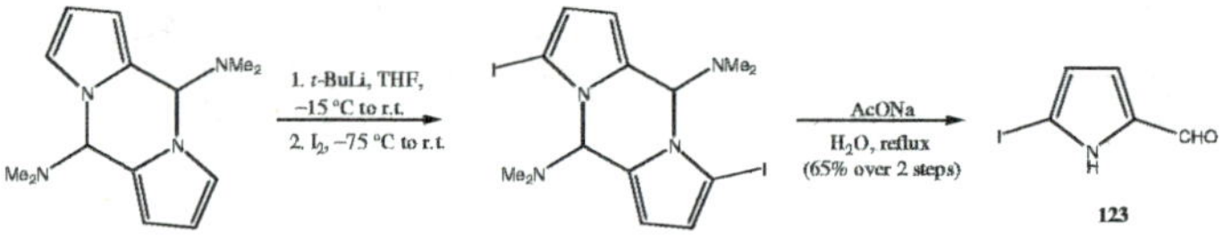

Scheme 13. Synthesis of the intermediate **123**.

3.2.2. Rhazinicine

The first total synthesis of the racemic *rac*-(±)-rhazinicine (**6**) was reported by Gaunt and co-workers in 2008.[63] The key steps of this convergent synthesis were the one-pot Ir^I-catalyzed C-H bond borylation,[64,65] the subsequent Suzuki coupling reaction[66] as well as the oxidative Pd^{II}-catalyzed pyrrole C-H bond cyclization. The synthesis started with the preparation of the two intermediates **138** and **144** (Scheme 14). For the synthesis of the biaryl compound **138**, a temporary blocking group at the C5 position of *N*-Boc-pyrrole had to be installed to avoid potential side reactions and to control the selectivity of the next reaction step. Deprotonation of *N*-Boc-pyrrole **135** with LDA followed by quench with TMS-chloride provided TMS-protected pyrrole **136** in 91% yield. Compound **136** was used in a one-pot Ir^I-catalyzed borylation and Suzuki coupling to form 3-arylated pyrrole **137** in 78% yield. Removal of Boc-protecting group under heating afforded the pyrrole **138** in 91% yield.

The alkene monoester intermediate **144** was prepared from the commercially available diethyl-4-oxo-pimelate **139**. The Wittig ethenylation of compound **139** used triphenylphosphonium ethyl bromide to give the diester **140** in 80% yield. The hydrolysis of the diester **140** produced quantitatively the diacid **141**. Heating of the diacid **141** in acetic anhydride and subsequent treatment of formed anhydride **142** with trimethylsilyl-ethanol **143** produced the intermediate **144** in 55% yield over two steps. The *N*-acylated pyrrole **145** was then obtained in 76% yield by reaction of lithium pyrrolate anion of **138** with acid chloride of **144** (Scheme 15). Intermediate **145**, containing all the carbons present in the target molecule, was treated with $Pd(TFA)_2$ catalyst and *t*-butylperoxybenzoate, resulting in the cyclization step and so producing the pivotal tetrahydroindolizine **146** in 53% yield. Simultaneous reduction of the nitro and alkene group of compound **146** afforded quantitatively the corresponding aniline **147**. $AlCl_3$-mediated removal of the 2-trimethylsilylethyl and trimethylsilyl protecting groups delivered the carboxylic acid **148** in 90% yield. Lactamization of this compound under Mukaiyama conditions produced the *rac*-(±)-rhazinicine (**6**) in 82% yield. The synthesis proceeds through eight steps starting from pyrrole **135** in 1.9% overall yield (through nine steps starting from diester **139** in 1.3% overall yield).

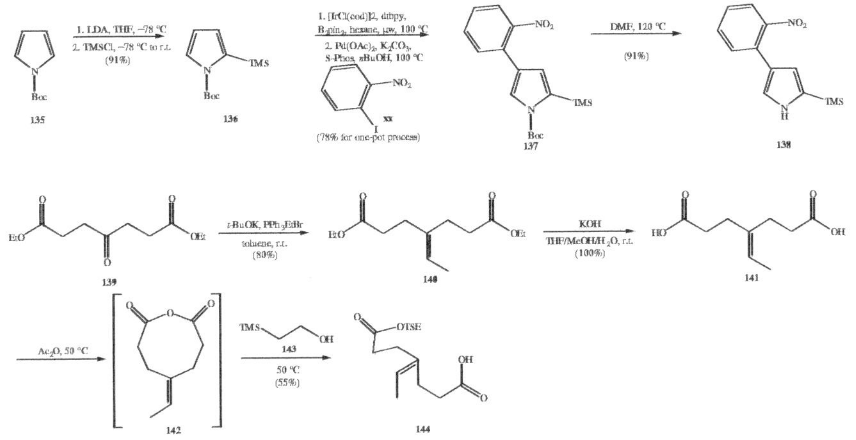

Intermediate138. 3 steps starting from pyrrole 135: 65% overall yield
Intermediate 144. 4 steps starting from diester 139: 44% overall yield

Scheme 14. Synthesis of intermediates **138** and **144**.

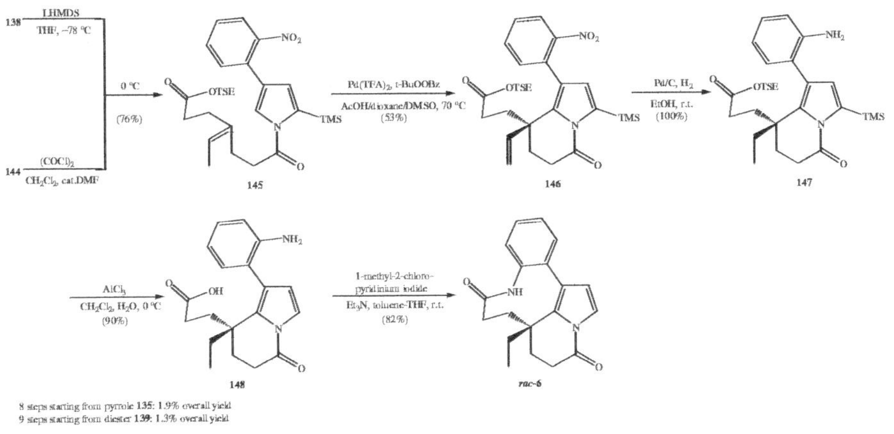

8 steps starting from pyrrole 135: 1.9% overall yield
9 steps starting from diester 139: 1.3% overall yield

Scheme 15. Total synthesis of *rac*-rhazinicine of Gaunt.

3.2.3. Leuconolam and *epi*-leuconolam

In 2006, Banwell and co-workers reported the first total syntheses of (–)-leuconolam (**8**) achieved in nineteen steps with 1.2% overall yield and (+)-*epi*-leuconolam (**9**)[56] achieved in nineteen steps with 2.0%

overall yield. Both syntheses involved as a key step an intramolecular Michael addition reaction. They were described in previous chapters (Scheme 11).

4. Summary and conclusions

The synthesis of rhazinilam has been studied for more than 35 years. Seven different syntheses have been reported so far. The development of synthetic methodologies since 1973 has been spectacular and it seems almost unfair to compare the first synthesis published in 1973 with the recent syntheses. In order to have a fair view on the different syntheses reported, we have tried to go back to commercial starting materials in all cases. If one tries to compare the overall yields and the number of synthetic steps for all seven syntheses reported, one has to take into account the following remarks. In some cases we had to go back to literature reports cited by the authors to identify the commercial starting materials. In these cases we assumed that the authors have used the procedure reported in the cited literature and we have taken the yields of these cited procedure for our calculation, at least if the authors of the total synthesis have not reported a modified procedure or an optimized transformation with better yields.

Four publications report on the synthesis of *rac*-rhazinilam and three publications describe the synthesis of *R*-(−)-rhazinilam. The number of steps from commercial starting material varies from 9 to 20 steps. The lowest overall yield is 1.6% and the best 19.8%. The average yield per step is more than 70% and for the highest yielding synthesis the average yield is 90%. Looking at these syntheses from the standpoint of overall yield, the early syntheses are surprisingly efficient. In comparing the four synthesis of *rac*-rhazinilam, the shortest synthesis with only 9 steps has been reported by Magnus. The three enantioselective syntheses are all considerably longer, 15 to 18 steps. In one case a chiral auxiliary was used, which had to be synthesized and to be resolved. In the enantioselective synthesis by Nelson and by Banwell, an enantioselective catalytic process is applied. Despite the use of the catalytic approach, the overall length of these two synthesis is 15 respectively 18 steps. Based on the overall yield, the synthesis reported by the Nelson group stands out with an overall yield, which is three times higher than the Magnus synthesis and at least a factor of five better than most of the other synthesis. Another observation concerns the use of a "protecting" or stabilizing substituent on the pyrrole ring. With one exception, all syntheses have to introduce or to carry on an electron attracting substituent on the pyrrole ring. These substituents have to be removed towards the end of the synthesis. Even if it is not formally stated in most of the publications, the function of these electron attracting substituents is clearly to make the intermediates stable enough that they can be manipulated. The only synthesis, where no electron attracting subtituent has been introduced, is the Magnus synthesis. Despite the fact that the Magnus' group uses an intermediate, which is identical with the one reported in the first synthesis of Sames, the Magnus group was able to avoid the use of an electron attracting group directly linked to the pyrrole ring.

Another important feature distinguishing the different reported syntheses is the use of a commercial precursor containing the intact pyrrole ring. The groups of Sames and of Magnus have applied a retrosynthesis where the pyrrole ring is constructed during the preparation of rhazinilam. All the other groups use an intact pyrrole ring as one of the key starting materials. In summary, the synthesis of rhazinilam has stayed a challenging synthetic problem over the whole period of 35 years which passed since the first synthesis has been reported in the literature. The syntheses reported are attractive and incorporate different interesting key steps. Despite the spectacular developments of organic synthesis, even the very first reported synthesis is still a valid approach to this fascinating natural product.

Acknowledgments

The writing of this review has been greatly facilitated by the contributions of Olivier Vallat[67] and Ana-Maria Buciumas.[68] The information contained in their respective PhD thesis has been of a great help. We would like to thank both of them for their contribution to the "rhazinilam project" and for the possibility to use their PhD thesis as a basis for our manuscript.

References

1. Gossauer, A. *Fortschr. Chem. Org. Naturst.* **2003**, *86*, 1.
2. Cookson, G. H.; Rimington, C. *Biochem. J.* **1954**, *57*, 476.
3. Weiss, M. J.; Webb, J. S.; Smith, J. M., Jr. *J. Am. Chem. Soc.* **1957**, *79*, 1266.
4. Waller, C. W.; Wolf, C. F.; Stein, W. J.; Hutchings, B. L. *J. Am. Chem. Soc.* **1957**, *79*, 1265.
5. Schroeder, F.; Franke, S.; Francke, W.; Baumann, H.; Kaib, M.; Pasteels, J. M.; Daloze, D. *Tetrahedron* **1996**, *52*, 13539.
6. Linde, H. H. A. *Helv. Chim. Acta* **1965**, *48*, 1822.
7. Banerji, A.; Majumder, P. L.; Chatterjee, A. *Phytochemistry* **1970**, *9*, 1491.
8. De Silva, K. T.; Ratcliffe, A. H.; Smith, G. F.; Smith, G. N. *Tetrahedron Lett.* **1972**, 913.
9. Lyon, R. L.; Fong, H. H. S.; Farnsworth, N. R.; Svoboda, G. H. *J. Pharm. Sci.* **1973**, *62*, 218.
10. Abraham, D. J.; Rosenstein, R. D.; Lyon, R. L.; Fong, H. H. S. *Tetrahedron Lett.* **1972**, 909.
11. Benoit, P. S.; Angry, G.; Lyon, R. L.; Fong, H. H. S.; Farnsworth, N. R. *J. Pharm. Sci.* **1973**, *62*, 1889.
12. Robert, G. M. T.; Ahond, A.; Poupat, C.; Potier, P.; Jolles, C.; Jousselin, A.; Jacquemin, H. *J. Nat. Prod.* **1983**, *46*, 694.
13. Goh, S. H.; Ali, A. R. M. *Tetrahedron Lett.* **1986**, *27*, 2501.
14. Thoison, O.; Guenard, D.; Sevenet, T.; Kan-Fan, C.; Quirion, J. C.; Husson, H. P.; Deverre, J. R.; Chan, K. C.; Potier, P. *C. R. Acad. Sci., Ser. 2* **1987**, *304*, 157.
15. Goh, S. H.; Ali, A. R. M.; Wong, W. H. *Tetrahedron* **1989**, *45*, 7899.
16. Varea, T.; Kan, C.; Remy, F.; Sevenet, T.; Quirion, J. C.; Husson, H. P.; Hadi, H. A. *J. Nat. Prod.* **1993**, *56*, 2166.
17. Zeches, M.; Mesbah, K.; Richard, B.; Moretti, C.; Nuzillard, J. M.; Men-Olivier, L. L. *Planta Med.* **1995**, *61*, 89.
18. Sheludko, Y.; Gerasimenko, I.; Platonova, O. *Planta Med.* **2000**, *66*, 656.
19. Lim, K.-H.; Hiraku, O.; Komiyama, K.; Koyano, T.; Hayashi, M.; Kam, T.-S. *J. Nat. Prod.* **2007**, *70*, 1302.
20. Wu, Y.; Suehiro, M.; Kitajima, M.; Matsuzaki, T.; Hashimoto, S.; Nagaoka, M.; Zhang, R.; Takayama, H. *J. Nat. Prod.* **2009**, *72*, 204.
21. Jewers, K.; Pusey, D. F. G.; Sharma, S. R.; Ahmad, Y. *Planta Med.* **1980**, *38*, 359.
22. Ratcliffe, A. H.; Smith, G. F.; Smith, G. N. *Tetrahedron Lett.* **1973**, 5179.
23. Szabo, L. F. *Arkivoc* **2008**, 167.
24. Morita, H.; Awang, K.; Hadi, A. H. A.; Takeya, K.; Itokawa, H.; Kobayashi, J. *Bioorg. Med. Chem. Lett.* **2005**, *15*, 1045.
25. David, B.; Sevenet, T.; Thoison, O.; Awang, K.; Pais, M.; Wright, M.; Guenard, D. *Bioorg. Med. Chem. Lett.* **1997**, *7*, 2155.
26. Dupont, C.; Guenard, D.; Tchertanov, L.; Thoret, S.; Gueritte, F. *Bioorg. Med. Chem.* **1999**, *7*, 2961.
27. Kreher, R.; Seubert, J. *Angew. Chem.* **1964**, *76*, 682.
28. Subramaniam, G.; Hiraku, O.; Hayashi, M.; Koyano, T.; Komiyama, K.; Kam, T.-S. *J. Nat. Prod.* **2008**, *71*, 53.
29. Kam, T. S.; Subramaniam, G. *Nat. Prod. Lett.* **1998**, *11*, 131.
30. Gerasimenko, I.; Sheludko, Y.; Stoeckigt, J. *J. Nat. Prod.* **2001**, *64*, 114.

31. Zhou, H.; He, H.-P.; Kong, N.-C.; Wang, Y.-H.; Liu, X.-D.; Hao, X.-J. *Helv. Chim. Acta* **2006**, *89*, 515.

32. Subramaniam, G.; Hiraku, O.; Hayashi, M.; Koyano, T.; Komiyama, K.; Kam, T.-S. *J. Nat. Prod.* **2007**, *70*, 1783.

33. Aimi, N.; Uchida, N.; Ohya, N.; Hosokawa, H.; Takayama, H.; Sakai, S.; Mendonza, L. A.; Polz, L.; Stoeckigt, J. *Tetrahedron Lett.* **1991**, *32*, 4949.

34. Goh, S. H.; Wei, C.; Ali, A. R. M. *Tetrahedron Lett.* **1984**, *25*, 3483.

35. Abe, F.; Yamauchi, T. *Phytochemistry* **1994**, *35*, 169.

36. Atta ur, R.; Abbas, S.; Zaman, K.; Qureshi, M. M.; Nighat, F.; Muzaffar, A. *Nat. Prod. Lett.* **1995**, *5*, 245.

37. Kam, T.-S.; Tee, Y.-M.; Subramaniam, G. *Nat. Prod. Lett.* **1998**, *12*, 307.

38. Mosettig, E. *Org. React.* **1954**, 218.

39. Johnson, J. A.; Sames, D. *J. Am. Chem. Soc.* **2000**, *122*, 6321.

40. Johnson, J. A.; Li, N.; Sames, D. *J. Am. Chem. Soc.* **2002**, *124*, 6900.

41. Magnus, P.; Rainey, T. *Tetrahedron* **2001**, *57*, 8647.

42. Liebowitz, S. M.; Belair, E. J.; Witiak, D. T.; Lednicer, D. *Eur. J. Med. Chem.-Chim. Ther.* **1986**, *21*, 439.

43. Crigg, R.; Myers, P.; Somasunderam, A.; Sridharan, V. *Tetrahedron* **1992**, *48*, 9735.

44. Hill, G. S.; Irwin, M. J.; Levy, C. J.; Rendina, L. M.; Puddephatt, R. J. *Inorg. Synth.* **1998**, *32*, 149.

45. Parikh, J. R.; Doering, W. v. E. *J. Am. Chem. Soc.* **1967**, *89*, 5505.

46. Bowie, A. L., Jr.; Hughes, C. C.; Trauner, D. *Org. Lett.* **2005**, *7*, 5207.

47. Harris, M. C.; Geis, O.; Buchwald, S. L. *J. Org. Chem.* **1999**, *64*, 6019.

48. Liu, Z.; Wasmuth, A. S.; Nelson, S. G. *J. Am. Chem. Soc.* **2006**, *128*, 10352.

49. Nelson, S. G.; Mills, P. M.; Ohshima, T.; Shibasaki, M. *Org. Synth.* **2005**, *82*, 170.

50. Nelson, S. G.; Wan, Z. *Org. Lett.* **2000**, *2*, 1883.

51. Nelson, S. G.; Peelen, T. J.; Wan, Z. *J. Am. Chem. Soc.* **1999**, *121*, 9742.

52. Zhu, C.; Shen, X.; Nelson, S. G. *J. Am. Chem. Soc.* **2004**, *126*, 5352.

53. Banwell, M. G.; Edwards, A. J.; Jolliffe, K. A.; Smith, J. A.; Hamel, E.; Verdier-Pinard, P. *Org. Biomol. Chem.* **2003**, *1*, 296.

54. Miyaura, N.; Suzuki, A. *Chem. Rev.* **1995**, *95*, 2457.

55. Barder, T. E.; Walker, S. D.; Martinelli, J. R.; Buchwald, S. L. *J. Am. Chem. Soc.* **2005**, *127*, 4685.

56. Banwell, M. G.; Beck, D. A. S.; Willis, A. C. *Arkivoc* **2006**, 163.

57. Paras, N. A.; MacMillan, D. W. C. *J. Am. Chem. Soc.* **2001**, *123*, 4370.

58. Li, J. H.; Snyder, J. K. *J. Org. Chem.* **1993**, *58*, 516.

59. Banwell, M.; Smith, J. *Synth. Commun.* **2001**, *31*, 2011.

60. Fatiadi, A. J. *Synthesis* **1987**, 85.

61. Bowie, A. L., Jr.; Trauner, D. *J. Org. Chem.* **2009**, *74*, 1581.

62. Freeman, F.; Kim, D. S. H. L.; Rodriguez, E. *J. Org. Chem.* **1992**, *57*, 1722.

63. Beck, E. M.; Hatley, R.; Gaunt, M. J. *Angew. Chem. Int. Ed.* **2008**, *47*, 3004.

64. Cho, J.-Y.; Tse, M. K.; Holmes, D.; Maleczka, R. E., Jr.; Smith, M. R., III *Science* **2002**, *295*, 305.

65. Ishiyama, T.; Takagi, J.; Ishida, K.; Miyaura, N.; Anastasi, N. R.; Hartwig, J. F. *J. Am. Chem. Soc.* **2002**, *124*, 390.

66. Walker, S. D.; Barder, T. E.; Martinelli, J. R.; Buchwald, S. L. *Angew. Chem. Int. Ed.* **2004**, *43*, 1871.

67. Vallat, O., Ph.D. thesis, Université de Neuchâtel, **2004**.

68. Buciumas, A.-M., Ph.D. thesis, Université de Neuchâtel, **2008**.

METAL-CATALYZED ELECTROPHILIC CYCLIZATION REACTIONS IN THE SYNTHESIS OF HETEROCYCLES

Félix Rodríguez and Francisco J. Fañanás

Instituto Universitario de Química Organometálica "Enrique Moles", Universidad de Oviedo, Julián Clavería 8, E-33006 Oviedo, Spain (e-mail: frodriguez@uniovi.es; fjfv@uniovi.es)

Abstract. *The metal-catalyzed electrophilic cyclization reactions have become an efficient way of constructing heterocyclic systems. These processes allow the synthesis of a wide variety of heterocyclic derivatives from usually simple precursors. In this review, we present the most significant developments in the field focusing our attention to reactions directed to the synthesis of oxygen- and nitrogen-containing heterocycles. Primarily guided by the nucleophile counterpart (oxygen or nitrogen), the reactions have been categorized by the nature of functional group of the nucleophile and by the kind of unsaturation (alkyne, alkene or allene). Additionally, some recent syntheses of heterocycles by metal-catalyzed cascade reactions are discussed.*

Contents

1. Introduction

Electrophilic cyclizations are those reactions in which an external electrophilic reagent activates the double (or triple) bond of an alkene (or alkyne) favouring a subsequent intramolecular addition of an internal nucleophile (Scheme 1). These processes result in the formation of cyclic products (carbocycles from carbon-centred nucleophiles and heterocycles from heteroatom-centred nucleophiles). The electrophilic cyclization reactions initially provide compounds with a functionality derived from the original electrophilic reagent which may be further transformed.

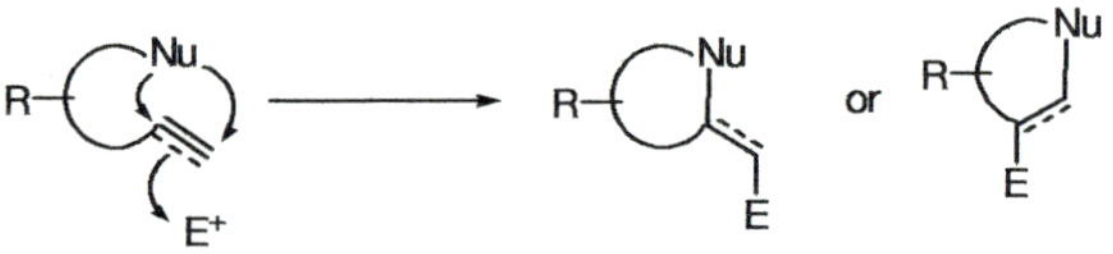

Scheme 1

The work developed by Bougault at the turn of the 20th century on the iodocyclization of unsaturated acids may be considered the birth of the electrophilic cyclization reactions.[1] Since then, the interest in this type of reactions has undergone incredible growth especially over the last decades. This growth is exemplified by the number of reviews on this subject that have appeared.[2]

Numerous electrophiles have been used to accomplish electrophilic cyclizations. Probably, the most common are the halonium ions (I^+, Br^+ and Cl^+) generated from their molecular form or from more complex reagents. Selenium-derived reagents have also widely used as electrophiles in this kind of cyclizations. In a lesser extension sulphur-derived reagents or a simple proton have also been used to activate the double or triple bond. The notorious ability of several metallic complexes to activate carbon-carbon double and triple bonds has resulted in the development of an impressive array of electrophilic cyclization reactions. In this context, those reactions where the metallic complex is used in catalytic amounts are of significant interest. In particular, much attention has been paid to the synthesis of heterocyclic compounds by means of metal-catalyzed electrophilic cyclizations.[3] The goal of this chapter is precisely to show selective, representative, illustrative and recent examples of this kind of reactions rather than a complete or comprehensive view. The reactions discussed have been categorized according to the nature of the nucleophile (mainly oxygen or nitrogen) and the unsaturated counterpart (alkenes, alkynes or allenes).

2. Mechanism and general aspects of the metal-catalyzed electrophilic cyclization

Scheme 2 shows the generally accepted mechanism for the metal-catalyzed electrophilic cyclization reaction. It starts with the coordination of the metal complex [M] to the unsaturated double or triple bond. This coordination induces the attack of the nucleophile to deliver an alky- or alkenyl-metallic complex. The integration of the metal-induced cyclizations into a productive catalytic cycle requires the eventual cleavage of the newly generated carbon-metal bond. The most usual processes for this cleavage involve a simple reaction with an electrophile E (a protodemetallation reaction is a typical process). However, other reactions such as β-hydride eliminations or rearrangement processes may be occur in order to close the catalytic cycle and to regenerate the metallic complex. The metallic complexes most widely used for this type of reactions are those defined as π-acids (the term π-acid has been recently defined by Fürstner and Davies).[4] In general, these complexes are derived from metals located in Groups 9–12 of the Periodic Table in an oxidation state that allows a d^8 or d^{10} electronic configuration. Normally, cationic complexes are more appropriate than neutral ones because in the cationic complexes the electrostatic interaction between the metal and the ligand (alkene or alkyne) is maximized and the back-donation from the metal to the ligand is minimized. Also as a general trend and due to polarizability effects, those catalysts derived from metals placed in lower part of a particular Group of the Periodic Table are more carbophilic and so, better catalysts to perform electrophilic cyclization reactions. Particularly appropriate are those catalysts derived from metals from the 6th Period of the Periodic Table as in this Period the relativistic effects become very important.[5]

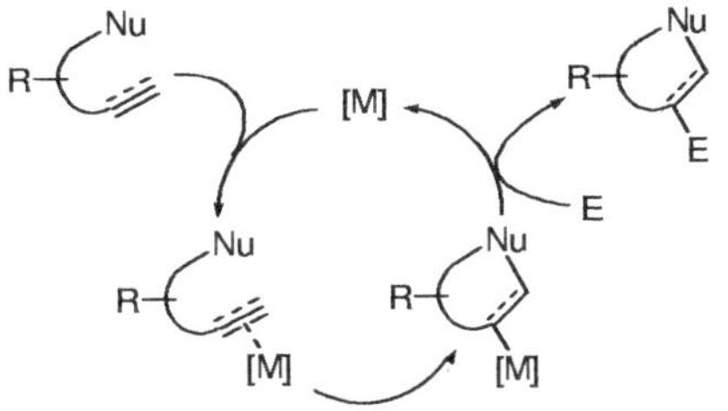

Scheme 2

The cyclization generally occurs following Baldwin ring closure rules. Thus, for 3- to 7-membered ring closures *exo*-trig processes (or *exo*-dig processes when triple bonds are implied) are usually favoured. On the contrary, 3- to 5-*endo* ring closures are not favoured and 6- and 7-*endo* are possible processes. However, the substitution of the double bond (or triple bond), the nature of the catalyst as well as other structural issues can influence this regiochemistry by electronic and/or steric effects.

3. Oxygen-centred nucleophiles

3.1. Cyclization of alcohols

3.1.1. Intramolecular addition to alkynes

The 5-*endo* cyclization reaction of *o*-alkynylphenols represents a very convenient method for the synthesis of benzofurans (Scheme 3).

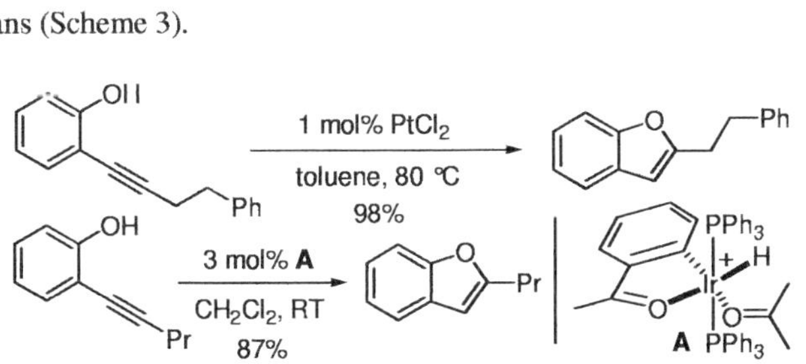

Scheme 3

As shown, the use of PtCl$_2$ as catalyst requires the heating of the toluene solution to 80 °C in order to get an almost quantitative yield of the benzofuran derivative.[6] However, when an iridium-derived catalyst is used, the reaction proceeds at room temperature.[7]

A rhodium-catalyzed formal 5-*endo* cyclization reaction has also been developed by B. M. Trost and co-workers for the synthesis of dihydrofuran derivatives.[8] The same group has also reported the synthesis of dihydropyran derivatives through a ruthenium-catalyzed formal 6-*endo* hydroalkoxylation reaction.[9] However, these reactions do not proceed through a typical electrophilic cyclization mechanism and the authors propose the initial formation of a metal-vinylidene intermediate. This kind of cyclization reactions are beyond the scope of this chapter.

Furan derivatives can also be easily obtained by cyclization of (Z)-pent-2-en-ynols (Scheme 4). Several catalysts derived from gold(I) and (III),[10] palladium(II),[11] silver(I),[12] rhodium(I),[13] iridium(I)[13b] and ruthenium(II)[14] have been reported to promote this transformation. The reaction proceeds through a 5-*exo* hydroalkoxylation process followed by an isomerization of the initially formed exocyclic enol ether to the corresponding endocyclic isomer. Representative examples of these reactions are shown in Scheme 4.

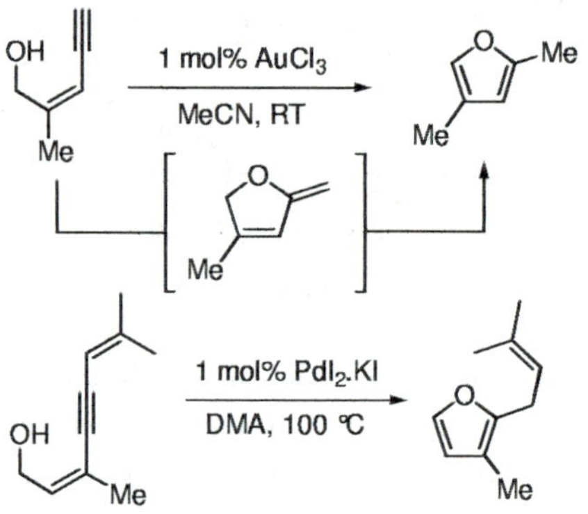

Scheme 4

The *exo*-hydroalkoxylation reaction is a powerful method for the synthesis of otherwise difficult to construct exocyclic enol ethers. However, as shown in the examples of Scheme 4, the thermodynamically favoured *exo→endo* isomerization process normally avoids the isolation of such exocyclic enol ethers. However, in some cases, the appropriate substitution of the starting ω-alkynol derivative avoids the *exo→endo* isomerization (Scheme 5).[15] In the first example of the Scheme 5, the gold(III) catalyzed cycloisomerization reaction of a (Z)-pent-2-en-ynol derivative in which the hydroxyl-containing carbon atom is quaternary leads to the corresponding exocyclic enol ether in high yield. By contrast to the examples discussed in Scheme 4, in this case the isomerization of the exocyclic double bond is not possible. In a similar way, in the second and third examples of the Scheme 5, the propargylic carbon of the starting materials are quaternary and so the corresponding cycloisomerization reaction furnishes exocyclic enol ethers where the isomerization of the double bond is not viable.

Exocyclic enol ethers could also be isolated in the tungsten-catalyzed reaction of ω-alkynols (Scheme 6).[16] In this reaction, the catalytic species are generated *in situ* under photochemical conditions from tungsten hexacarbonyl in the presence of an excess of triethylamine. Although in this case the structure of the starting material allows the *exo→endo* isomerization, the basic reaction media avoids such a process.

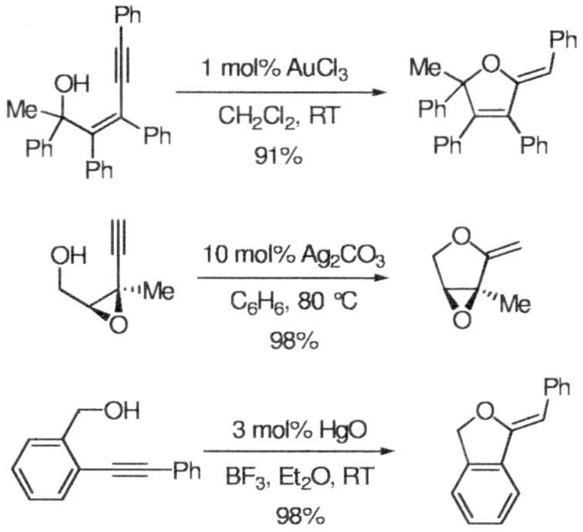

Scheme 5

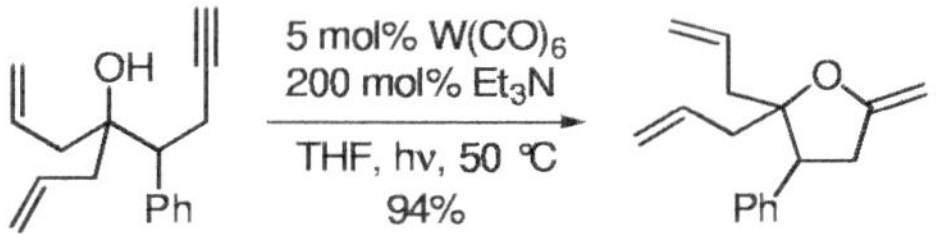

Scheme 6

K. Kato and co-workers have also developed an interesting method for the synthesis of exocyclic enol ethers consisting in a palladium-catalyzed cascade reaction initiated by an alkoxycyclization reaction followed by a methoxycarbonylation process (Scheme 7).[17] A vinylpalladium intermediate is proposed in this reaction. In this case, the *exo→endo* isomerization is disfavoured because the exocyclic double bond is conjugated with the carbonyl group.

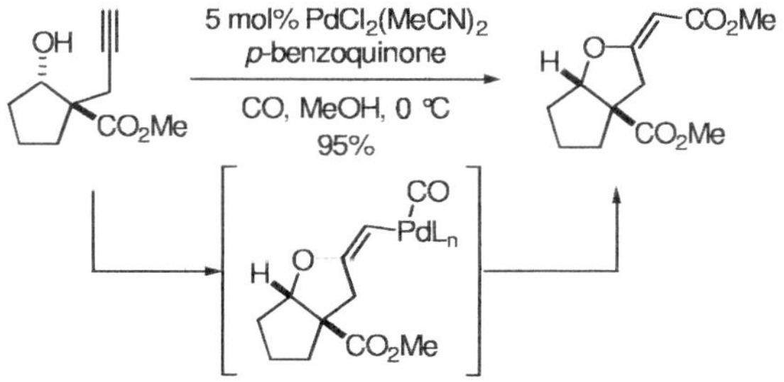

Scheme 7

The rich reactivity of enol ethers has allowed the development of several cascade reactions based on an initial hydroalkoxylation reaction. The formed enol ether can be activated by the metallic catalyst favouring

277

inter- or intramolecular reaction with a second nucleophile. For example, if this second nucleophile is another alcohol, the global reaction gives rise to functionalized acetals. Some examples of this strategy for the synthesis of acetals from ω-alkynols are shown in Scheme 8. In the first one, an iridium complex is used as catalyst and an initial 5-*exo* hydroalkoxylation reaction is proposed.[18] In the second example, a gold-derived catalyst promotes the initial 5-*endo* hydroalkoxylation reaction and the subsequent reaction of the enol ether formed with ethanol.[19] As shown in the third example of Scheme 8, not only five-membered but also six-membered cyclic acetals may be synthesized by applying this methodology.[20]

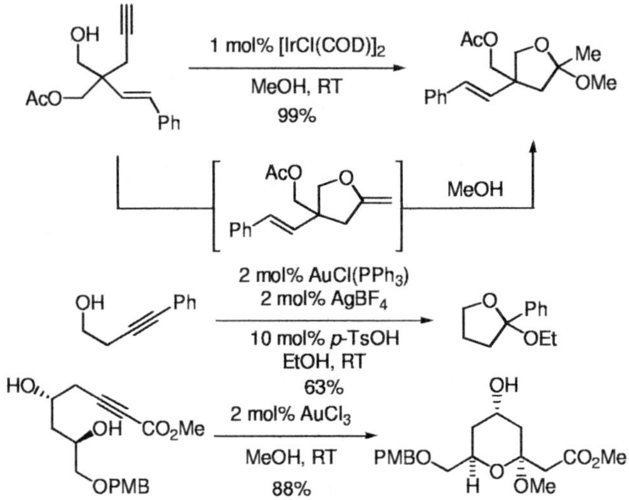

Scheme 8

The intramolecular version of the above commented process supposes the reaction of an alkyndiol derivative to give bicyclic acetals.[21] Several metallic complexes derived from gold(I),[22] platinum(IV),[23] palladium(II)[24] and iridium(I)[25] were shown to be appropriate to accomplish this reaction (Scheme 9).

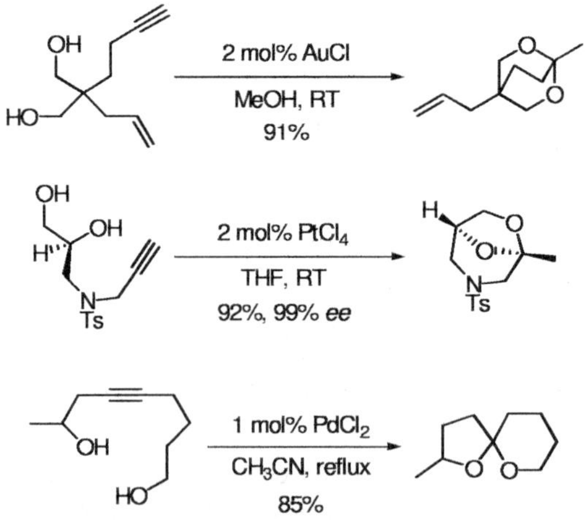

Scheme 9

Particularly interesting is the application of this reaction for the synthesis of spiroacetal derivatives.[21] The intramolecular double hydroalkoxylation reaction of an alkyne has been used in the key step of the total synthesis of several natural products.[26]

In the previous examples, two hydroxyl groups attack the triple bond to form acetal derivatives. However, the nature of the second nucleophile attacking the enol ether intermediate may be changed and for example an electron-rich aromatic ring may be incorporated (Scheme 10).[27] The global sequence may be seen as a cascade hydroalkoxylation/hydroarylation reaction of a carbon-carbon triple bond. Both gold- and platinum-derived catalysts have been shown to be the most active for this kind of cyclization processes. As shown in Scheme 10, the final hydroarylation reaction can be performed in an inter- or intramolecular fashion to give cyclic or bicyclic core structures, respectively. Interestingly, the intramolecular version of this process has been used in the key step of the total synthesis of *bruguierol A*.[28]

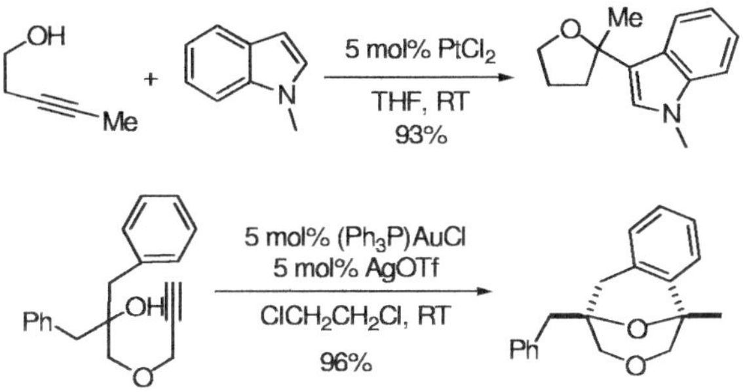

Scheme 10

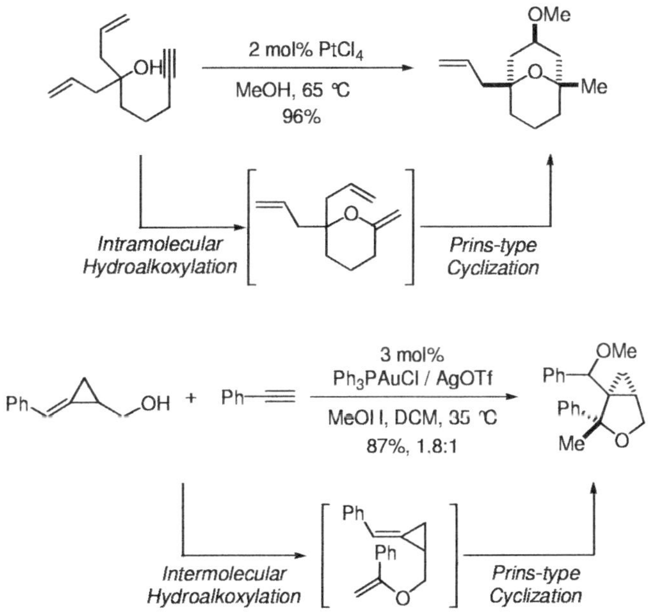

Scheme 11

A highly efficient method for the diastereoselective synthesis of 9-oxabicyclo[3.3.1]nonane derivatives from easily available alkynol derivatives has been recently reported (Scheme 11; equation 1).[29] The reaction is based on a cascade process that involves an intramolecular hydroalkoxylation of a triple bond followed by a Prins-type cyclization. Although several metallic complexes have been found to catalyze this transformation, those derived from gold and platinum have been found to be the most efficient ones. A similar process involving an intermolecular hydroalkoxylation reaction of an alkyne followed by a Prins-type cyclization to give furan derivatives has been developed (Scheme 11; equation 2).[30]

3.1.2. Intramolecular addition to alkenes

The palladium(II)-catalyzed oxidative transformations of alkenol derivatives have evolved into a useful methodology in organic synthesis. The intramolecular version of the Wacker reaction provides cyclic ethers (Scheme 12).[31] This reaction proceeds through an initial nucleophilic attack of the hydroxyl group to the activated carbon-carbon double bond to afford an alkyl-palladium(II) intermediate that further evolves by a β-elimination process. Asymmetric versions of this reaction by the use of chiral non racemic palladium complexes as catalysts have been published.[32] It should be noted that recent studies reveal that these reactions do not proceed through a Wacker-type mechanism and a *syn*-oxypalladation reaction of the carbon-carbon double bond seems more likely.[33]

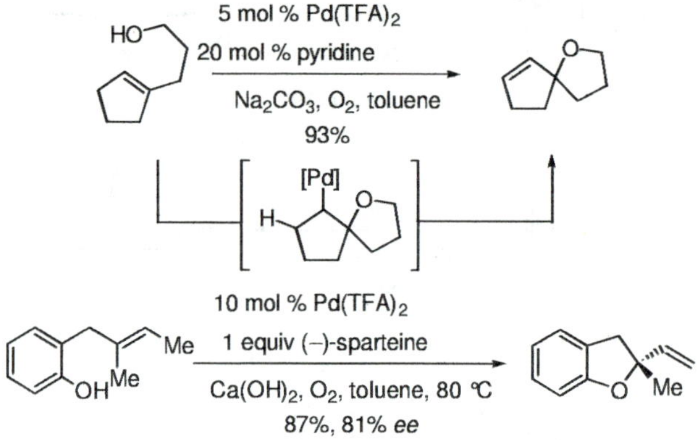

Scheme 12

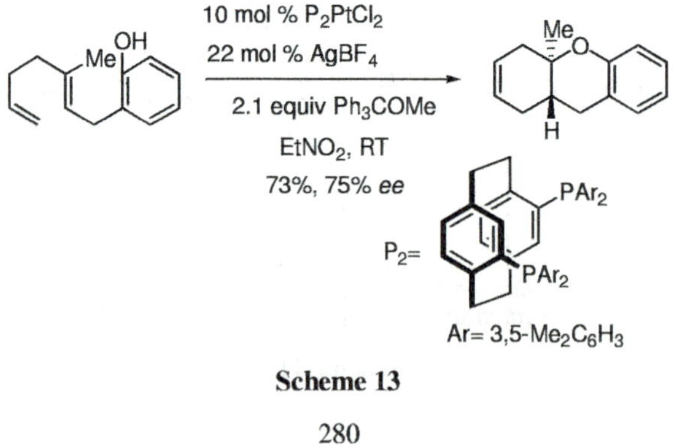

Scheme 13

The elegant asymmetric oxidative cation/olefin cyclization reaction of polyenes catalyzed by chiral platinum complexes developed by M. R. Gagné and co-workers should be remarked (Scheme 13).[34] Interestingly, in this oxidative cyclization, the catalyst turnover is achieved by using a trityl cation to abstract a hydride from a platinum hydride complex intermediate.

A diastereoselective synthesis of tetrahydrofurans *via* the palladium-catalyzed reaction of aryl bromides with γ-hydroxy alkenes has been reported (Scheme 14).[35] This carboetherification reaction is a powerful tool for the construction of these heterocycles as it is convergent and may allow access to a variety of analogous from a single alkenol starting material. By contrast to the well established palladium(II)-catalyzed carbonylative cyclization of alkenol derivatives, this reaction do not proceed through a Wacker-type mechanism and an intramolecular insertion of the olefin into a Pd(Ar)(OR) intermediate is invoked.[36]

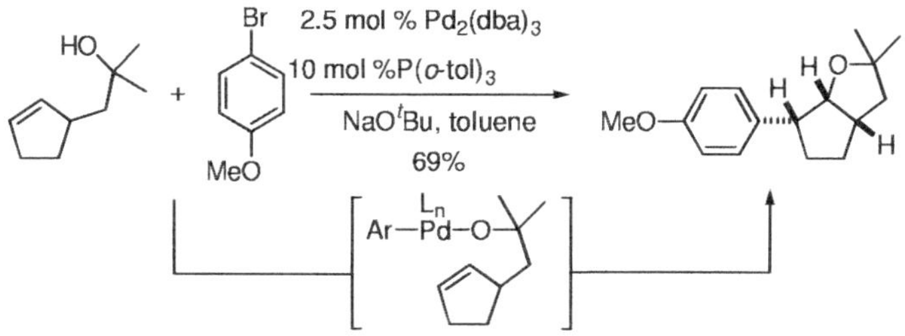

Scheme 14

The simple intramolecular hydroalkoxylation reaction of alkenes to give the corresponding saturated cyclic ethers has been achieved by using metallic complexes derived from mercury,[37] ruthenium,[38] tin,[39] platinum,[40] gold,[41] silver[42] and rhodium.[43] Representative examples are shown in Scheme 15.

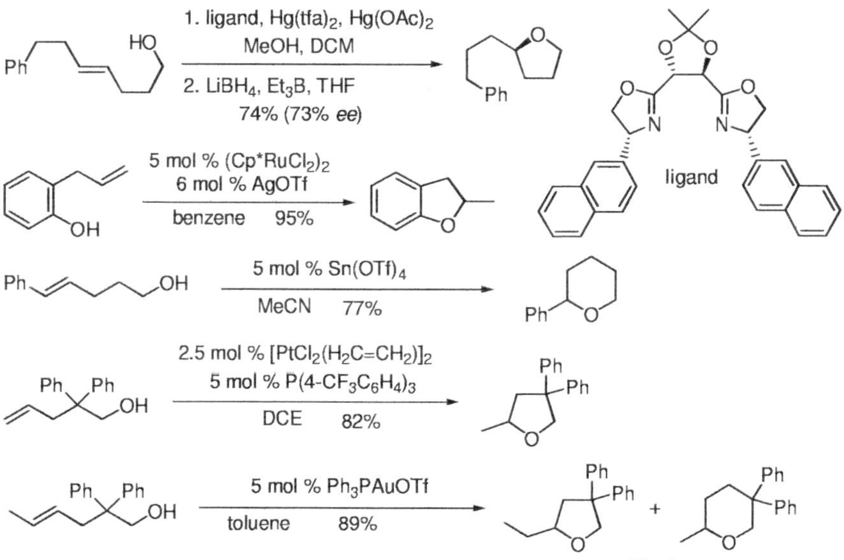

Scheme 15

281

In this context, the enantioselective catalytic mercuriocyclization reaction of γ-hydroxy-*cis*-alkenes employing $Hg(OAc)_2$ in the presence of chiral bisoxazoline-mercury complexes should be remarked (see the first reaction in Scheme 15).[37a] It should be noted that J. F. Hartwig and co-workers have demonstrated that simple Brønsted acids catalyze the addition of alcohols to olefins and so one must consider carefully whether a hydroalkoxylation of a carbon-carbon double bond is a true metal-catalyzed process or if the metallic compound simply generates a protic acid.[44]

3.1.3. Intramolecular addition to allenes

The cyclization of hydroxyallenes is a particularly useful reaction for the stereoselective synthesis of cyclic ethers since it normally occurs with efficient axis-to-centre chirality transfer. Besides palladium complexes,[45] more recently, other metallic compounds derived from silver,[46] gold[47] and platinum[48] have been found to efficiently catalyze this type of cyclization. Especially interesting is the recent development of new asymmetric intramolecular hydroalkoxylation reactions of non-chiral allenes by the use of chiral non-racemic catalysts (Scheme 16).[49]

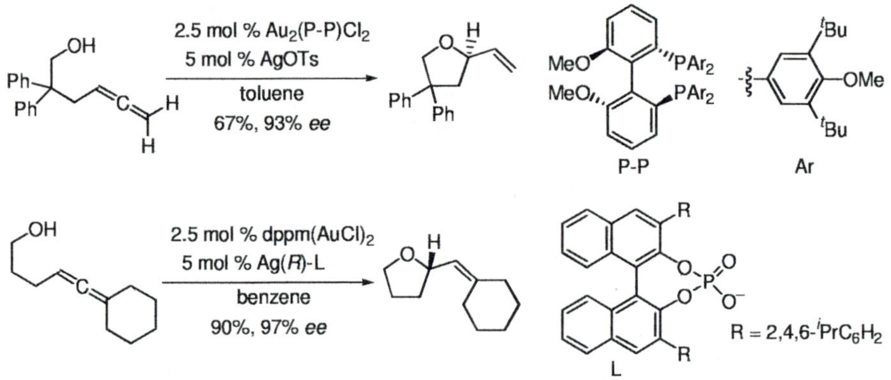

Scheme 16

3.2. Cyclization of carboxylic acids

3.2.1. Intramolecular addition to alkynes

Five- and six-membered lactones are readily available by the corresponding *endo-* or *exo*-cyclization of the corresponding alkynoic acids in the presence of metallic complexes derived from mercury,[50] palladium,[51] rhodium,[52] ruthenium,[53] silver,[54] gold[55] and catalytic systems of palladium-copper[56] or molybdenum-nickel.[57] For representative examples, see Scheme 17.

An interesting dichotomy between gold(I) and gold(III) catalyst in the cyclization of alkynoic acids has been recently discovered.[58] So, the gold(I)-catalyzed cyclization of γ- and δ-acetylenic acids provided the expected γ- and δ-alkylidene lactones. Interestingly, whatever the substitution at the terminal acetylenic end, only the *exo-dig* product was formed in the presence of gold(I) chloride and potassium carbonate (Scheme 18). In sharp contrast, the treatment of γ- and δ-acetylenic acids with gold(III) chloride and potassium carbonate offered a direct access to dimeric methylene lactones as a single regio- and stereoisomer (Scheme 18). The mechanism of this interesting dimerization process is unclear.

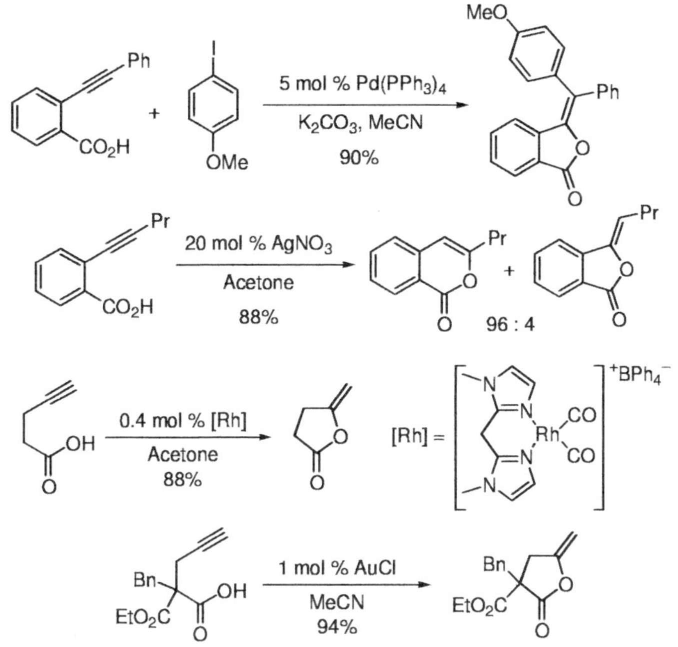

Scheme 17

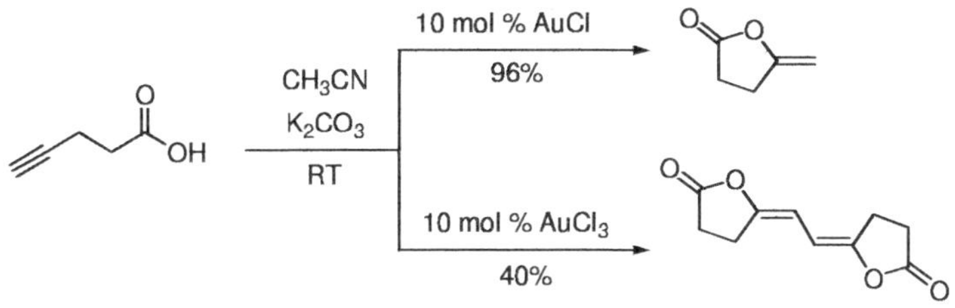

Scheme 18

3.2.2. Intramolecular addition to alkenes

Palladium complexes have been used to catalyze intramolecular addition reaction of carboxylic acids to olefins.[59] These reactions seem to proceed through an initial *anti*-oxypalladiation followed by the subsequent β-hydride elimination reaction.

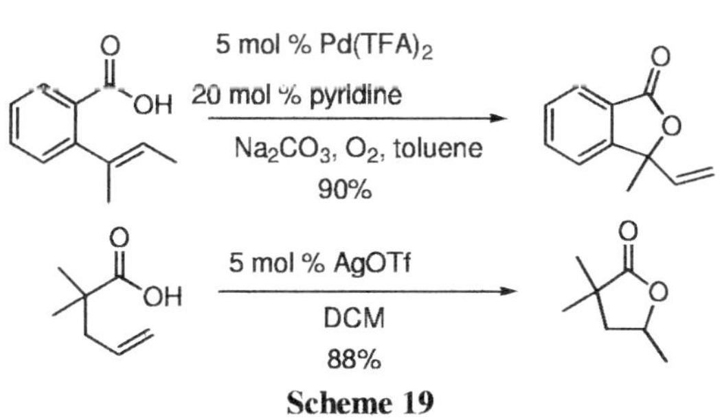

Scheme 19

So, lactones containing a carbon-carbon double bond in their structure are usually obtained (Scheme 19; first equation). However, the typical hydroalkoxylation reaction of the alkene to give the corresponding saturated lactone may be achieved by using other catalysts such as silver(I) triflate (Scheme 19; second equation).[42] In this example, the possibility of a Brønsted-acid catalyzed process instead of a metallic-catalyzed one should not be ruled out.[44]

3.2.3. Intramolecular addition to allenes

Gold(III) chloride catalyzes the cyclization of allene-substituted esters to give unsaturated lactone derivatives (Scheme 20; first equation).[60] When *tert*-butyl esters are used, the reaction is proposed to proceed through an initial cleavage of the *tert*-butyl group followed by gold-catalyzed cyclization of the *in situ* generated allenoic acid. The asymmetric intramolecular hydrocarboxylation reaction of allenoic acids promoted by a gold(I)-derived catalyst has been recently achieved (Scheme 20; second equation).[49b] As shown in the Scheme, the first example formally supposes the 5-*endo* cyclization of the starting material in such a way that the carboxylic group attacks the "terminal" carbon of the allene moiety. However, in the second example, a 5-*exo* cyclization occurs and the carboxylic group attacks the "internal" carbon of the allene.

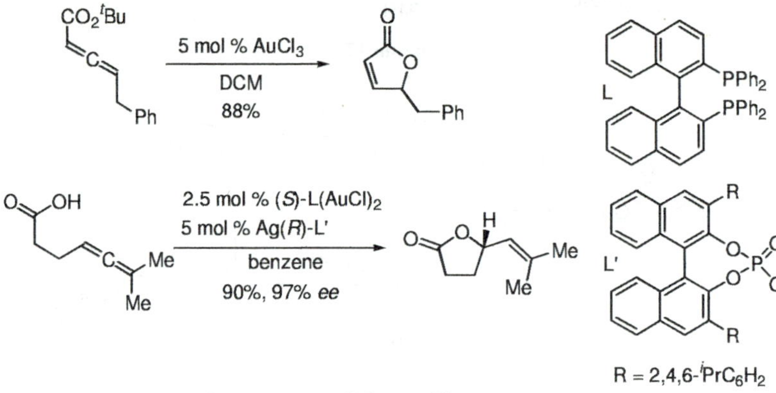

Scheme 20

3.3. Cyclization of amides, carbonates and carbamates

Amides, carbonates and carbamates containing in their structure an alkyne at appropriate disposition react through electrophilic cyclization processes by nucleophilic addition of the oxygen atom of the carbonyl group to the carbon-carbon triple bond. Thus, for example it has been found that *N*-propargylamides react in the presence of 5 mol% of $AuCl_3$ and 5 mol% of $AgSbF_6$ to give the corresponding 5-alkylidene-1,3-oxazolidine (Scheme 21).[61]

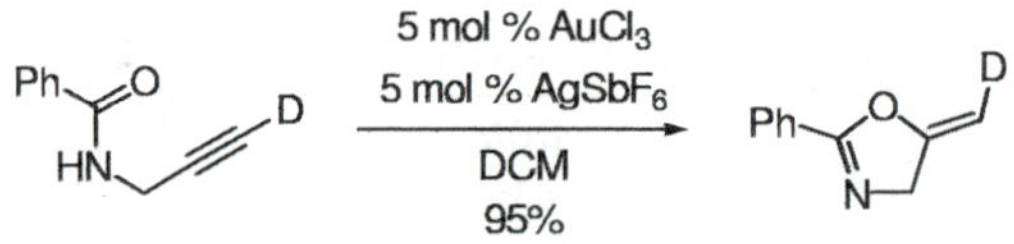

Scheme 21

As shown, the intramolecular 5-*exo* addition of the oxygen atom to the alkyne is stereoselective. Globally, this transformation may be considered as an intramolecular hydroalkoxylation reaction. However, it should be noted that, if possible, amides may also react through hydroamination processes (see Section 4.2.).

Gold[62] and mercury[63] complexes have been shown to be powerful catalysts for the cyclization of alkynyl *tert*-butylcarbonates under mild conditions. By this strategy, both 1,3-dioxan-2-ones and 1,3-dioxolan-2-ones are easily synthesized. In a similar way, *N*-Boc-protected alkynylamines are converted into the corresponding cyclic carbamate derivatives by reaction with catalytic amounts of gold complexes.[64] Both, 5-*exo* and 5-*endo* cyclization processes have been reported. It should be noted that, in all the above commented reactions, a molecule of isobutylene is released. Representative examples of all these processes are shown in Scheme 22.

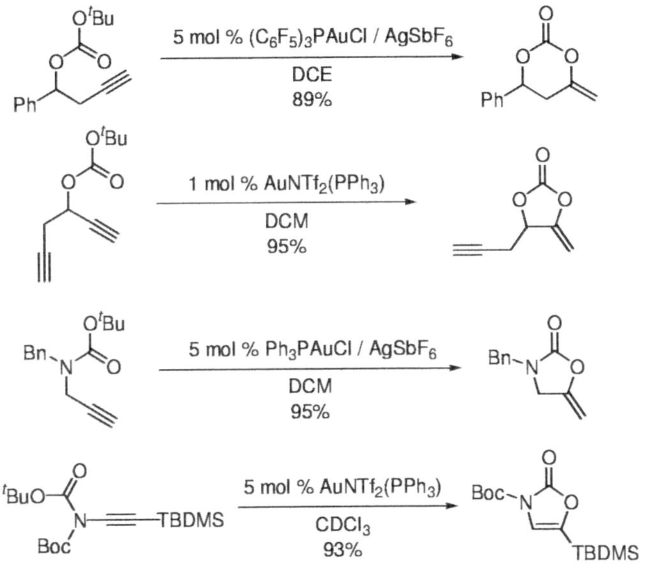

Scheme 22

3.4. Cyclization of aldehydes and ketones

3.4.1. Intramolecular addition to alkynes

The gold-catalyzed reaction of *o*-alkynylbezaldehydes and trifluoroacetic acid supposes a very appropriate strategy to access to isochromenylium salts (Scheme 23).[65] The reaction proceeds by addition of oxygen of the aldehyde to the carbon-carbon triple activated bond by means of the gold catalyst to give a gold-benzopyrilium intermediate. Finally, a protodemetallation reaction gives rise to the isochromenylium salt.

When the above commented reaction is performed in the presence of an olefin or alkyne instead of triflic acid, the metal-isochromenylium intermediate evolves through a cycloaddition reaction to finally give different carbocyclic compounds.[66] These reactions are far beyond the scope of this review as non-heterocyclic compounds are the final products of these processes. However, if the reaction is carried out in the presence of nucleophiles such as alcohols or terminal alkynes, the metal-isochromenylium intermediate may be intercepted and enol acetals or isochromene derivatives are obtained, respectively. Thus, it has been reported that the silver-catalyzed reaction of *o*-alkynylbezaldehydes in methanol leads to five- or six-membered enol acetals (*exo*- or *endo*-addition) depending on the nature of the catalyst used (Scheme 24).[67] Similar results were obtained by using copper(I)[68] and palladium(II)[69] catalysts.

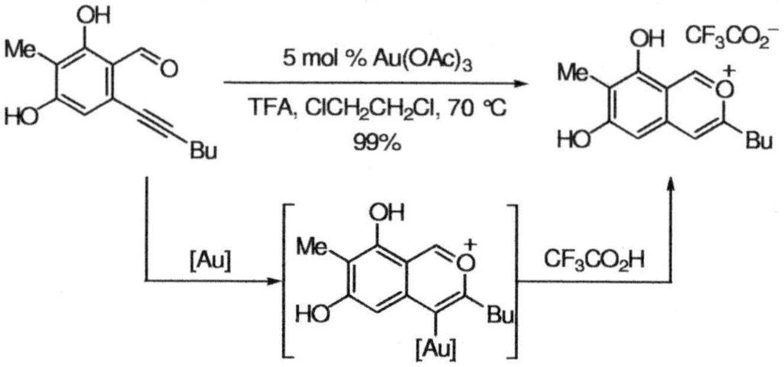

Scheme 23

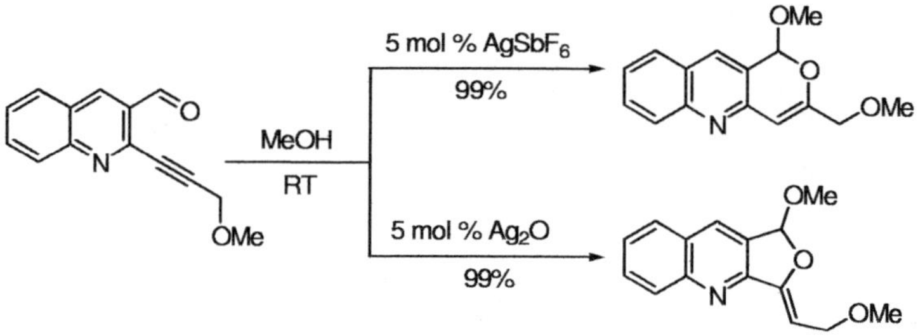

Scheme 24

As previously stated, this reaction may also be carried out in the presence of a terminal alkyne. This alkyne acts as a nucleophile trapping the metal-isochromenylium intermediate. As shown in the example of Scheme 25, the gold(I)-catalyzed reaction of an *o*-alkynylbenzaldehyde and phenylacetylene leads to the formation of a 1-alkynylisochromene derivative in high yield.[70]

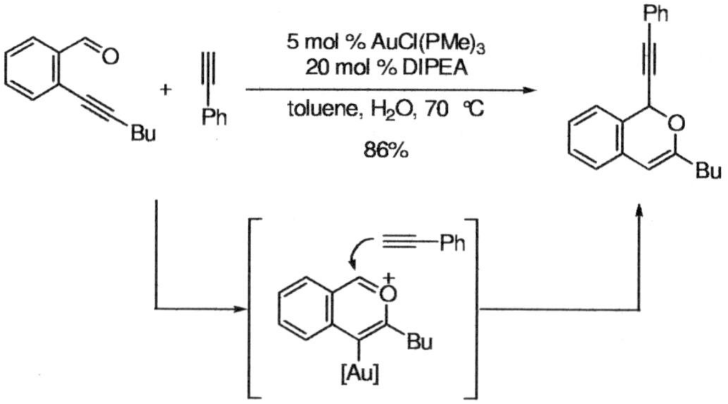

Scheme 25

Interesting versions of the above commented reactions have been developed with α,β-unsaturated ketones instead of aldehydes (Scheme 26).[71] These reactions proceed through a cascade sequence involving an initial cycloisomerization reaction to form the corresponding metal-isochromenylium intermediate followed by a Michael-type addition of an alcohol. As shown in Scheme 26, this reaction has been used for the synthesis of functionalized isochromene and furan derivatives.

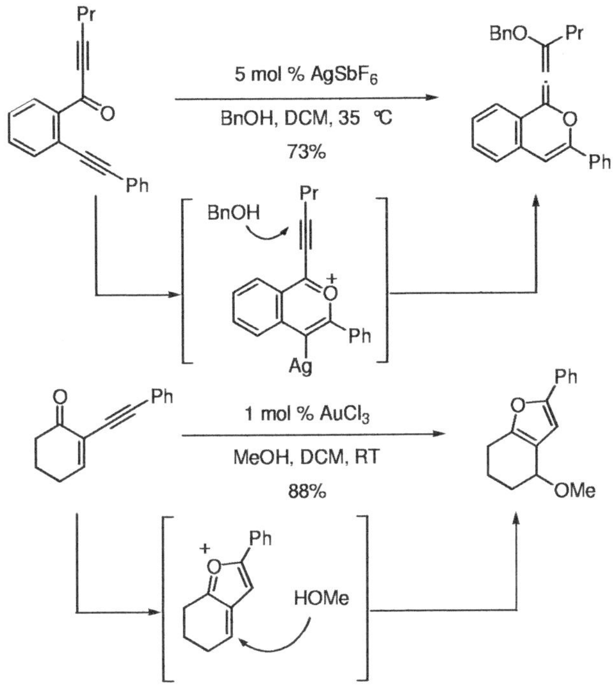

Scheme 26

A mild and efficient method for the synthesis of multiply substituted furans from 1-(1-alkynyl)-cyclopropyl ketones has been developed.[72] This gold-catalyzed process involves an initial cycloisomerization reaction to form the corresponding gold-isochromenylium intermediate that further reacts with an alcohol or an indole to finally form the furan derivative through a cyclopropane ring opening process (Scheme 27). Particularly remarkable is the structural complexity of the heterocyclic compounds obtained from the reactions performed with indoles.

The reaction of 2-hydroxy-2-alkynylcarbonyl compounds with catalytic amounts of platinum or gold complexes affords 3(2*H*)-furanone derivatives through a tandem reaction involving an initial cyclization reaction followed by a 1,2-alkyl shift (first equation in Scheme 28).[73] Moreover, acetylenic dicarbonyl compounds have been cyclized by metallic catalysts.[74] In the example shown in the second equation of Scheme 28, 4-propargyl-1,3-cyclohexanedione is treated with 5 mol% of palladium diacetate to give a 7-oxabicyclo[4.3.0]nonane derivative in high yield. Other metallic complexes derived from platinum, tungsten and ruthenium have been shown to efficiently catalyze the process. However, the regiochemistry of the addition depends on the catalyst used and also on the structure of the starting diketone derivative.[74a]

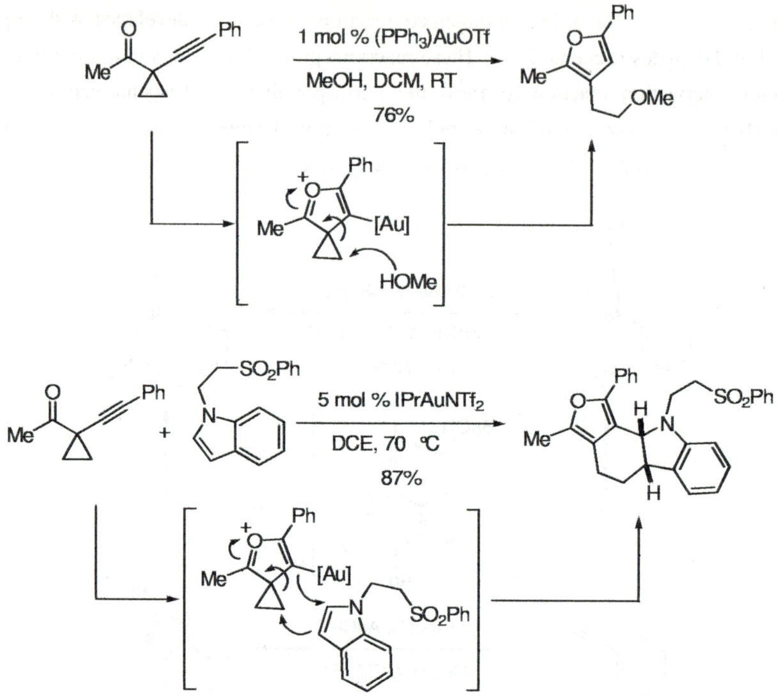

Scheme 27

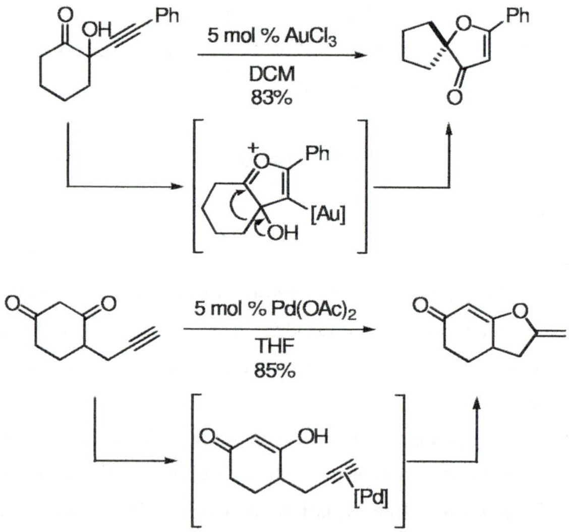

Scheme 28

3.4.2. Intramolecular addition to allenes

Transition-metal catalyzed cycloisomerization of allenyl ketones is an efficient approach for the assembly of the furan ring.[75] These reactions have been shown to proceed with catalysts derived from silver, gold, palladium and copper. The process can be seen as a cascade sequence involving a formal cycloisomerization reaction followed by a [1,2]-shift of a hydrogen atom, an alkyl group, a sulphur-group or a halogen. An interesting example of this kind of reactions is shown in Scheme 29. By using a gold(I) catalyst, the reaction delivers a 2-bromofuran derivative. However, by using the same starting material but a gold(III) catalyst a 3-bromofuran derivative is obtained.[75j] DFT computational and experimental studies reveal that both gold(I) and gold(III) catalysts activate the distal double bond of the allene to produce cyclic zwitterionic intermediates which undergo a kinetically favoured 1,2-Br migration. However, in the case of $(PEt_3)AuCl$ catalyst, the counterion-assisted H-shift is the major process indicating that the regioselectivity of the gold-catalyzed 1,2-H *vs* 1,2-Br migration is ligand dependent.[75n]

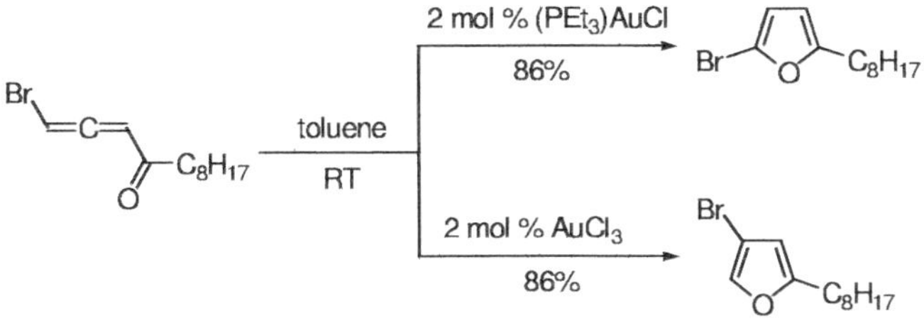

Scheme 29

3.5. Cyclization of ethers and acetals. Intramolecular addition to alkynes

The oxygen atom of an ether or acetal can also intramolecularly interact with an electrophilically activated alkyne. In general, these processes lead to the initial formation of an enoloxonium intermediate that evolves by [1,3]-shift of an appropriate migrating group. This shift is favoured when the migrating group can stabilize a positive charge. Thus, allyl cations (first equation of Scheme 30) and oxonium cations (second equation in Scheme 30) easily migrate in this kind of reactions. Platinum-derived catalysts have been usually employed for these cyclizations.[76]

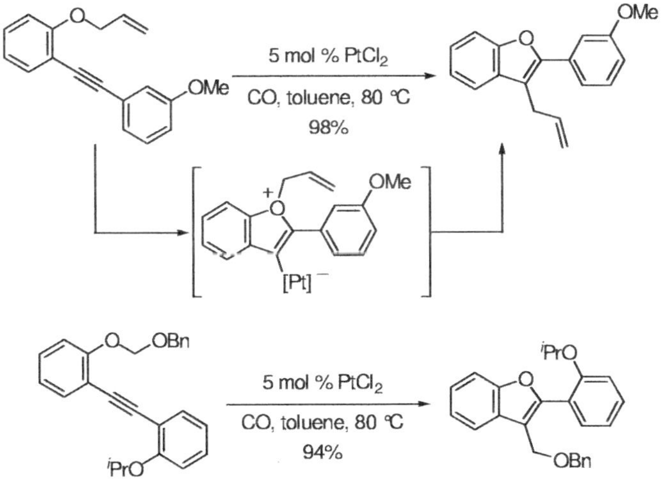

Scheme 30
289

4. Nitrogen-centred nucleophiles

4.1. Cyclization of amines

The catalytic addition of an organic amine N-H bond to alkenes and alkynes (hydroamination) to give nitrogen-containing molecules is of great interest to both academia and industrial researchers. Group 4[77] and especially lanthanide[78] metal complexes are highly efficient catalysts for the hydroamination of various carbon-carbon bonds. However, in general, the mechanism proposed for these processes can not strictly be considered as electrophilic cyclizations. For this reason, in the following sections we will focus on those processes where the metal complex activates the unsaturated carbon-carbon bond favouring the subsequent nucleophilic attack of the nitrogen atom.

4.1.1. Intramolecular addition to alkynes

Simple alkynylamines react in the presence of catalytic amounts of several metallic complexes to afford the corresponding cyclic imine derivatives. Thus, palladium,[79] copper,[80] silver,[81] gold,[82] zinc,[83] rhenium,[84] rhodium[85] and iridium[86] complexes have been used in this context. Two representative examples are shown in Scheme 31. The first palladium-catalyzed reaction implies an initial 5-*endo* hydroamination reaction followed by an isomerization of the enamine formed to the corresponding imine. The second example is a gold-catalyzed process initiated by a 6-*exo* hydroamination reaction and then an enamine to imine isomerization occurs.

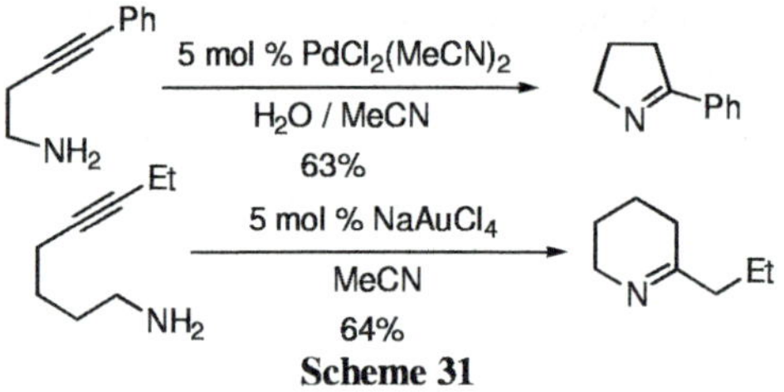

Ph   5 mol % PdCl$_2$(MeCN)$_2$ H$_2$O / MeCN   63%   N–Ph

NH$_2$ Et

5 mol % NaAuCl$_4$ MeCN   64%   N–CH$_2$Et

NH$_2$

Scheme 31

Indoles are easily synthesized from 2-(1-alkynyl)anilines by reaction with catalytic amounts of palladium,[87] copper,[88] gold,[89] platinum,[90] rhodium,[91] iridium,[92] molybdenum[93] and indium complexes.[94] A few examples of this kind of hydroamination reactions are shown in Scheme 32.

All reactions above commented may be considered as intramolecular hydroamination reactions. All the starting materials (amines or amides) contain an N-H function. However, tertiary amines or amides can also participate in electrophilic cyclization processes. For example, tertiary amides derived from 2-alkynyl-anilines provide an easy access to functionalized indoles.[95] In particular, the platinum-catalyzed reaction with acetamides furnishes 3-acylindole derivatives through a cyclization/[1,3]-acyl shift sequence (first equation in Scheme 33).[95a] However, a different reaction is observed when lactams derived from 2-alkynyl-anilines are used as starting material (second equation in Scheme 33).[95b] In this case, the reaction proceeds through an initial aminometallation reaction to furnish a metal-containing ammonium ylide that evolves by successive 1,2-acyl and 1,2-methyl migrations to finally give the corresponding pyrido[1,2-*a*]indol-6-one derivative. Both PtCl$_2$ and PtCl$_4$ have shown to be active catalysts in this process. The authors also observed that the reaction could be improved in terms of yield and reaction time by performing the process under an atmosphere of oxygen.

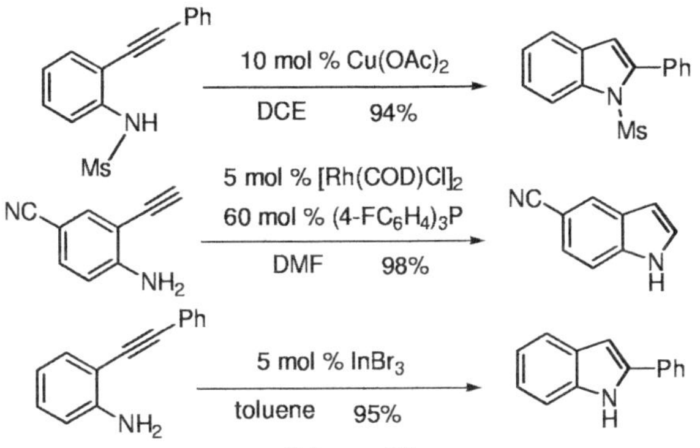

Scheme 32

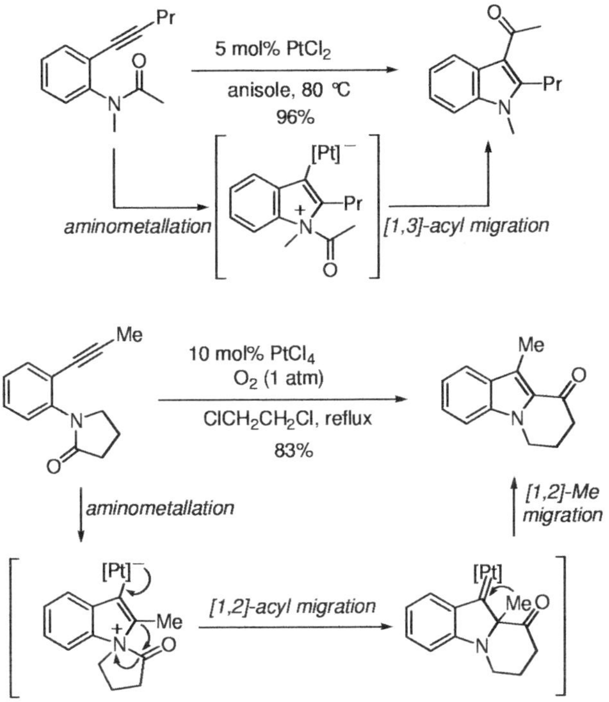

Scheme 33

An interesting strategy for the synthesis of N-fused pyrroloheterocycles from propargyl N-containing heterocycles has been developed.[96] The reaction involves a cyclization/1,2-migration sequence catalyzed by gold, copper or platinum complexes. A representative example where a platinum catalyst is used is shown in Scheme 34.[96b]

The intramolecular hydroamination reaction of carbon-carbon triple bonds catalyzed by metallic complexes have also been applied for the synthesis of other nitrogen-containing heterocycles such as pyrazoles [166],[97] benzimidazoles [167],[98] 2,5-dihydroisoxazoles [168],[99] 1,2-dihydroisoquinolines [169].[100]

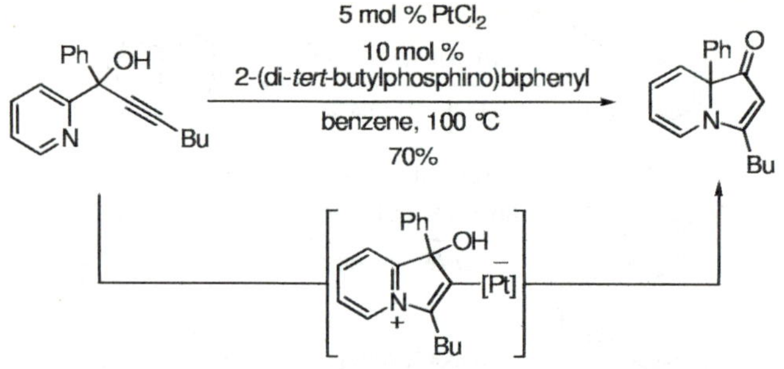

Scheme 34

4.1.2. Intramolecular addition to alkenes

The catalytic hydroamination of inactivated alkenes remains a challenge in organic synthesis. However, some recent advances in this field should be highlighted. In the context of palladium chemistry, the major drawback of the reaction is the formation of oxidative amination products rather than the desired hydroamination compounds. This problem has been wisely solved by the use of tridentate ligands on palladium that effectively inhibit the β-hydride elimination (Scheme 35).[101]

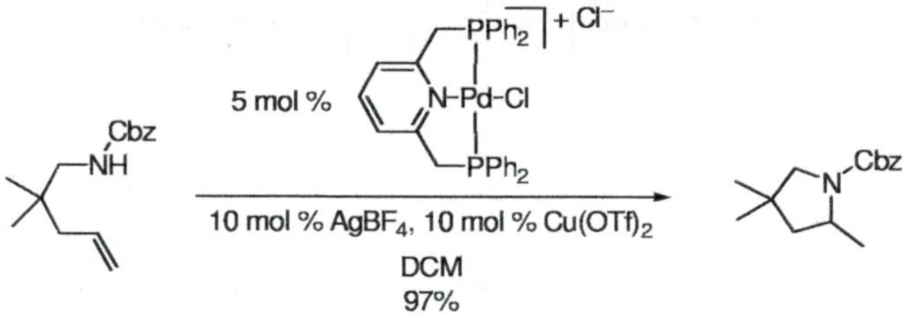

Scheme 35

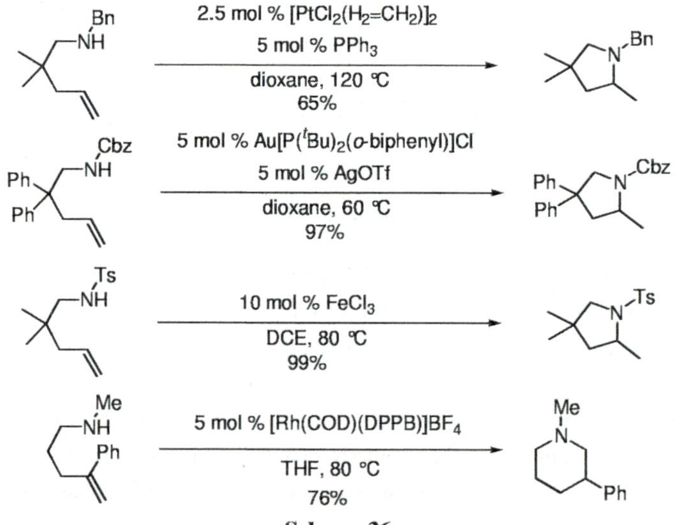

Scheme 36

Platinum[102] and gold[103] complexes have also shown to be appropriate catalysts to perform the hydroamination of unactivated alkenes (first and second equations of Scheme 36). These reactions are particularly interesting for the synthesis of pyrrolidine and piperidine derivatives through 5-*exo* and 6-*exo* hydroamination reactions, respectively. The discovery of an important catalytic activity by the use of simple iron salts should be remarked (third equation of Scheme 36).[104] Particularly unusual is the rhodium-catalyzed intramolecular, *anti*-Markovnikov hydroamination of vinylarenes shown in the last equation of Scheme 36. This process occurs through a formal 6-*endo* hydroamination proccess.[105]

4.1.3. Intramolecular addition to allenes

A series of nitrogen-containing heterocycles can be constructed using palladium,[106] gold[107] and silver[108] catalysts by annulation of allenes with pendant amines. Remarkable are the recent advances in the asymmetric hydroamination reaction of allenes achieved by using a chiral ligand on the gold complex (first example in Scheme 37)[109] or a chiral counterion (second example in Scheme 37).[49b]

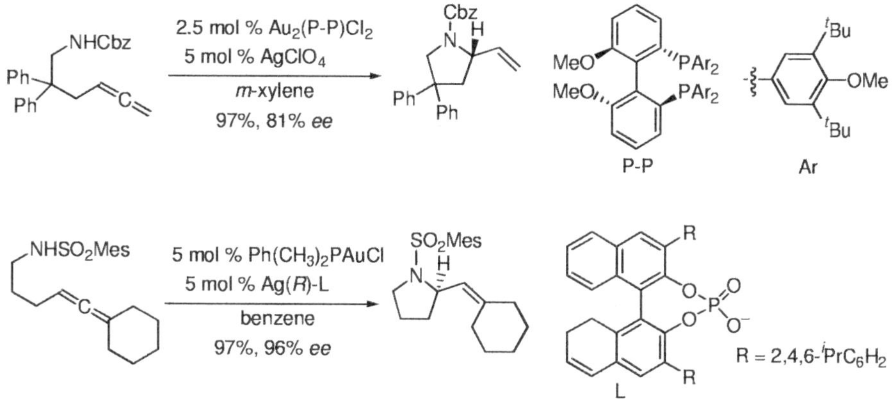

Scheme 37

4.2. Cyclization of amides, carbamates and trichloroacetimidates

Amide and carbamate functional groups contain two nucleophilic positions: the oxygen and the nitrogen atoms. So, when an electrophilic cyclization of products containing such functional groups is performed, two possible compounds may be obtained depending on which atom acts as nucleophile. In Section 3.3., we have shown several examples of reactions where amides and carbamates react through electrophilic cyclization processes where the oxygen atom was the nucleophilic counterpart. In the next sections, we will focus on reactions where the nitrogen atom of amides, carbamates and trichloroacetimidates acts as nucleophile.

4.2.1. Intramolecular addition to alkynes

Isoindolinone and isoquinolinone derivatives are easily available by palladium catalyzed cyclization of 2-(1-alkynyl)arenecarboxamides.[110] In a similar way, bicyclic lactams have been prepared.[111] On treatment with catalytic amounts of gold(I) chloride and a base, *O*-propargyl carbamates smoothly undergo cyclization to afford 4-methylene-2-oxazolidinone derivatives.[112] The intramolecular hydroamination reaction of trichloroacetimidates derived from propargyl and homopropargyl alcohols under gold catalyst conditions has also been described.[113] Representative examples of all these reactions are shown in Scheme 38.

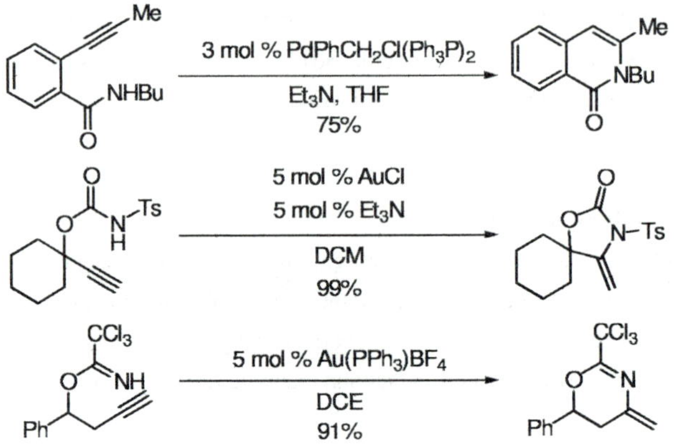

Scheme 38

4.2.2. Intramolecular addition to allenes

Allene containing lactams and oxazolidinone derivatives have been reacted in the presence of palladium[114] or gold[115] catalysts to give the corresponding bicyclic systems (Scheme 39). In the first example of Scheme 39 an *in situ* reduction of Pd(II) to Pd(0) is proposed. Further reaction of these species with allyl bromide leading to the formation of a π-allylpalladium complex is proposed. This palladium(II) π-complex activates the allene moiety favouring the intramolecular attack of the nitrogen atom to form a vinyl(allyl)palladium(II) intermediate. A final reductive elimination delivers the final product and regenerates Pd(0). The second example of the Scheme 39 can be considered a typical hydroamination reaction.

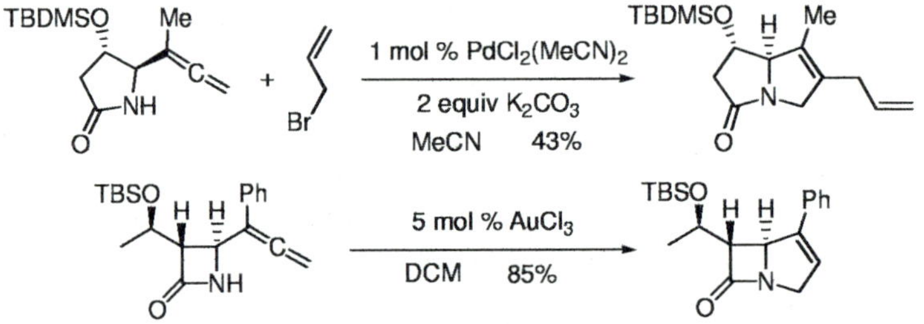

Scheme 39

4.3. Cyclization of imines

Imines derived from 2-(1-alkynyl)anilines react with alcohols and catalytic amounts of copper(I) chloride to give indole derivatives.[116] The same imines have been used to obtain pyrrolo[1,2-*a*]indole derivatives by reaction with enol ethers and catalytic amounts of gold trichloride (Scheme 40).[117]

A wide variety of isoquinoline derivatives have been prepared from *tert*-butyl imines derived from 2-(1-alkynyl)benzaldehydes under copper and silver catalyzed conditions.[118] 3,4-Disubstituted isoquinoline derivatives have been obtained by palladium catalyzed cyclization/cross-coupling reactions.[119] Representative examples of these reactions are shown in Scheme 41.

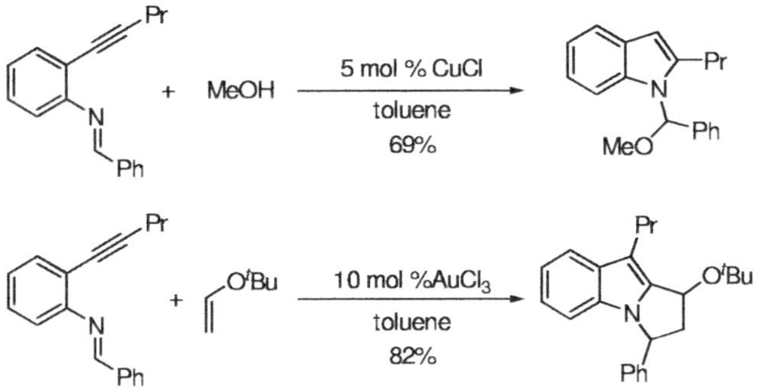

Scheme 40

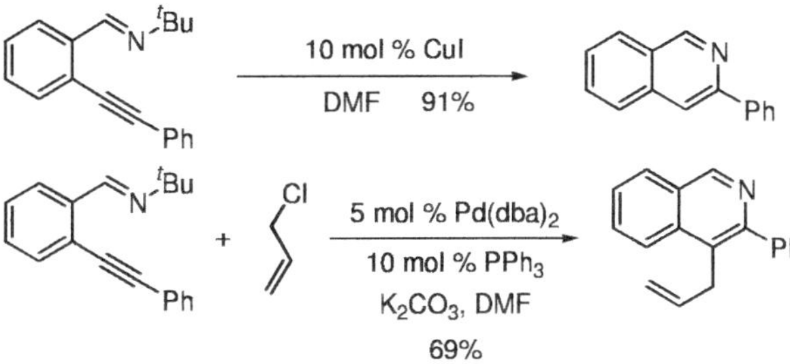

Scheme 41

5. Cascade reactions initiated by a metal-catalyzed electrophilic cyclization

The development of catalytic cascade electrophilic cyclization reactions for the synthesis of policyclic compounds is highly attractive as molecules with complex architectures are easily formed from simple starting materials. In these reactions, the metallic catalyst should promote two or more consecutive cyclizations. For example, a one-pot multicatalytic and multicomponent cascade reaction that allows the synthesis of spirofuranquinoline derivatives from alkynol derivatives, aldehydes and aromatic amines by the cooperative effect of two catalysts– a platinum complex and a Brønsted acid– has been developed (Scheme 42).[120]

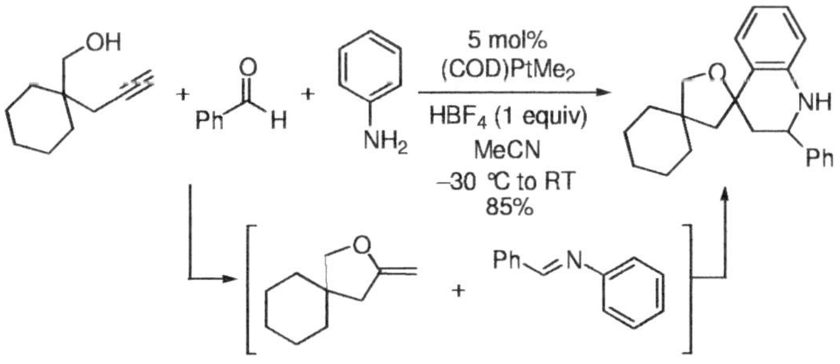

Scheme 42

This reaction may be considered a Povarov reaction between two reagents, an exocyclic enol ether and an imine, catalytically formed *in situ*. Moreover, this reaction can be formally considered as the first Povarov reaction performed with exocyclic enol ethers. A related platinum-catalyzed one-pot three-component coupling reaction of alkynol derivatives and imines for the diastereoselective synthesis of functionalized furo[3,2-*c*]quinoline derivatives has been reported.[121]

A related version of the above commented reaction has been applied for the diastereoselective synthesis of chroman spiroacetals (Scheme 43).[122] The reaction supposes a palladium(II)-catalyzed one pot three-component coupling reaction between an alkynol derivative, a salicylaldehyde and an amine or an orthoester. As shown, depending on the nature of the third component of the reaction (an amine or an orthoester) nitrogen- or oxygen-substituted spiroacetals can be synthesized.

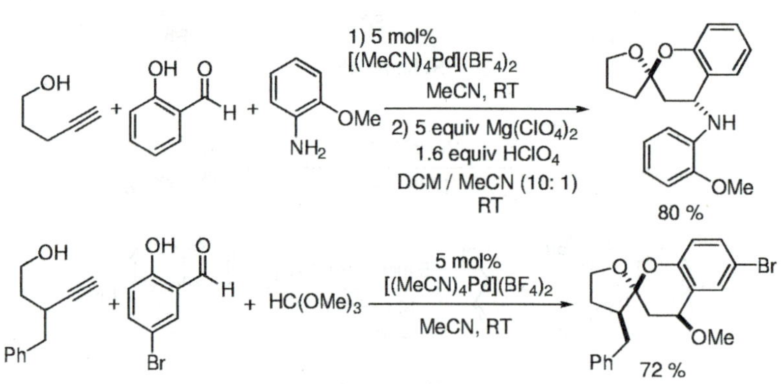

Scheme 43

An interesting synthesis of pyrazine derivatives by means of a cascade reaction based on an initial intramolecular addition of a carboxylic acid to an alkyne has been recently developed by D. J. Dixon and co-workers (Scheme 44).[123] Thus, for instance, 5-hexynoic acid reacts with 1-(aminoethyl)pyrrole in the presence of catalytic amounts of an *in situ* formed cationic gold(I) complex to give 11a-methyl-9,10,11,11a-tetrahydro-5*H*-pyrido[1,2-*a*]pyrrolo[2,1-*c*]pyrazin-8(6*H*)-one. The cascade sequence involves the 6-*exo* cyclization of the carboxylic acid to form a methylenelactone derivative which reacts with the amine to give an acyliminium intermediate. Further Friedel-Crafts cyclization leads to the final heterocycle.

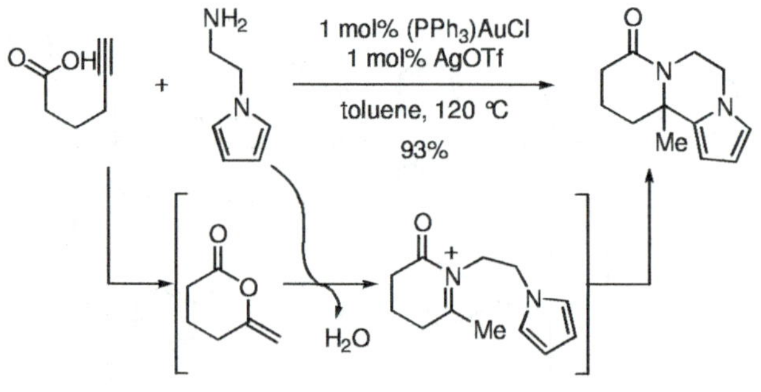

Scheme 44

296

An interesting protocol for the synthesis of pyrrolo[1,2-*a*]quinolines has been recently reported (Scheme 45).[124] This reaction is performed with *N*-aryl-4-alkynamines and alkynes in water and in the presence of a gold catalyst and features the formation of two new carbon-carbon bonds and one carbon-nitrogen bond to make two rings in a one-pot process. The reaction pathway may be understood as a cascade process initiated by an intramolecular hydroamination reaction of the 4-alkynamine followed by addition of the alkyne and a final hydroarylation step. The method has been applied in gram-scale synthesis of substituted pyrrolo[1,2-*a*]quinolines and some of these compounds have exhibited a significant cytotoxicity against the cervical epithelioid carcinoma cell line suggesting a potential application of this class of compounds as anticancer drugs.

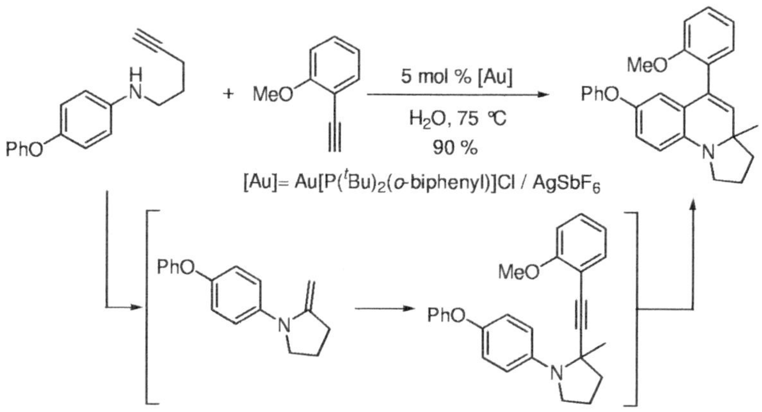

Scheme 45

An efficient platinum(II)-catalyzed cascade reaction for the synthesis of indoline and quinoline derivatives from simple secondary alkynylamines and 1,3-diketones has been recently developed (Scheme 46).[125] The reaction is believed to proceed through the condensation of an enamine derivative, *in situ* formed by hydroamination of the alkynylamine and the 1,3-diketone.

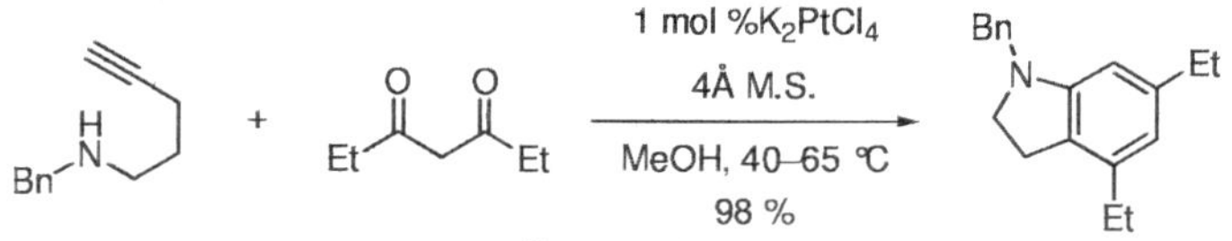

Scheme 46

6. Conclusions

As shown along this review, recent years have witnessed remarkable growth in the number of reactions directed to the synthesis of heterocyclic systems based on metal-catalyzed electrophilic cyclizations. In this review, we have focused on the most recent representative reactions reported in the literature for the synthesis of oxygen-and nitrogen-containing heterocycles. We hope that this compilation has provided an appropriate background for the present topic and exciting developments in the field are expected in the future.

Acknowledgments

Financial support from the Spanish Ministerio de Educación y Ciencia (CTQ2007-61048/BQU) and Principado de Asturias (IB08-088) is gratefully acknowledged. We are also indebt to all our collaborators and in particular to Professor José Barluenga for his invaluable assistance.

References
1. Bougault, M. J. *C. R. Acad. Sci.* **1904**, *139*, 864.
2. (a) Dowle, M. D.; Davies, D. I. *Chem. Soc. Rev.* **1979**, 171. (b) Barlett, P. A. *Asymmetric Synthesis*; Morrison, J. D., Ed.; Academic Press: San Diego, 1984; Vol. 3. Chap. 6. (b) Cardillo, G.; Orena, M. *Tetrahedron* **1990**, *46*, 3321. (c) Harding, K. E.; Tiner, T. H. *Comprehensive Organic Synthesis*; Trost, B. M., Ed.; Pergamon Press: New York, 1991; Vol. 4, p. 363. (d) Robin, S.; Rousseau, G. *Tetrahedron* **1998**, *54*, 13681. (e) Ranganathan, S.; Muraleedharan, K. M.; Vaish, N. K.; Jayaraman, N. *Tetrahedron* **2004**, *60*, 5273. (f) Larock, R. C. *Acetylene Chemistry: Chemistry, Biology, and Material Science*; Diederich, F.; Stang, P. J.; Tykwinski, R. R., Eds.; Wiley-VCH: New York, 2005; p. 51. (g) French, A. N.; Bissmire, S.; Wirth, T. *Chem. Soc. Rev.* **2004**, *33*, 354.
3. (a) Kirsch, S. F. *Synthesis* **2008**, 3183. (b) Patil, N. T.; Yamamoto, Y. *Chem. Rev.* **2008**, *108*, 3395. (c) Shen, H. C. *Tetrahedron* **2008**, *64*, 3885. (d) Zenit, G.; Larock, R. C. *Chem. Rev.* **2006**, *106*, 4644.
4. Fürstner, A.; Davies, P. W. *Angew. Chem. Int. Ed.* **2007**, *46*, 3410.
5. Gorin, D. J.; Toste, F. D. *Nature* **2007**, *446*, 395.
6. Fürstner, A.; Davies, P. W. *J. Am. Chem. Soc.* **2005**, *127*, 15024.
7. Li, X.; Chianese, A. R.; Vogel, T.; Crabtree, R. *Org. Lett.* **2005**, *7*, 5437.
8. Trost, B. M.; Rhee, Y. H. *J. Am. Chem. Soc.* **2003**, *125*, 7482.
9. Trost, B. M.; Rhee, Y. H. *J. Am. Chem. Soc.* **2002**, *124*, 2528.
10. (a) Hashmi, A. S. K.; Schwarz, L.; Choi, J.-H.; Frost, T. M. *Angew. Chem. Int. Ed.* **2000**, *39*, 2285. (b) Liu, Y.; Song, F.; Song, Z.; Liu, M.; Yan, B. *Org. Lett.* **2005**, *7*, 5409.
11. (a) Gabriele, B.; Salerno, G.; Costa, M. *Synlett* **2004**, 2468. (b) Cadierno, V.; Díez, J.; García-Alvarez, J.; Gimeno, J.; Nebra, N.; Rubio-García, J. *Dalton Trans.* **2006**, 5593.
12. Pale, P.; Chuche, J. *Eur. J. Org. Chem.* **2000**, 1019.
13. (a) Elgafi, S.; Field, L. D.; Messerle, B. A. *J. Organomet. Chem.* **2000**, *607*, 97. (b) Díaz-Alvarez, A. E.; Crochet, P.; Zablocka, M.; Duhayon, C.; Cadierno, V.; Gimeno, J.; Majoral, J.-P. *Adv. Synth. Cat.* **2006**, *348*, 1671.
14. (a) Çetinkaya, B.; Özdemir, I.; Bruneau, C.; Dixneuf, P. H. *Eur. J. Inorg. Chem.* **2000**, 29. (b) Albers, J.; Cadierno, V.; Crochet, P.; García-Garrido, S. E.; Gimeno, J. *J. Organomet. Chem.* **2007**, *692*, 5234.
15. (a) Pale, P.; Chuche, J. *Eur. J. Org. Chem.* **2000**, 1019. (b) Riediker, M.; Schwartz, J. *J. Am. Chem. Soc.* **1982**, *104*, 5842. (c) Gabriele, B.; Salerno, G.; Fazio, A.; Pittelli, R. *Tetrahedron* **2003**, *59*, 6251.
16. Barluenga, J.; Diéguez, A.; Rodríguez, F.; Fañanás, F. J.; Sordo, T.; Campomanes, P. *Chem. Eur. J.* **2005**, *11*, 5735.
17. (a) Kato, K.; Nishimura, A.; Yamamoto, Y.; Akita, H. *Tetrahedron Lett.* **2001**, *42*, 4203. (b) Kato, K.; Tanaka, M.; Yamamoto, Y.; Akita, H. *Tetrahedron Lett.* **2002**, *43*, 1511. See also: (c) Gabriele, B.; Salerno, G.; de Pascali, F.; Costa, M.; Chiusoli, G. P. *J. Organomet. Chem.* **2000**, *593–594*, 409.
18. Genin, E.; Antoniotti, S.; Michelet, V.; Gênet, J.-P. *Angew. Chem. Int. Ed.* **2005**, *44*, 4949.
19. Belting, V.; Krause, N. *Org. Lett.* **2006**, *8*, 4489.
20. Diéguez-Vázquez, A.; Tzschucke, C. C.; Crecente-Campo, J.; McGrath S.; Ley, S. V. *Eur. J. Org. Chem.* **2009**, 1698.
21. Utimoto, K. *Pure Appl. Chem.* **1983**, *55*, 1845.
22. (a) Antoniotti, S.; Genin, E.; Michelet, V.; Gênet, J.-P. *J. Am. Chem. Soc.* **2005**, *127*, 9976. (b) Liu, B.; De Brabander, J. K. *Org. Lett.* **2006**, *8*, 4907.
23. Diéguez-Vázquez, A.; Tzschucke, C. C.; Lam, W. Y.; Ley, S. V. *Angew. Chem. Int. Ed.* **2008**, *47*, 209. See also reference 22b.
24. Ramana, C. V.; Mallik, R.; Gonnade, R. G. *Tetrahedron* **2007**, *64*, 219. See also reference 21.
25. Messerle, B. A.; Vuong, K. Q. *Organometallics* **2007**, *26*, 3031.

26. (a) Trost, B. M.; Home, D. B.; Woltering, M. J. *Angew. Chem. Int. Ed.* **2003**, *42*, 5987. (b) Li, Y.; Zhou, F.; Forsyth, C. J. *Angew. Chem. Int. Ed.* **2007**, *46*, 279. (c) Trost, B. M.; Weiss, A. H. *Angew. Chem. Int. Ed.* **2007**, *46*, 7664. (d) Ramana, C. V.; Induvadana, B. *Tetrahedron Lett.* **2009**, *50*, 271. (e) Ramana, C. V.; Suryawanshi, S. B.; Gonnade, R. G. *J. Org. Chem.* **2009**, *74*, 2842.

27. (a) Ferrer, C.; Amijs, C. H. M.; Echavarren, A. M. *Chem. Eur. J.* **2007**, *13*, 1358. (b) Bhuvaneswari, S.; Jeganmohan, M.; Cheng, C.-H. *Chem. Eur. J.* **2007**, *13*, 8285. (c) Barluenga, J.; Fernández, A.; Satrústegui, A.; Diéguez, A.; Rodríguez, F.; Fañanás, F. J. *Chem. Eur. J.* **2008**, *14*, 4153.

28. Fañanás, F. J.; Fernández, A.; Çevic, D.; Rodríguez, F. *J. Org. Chem.* **2009**, *74*, 932.

29. Barluenga, J.; Diéguez, A.; Fernández, A.; Rodríguez, F.; Fañanás, F. J. *Angew. Chem. Int. Ed.* **2006**, *45*, 2091.

30. Tian, G.-Q.; Shi, M. *Org. Lett.* **2007**, *9*, 4917.

31. Hegedus, L. S. *Organometallics in Synthesis*; Schlosser, M., Ed.; John Wiley and Sons: Chichester, 1994; p. 383.

32. (a) Arai, M. A.; Kuraishi, M.; Arai, T.; Sasai, H. *J. Am. Chem. Soc.* **2001**, *123*, 2907. (b) Trend, R. M.; Ramtohul, Y. K.; Ferreira, E. M.; Stoltz, B. M. *Angew. Chem. Int. Ed.* **2003**, *42*, 2892.

33. Trend, R. M.; Ramtohul, Y. K.; Stolz, B. M. *J. Am. Chem. Soc.* **2005**, *127*, 17778.

34. Mullen, C. A.; Campbell, A. N.; Gagné, M. R. *Angew. Chem. Int. Ed.* **2008**, *47*, 6011.

35. Wolfe, J. P.; Rossi, M. A. *J. Am. Chem. Soc.* **2004**, *126*, 1620.

36. Wolfe, J. P. *Eur. J. Org. Chem.* **2007**, 571.

37. (a) Kang, S. H.; Kim, M.; Kang, S. Y. *Angew. Chem. Int. Ed.* **2004**, *43*, 6177. (b) Kang, S. H.; Kim, M. *J. Am. Chem. Soc.* **2003**, *125*, 4684.

38. Hori, K.; Kitagawa, H.; Miyoshi, A.; Ohta, T.; Furukawa, I. *Chem. Lett.* **1998**, 1083.

39. Coulombel, L.; Favier, I.; Duñach, E. *Chem. Commun.* **2005**, 2286.

40. Qian, H.; Han, X.; Widenhoefer, R. A. *J. Am. Chem. Soc.* **2004**, *126*, 9536.

41. Yang, C.-G.; He, C. *J. Am. Chem. Soc.* **2005**, *127*, 6966.

42. Yang, C.-Q.; Reich, N. W.; Shi, Z.; He, C. *Org. Lett.* **2005**, *7*, 4553.

43. Zhao, P.; Incarvito, C. P.; Hartwig, J. F. *J. Am. Chem. Soc.* **2006**, *128*, 9642.

44. (a) Rosenfeld, D. C.; Shekhar, S.; Takemiya, A.; Utsunomiya, M.; Hartwig, J. F. *Org. Lett.* **2006**, *8*, 4179. See also: (b) Kelly, B. D.; Allen, J. M.; Tundel, R. E.; Lambert, T. H. *Org. Lett.* **2009**, *11*, 1381.

45. (a) Ma, S.; Gao, W. *Tetrahedron Lett.* **2000**, *41*, 8933. (b) Ma, S.; Gao, W. *J. Org. Chem.* **2002**, *67*, 6104.

46. (a) Olsson, L. I.; Claesson, A. *Synthesis* **1979**, 793. (b) Marshall, J. A.; Wang, X.-J. *J. Org. Chem.* **1991**, *56*, 4913. (c) Marshall, J. A.; Pinney, K. G. *J. Org. Chem.* **1993**, *58*, 7180. (d) Aurrecoechea, J. M.; Solay, M. *Tetrahedron* **1998**, *54*, 3851. (e) VanBrunt, M. P.; Standaert, R. F. *Org. Lett.* **2000**, *2*, 705. (f) Xu, D.; Li, Z.; Ma, S. *Chem. Eur. J.* **2002**, *8*, 5012. (f) Lepage, O.; Kattnig, E.; Fürstner, A. *J. Am. Chem. Soc.* **2004**, *126*, 15970.

47. (a) Hoffmann-Röder, A.; Krause, N. *Org. Lett.* **2001**, *3*, 2537. (b) Krause, N.; Hoffmann-Röder, A.; Canisius, J. *Synthesis* **2002**, 1759. (c) Gockel, B.; Krause, N. *Org. Lett.* **2006**, *8*, 4485. (d) Deutsch, C.; Gockel, B.; Hoffmann-Röder, A.; Krause, N. *Synlett* **2007**, 1790. (e) Volz, F.; Wadman, S. H.; Hoffmann-Röder, A.; Krause, N. *Tetrahedron* **2008**, *65*, 1902. (f) Poonoth, M.; Krause, N. *Adv. Synth. Catal.* **2009**, *351*, 117.

48. Kong, W.; Cui, J.; Yu, Y.; Chen, G.; Fu, C.; Ma, S. *Org. Lett.* **2009**, *11*, 1213.

49. (a) Zhang, Z.; Widenhoefer, R. *Angew. Chem. Int. Ed.* **2007**, *46*, 283. (b) Hamilton, G. L.; Kang, E. J.; Mba, M.; Toste, F. D. *Science* **2007**, *317*, 496.

50. See for example: (a) Amos, R. A.; Katzenellenbogen, J. A. *J. Org. Chem.* **1978**, *43*, 560. (b) Yamamoto, M. *J. Chem. Soc., Perkin Trans. 1* **1981**, 582. (c) Jellal, A.; Grimaldi, J.; Santelli, M. *Tetrahedron Lett.* **1984**, *25*, 3179.

51. See for example: (a) Tsuda, T.; Ohashi, Y.; Nagahama, N.; Sumiya, R.; Saegusa, T. *J. Org. Chem.* **1988**, *53*, 2650. (b) Wang, Z.; Lu, X. *J. Org. Chem.* **1996**, *61*, 2254. (c) Arcadi, A.; Burini, A.; Cacchi, S.; Delmastro, M.; Marinelli, F.; Pietroni, B. R. *J. Org. Chem.* **1992**, *57*, 976. (d) Bouyssi, D.; Goré, J.; Balme, G. *Tetrahedron Lett.* **1992**, *33*, 2811. (e) Rossi, R.; Bellini, F.; Biagetti, M.; Catanese, A.; Mannina, L. *Tetrahedron Lett.* **2000**, *41*, 5281.

52. Elgafi, S.; Field, L. D.; Messerle, B. A. *J. Organomet. Chem.* **2000**, *607*, 97.

53. Jiménez-Tenorio, M.; Puerta, M. C.; Valerga, P.; Moreno-Dorado, F. J.; Guerra, F. M.; Massanet, G. M. *Chem. Commun.* **2001**, 2324.

54. (a) Ogawa, Y.; Maruno, M.; Wakamatsu, T. *Synlett* **1995**, 871. (b) Dalla, U.; Pale, P. *New J. Chem.* **1999**, *23*, 803. (c) Bellina, F.; Ciucci, D.; Vergamini, P.; Rossi, R. *Tetrahedron* **2000**, *56*, 2533.

55. (a) Genin, E.; Toullec, P. Y.; Antoniotti, S.; Brancour, C.; Genêt, J.-P.; Michelet, V. *J. Am. Chem. Soc.* **2006**, *128*, 3112. (b) Marchal, E.; Uriac, P.; Legouin, B.; Toupet, L.; van de Waghe, P. *Tetrahedron* **2007**, *63*, 9979. (c) Genin, E.; Toullec, P. Y.; Marie, P.; Antoniotti, S.; Brancour, C.; Genêt, J.-P.; Michelet, V. *Arkivoc* **2007**, *v*, 67.

56. Subramanian, V.; Batchu, V. R.; Barange, D.; Pal, M. *J. Org. Chem.* **2005**, *70*, 4778.

57. Takei, I.; Wakebe, Y.; Suzuki, K.; Enta, Y.; Suzuki, T.; Mizobe, Y.; Hidai, M. *Organometallics* **2003**, *22*, 4639.

58. Harkat, H.; Dembelé, A. Y.; Weibel, J.-M.; Blanc, A.; Pale, P. *Tetrahedron* **2009**, *65*, 1871.

59. See for example references 32b and 33.

60. (a) Kang, J.-E.; Lee, E.-S.; Park, S.-I; Shin, S. *Tetrahedron Lett.* **2005**, *46*, 7431. (b) Piera, J.; Krumlinde, P.; Strübing, D.; Bäckvall, J.-E. *Org. Lett.* **2007**, *9*, 2235.

61. Hashmi, A. S. K.; Weyrauch, J. P.; Frey, W.; Bats, J. W. *Org. Lett.* **2004**, *6*, 4391.

62. (a) Kang, J.-E.; Shin, S. *Synlett* **2006**, 717. (b) Buzas, A.; Gagosz, F. *Org. Lett.* **2006**, *8*, 515.

63. Yamamoto, H.; Nishiyama, M.; Imagawa, H.; Nishizawa, M. *Tetrahedron Lett.* **2006**, *47*, 8369.

64. (a) Robles-Martín, R.; Adrio, J.; Carretero, J. C. *J. Org. Chem.* **2006**, *71*, 5023. (b) Buzas, A.; Gagosz, F. *Synlett* **2006**, 2727. (c) Hashmi, A. S. K.; Salathé, R.; Frey, W. *Synlett* **2007**, 1763.

65. (a) Zhu, J.; Germain, R.; Porco, J. A., Jr *Angew. Chem. Int. Ed.* **2004**, *43*, 1239. (b) Beeler, A. B.; Su, S.; Singleton, C. A.; Porco, J. A., Jr *J. Am. Chem. Soc.* **2007**, *129*, 1413.

66. For a review, see: Patil, N. T.; Yamamoto, Y. *Arkivoc* **2007**, *v*, 6.

67. Godet, T.; Vaxelaire, C.; Michel, C.; Milet, A.; Belmont, P. *Chem. Eur. J.* **2007**, *13*, 5632.

68. Patil, N. T.; Yamamoto, Y. *J. Org. Chem.* **2004**, *69*, 5139.

69. Asao, N.; Nogami, T.; Takahashi, K.; Yamamoto, Y. *J. Am. Chem. Soc.* **2002**, *124*, 764.

70. Yao, X.; Li, C.-J. *Org. Lett.* **2006**, *8*, 1953.

71. (a) Yao, T.; Zhang, X.; Larock, R. *J. Am. Chem. Soc.* **2004**, *126*, 11164. (b) Patil, N. T.; Wu, H.; Yamamoto, Y. *J. Org. Chem.* **2005**, *70*, 4531. (c) Yao, T.; Zhang, X.; Larock, R. *J. Org. Chem.* **2005**, *70*, 7679. (d) Patil, N. T.; Pahadi, N. K.; Yamamoto, Y. *J. Org. Chem.* **2005**, *70*, 10096. (e) Xiao, Y.; Zhang, J. *Angew. Chem. Int. Ed.* **2008**, *47*, 1903.

72. (a) Zhang, J.; Schmalz, H.-G. *Angew. Chem. Int. Ed.* **2006**, *45*, 6704. (b) Zhang, G.; Huang, X.; Li, G.; Zhang, L. *J. Am. Chem. Soc.* **2008**, *130*, 1814.

73. (a) Kirsch, S. F.; Binder, J. T.; Liébert, C.; Menz, H. *Angew. Chem. Int. Ed.* **2006**, *45*, 5878. (b) Crone, B.; Kirsch, S. F. *J. Org. Chem.* **2007**, *72*, 5435.

74. (a) Gulías, M.; Rodríguez, J. R.; Castedo, L.; Mascareñas, J. L. *Org. Lett.* **2003**, *5*, 1975. For palladium catalyzed reactions, see: (b) Arcadi, A.; Cacchi, S.; Larock, R. C.; Marinelli, F. *Tetrahedron Lett.* **1993**, *34*, 2831. (c) Arcadi, A.; Rossi, E. *Tetrahedron Lett.* **1996**, *37*, 6811. (d) Cacchi, S.; Fabrizi, G.; Moro, L. *J. Org. Chem.* **1997**, *62*, 5327. (e) Kato, K.; Yamamoto, Y.; Akita, H. *Tetrahedron Lett.* **2002**, *43*, 4915. (f) Arcadi, A.; Cacchi, S.; Fabrizi, G.; Marinelli, F.; Parisi, L. M. *Tetrahedron* **2003**, *59*, 4661.

75. (a) Marshall, J. A.; Robinson, E. D. *J. Org. Chem.* **1990**, *55*, 3450. (b) Marshall, J. A.; Wang, X.-J. *J. Org. Chem.* **1991**, *56*, 960. (c) Marshall, J. A.; Bartley, G. S. *J. Org. Chem.* **1994**, *59*, 7169. (d) Marshall, J. A.; Wallace, E. M. *J. Org. Chem.* **1995**, *60*, 796. (e) Marshall, J. A.; Sehon, C. A. *J. Org. Chem.* **1995**, *60*, 5966. (f) Hashmi, A. S. K. *Angew. Chem. Int. Ed. Engl.* **1995**, *34*, 1581. (g) Hashmi, A. S. K.; Schwarz, L.; Choi, J.-H.; Frost, T. M. *Angew. Chem. Int. Ed.* **2000**, *39*, 2285. (h) Kim, J. T.; Kel'in, A. V.; Gevorgyan, V. *Angew. Chem. Int. Ed.* **2003**, *42*, 98. (i) Sromek, A. W.; Kel'in, A. V.; Gevorgyan, V. *Angew. Chem. Int. Ed.* **2004**, *43*, 2280. (j) Sromek, A. W.; Rubina, M.; Gevorgyan, V. *J. Am. Chem. Soc.* **2005**, *127*, 10500. (k) Zhou, C.-Y.; Chan, P. W. H.; Che, C.-M. *Org. Lett.* **2006**, *8*, 325. (l) Dudnik, A. S.; Gevorgyan, V. *Angew. Chem. Int. Ed.* **2007**, *46*, 5195. (m) Dudnik, A. S.; Sromek, A. W.; Rubina, M.; Kim, J. T.; Kel'in, A. V.; Gevorgyan, V. *J. Am. Chem. Soc.* **2008**, *130*, 1440. (n) Xa, Y.; Dudnik, A. S.; Gevorgyan, V.; Li, Y. *J. Am. Chem. Soc.* **2008**, *130*, 6940.

76. (a) Nakamura, I.; Mizushima, Y.; Yamamoto, Y. *J. Am. Chem. Soc.* **2005**, *127*, 15022. (b) Fürstner, A.; Davies, P. W. *J. Am. Chem. Soc.* **2005**, *127*, 15024. (c) Fürstner, A.; Heilmann, E. K.; Davies, P. W. *Angew. Chem. Int. Ed.* **2007**, *46*, 4760. (d) Nakamura, I.; Chan, C.-S.; Araki, T.; Terada, M.; Yamamoto, Y. *Org. Lett.* **2008**, *10*, 309.

77. Pohlki, F.; Doye, S. *Chem. Soc. Rev.* **2003**, *32*, 104.

78. Hong, S.; Marks, T. J. *Acc. Chem. Res.* **2004**, *37*, 673.

79. (a) Fukuda, Y.; Matsubara, S.; Utimoto, K. *J. Org. Chem.* **1991**, *56*, 5812. (b) Hiemstra, L.; Schoemaker, H. E.; Rutjes, F. P. J. T. *Adv. Synth. Catal.* **2002**, *344*, 70. (c) Jacobi, P. A.; Liu, H. *J. Org. Chem.* **2000**, *65*, 7676. (d) Gabriele, B.; Salerno, G.; Fazio, A. *J. Org. Chem.* **2003**, *68*, 7853. See also: (d) Lutete, L. M.; Kadota, I.; Yamamoto, Y. *J. Am. Chem. Soc.* **2004**, *126*, 1622.

80. Müller, T. E.; Pleier, A.-K. *J. Chem. Soc., Dalton Trans.* **1999**, 583.

81. van Esseveldt, B. C. J.; Vervoort, P. W. H.; van Delft, F. L.; Rutjes, F. P. J. T. *J. Org. Chem.* **2005**, *70*, 1791.

82. Fukuda, Y.; Utimoto, K. *Synthesis* **1991**, 975.

83. Zulys, A.; Dochnahl, M.; Hollmann, D.; Löhnwitz, K.; Herrmann, J.-S.; Roesky, P. W.; Blechert, S. *Angew. Chem. Int. Ed.* **2005**, *44*, 7794.

84. Ouh, L. L.; Müller, T. E.; Yan, Y. K. *J. Organomet. Chem.* **2005**, *690*, 3774.

85. (a) Campi, E. M.; Jackson, W. R. *J. Organomet. Chem.* **1996**, *523*, 205. (b) Field, L. D.; Messerle, B. A.; Vuong, K. Q.; Turner, P. *Organometallics* **2005**, *24*, 4241. (c) Field, L. D.; Messerle, B. A.; Vuong, K. Q.; Turner, P.; Failes, T. *Organometallics* **2007**, *26*, 2058. (d) Burling, S.; Field, L. D.; Messerle, B. A.; Rumble, S. L. *Organometallics* **2007**, *26*, 4335.

86. See references 85b–d.

87. (a) Iritani, K.; Matsubara, S.; Utimoto, K. *Tetrahedron Lett.* **1988**, *29*, 1799. (b) Arcadi, A.; Cacchi, S.; Marinelli, F. *Tetrahedron Lett.* **1989**, *30*, 2581. (c) Cacchi, S.; Carnicelli, V.; Marinelli, F. *J. Organomet. Chem.* **1994**, *475*, 289. (d) Cacchi, S.; Fabrizi, G.; Marinelli, F.; Moro, L.; Pace, P. *Synlett* **1997**, 1363. (e) Battistuzzi, G.; Cacchi, S.; Fabrizi, G. *Eur. J. Org. Chem.* **2002**, 2671. (f) van Esseveldt, B. C. J.; van Delft, F. L.; de Gelder, R.; Rutjes, F. P. J. T. *Org. Lett.* **2003**, *5*, 1717.

88. (a) Hiroya, K.; Itoh, S.; Sakamoto, T. *J. Org. Chem.* **2004**, *69*, 1126. (b) Ohno, H.; Ohta, Y.; Oishi, S.; Fujii, N. *Angew. Chem. Int. Ed.* **2007**, *46*, 2295.

89. See reference 87a and also: Nakamura, I.; Yamagishi, U.; Song, D.; Konta, S.; Yamamoto, Y. *Angew. Chem. Int. Ed.* **2007**, *46*, 2284.

90. Cariou, K.; Ronan, B.; Mignani, S.; Fensterbank, L.; Malacria, M. *Angew. Chem. Int. Ed.* **2007**, *46*, 1881.

91. Trost, B. M.; McClory, A. *Angew. Chem. Int. Ed.* **2007**, *46*, 2074.

92. Li, X.; Chianese, A. R.; Vogel, T.; Crabtree, R. H. *Org. Lett.* **2005**, *7*, 5437.

93. McDonald, F. E.; Chatterjee, A. J. *Tetrahedron Lett.* **1997**, *38*, 7687.

94. Sakai, N.; Annaka, K.; Konakahara, T. *Org. Lett.* **2004**, *6*, 1527. See also reference 89.

95. (a) Shimada, T.; Nakamura, I.; Yamamoto, Y. *J. Am. Chem. Soc.* **2004**, *126*, 10564. (b) Nakamura, I.; Sato, Y.; Konta, S.; Terada, M. *Tetrahdron Lett.* **2009**, *50*, 2075. (c) Li, G.; Huang, X.; Zhang, L. *Angew. Chem. Int. Ed.* **2008**, *47*, 346.

96. (a) Seregin, I. V.; Gevorgyan, V. *J. Am. Chem. Soc.* **2006**, *128*, 12050. (b) Smith, C. R.; Bunnelle, E. M.; Rhodes, A. J.; Sarpong, R. *Org. Lett.* **2007**, *9*, 1169. (c) Yan, B.; Liu, Y. *Org. Lett.* **2007**, *9*, 4323. (d) Yan, B.; Zhou, Y.; Zhang, H.; Chen, J.; Liu, Y. *J. Org. Chem.* **2007**, *72*, 7783.

97. Cacchi, S.; Fabrizi, G.; Carangio, A. *Synlett* **1997**, 959.

98. Sun, Q.; LaVoie, E. J. *Heterocycles* **1996**, *43*, 737.

99. Yeom, H.-S.; Lee, E.-S.; Shin, S. *Synlett* **2007**, 2292.

100. Yanada, R.; Obika, S.; Kono, H.; Takemoto, Y. *Angew. Chem. Int. Ed.* **2006**, *45*, 3822.

101. (a) Michael, F. E.; Cochran, B. M. *J. Am. Chem. Soc.* **2006**, *128*, 4246. (b) Michael, F. E.; Cochran, B. M. *Org. Lett.* **2008**, *10*, 329.

102. Bender, C. F.; Widenhoefer, R. A. *J. Am. Chem. Soc.* **2005**, *127*, 1070.

103. (a) Han, X.; Widenhoefer, R. A. *Angew. Chem. Int. Ed.* **2006**, *45*, 1747. (b) Bender, C. F.; Widenhoefer, R. A. *Org. Lett.* **2006**, *8*, 5303. (c) Zhang, J.; Yang, C.-G.; He, C. *J. Am. Chem. Soc.* **2006**, *128*, 1798. (d) Bender, C. F.; Widenhoefer, R. A. *Chem. Commun.* **2008**, 2741.

104. Komeyama, K.; Morimoto, T.; Takaki, K. *Angew. Chem. Int. Ed.* **2006**, *45*, 2938.

105. Takemiya, A.; Hartwig, J. *J. Am. Chem. Soc.* **2006**, *128*, 6042.

106. For recent examples, see: (a) Ma, S.; Yu, F.; Gao, W. *J. Org. Chem.* **2003**, *68*, 5943. (b) Ohno, H.; Anzai, M.; Toda, A.; Ohishi, S.; Fujii, N.; Tanaka, T.; Takemoto, Y.; Ibuka, T. *J. Org. Chem.* **2001**, *66*, 4904. (c) Kang, S.-K.; Kim, K.-J. *Org. Lett.* **2001**, *3*, 511.

107. (a) Morita, N.; Krause, N. *Eur. J. Org. Chem.* **2006**, 4634. (b) Nishina, N.; Yamamoto, Y. *Angew. Chem. Int. Ed.* **2006**, *45*, 3314. (c) Zhang, Z.; Liu, C.; Kinder, R. E.; Han, X.; Qian, H.; Widenhoefer, R. A. *J. Am. Chem. Soc.* **2006**, *128*, 9066.

108. (a) Dieter, R. K.; Yu, H. *Org. Lett.* **2001**, *3*, 3855. (b) Mitasev, B.; Brummond, K. M. *Synlett* **2006**, 3100.

109. (a) Zhang, Z.; Bender, C. F.; Widenhoefer, R. A. *Org. Lett.* **2007**, *9*, 2887 (b) LaLonde, R. L.; Sherry, B. D.; Kang, E. J.; Toste, F. D. *J. Am. Chem. Soc.* **2007**, *129*, 2452. (c) Zhang, Z.; Bender, C. F.; Widenhoefer, R. A. *J. Am. Chem. Soc.* **2007**, *129*, 14148. See also reference 49b.

110. (a) Nagarajan, A.; Balasubramanian, T. R. *Indian J. Chem.* **1989**, *28B*, 67. (b) Sashida, H.; Kawamukai, A. *Synthesis* **1999**, 1145. (c) Kundu, N. G.; Khan, M. W. *Tetrahedron* **2000**, *56*, 4777.

111. Karsten, W. F. J.; Stol, M.; Rutjes, F. P. J. T.; Kooijman, H.; Spek, A. L.; Hiemstra, H. *J. Organomet. Chem.* **2001**, *624*, 244.

112. (a) Ritter, S.; Horino, Y.; Lex, J.; Schmalz, H.-G. *Synlett* **2006**, 3309. Related cyclizations using copper, silver and palladium catalysts have been described: (b) Tamaru, Y.; Kimura, M.; Tanaka, S.; Kure, S.; Yoshida, Z. *Bull. Chem. Soc. Jpn.* **1994**, *67*, 2383. (c) Ohe, K.; Matsuda, H.; Ishihara, T.; Chatani, N.; Kawasaki, Y.; Murai, S. *J. Org. Chem.* **1991**, *56*, 2267. (d) Lei, A.; Lu, X. *Org. Lett.* **2000**, *2*, 2699.

113. Kang, J.-E.; Kim, H.-B.; Lee, J.-W.; Shin, S. *Org. Lett.* **2006**, *8*, 3537.

114. Karstens, W. F. J.; Klomp, D.; Rutjes, F. P. J. T.; Hiemstra, H. *Tetrahedron* **2001**, *57*, 5123.

115. Lee, P. H.; Kim, H.; Lee, K.; Kim, M.; Noh, K.; Kim, H.; Seomoon, D. *Angew. Chem. Int. Ed.* **2005**, *44*, 1840.

116. Kamijo, S.; Sasaki, Y.; Yamamoto, Y. *Tetrahedron Lett.* **2004**, *45*, 35.

117. Kusama, H.; Miyashita, Y.; Takaya, J.; Iwasawa, N. *Org. Lett.* **2006**, *8*, 289.

118. (a) Roesch, K. R.; Larock, R. C. *Org. Lett.* **1999**, *1*, 553. (b) Zhang, H.; Larock, R. C. *J. Org. Chem.* **2002**, *67*, 7048.

119. (a) Dai, G.; Larock, R. C. *J. Org. Chem.* **2002**, *67*, 7042. (b) Dai, G.; Larock, R. C. *J. Org. Chem.* **2003**, *68*, 920. (c) Huang, Q.; Larock, R. C. *J. Org. Chem.* **2003**, *68*, 980.

120. Barluenga, J.; Mendoza, A.; Rodríguez, F.; Fañanás, F. J. *Angew. Chem. Int. Ed.* **2008**, *47*, 7044.

121. Barluenga, J.; Mendoza, A.; Rodríguez, F.; Fañanás, F. J. *Chem. Eur. J.* **2008**, *14*, 10892.

122. Barluenga, J.; Mendoza, A.; Rodríguez, F.; Fañanás, F. J. *Angew. Chem. Int. Ed.* **2009**, *48*, 1644.

123. Yang, T.; Campbell, L.; Dixon, D. J. *J. Am. Chem. Soc.* **2007**, *129*, 12070.

124. Liu, X.-Y.; Che, C.-M. *Angew. Chem. Int. Ed.* **2008**, *47*, 3805.

125. Li, X.-Y.; Che, C.-M. *Angew. Chem. Int. Ed.* **2009**, *48*, 2367.

STEREOSELECTIVE SYNTHESIS OF OPTICALLY ACTIVE PYRIDYL ALCOHOLS. PART I: PYRIDYL *SEC*-ALCOHOLS

Giorgio Chelucci

Dipartimento di Chimica, Università di Sassari, Via Vienna 2, I-07100 Sassari, Italy

(e-mail: chelucci@uniss.it)

Abstract. *This account deals with the stereoselective synthesis of optically active pyridyl sec-alcohols, through the description of a variety of methods allowing to generate a sec-carbinol chiral centre bound to a pyridine ring.*

Contents

1. Introduction
2. Pyridyl *sec*-alcohols
    2.1. Addition of 2-pyridyllithium derivatives to chiral aldehydes
    2.2. Addition of chiral 2-pyridyllithium derivatives to aldehydes
    2.3. Stereoselective addition of organometallic reagents to pyridyl carboxaldehydes
        2.3.1. Enantioselective addition
        2.3.2. Diasteroselective addition
    2.4. Rearrangement of pyridine *N*-oxides
    2.5. Reduction of pyridyl ketones
        2.5.1. Stoichiometric enantioselective reduction
        2.5.2. Catalytic enantioselective reduction
            2.5.2.1. Borane reduction
            2.5.2.2. Hydrogenation transfer reactions
            2.5.2.3. Hydrogenation
        2.5.3. Stoichiometric diasteroselective reduction
        2.5.4. Catalytic diasteroselective reduction
    2.6. Cyclotrimerization of chiral 2-hydroxynitriles
    2.7. Asymmetric dihydroxylation
    2.8. Chiral aziridine and epoxide ring opening
3. Conclusion
4. Note added in proof
Acknowledgments
References

1. Introduction

Chiral pyridyl alcohols have seen a number of applications as privileged catalysts and ligands for enantioselective synthesis and catalysis.[1] Examples amenable to catalysis with pyridyl alcohols include the epoxidation of allylic alcohols,[1a] the conjugate addition of dialkylzinc reagents to α,β-unsaturated ketones,[1b–d] the nucleophilic addition of dialkylzinc reagents to aldehydes,[1e–i] Strecher-type reaction,[1j] ring-opening reaction[1k–m] and etc. Moreover, phosphorus ligands derived from pyridyl alcohols have been recently applied with success to iridium-catalyzed asymmetric hydrogenation.[1n–q] Many enantiomerically pure pyridyl alcohols are also biologically relevant compounds and key intermediates for commercial drugs such as ceramide,[2a–e] (*S*)-carbinoxamine,[2f,g] *allo*-heteroyohimbine,[2h] (*S*)-naproxen,[2i,j] (*R,S*)-mefloquine,[2k] etc. Moreover, this skeleton is present in a number of compounds that have been described in recent patents reporting their herbicide and fungicide properties.[2l,m]

We discuss the subject of the synthesis optically active pyridyl alcohols in three accounts; the first two deal with chemical syntheses, whereas the third one talks about enzymatic methods. The first part, namely this account, is dedicated only to pyridyl *sec*-alcohols, whereas the second one is devoted to the residual topic (pyridyl *tert*-alcohols, pyridyl β-carbinols, pyridines bound to a phenolic ring, etc.). In particular, this report considers the synthesis of a range of optically active pyridyl alcohols through the description of a variety of approaches allowing to obtain this kind of compounds with a stereogenic carbon bound to both a hydroxy group and a pyridine ring, although the synthesis of some examples of pyridyl *sec*-alcohols without this topological element are explained too.

2. Pyridyl *sec*-alcohols

2.1. Addition of 2-pyridyllithium derivatives to chiral aldehydes

The easiest entry to secondary pyridyl alcohols involves the trapping of 2-pyridyllithium derivatives with optically active aldehydes. Chiral pyridyl and bipyridyl alcohols have been prepared using as a starting point (*R*)-2,3-*O*-isopropylideneglyceraldehyde **2** obtained from D-mannitol (Scheme 1).[3]

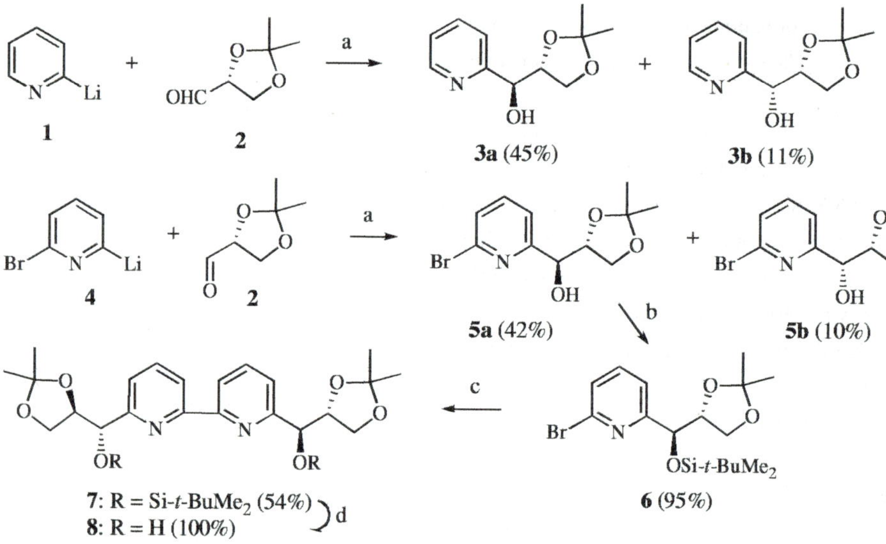

a: Et₂O, –78 °C to r.t.; b: TBDMSCl, imidazole, DMF, r.t.; c: PPh₃, NiCl₂, Zn, DMF, 70 °C; d: TBAF, THF, r.t.

Scheme 1

Thus, reaction of 2-pyridyllithium **1**, obtained by lithiation of 2-bromopyridine with *n*-BuLi at –78 °C, with **2** gave the epimeric alcohols **3a** and **3b** that were separated by chromatography (45 and 11% yields, respectively). When (6-bromopyridin-2-yl)lithium **4**, prepared from 2,6-dibromopyridine with *n*-BuLi, was treated with **2**, the epimeric alcohols **5a** and **5b** were obtained in 52% combined yield and in a ratio of about 4:1, respectively. These alcohols were separated and the major isomer **5a** was converted to the silyl ether **6** using *tert*-butyldimethylsilyl chloride (54% yield). Finally, compound **6** was subjected to nickel-catalyzed homocoupling to produce the bipyridine **14** (60% yield) that was then quantitatively desylilated to dialcohol **8**.

Bittman *et al.* reported an efficient and stereoselective method for the preparation of the heteroaryl ceramide analogue **13** (Scheme 2).[4] Lithiation of bromopyridine derivative **9**, followed by reaction with Garner aldehyde **14**, gave a 5:1 mixture of *erythro*-**10** and *threo*-**11**, which however could not be separated by column chromatography. Hydrolysis of **10** and **11** (1M HCl, THF, 70 °C, 5 hours) resulted in the formation of diastereomeric sphingoid alcohols **12**. Since chromatographic separation of these diastereomers at this stage was still difficult, the mixture of sphingosine analogue **12** was *N*-acylated with *p*-nitrophenyl butyrate. Fortunately, the ceramide diastereomers were now separable by column chromatography and thus, the desired D-erythro stereoisomer **13** was obtained in 70% yield.

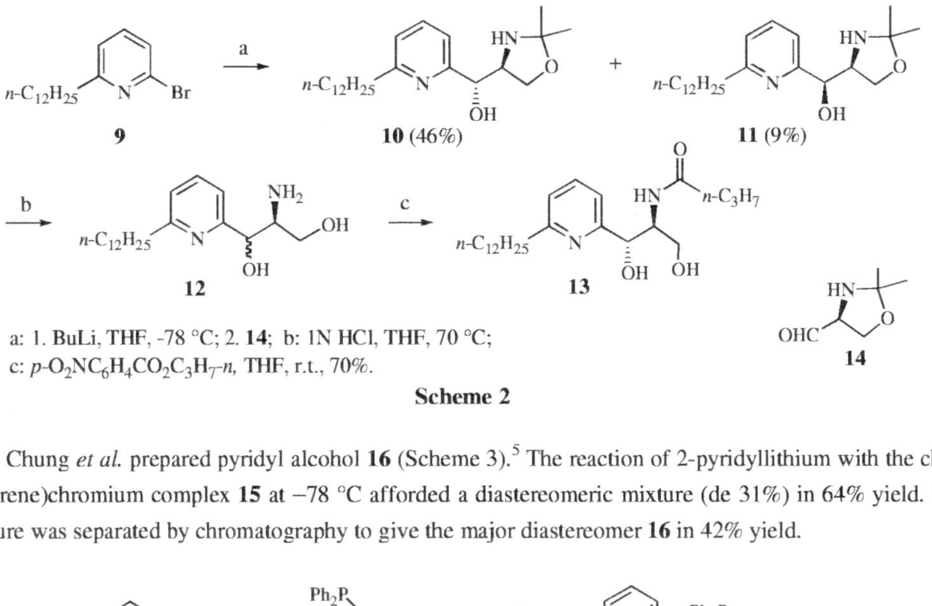

a: 1. BuLi, THF, -78 °C; 2. **14**; b: 1N HCl, THF, 70 °C;
c: *p*-O$_2$NC$_6$H$_4$CO$_2$C$_3$H$_7$-*n*, THF, r.t., 70%.

Scheme 2

Chung *et al.* prepared pyridyl alcohol **16** (Scheme 3).[5] The reaction of 2-pyridyllithium with the chiral (η^6-arene)chromium complex **15** at −78 °C afforded a diastereomeric mixture (de 31%) in 64% yield. This mixture was separated by chromatography to give the major diastereomer **16** in 42% yield.

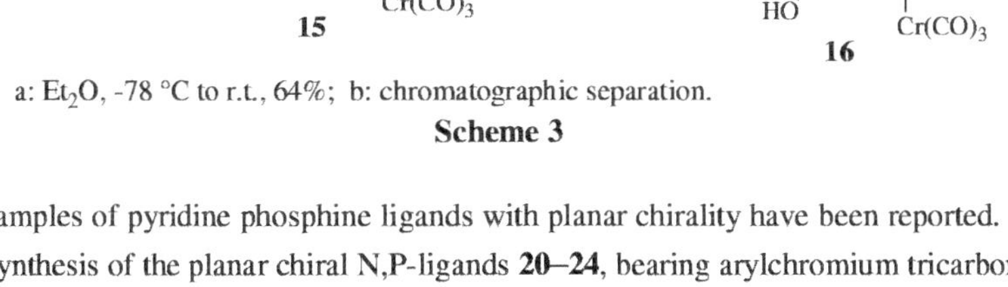

a: Et$_2$O, -78 °C to r.t., 64%; b: chromatographic separation.

Scheme 3

Three examples of pyridine phosphine ligands with planar chirality have been reported. Chung *et al.*[6,7] described the synthesis of the planar chiral N,P-ligands **20–24**, bearing arylchromium tricarbonyl, phosphine and pyridine moieties (Scheme 4). The synthesis of these ligands is based on the addition of 2-pyridyllithium to the optically active benzaldehyde derivative **19** that was accessible starting from the planar chiral known compound **17**. The diastereomeric ratio of **20** and **21** was slightly dependent upon the substituent in the aryl phosphine moiety. The relative amount of **20** raises as the number of electron-withdrawing substituents at phosphorus increases. The complexes **22** and **23** were obtained by deprotonation of **20** and **21** with NaH in THF, followed by the addition of MeI.[6] The *O*-benzyl and *O*-4-methoxybenzyl derivatives (**25** and **26**, respectively) of the complex **1a** were prepared by deprotonation of this complex, followed by treatment with benzyl bromide or 4-methoxybenzylchloride.[7] In order to understand the role of the $Cr(CO)_3$ moiety in the planar chiral ligand, Chung *et al.* carried out the demetallation of **22c**. However, treatment of **22c** with an oxidizing reagent (I_2, THF) led to the isolation of the phosphine oxide **24** (Scheme 4).[6]

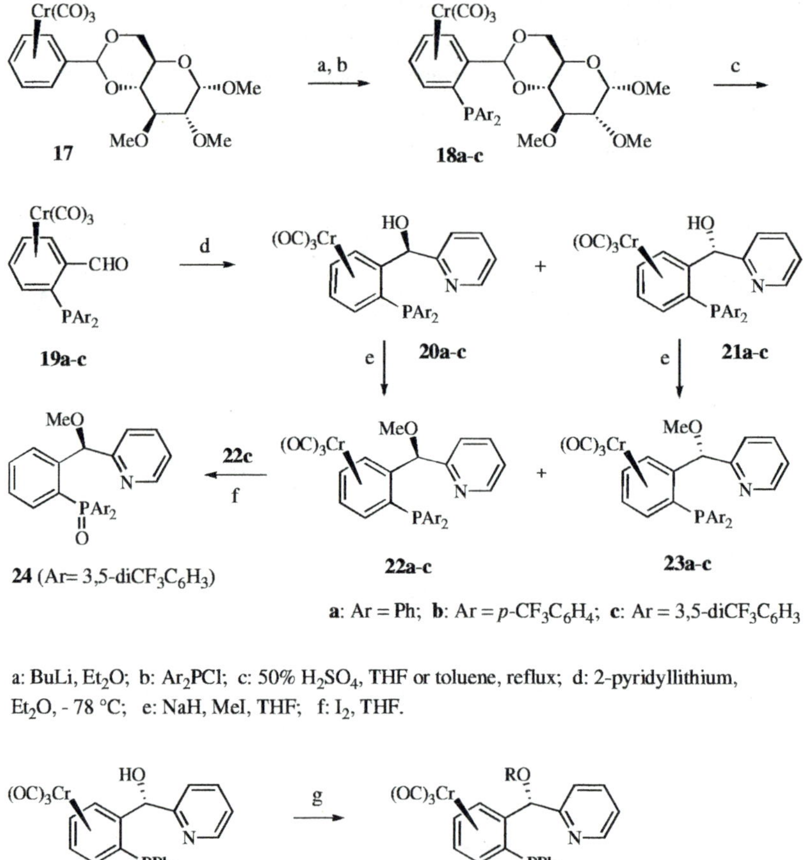

a: Ar = Ph;   b: Ar = p-CF$_3$C$_6$H$_4$;   c: Ar = 3,5-diCF$_3$C$_6$H$_3$

a: BuLi, Et$_2$O;   b: Ar$_2$PCl;   c: 50% H$_2$SO$_4$, THF or toluene, reflux;   d: 2-pyridyllithium, Et$_2$O, - 78 °C;   e: NaH, MeI, THF;   f: I$_2$, THF.

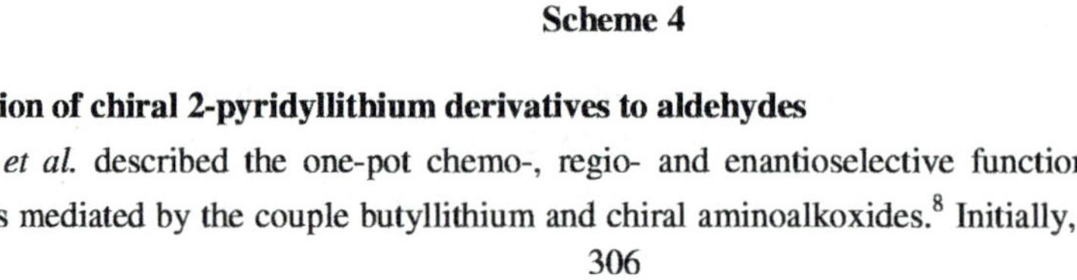

g: NaH, BnBr or 4-MeOC$_6$H$_4$CH$_2$Cl, THF.

Scheme 4

2.2. Addition of chiral 2-pyridyllithium derivatives to aldehydes

Fort *et al.* described the one-pot chemo-, regio- and enantioselective functionalization of pyridine compounds mediated by the couple butyllithium and chiral aminoalkoxides.[8] Initially, 2-chloropyridine was metallated by the superbases generated by butyllithium and several chiral aminoalcohols and then treated with benzaldehyde. As expected, the superbase promoted the clean C-6 functionalization of 2-chloropyridine,[9] but the reaction course was critically dependent on the aminoalcohol structure, except with (S)-(1-methylpyrrolidin-2-yl)methanol **28**, which led to (6-chloropyridin-2-yl)(phenyl)methanol in 59% yield and 58% ee. (Scheme 5). After optimization of the reaction conditions, the versatility and synthetic value of the reaction for the preparation of chiral polyfunctional pyridines was inspected. The reaction of various substituted pyridines with aldehydes in the presence of the couple BuLi/**28** afforded the related pyridyl alcohols in moderate to good yields (47–74%), but low enantioselectivities (30–45% ees) (Scheme 5).

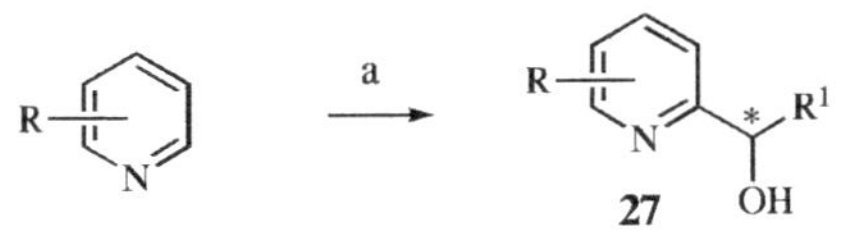

a: 1. BuLi/**28**, hexane, -78 °C; 2. R^{1}CHO, THF, -78 °C.

27	R	R^1	yield(%)	ee(%)
a	6-Cl	*t*-Bu	62	35
b	6-Cl	4-MeOC$_6$H$_4$	61	45
c	6-Cl	4-ClC$_6$H$_4$	56	23
d	5-Me	Ph	47	39
e	6-F	Ph	74	30
f	6-Cl	Ph	59	58

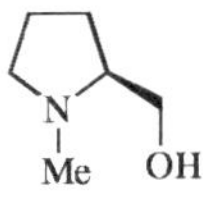

Scheme 5

2.3. Stereoselective addition of organometallic reagents to pyridyl carboxaldehydes

2.3.1. Enantioselective addition

The addition of dialkylzinc reagents to pyridinecarboxaldehydes catalyzed by chiral ligands consents an easy entry to optically active pyridyl alcohols.[10] However, this reaction can be problematic since the pyridylalkanol formed *in situ* acts as a catalyst eroding the enantioselectivity (*vide infra*). Notwithstanding that, Hoshino *et al.* reported highly enantioselective and rapid addition (30 seconds) of dialkylzinc reagents to pyridinecarboxaldehydes using catalytic amount (10 mol%) of the ligand **31** at room temperature (Scheme 6).[1g] Reaction of pyridine 3- and 4-carboxaldehydes with diethylzinc afforded, in the presence of **31**, the pyridyl alcohols **30b** and **30c** in 88 and 83% ees, respectively. On the other hand, the isomer **30a** was obtained in quasi-racemic form because of the facile intramolecular chelation of diethylzinc between the nitrogen atom and the carbonyl group in **29a** that accelerates non-catalyzed reactions. The validity of this assumption was demonstrated submitting 2-bromo-6-carboxypyridine **29d** to the addition reaction. The bulky as well as electron-withdrawing bromine atom (**29d**) allowed to obtain the corresponding pyridyl alcohol **30d** in 81% yield and 70% ee. Radical mediated reduction of **30d** afforded **30a** in 86% yield without loss of optical purity.

In contrast to the above findings, Braga and co-workers have recently found that the amino diselenide (*R,R*)-**32** (Scheme 6) catalyzes the enantioselective addition of diethylzinc to pyridine-2-carboxaldehyde **29a** to give the corresponding alcohol **30a** with high enantiomeric excess (90%) and yield (85%).[11]

Soai and co-workers carried out an extensive study on the enantioselective autocatalysis of pyridyl and quinolinyl alcohols.[12] This group reported the first asymmetric autocatalysis using pyridyl alcohols as asymmetric autocatalysts. Thus, when pyridine-3-carboxaldehyde was treated with diisopropylzinc using (−)-2-methyl-1-(pyridin-3-yl)propan-1-ol (86% ee), as a chiral catalyst, the same alcohol was produced with retention of configuration in 47% enantiomeric excess, indicating that it acts as a chiral self-catalyst (Scheme 7).[12b] In a similar manner, other chiral non-racemic 3-pyridylalkyl alcohols (42–56% ees) afforded themselves (6–14% ees) in the enantioselective addition of (alkyl)$_2$Zn (Scheme 7).[12b]

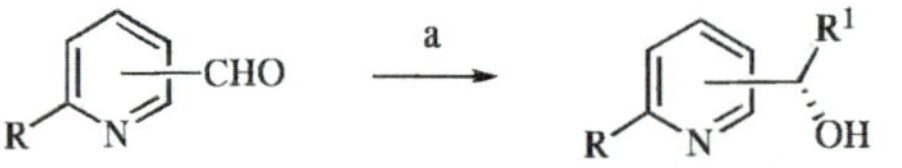

29a: 2-formyl, R= H
29b: 3-formyl, R= H
29c: 4-formyl, R= H
29d: 2-formyl, R= Br

(*S*)-**30a**: 1-(pyridin-2-yl)propanol: 47%, 10% ee
(*S*)-**30b**: 1-(pyridin-3-yl)propanol: 77%, 88% ee
(*S*)-**30c**: 1-(pyridin-4-yl)propanol: 73%, 83% ee
(*S*)-**30d**: 1-(6-bromopyridin-2-yl)propanol: 81%, 70% ee
(*S*)-**30e**: 1-(pyridin-2-yl)ethanol: 70%, 85% ee

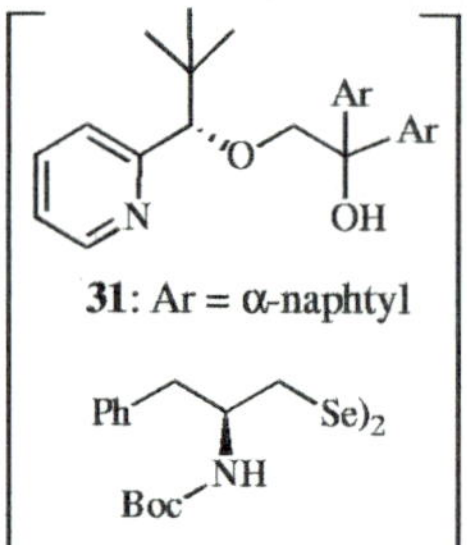

31: Ar = α-naphtyl

(*R,R*)-**32**

a: Et$_2$Zn or Me$_2$Zn, **31** (10 mol %), toluene-hexane (1:1), r.t., 30 sec (1 h for **30e**).

Scheme 6

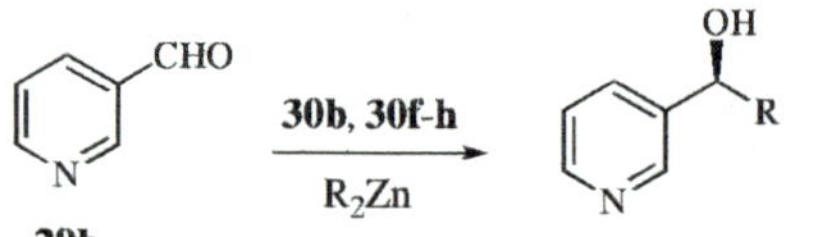

29b

30b, 30f-h, R$_2$Zn

30b, 30f-h

(*S*)-**30b**: R = Et (56% ee) ⟶ (*S*)-**30b**: 14% ee
(*S*)-**30f**: R = Me (42% ee) ⟶ (*S*)-**30f**: 7% ee
(*S*)-**30g**: R = *n*-Bu (47% ee) ⟶ (*S*)-**30g**: 6% ee
(*S*)-**30h**: R = *i*-Pr (86% ee) ⟶ (*S*)-**30h**: 47% ee

Scheme 7

The same group reported the enantioselective preparation of 5-carbamoylpyridin-3-yl alcohols (*S*)-**34a–d** (88–94% ees) by addition of diisopropylzinc to the related 3-carboxaldehydes **33a–d** in the presence of a catalytic amount of (1*S*,2*R*)-*N,N*-dipropylnorephedrine **35** (Scheme 8).[12c] Interestingly, a catalytic amount of the zinc alkoxide of these new alcohols (*S*)-**34a–c** (88–94% ees) catalyzed the enantioselective alkylation of the 3-carboxaldehydes **33a–c** by diisopropylzinc to afford itself with the same configuration as the catalyst and in enantiomeric excesses up to 86% depending on the structure on the nitrogen atom of the amide.

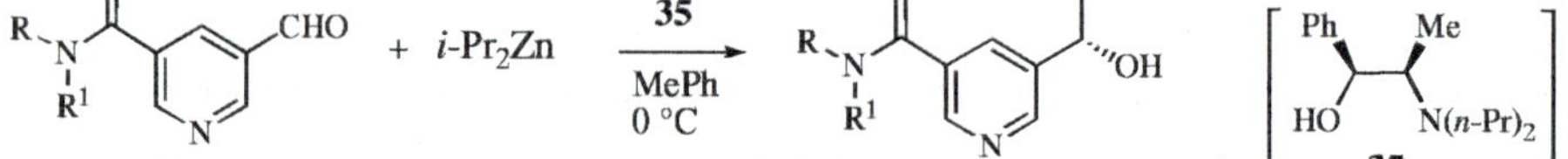

33a-d + *i*-Pr$_2$Zn → (via **35**, MePh, 0 °C) → (*S*)-**34a-d**

35

33a: R = OMe, R^1= Me; **33b**: R = R^1= *i*-Pr
33c: R = R^1= *i*-Bu; **33d**: R = R^1= *n*-Bu

34a: 56%, 90% ee; **34b**: 56%, 94% ee
34c: 82, 88% ee; **34d**: 70, 89% ee

Scheme 8

Soai and co-workers have recently described the enantioselective synthesis of 2-, 5-, 6-, 7- and 8-substituted 3-quinolyl alkanols and their behaviour as asymmetric autocatalysts in the addition of diisopropylzinc to the corresponding quinoline-3-carboxaldehydes.[12d] The newly formed products have the same structure and absolute configuration as those of the initially used asymmetric autocatalyst. In addition, low enantiomeric excess has been amplified up to 97% ee by consecutive asymmetric autocatalysis without the intervention of any other chiral substance.

Pagenkopf and Carreira reported that a titanium(IV) catalyst prepared *in situ* from TiF$_4$ and (*R,R*)-1,2-bis(diphenylmethanol)cyclohexane catalyzes the asymmetric addition of trimethylaluminium to aldehydes. Under these conditions pyridine-3-carboxaldehyde **29b** was methylated to give the 3-pyridyl-ethanol (*R*)-**30e** in 83% yield and 85% ee (Scheme 9).[13]

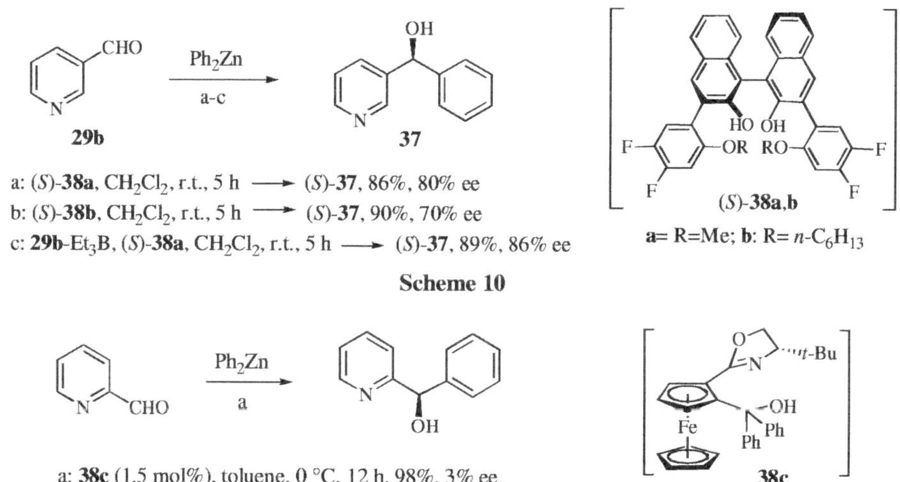

a: Me$_3$Al (1.4 equiv), TiF$_4$ (14 mol%), **36** (15 mol%), THF, -10 (12 h) to 0 °C.

Scheme 9

Pu and Huang studied the use of the 1,1-binaphthyl derivative (*S*)-**38a,b** as catalysts in the asymmetric addition of diphenylzinc to pyridine-3-carboxaldehyde (Scheme 10).[14] Ligand (*S*)-**38a** afforded the corresponding pyridyl phenyl carbinol (+)-**37** with a better steroselectivity than (*S*)-**38b** (80 *vs* 70% ee), but lower yield (80 *vs* 90%). When the nitrogen of the pyridine was protected with triethylborane, its reaction with diphenylzinc in the presence of (*S*)-**38a** gave 86% ee and 89% yield (Scheme 10).

a: (*S*)-**38a**, CH$_2$Cl$_2$, r.t., 5 h ⟶ (*S*)-**37**, 86%, 80% ee
b: (*S*)-**38b**, CH$_2$Cl$_2$, r.t., 5 h ⟶ (*S*)-**37**, 90%, 70% ee
c: **29b**-Et$_3$B, (*S*)-**38a**, CH$_2$Cl$_2$, r.t., 5 h ⟶ (*S*)-**37**, 89%, 86% ee

a= R=Me; b: R= *n*-C$_6$H$_{13}$

Scheme 10

a: **38c** (1.5 mol%), toluene, 0 °C, 12 h, 98%, 3% ee.

Scheme 11

Bolm and Muñiz reported that the asymmetric addition of diphenylzinc to aldehydes in the presence of catalytic amounts of the planar chiral ferrocene based hydroxy oxazoline **38c** afforded alcohols with enantiomeric excesses of up to 96%.[15] Unfortunately, the arylation of pyridine-2-carboxaldehyde proceeded with good yield (98%), but with very low stereoselectivity (3% ee) (Scheme 11). Chiral terpenyl-based allylborane reagents, *B*-allyldiisopinocampheylborane (Ipc$_2$Ball) and *B*-allyl-bis(4-isocaranyl)borane (Icr$_2$Ball) undergo facile condensation with pyridylcarboxaldeides **29a–c** at −100 °C to afford the corresponding terpenylborinate **39a–c** (Scheme 12).[16] These intermediates were not isolated, but directly converted into the homoallylic alcohols **40a–c** by treatment with a methanolic solution of 8-hydroxy-quinoline. Good overall yields (78–85%) and enantiomeric purities approaching 100% were achieved. The reagent Ipc$_2$Ball was next used with success to synthesize the optically pure (−)-(*S*,*S*)-α,α'-diallyl-2,6-pyridinedimethanol **42** from the pyridine-2,6-dicarboxaldehyde **41** (Scheme 12).[17]

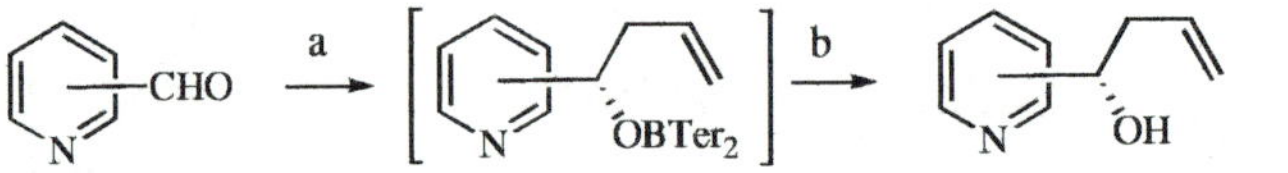

29a-c　　　　　**39a-c**

40a: 85%, >99% ee
40b: 84%, >96% ee
40c: 78%, >99% ee

a: 2-pyridyl, **b**: 3-pyridyl, **c**: 4-pyridyl

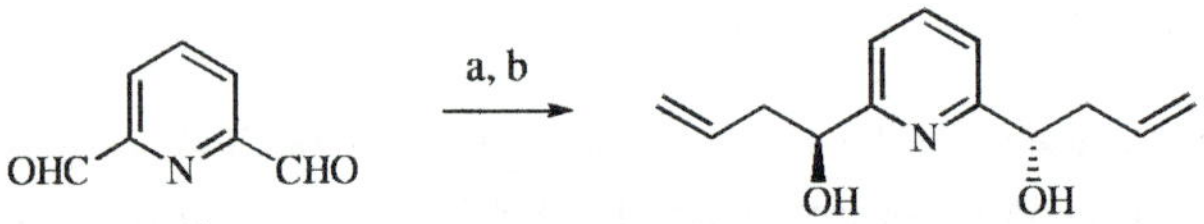

41

42 (94%, 92% ee)

a: Ipc$_2$BAll (for **29a,b** and **41**) or Icr$_2$BAll (for **29c**), Et$_2$O
(or THF for **41**), -100 °C, 1 h; b: 8-hydroxyquinoline,
MeOH, 25 °C, 12 h.

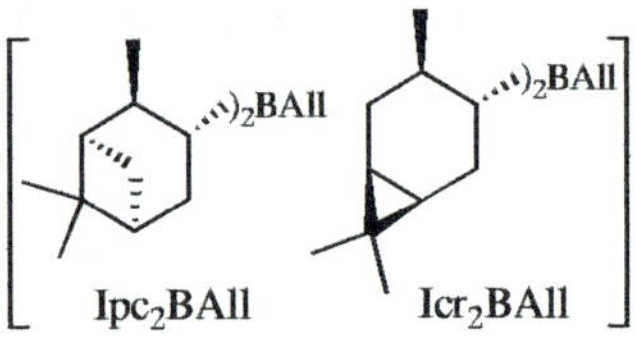

Ipc$_2$BAll　　　Icr$_2$BAll

Scheme 12

For the synthesis of 2-pyridyl carbinols, the addition of organo-magnesium reagents to pyridine-2-carboxaldehyde is a valuable alternative to the addition of 2-pyridyllithium to carbonyl compounds. Cunningham *et al.*[18] prepared from (−)-menthol the (−)-menthyl chloride **43** (containing less of 10% of its axial neomenthyl isomer) which was converted into the related Grignard reagent (Scheme 13).

Treatment of this magnesium derivative with pyridine-2-carboxaldehyde gave all four possible product alcohols **44a–d** (65% combined yield) and a significant amount of the epimeric ketones **45a** and **45b** (Scheme 13). It was impossible to obtain reliable ratios for these alcohols using ^{1}H NMR spectroscopy, but integration of the ^{13}C-signals for their α-carbinyl carbon atoms suggested the relative proportions of **44a:44b:44c:44d** to be 1.8:2.3:1:1. Pure samples of each of these four alcohols were obtained by careful chromatography over silica gel.

The alcohol **44a** was next obtained more conveniently by reduction of the ketone **45a**, obtained in turn by oxidation of a mixture of **44a–d** (*vide infra*, Scheme 41). Moreover, it was possible to obtain the alcohol **44b** *via* Mitsunobu inversion of **44a**. Thus, this alcohol was first efficiently converted into the nitrobenzoate **46**, which was then hydrolyzed to provide **44b** in 69% overall yield from **44a**.

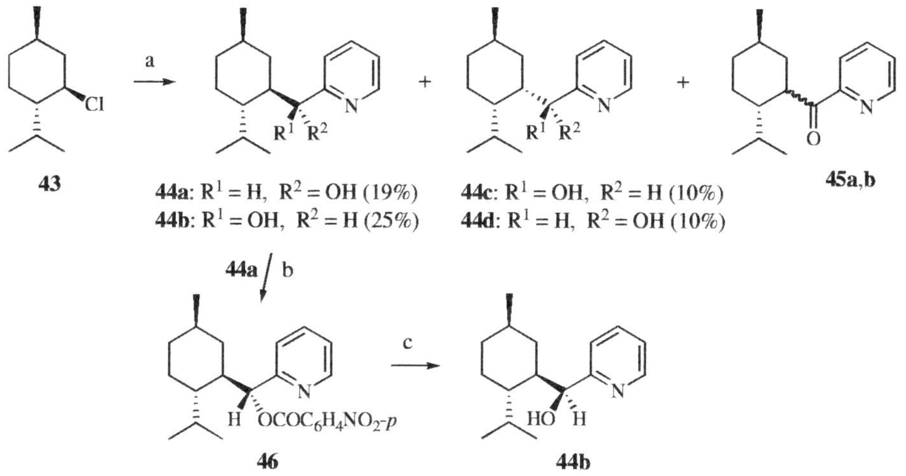

43

44a: R^1 = H, R^2 = OH (19%) **44c**: R^1 = OH, R^2 = H (10%) **45a,b**
44b: R^1 = OH, R^2 = H (25%) **44d**: R^1 = H, R^2 = OH (10%)

44a / b

H OCOC$_6$H$_4$NO$_2$-p HO H

46 **44b**

a: 1. Mg, THF, I$_2$; 2. pyridine-2-carboxaldehyde, reflux, 12 h; b: PPh$_3$, HOCOC$_6$H$_4$NO$_2$-p, diethyl azidodicarboxylate, THF, r.t., 12 h, 75%; c: NaOH (10%), 95%.

Scheme 13

Knochel and Cheemala have prepared ferrocenyl N,P-ligands based on a pyridyl alcohol framework (Scheme 14).[19] The synthesis starts from the enantiopure sulfoxide **47** that, by submission to sulfoxide/lithium exchange using phenyllithium (THF, −78 °C, 10 minutes) followed by the addition of 2-pyridinecarboxaldehyde, gave a 3:2 mixture of two diastereomeric ferrocenyl alcohols **48** in 72% yield. This inseparable mixture of alcohols was alkylated with MeI or PhCH$_2$Br, leading to readily separable ferrocenyl ethers **49a** (54% yield) and **49b** (35% yield) or **50a** (56% yield) and **50b** (36% yield), respectively (Scheme 14).

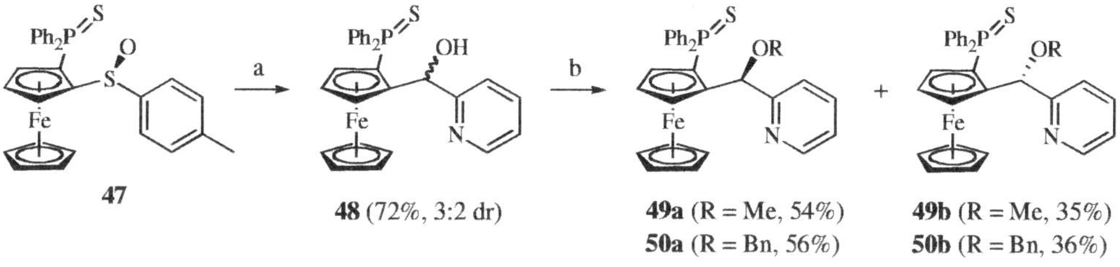

47 **48** (72%, 3:2 dr) **49a** (R = Me, 54%) **49b** (R = Me, 35%)
50a (R = Bn, 56%) **50b** (R = Bn, 36%)

a: 1. PhLi, Et$_2$O, -78 °C, 10 min; 2. 2-PyCHO, -78 °C, 1.5 h; b: 1. KH, THF, 0 °C, 1 h; 2. MeI or BnBr, 0 °C, 15 min.

Scheme 14

2.3.2. Diasteroselective addition

Ruzziconi *et al.* prepared diastereomeric quinolinophanylcarbinols (R_p,R)- and (R_p,S)-**52–55** with both central and planar chirality by addition of alkyl and aryl Grignard reagents to the quinolinophane-2-carboxyaldehyde (R_p)-**51** (Scheme 15).[20] Carbinols (R_p,R)-**55** and (R_p,S)-**55** had to be converted into the corresponding benzoate esters before being separated by HPLC; (R_p,R)-**54** and (R_p,S)-**54** were directly separated by HPLC; while all the other diastereomeic mixture were easily resolved by simple chromatography. The absolute configurations of the new stereogenic centres were assigned by NOESY experiments.

2.4. Rearrangement of pyridine *N*-oxides

The Bockelheide rearrangement of pyridine *N*-oxides represents a valuable method for the introduction of a hydroxy group on the α-position of 2-alkyl substituted pyridines.[21]

Exploiting this method, Von Zelewsky *et al.* prepared the two epimeric chiral 7-hydroxy-5,6,7,8-tetra-hydroquinolines (8*S*)-**61** and (8*R*)-**61** from (−)-pinocarvone **56** (Scheme 16).[22] Kröhnke type aza-annulation[23] of the pyridinium salt **57** with **56** gave the quinoline **58** (60% yield) which was oxidized to the corresponding *N*-oxide **59** with peracetic acid. Bockelheide rearrangement of **59** with isobutyrric anhydride afforded a mixture of epimeric esters (8*S*)-**60** and (8*R*)-**60** in a ratio of 65/35. This mixture was separated and the single epimers were hydrolysed to give pure alcohols (8*S*)-**61** and (8*R*)-**61**.

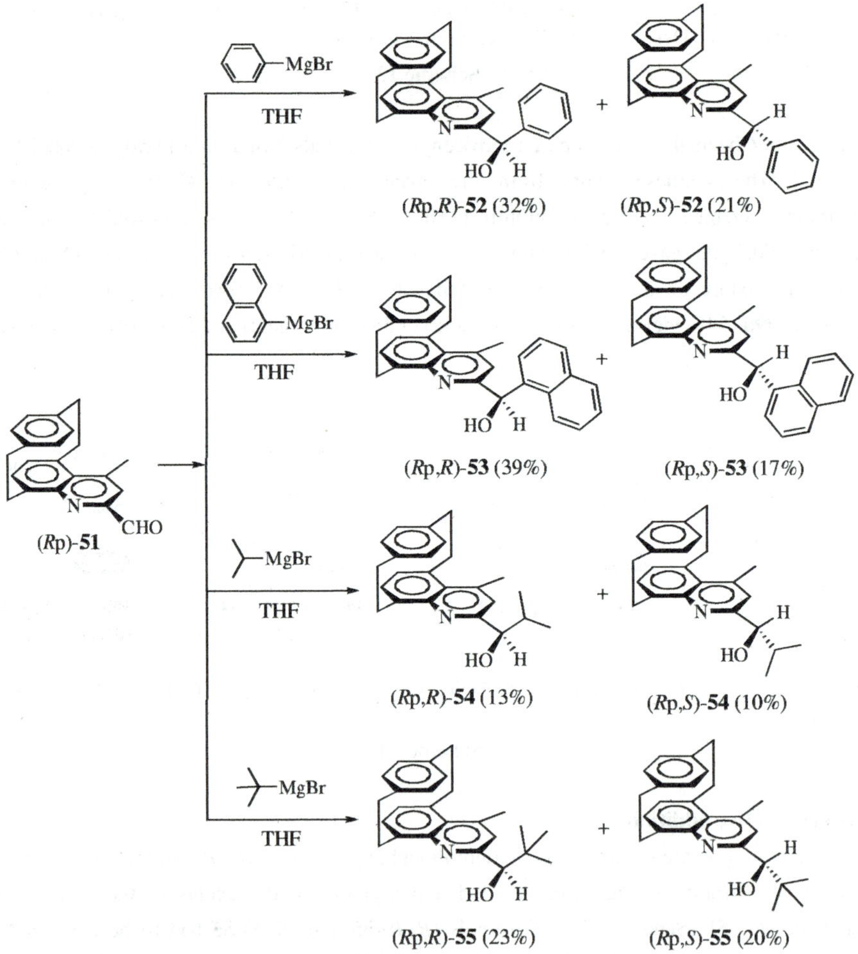

Scheme 15

312

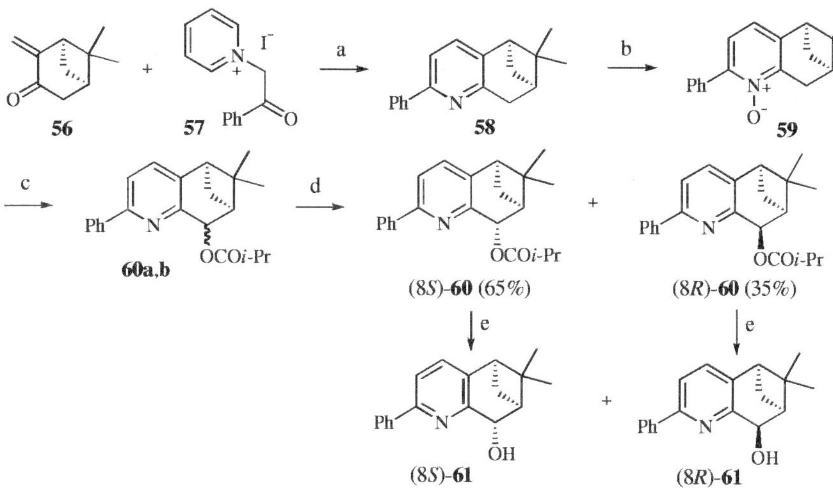

a: AcOH, AcONH$_4$, 100 °C, 20 h, 60%; b: AcOH, H$_2$O$_2$, 70 °C, 24 h; c: isobutyric anhydride, 100 °C, 8 h, 65%; d: crystallization and column chromatography; e: NaOH, MeOH-H$_2$O, reflux, 1 h, 95%.

Scheme 16

A notable example of application of the Bockelheide reaction to natural products synthesis has been described by Nicolau *et al.* who prepared a 7-hydroxy-5,6,7,8-tetrahydroquinoline as a key intermediate in the total synthesis of the antibiotic thiostrepton (see Scheme 52).[24]

2.5. Reduction of pyridyl ketones

2.5.1. Stoichiometric enantioselective reduction

A variety of organoborane reagents have been developed for the asymmetric reduction of a variety of pyridyl ketones. Midland and McLoughlin reported that (*S*)-1-(pyridin-3-yl)ethanol was obtained in 67% yield and 92% eanatiomeric excess when the reduction of acetylpyridine was carried out using B-3-pinanyl-9-borabicyclo[3.3.1]nonane (Alpine-Borane) in THF at 6000 atm, whereas the reaction at 1.0 atm gave lower enantiomeric excess (90%) and conversion (Scheme 17).[25] Brown *et al.* examined, among a series of aryl alkyl ketones, the reduction of 3-acetylpyridine using (−)-β-chlorodiisopinocampheylborane (−)-Ipc$_2$BCl[26a] and B-(*iso*-2-ethylapopinocampheyl-9-borabicyclo[3.3.1] nonane (Eapine-Borane)[26b] (Scheme 17). 3-Acetylpyridine was reduced using 2.0 equivalents of Ipc$_2$BCl in 92% ee and 67% yield, whereas no reaction occurred when 1.0 equivalent of the reagent was employed.[26a] Higher yield (93%) and enantioselectivity (96% ee) were obtained using 3.0 equivalents of Eapine-Borane, albeit at a slower rate (15 days).[26b]

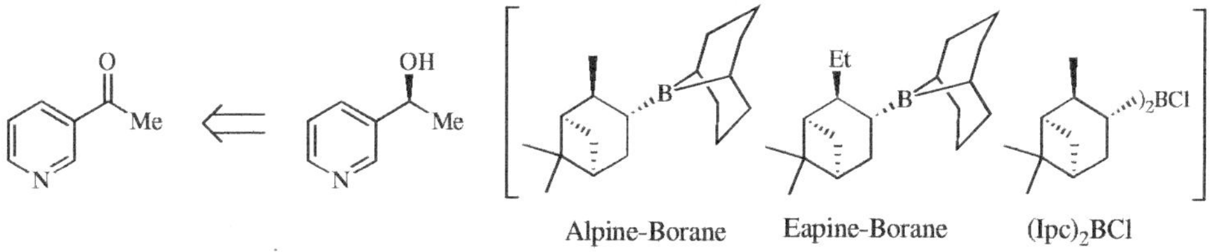

Scheme 17

313

Chelucci and Soccolini described the reduction of pivaloylpyridine **62** with (−)-Ipc$_2$BCl (Scheme 18).[1i] The reaction was carried out for 25 days at −25 °C in THF to give (*R*)-2,2-dimethyl-1-(2-pyridyl)propan-1-ol in 70% yield and 91% ee. More recently, this protocol has been applied to the synthesis of the 4-dimethyl-amino substituted analogue of **62** to give the related alcohol in 71% yield and 96% ee, which after recrystallization afforded material with >99% ee (Scheme 18).[27]

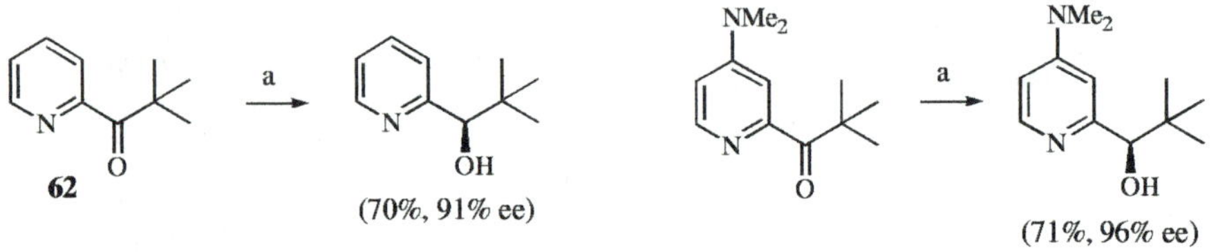

a: (-)-Ipc$_2$BCl, THF, -25 °C, 15 d.

Scheme 18

Bolm *et al.* reported the synthesis of optically pyridine and bipyridine alcohols, whose synthesis has the asymmetric reduction of prochiral pyridyl ketones as a key step.[28] Table 1 summarized the results obtained in the enantioselective reduction of ketones **63a–c** and **65a–e** (Scheme 19) with (+)- and (−)-Ipc$_2$BCl. The alcohols were obtained in 32–88% yields and 82–92% ees. The optical purity of (*R*)-**64a** and (*R*)-**64b** was easily raised by a single recrystallization of the corresponding camphanic acid esters followed by hydrolysis to give the alcohols in ee >99%. Ketones **63a,b** and **65a,b,d,e** (Scheme 19) were prepared by metal-halogen exchange of 3-bromopyridine or 2,6- and 2,5-dibromopyridine with *n*-BuLi in ether, followed by treatment with pivalonitrile or *N,N*-dimethylacetamide. Ketones **63c** and **65c** (Scheme 19) were in turn obtained by cross-coupling of the bromopyridines **63a** and **65c** with phenylboronic acid in the presence of tetrakis(triphenylphosphine)palladium(0) [Pd(PPh$_3$)$_4$].

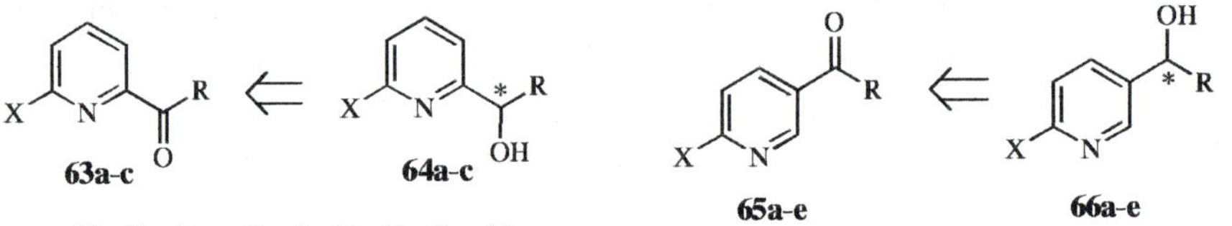

63a-c   **64a-c**

a: X = Br, R = *t*-Bu; b: X = Br, R = Me;
c: X = Ph, R = *t*-Bu

65a-e   **66a-e**

a: X = Br, R = *t*-Bu; b: X = Br, R = Me;
c: X = Ph, R = *t*-Bu; d: X = H, R = *t*-Bu;
e: X = H, R = Me

Scheme 19

In Schemes 20 and 21, many of these transformations are exemplified in the case of compounds substituted with the *tert*-butyl group. A nickel(0)/triphenylphosphine mediated reaction of (*R*)-**64a** afforded the C$_2$-symmetric bipyridine **69** (55% yield), while the heterocoupling of (*R*)-**64a** with 2-pyridylzinc chloride [2-pyZnCl] or 2-pyridyl-tri-*n*-butylstannane [2-pySnBu$_3$] in the presence of Pd(PPh$_3$)$_4$ provided the C$_1$-symmetric bipyridine **68** in about 50% yield.

Moreover, (*R*)-**64a** was transformed into pyridyl acetylene **70** by Sonogashira coupling with trimethylsilylacetylene, followed by desilylation with Bu$_4$NF. Alcohol (*R*)-**64a** was also hydrodehalogenated through radical debromination using Bu$_4$SnH/AIBN in toluene to give the alcohol (*R*)-**71**. Finally, following the same procedure used for **69**, starting from (*R*)-**64b**, the C_2-symmetric bipyridine **72** was prepared in 55% yield (Scheme 21).

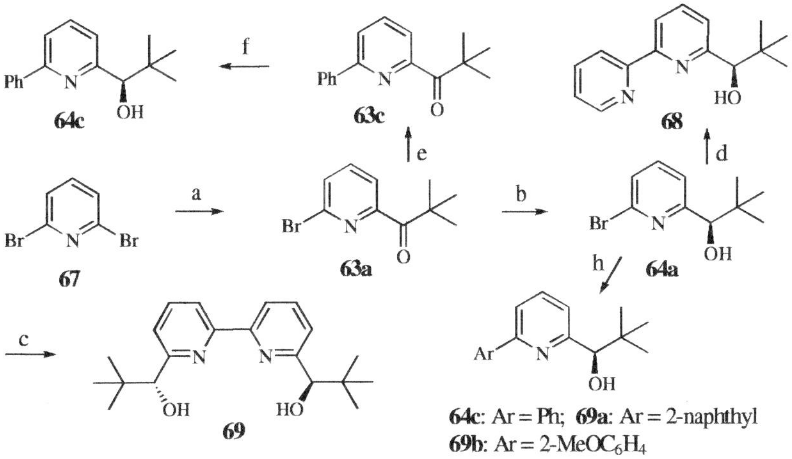

a: 1. BuLi, THF, -78 °C; 2. *t*-BuCN, -78 °C to r.t. 3.5 h, 80%; b: 1. (-)-Ipc$_2$BCl, neat, r.t., 2 d; 2. iminodiethanol, Et$_2$O, 3 h, 59%; c: NiCl$_2$-6H$_2$O, Zn, PPh$_3$, DMF, 72 °C, 3.5 h, 55%; d: 2-PyZnCl (53 %) or PySnBu$_3$ (50 %), Pd(PPh$_3$)$_4$; e: PhB(OH)$_2$, Pd(PPh$_3$)$_4$, Na$_2$CO$_3$.; f: 1. (-)-Ipc$_2$BCl, neat, r.t., 5 d; 2. iminodiethanol, Et$_2$O, 3 h, 62%; g: ArB(OH)$_2$, Pd(PPh$_3$)$_4$, Na$_2$CO$_3$.

Scheme 20

Table 1. Stereoselective reduction of pyridyl ketones (Scheme 19 by using Ipc$_2$BCl).

ketone	alcohol	enantiomer of the borane	yield	ratio of (*R*)/(*S*)
63a	**64a**	(−)	61	95:5
63b	**64b**	(−)	85	4:96
63c	**64c**	(−)	75	93:7
65a	**66a**	(−)	32	91:9
65b	**66b**	(+)	88	96:4
65c	**66c**	(−)	75	93:7
65d	**66d**	(+)	34	7:93
65e	**66e**	(−)	65	4:96

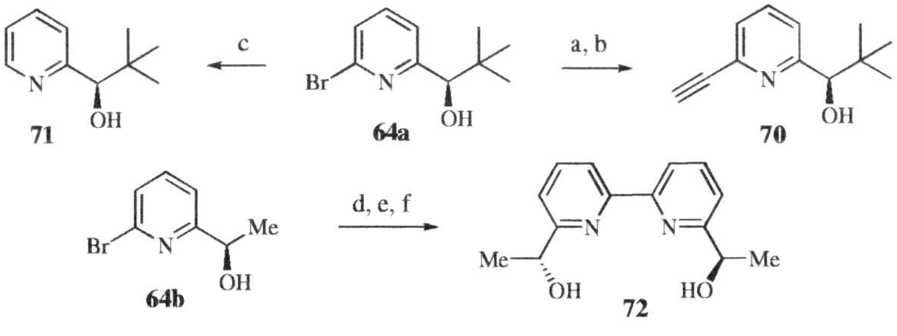

a: trimethylsilylacetylene, Pd(PPh$_3$)$_4$, CuI, NEt$_3$; b: Bu$_4$NF; c: Bu$_4$SnH, AIBN, toluene; d: Ac$_2$O, Py; e: NiCl$_2$-6H$_2$O, Zn, PPh$_3$, DMF, 72 °C, 3.5 h; f: K$_2$CO$_3$, MeOH.

Scheme 21

315

Reduction with Ipc_2BCl has been also the key step in the synthesis of other C_2-symmetric bipyridines **73a–c** and terpyridine **74** (Figure 1).[29]

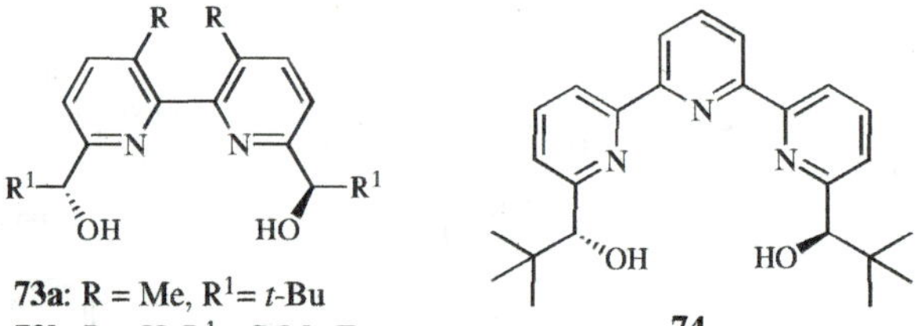

73a: R = Me, R^1= *t*-Bu
73b: R = H, R^1= C(Me)Et$_2$
73c: R = H, R^1= CMe$_2$CEt$_2$

74

Figure 1

Hoshino *et al.* examined the preparation of both enantiomer of the pyridyl diol **76** by asymmetric reduction of 2,6-dipivaloylpyridine **75** with (+)- or (−)-Ipc_2BCl (Scheme 22).[1e] This asymmetric synthesis was pursued since the previously reported method by a resolution procedure did not seem suitable for a large scale preparation.[1a,30] Reduction of with (+)- or (−)-Ipc_2BCl gave (*R*,*R*)- and (*S*,*S*)-**76**, whose enantiomeric excesses were determined to be both 100% by HPLC analysis using a chiral column. These findings are different from those previously reported for **75**, but in this case the very good result might be due to the stereospecific reduction of keto alcohols, which would be formed by a first highly stereoselective reduction of **75** with (+)- or (−)-Ipc_2BCl.

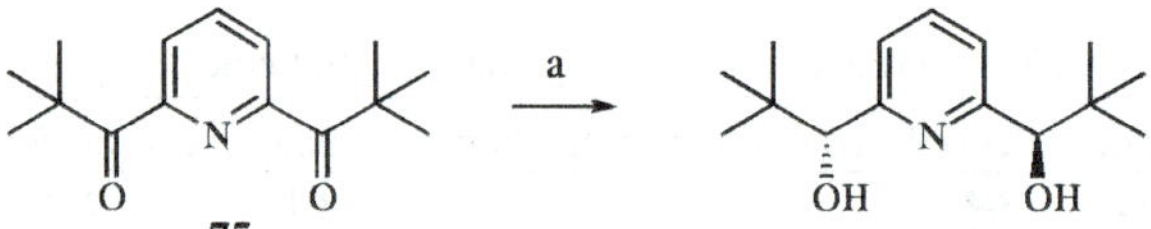

75

(*R*,*R*)-**76** (>99% ee)

a: 1. (+)-(Ipc)$_2$BCl, neat, r.t., 9 d; 2. iminodiethanol, Et$_2$O, 43%.

Scheme 22

Two years later, Brown and co-workers reported a systematic study to obtain C_2-symmetric diols by reduction of diacylaromatic ketones with Ipc_2BCl.[31] Among them, pyridyl ketones **77a–d** were reduced to the corresponding enantiopure alcohols (≥99% ee) (Scheme 23). Unlike Hoshino, who used 5 equivalents of Ipc_2BCl for the reduction of **75**, Brown employed only 2.5 equivalents of Ipc_2BCl to obtain complete conversion of the starting material. Only the 2,6-diacetylpyridine **77a** could not be completely reduced affording the hydroxyketone **78a** (36-44% yields and ≥95% ee) with the chiral diol **79a** (45–54% yields, ≥99% ee) and the *meso* diol (ca 5%).

Zhang *et al.*, in an independent work, obtained similar results for the reduction of pyridyl ketones **77a** and **75** with Ipc_2BCl.[32] Thus asymmetric reduction of **77a** with (−)-Ipc_2BCl (THF, −18 °C) afforded the diol (*S*,*S*)-**79a** in 50% yield and >98% ee along with 5–10% of the *meso* diol. This chiral diol was then obtained enantiomerically pure by recrystallization of the corresponding di-*p*-bromobenzoate ester obtained from the diol mixture. In an analogous manner, the diol **75** was reduced after 7 days at room temperature in 42% yield and 100% ee.

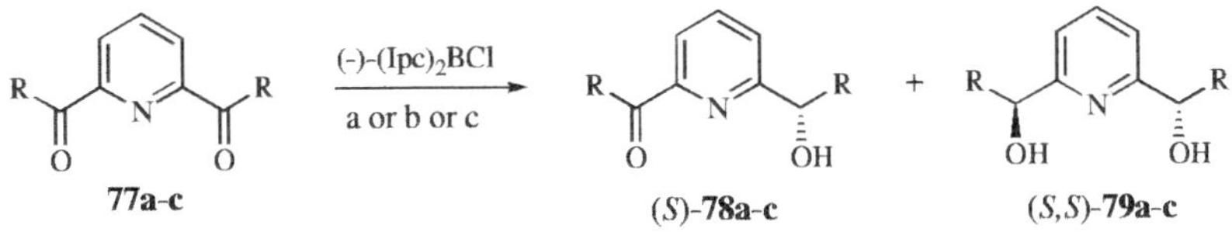

a: R= Me; **b:** R= CF$_3$; **c:** R= C$_2$F$_5$

a: **77a**, Et$_2$O, -25 °C, 17 h ⟶ (*S*)-**78a**: 36%, 95% ee + (*S,S*)-**79a**: 54%, 99% ee
b: **77b**, Et$_2$O, 25 °C, 3d ⟶ (*S,S*)-**79b**: 71%, 99% ee
c: **77c**, THF, 25 °C, 7d ⟶ (*S,S*)-**79c**: 72%, 99% ee

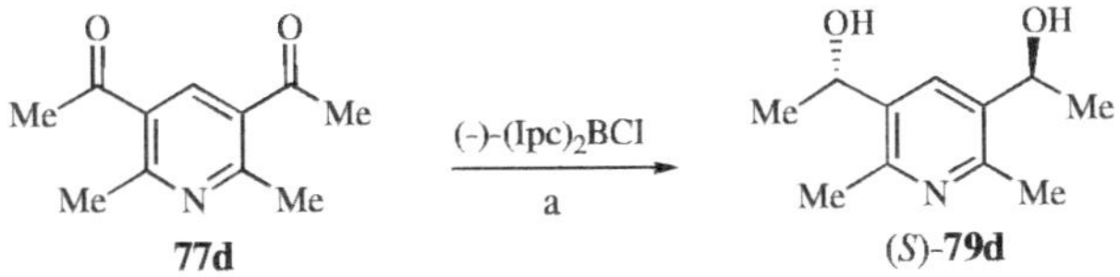

a: **77d**, Et$_2$O, -25 °C, 5 h ⟶ (*S,S*)-**79d**: 94%, 99% ee

Scheme 23

Pfaltz *et al.* reported the optically active pyridyl phosphane **82** (Scheme 24)[33] in which the enantioselective reduction of the ketone **80** with (−)-β-chlorodiisopinocampheyl borane was exploited to obtain the alcohol **81** in 75% yield and 92% ee.

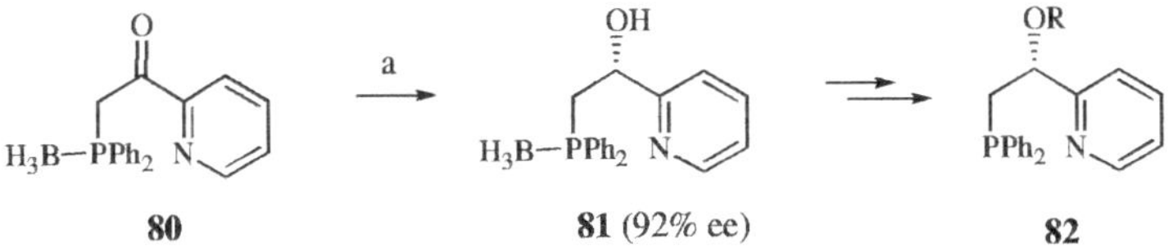

80   **81** (92% ee)   **82**

a: (-)-Ipc$_2$BCl (2.5 equiv), THF, 0 °C, 19 h, 75%.

Scheme 24

Efficient stereoselective total syntheses of furo[2,3-c]pyridine thiopyrimidine HIV-1 reverse transcriptase inhibitors **PNU-142721** and **PNU-109886** (Scheme 25) has been developed.[34] In both syntheses the key step is the preparation of the optically active furo[2,3-c]pyridyl alcohol **83**.

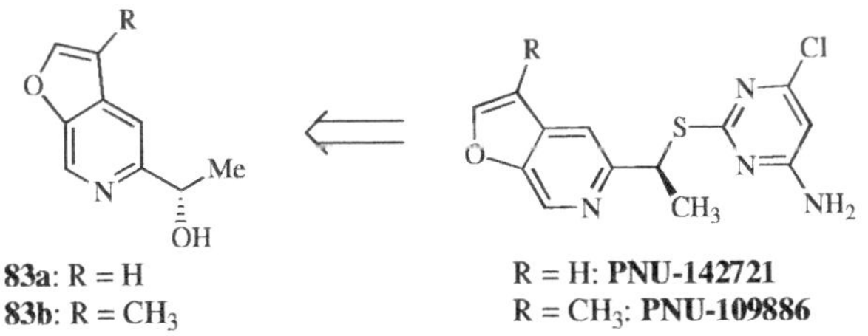

83a: R = H
83b: R = CH$_3$

R = H: **PNU-142721**
R = CH$_3$: **PNU-109886**

Scheme 25

To obtain **PNU-109886**, the stereoselective reduction of 1-(7-chloro-3-methylfuro[2,3-c]pyridin-5-yl)-ethanone **84** by using (−)-Ipc$_2$BCl was pursued (Scheme 26). Reduction was carried out in THF at room temperature overnight to give (*S*)-**85** in 94% yield and 92% ee, which was increased to 99% by recrystallization. The subsequent hydrodechlorination of (*S*)-**85** under transfer hydrogenolysis conditions afforded (*S*)-**83b** in >99% ee. The related reduction of the 3-methyldihydrofuropyridine derivative **86** afforded **87** as a mixture of diastereomers (99% yield) that was converted (hydrodechlorination, acetylation and dehydrogenation) into (*S*)-**83b** in 79% yield and 89% ee.

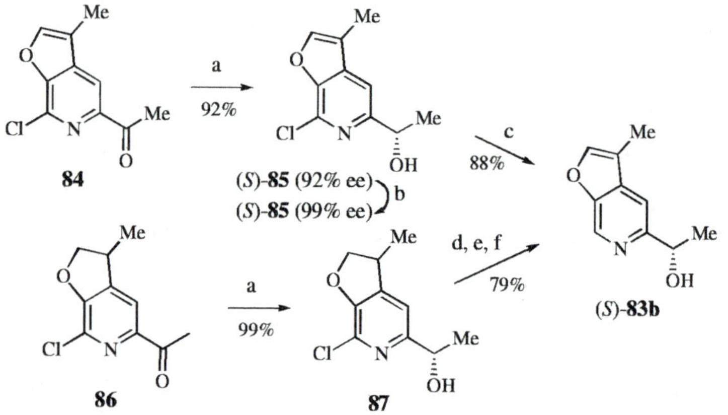

a: (-)-Ipc$_2$BCl, THF, -18 ¡C to r.t., overnight, 94%; b: recrystallization (ether/hexane), 78%; c: Pd(OH)$_2$/C, 1,4-cyclohexadiene, EtOH, reflux; d: H$_2$, 20% Pd(OH)$_2$/C; e: Ac$_2$O; f: 1. chloranil; 2. NaOH.

Scheme 26

The asymmetric reduction of pyridyl ketones was carried out using chiral 1,4-dihydronicotinamide (NADH) derivatives carrying polar groups in the chiral 3-carbamoyl moiety (Figure 2).[35–38] The reactions were performed employing 1–1.5 equivalents of the NADH model in the presence of 1–1.5 equivalents of magnesium perchlorate or bromide in dry acetonitrile or acetonitrile/chloroform (3/1) or THF.

Ohno *et al.* described the reduction of alkyl and fluorinated alkyl 2-pyridyl ketones in low to moderate enantiomeric excesses with the NADH model **88a**, observing that the optical yield decreases in the order of the substituent: Me> Et > *t*-Bu> *i*-Pr and CH$_3$> CH$_2$F> CHF$_2$> CF$_3$ (Table 2).[35] These results were interpreted in terms of conformation of the substrates for the alkyl 2-pyridyl ketones and of electronic competition effect between two substituents for fluorinated alkyl 2-pyridyl ketones.

Inouye *et al.* reported the reduction of 2-acetyl- and 2-benzoylpyridine by bis(NADH) model compounds **88b** and **88c** in which two chiral 1,4-dihydronicotinamide units, carrying L-prolinamide as the asymmetric centre, were spanned with a *p*-xylylene and a hexamethylene bridge, respectively.[36,37] With these reagents, very high stereoselectivities were achieved (Table 2). The origin of the enantiospecificity was tentatively ascribed to the specific blockage of diastereotopic faces of the dihydronicotinamide nuclei due to the probable C$_2$-conformation that the model adopts in association with the magnesium catalyst.

Levacher *et al.* examined the reduction of benzoylpyridine by quinolines and pyrido[2,3-b]indoles based NADH mimics. In these reducing agents, the 5,6-double bond was included in an electron rich homo or hetero aromatic ring that should favour the hydride equivalent departure through the electron donating effect of the aromatic ring and so open the possibility of using different metal ions. Benzoylpyridine was reduced quantitatively with **88d** (Figure 2) in the presence either of magnesium perchlorate in refluxing acetonitrile or magnesium bromide in refluxing THF, but the stereoselectivity was higher with the former system (78 and 47% ee, respectively) (Table 2).[38]

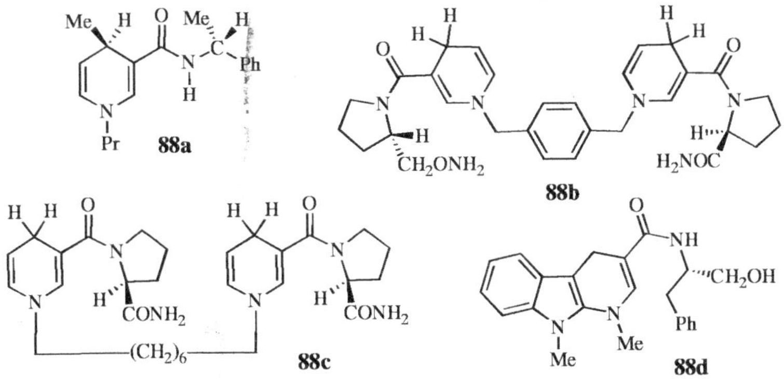

Figure 2

Table 2. Asymmetric reduction of 2-pyridyl ketones by chiral NADH model compounds.

R	NADH	yield (%)	conversion (%)	ee (%)	configuration	reference
Me	**88a**[a]	61	100	63	(R)	35
Et	**88a**	85	76	52	(R)	35
i-Pr	**88a**	75	92	0	--	35
t-Bu	**88a**	32	47	43	(R)	35
CH$_2$F	**88a**	88	98	53	(S)	35
CHF$_2$	**88a**	75	100	30	(S)	35
CF$_3$	**88a**	34	100	16	(S)	35
Me	**88b**[b]	67	100	90	(R)	36
Ph	**88b**	72	100	100	(R)	36
Me	**88c**[b]	84	100	67	(R)	37
Ph	**88c**	67	100	93	(R)	37
Ph	**88d**[c]	--	100	47	--	38
Ph	**88d**[d]	--	100	68	--	38

[a]Procedure: **88a** (1.0 equiv), Mg(ClO$_4$)$_2$ (1.0 equiv), MeCN, 25 °C.

[b]Procedure: **88b** (1.0 equiv), Mg(ClO$_4$)$_2$ (1.0 equiv), MeCN/CHCl$_3$ (3/1), 25 °C, 1–17 h.

[c]Procedure: **88c** (1.5 equiv), MgBr$_2$ (1.5 equiv), THF, reflux, 24 h.

[d]Procedure: **88d** (1.5 equiv), Mg(CO$_4$)$_2$ (1.5 equiv), MeCN, reflux, 24 h.

Optically active pyridyl ethanols were obtained in moderate to good enantiomeric excesses from the enantioselective reduction of acetylpyridines using lithium borohydride modified with (R,R)-N,N'-dibenzoyl-cystine **89** and ethanol (Scheme 27).[39]

1-(pyridin-2-yl)ethanol: 58%, 56% ee
1-(pyridin-3-yl)ethanol: 73%, 87% ee
1-(pyridin-4-yl)ethanol: 72%, 79% ee

a: (*R*,*R*)-**89** (1.1 equiv), LiBH$_4$ (3.6 equiv), EtOH (1.6 equiv), THF, -78 °C, 5 h.

Scheme 27

The reduction of 2-, 3- and 4-acylpyridines with (−)-bornan-2-*exo*-yloxyaluminium dichloride and LiAlH$_4$ in diethyl ether afforded the corresponding (*R*)-pyridyl alkanols with fairly good yields (60–75%), but low stereoselectivities (11–37% ees) (Scheme 28).[40]

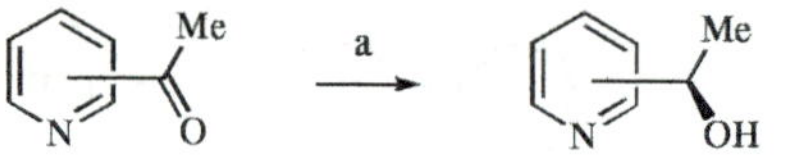

1-(pyridin-2-yl)ethanol: 75%, 11% ee
1-(pyridin-3-yl)ethanol: 60%, 37% ee
1-(pyridin-4-yl)ethanol: 70%, 15% ee

a: [bornyl]-OH (10 mmol), AlCl$_3$ (8.6 mmol), LiAlH$_4$ (2.5 mmol), Et$_2$O, 0 °C to r.t.

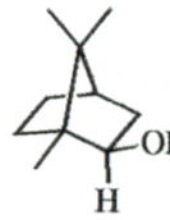

Scheme 28

The three isomeric acetylpyridines were reduced with sodium borohydride in the presence of amphiphilic dendrimer bearing sugar moieties at their ends [G(3)G] in THF at 0 °C. Yields (25–88%) and enantioselectivities (43–90% ees) were depending on the position of the carbonyl group relatively to the nitrogen atom. The best result was obtained with 2-acetylpyridine (88% yield, 90% ee), owing to possible hydrogen bonding with a hydroxylic group of the dendrimer.[41]

Complexes formed by reaction of chiral diols with LiAlH$_4$ were used to reduce prochiral ketones to optically active alcohols. Among them, 2-acetylpyridine was reduced to (*S*)-1-(pyridin-2-yl)ethanol in 37–47% yields and 13–21% ees using the complexes **90a** and **90b** (Scheme 29).[42]

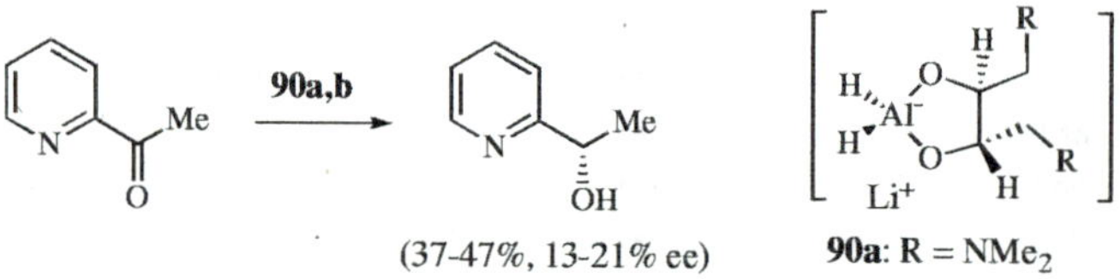

(37-47%, 13-21% ee)

90a: R = NMe$_2$
90b: R = 1-pyrrolidinyl

Scheme 29

Optically active pyridyl ethanols were produced from the reduction of acetylpyridines at a mercury electrode providing that a catalytic concentration of certain alkaloids were present.[43] A variety of chiral alkaloids were screened and among them, strychnine gave the highest optical yields. (+)-1-(Pyridin-2-yl)ethanol and (+)-1-(pyridin-4-yl)ethanol were obtained in 48 and 40% optical purity, respectively, whereas the reduction of 3-acetylpyridine gave optically inactive alcohol under all conditions employed. The 2,3-di-(pyridyl)butane-2,3-diols formed competitively in the reductions of these three ketones were optically inactive in all cases.

320

2.5.2. Catalytic enantioselective reduction

2.5.2.1. Borane reduction

Asymmetric reduction of prochiral ketones with chiral oxazaborolidines is an useful syntheyic methods to synthesize optically active secondary alcohols.[44]

Quallich and Woodall first reported the catalytic enantioselective reduction of pyridyl ketones with chiral oxazaborolidines in the presence of borane excess, notwithstanding, substrates with borane coordinating groups were not anticipated to afford high ees due to the possibile formation of borane complexes that not only decrease the borane source but also compete with the enantioselective carbonyl reduction.[45] Thus, 2-acetylpyridine was reduced with oxazaborolidine (R)-**92** and 1.7 equivalents of borane to yield the alcohol in 45% ees, whereas 3- and 4-acetylpyridine were reduced with higher enantioesectivities (62 and 65% ee, respectively) (Scheme 30). Azatetralone **90** gave the related alcohol with lower ee (40%). Using 1.0 equivalents of (R)-**92**, 3-acetylpyridine was reduced with enantioselectivity rise up to 80% ee.

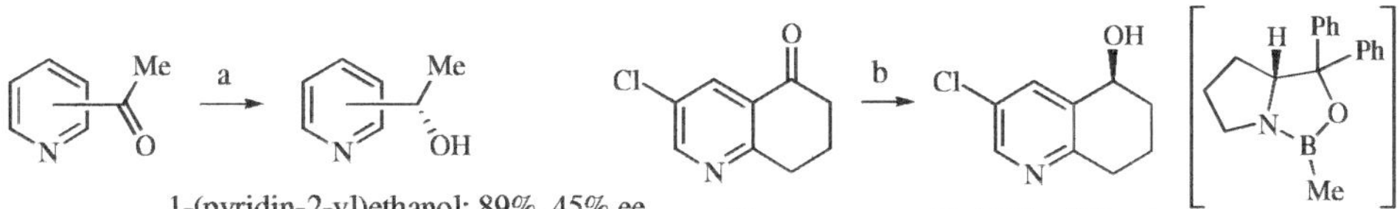

1-(pyridin-2-yl)ethanol: 89%, 45% ee
1-(pyridin-3-yl)ethanol: 92%, 62% ee
1-(pyridin-4-yl)ethanol: 93%, 65% ee

90

91 (85%, 40% ee)

(R)-**92**

a: (R)-**92** (0.2 equiv), $BH_3 \cdot Me_2S$ (1.7 equiv), THF, r.t.; b: (S)-**92** (5 mol%), $BH_3 \cdot Me_2S$ (1.7 equiv), THF, r.t.

Scheme 30

Next, a more efficient method for the asymmetric borane reduction with reagent prepared *in situ* from amino alcohols and trimethylborate was reported.[46] Among the examined amino alcohols, **93** and **94** were used in the reduction of pyridyl ketones (Scheme 31).

R = Me or Ph

93

94

a: **93** or **94**, $B(OMe)_3$ (1.2 equiv against the amino alcohol), $BH_3 \cdot Me_2S$ (7-20 equiv against the ketone), THF.

ketone	amino alcohol (equiv)	temp (°C)	yield (%)	% ee (conf.)
2-acetyl	**94** (0.2)	0-5	95	24 (S)
2-acetyl	**94** (0.5)	0-5	95	79 (S)
2-acetyl	**94** (1.0)	0-5	95	98 (S)
3-acetyl	**94** (0.1)	0-5	93	99 (S)
4-acetyl	**94** (0.1)	0-5	95	99 (S)
4-benzoyl	**93** (0.1)	25-30	99	69 (R)
4-benzoyl	**94** (0.1)	25-30	95	83 (S)

Scheme 31

Thus, 3- and 4-acetylpyridine were reduced in highly stereoselective manner (99% ee) by addition of borane-dimethyl sulfide complex, in the presence of catalytic amount of **94** and trimethylborate. The reduction of 2-acetylpyridine with 0.1 equivalents of **94**, provided 1-(pyridin-2-yl)ethanol low enantioselectivity, which rose up to 98% ee by increasing the amount of the amino alcohol. The reduction of 4-benzoylpyridine with amino alcohol **94** gave more satisfactory results relative to amino alcohol **93**.

A notable example of the application of asymmetric borane reduction using the system amino alcohols **94** and trimethylborate was described by Nicolau *et al.* in the construction of the antibiotic thiostrepton. In this circumstance, a 4-acetyl-5,6,7,8-tetrahydroquinoline was reduced to the corresponding alcohol in 95% yield and 90% ee (see Scheme 52).[24]

The oxazaborolidine *(S)*-**92** has been recently employed to catalyze the enantioselective reduction of cyclic pyridyl ketones **95** and **97** (Scheme 32).[47] Under optimal conditions, the catalytic enantioselective reduction of **95** with borane *(S)*-**92** generated *(R)*-alcohol **96** in quantitative yield with 93% ee, which was subjected to recrystallization to give enantiomerically pure chloroalcohol **96** with 50% yield. Using a similar procedure, its six-membered analogue, *(R)*-**98**, was also successfully prepared enantiomerically pure and in good yield from the ketone **97**.

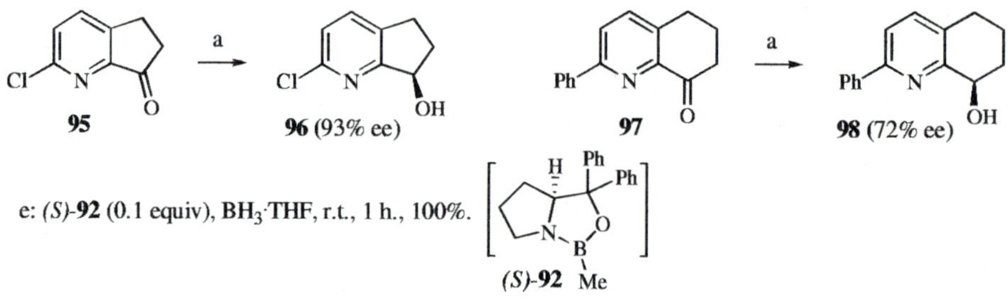

e: *(S)*-**92** (0.1 equiv), BH$_3$·THF, r.t., 1 h., 100%.

Scheme 32

Corey *et al.*, in order to prepare the *(S)*-carbinoxamine **102**, a medically useful histamine antagonist, developed a highly enantioselective catalytic reduction of aroylpyridines.[48]

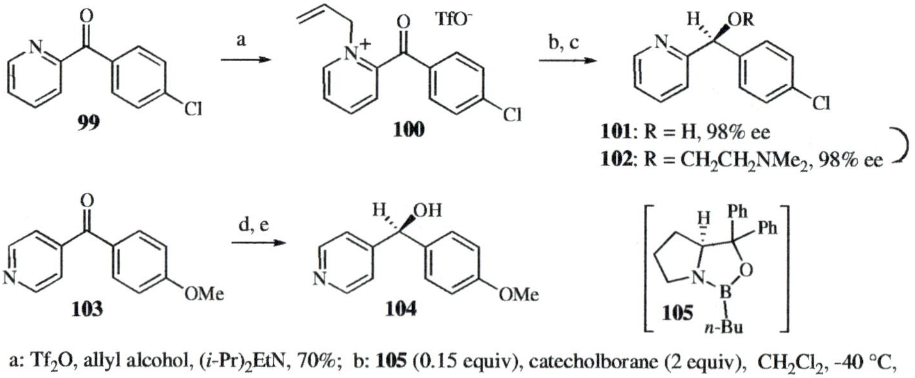

101: R = H, 98% ee

102: R = CH$_2$CH$_2$NMe$_2$, 98% ee

a: Tf$_2$O, allyl alcohol, (*i*-Pr)$_2$EtN, 70%; b: **105** (0.15 equiv), catecholborane (2 equiv), CH$_2$Cl$_2$, -40 °C, 36 h; c: *n*-BuNH$_3^+$ HCO$_2^-$ (0.15 equiv), Pd(PPh$_3$)$_4$ (0.05 equiv), 3 h, 78% (last two steps); d: BF$_3$·Et$_2$O, CH$_2$Cl$_2$; e: **105** (0.15 equiv), catecholborane (1.5 equiv), CH$_2$Cl$_2$, -78 °C, 17 h.

Scheme 33

Initial experiments on the oxazaborolidine(**105**)-catalyzed reduction of 2-benzoylpyridine with or without the addition of Lewis acid to coordinate the pyridine nitrogen, gave unsatisfactory results. Next, high enantiomeric induction was obtained using a readily removable *N*-allyl group both to protect the nitrogen and to provide spatial bias for the enantioselective reduction. Thus, treatment of **100** with the chiral oxazaborolidine **105** (0.15 equivalents) and catecholborane (2 equivalents) at −40 °C afforded a *N*-allyl alcohol intermediate that was not isolated, but it was treated with Pd(PPh₃)₄ and *n*-butyl ammonium formate[49] to give the pyridyl alcohol **101** in 78% yield and 98% enantiomeric excess (Scheme 33). Using a similar procedure, the reduction of the 4-(methoxybenzoyl)pyridine(**103**)-BF₃ adduct could be accomplished to afford the pyridyl alcohol **104** with an enantioselectivity of about 30:1 (Scheme 33).

Garcia *et al.* carried out the enantioselective reduction of a variety of ketones with borane and a catalytic amount of (*R*)-*B*-methyl-4,5,5-triphenyl-1,3,2-oxaborolidine **106** (Scheme 34).[50] Among the examined ketones, 3-acetylpyridine was reduced with excellent enantioselectivity to give (*S*)-1-(pyridin-3-yl)ethanol in 96% ee.

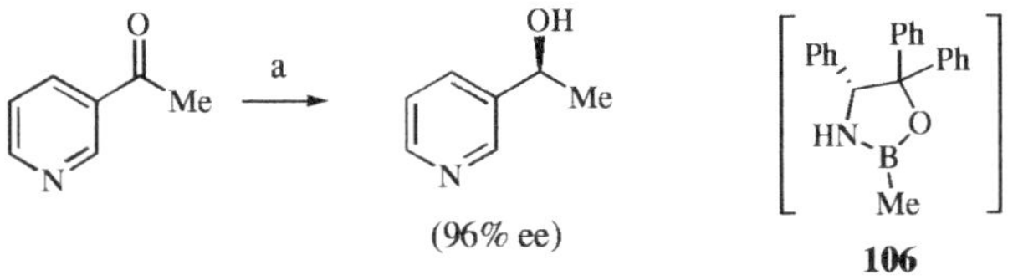

a: **106** (0.1 equiv), BH₃·Me₂S (1.2 equiv), THF, 0 °C, 1.5 h.

Scheme 34

Ortiz-Marciales and co-workers investigated the effectiveness of several spiroborate ester catalysts in the asymmetric borane reduction of pyridyl ketones under different reaction conditions.[51] Highly enantiomerically enriched pyridyl ethanols and phenyl(pyridin-3-yl)methanol were obtained using 1–10% catalytic loads of the spiroborate **107** derived from diphenylprolinol and ethylene glycol (Figure 3).

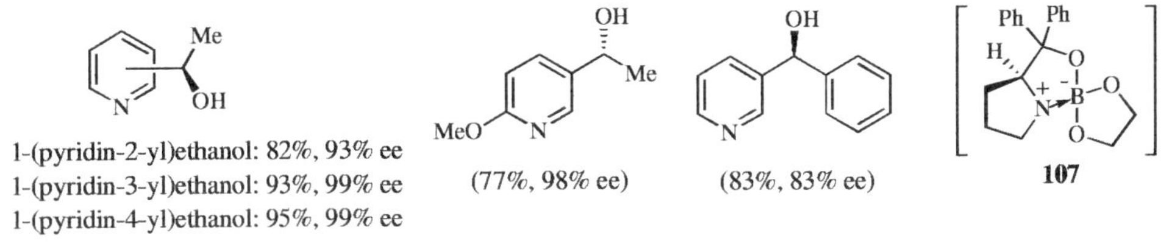

1-(pyridin-2-yl)ethanol: 82%, 93% ee
1-(pyridin-3-yl)ethanol: 93%, 99% ee
1-(pyridin-4-yl)ethanol: 95%, 99% ee

(77%, 98% ee)

(83%, 83% ee)

107

Reaction conditions: **107** (1-10 mol%), BH₃·Me₂S (1.6-2.0 equiv), THF, 25 °C, 1 h.

Figure 3

2.5.2.2. Hydrogenation transfer reactions

Owing to its operational simplicity and utilization of a cheap and safe hydrogen source, catalytic asymmetric hydrogen transfer reduction of ketones using a variety of ruthenium complexes with different hydrogen sources have been extensively investigated to access optically active alcohols.[52] Figure 4 shows a series of ruthenium based catalysts used in the asymmetric transfer hydrogenation of pyridyl ketones.

Ikariya and co-workers[53] have demonstrated that RuCl[(*S*,*S*)-*N*-(*p*-toluensulfonyl)-1,2-diphenyl-ethylenediamine](*p*-cymene) (1*S*,2*S*)-**108** (Figure 4), serves as an efficient catalyst for the asymmetric transfer hydrogenation of a series of pyridyl ketones (Figure 5) with a substrate/catalyst molar ratio of 1000 to 200 and with HCOOH as a hydrogen source. Yields and enantioselectivities are generally high (>85% ee), except in the case of the 2-benzoylpyridine **113f** that afforded the corresponding carbinol in only 9% ee (Table 3). The reduction of 1-(7-chlorofuro[2,3-c]pyridin-5-yl)ethanone **115b** gave the pyridyl ethanol **116b** (an intermediate of **PNU-142721**, a potent synthetic anti-HIV medicine) in almost quantitative yield and 86% ee. By contrast, the analogue acetylpyridine **115a** without the electron-withdrawing group, was reduced with the same enantioselectivity, but with a more less yield (36%). The 2,6-diacetylpyridine **77a** gave the diol **79a** in 91% yield and 99.6% ee besides the *meso*-diol (9%).

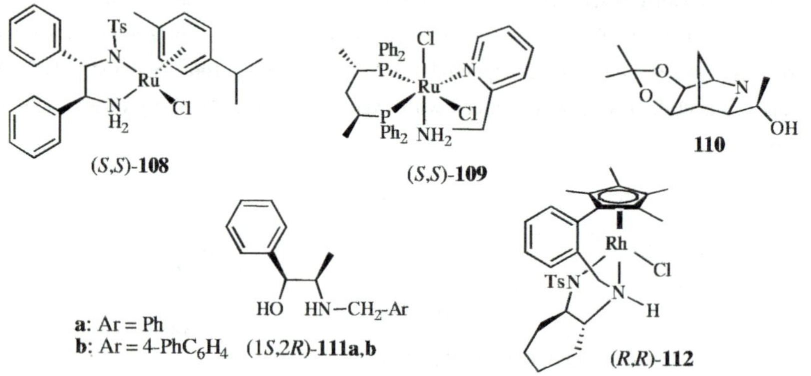

Figure 4

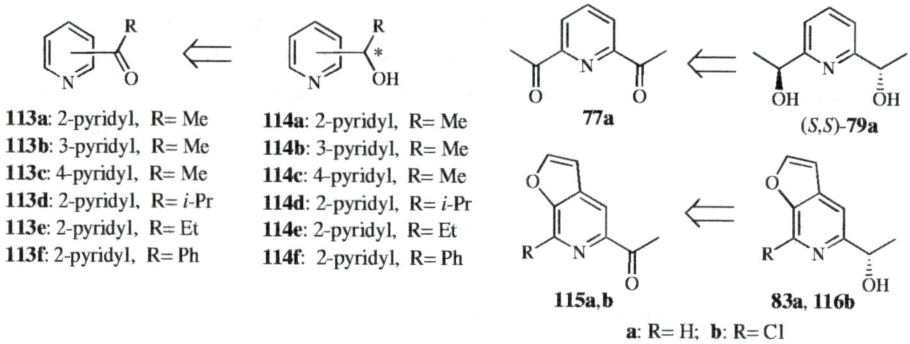

Figure 5

The system (1S,2S)-**108** with the couple HCOOH/NEt₃ as a hydrogen source was found to be very efficient (93–99% yields) for the reduction of a number of 6-bromopyridin-2-yl ketones **63a,b** and **117a–c** (Scheme 35).[54] However, a good stereoselectivity was obtained only in the case of methyl and *tert*-butyl ketones (about 90% ee). Optically pure alcohols were obtained by either recrystallization or column chromatography of the corresponding camphanates. Homo-coupling of bromo alcohols carried out in the presence of PdCl₂(PhCN)₂ and tetrakis(dimethylamino)ethylene (TDAE) afforded the corresponding 2,2'-bi-pyridines in moderate to good yields (58–93%), except in the case of the bromopyridine **119c** bearing a methoxy group (36% yield), which probably reduces the catalytic activity of palladium by decreasing the electron-donating ability nitrogen.

In contrast to (1*S*,2*S*)-**108**, the Ru catalyst (*S*,*S*)-**109** (Figure 4), obtained from (*S*,*S*)-Skewphos and 2-(aminomethyl)pyridine, allowed the rapid (TOF 1.5 10^5 h^{-1}) reduction of benzoylpyridine **113f** in 2-propanol/NaOH with high enantioselectivity (90% ee) (Table 3).[55]

The Ru catalyst [Ru(*p*-cymene)(**110**)] (Figure 4) effects highly enantioselective transfer hydrogenation of aromatic ketones at low catalyst loading and high rate.[56] Thus, the reduction of the three isomeric acetylpyridines **113a–c** was accomplished in less than 15 minutes with a substrate/catalyst ratio of 200 with about 90% enantiomeric excess (Table 3).

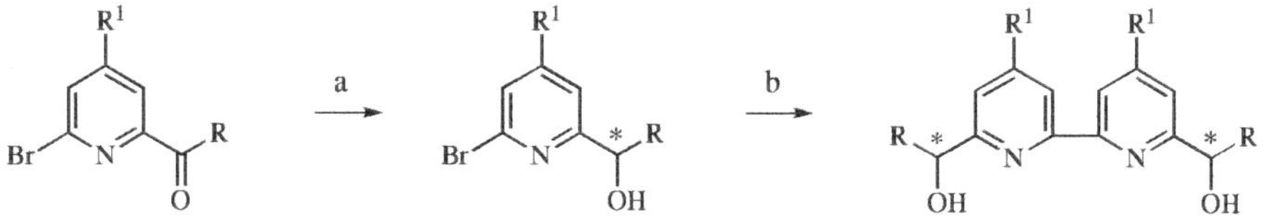

alcohol	R	R^1	yield (%)	ee (%)	conf.
64a	*t*-Bu	H	95	91	*S*
64b	Me	H	97	87	*S*
118a	*i*-Pr	H	93	18	*R*
118b	Ph	H	99	47	*R*
118c	*t*-Bu	OMe	94	90	*S*

a: (1*S*,2*S*)-**108** (0.01 equiv), HCOOH (4.3 equiv), NEt$_3$ (2.5 equiv), r.t.; b: PdCl$_2$(PhCN)$_2$ (0.05 equiv), TDAE (2 equiv), DMF, 50 °C, (*S*,*S*)-**69**: 93%, (*S*,*S*)-**79**: 79%, (*R*,*R*)-**119a**: 84%, (*R*,*R*)-**119b**: 58%, (*S*,*S*)-**119c**: 36%

Scheme 35

Table 3. Enantioselective transfer hydrogenation of pyridyl ketones (Figure 5).

ketone	catalyst	S/C	temp. (°C)	time (h)	alcohol	yield (%)	ee (%)	conf.	reference
113a	(*S*,*S*)-**108**	1000	27	24	**114a**	91	93	*S*	53
113b	(*S*,*S*)-**108**	200	27	24	**114b**	99	89	*S*	53
113c	(*S*,*S*)-**108**	200	27	24	**114c**	99	92	*S*	53
113e	(*S*,*S*)-**108**	200	27	24	**114d**	100	89	*S*	53
113f	(*S*,*S*)-**108**	200	27	24	**114b**	84	9	*S*	53
77a	(*S*,*S*)-**108**	200	27	24	**79a**	91	99.6	*S*,*S*	53
115a	(*S*,*S*)-**108**	200	10	24	**116a**	36	85	*S*	53
11b	(*S*,*S*)-**108**	200	10	24	**116b**	95	86	*S*	53
113f	(*S*,*S*)-**9**	2000	82	0.08	**114b**	84	90	*S*	55
113a	**110**/*p*-cymene	200	25	0.25	**114a**	90	91	-	56
113b	**110**/*p* cymene	200	25	0.07	**114b**	98	89		56
113c	**110**/*p*-cymene	200	25	0.05	**114c**	97	91	-	56
113a	**110**/*p*-cymene	100	20	16	**114a**	100	89	*R*	57
113a	**110**/*p*-cymene	200	20	6	**114a**	100	88	*R*	57
113c	(*R*,*R*)-**112**	200	28	1.5	**114c**	100	95	*R*	57

Significant catalytic activity and enantioselectivity (up to 89% ee) for the asymmetric transfer hydrogenation of 2-acetylpyridine was achieved by a combination of [RuCl$_2$(η^6-*p*-cymene)]$_2$ (0.5 mol%) and

N-substituted derivatives of (1*S*,2*R*)-norephedrine **111a** and **111b** (2.0 mol%) in 2-propanol containing *i*-PrOK (6.0 mol%) (Figure 4 and Table 3).[57] The complex Rh/tetramethylcyclopentadienyl complex containing tethered diamine (*R*,*R*)-**112** proved to be an excellent catalyst for the asymmetric reduction of a wide range of ketones in HCOOH/Et$_3$N media.[58] Reduction of 4-acetylpyridine was complete within 1.5 hours (for S/C=200) to give the corresponding alcohol in 95% ee (Table 3).

An effective method was developed to catalyze the asymmetric reduction of heteroaromatic ketones based on a highly activated form of a CuH ligated to the nonracemic biphosphine DTBM-SEGPHOS.[59] Thus, the catalyst **124** (Scheme 36) (generated *in situ* from DTBM-SEGPHOS, CuCl, *t*-BuONa and polymethylhydrosiloxane (PMHS) used as an inexpensive source of hydride) effected the reduction of acetyl pyridines **113a–c** with a substrate-to-ligand ratio on the order of 2000:1, in high yields (92–97%) and enantiomeric excesses (75–90%). Next, this protocol was extended to aryl pyridyl ketones (Scheme 36).[60] The hydrosilylation of 2-benzopyridines **120a** and **120b** gave the alcohols in modest ees (56 and 67%, respectively), whereas the parent ketone **113f** led to racemic material. On the other hand, 3-pyridyl ketone **122** gave the alcohol **123** in much higher ee (86%).

The same authors applied this type of reduction to the preparation of a number of known physiologically active compounds.[61] Among them, the ketone **115a** (Scheme 36) was reduced in 88% yield and 97% ee to the alcohol (*R*)-**83a** which is a precursor of the HIV-1 non-nucleoside transcriptase inhibitor **PNU-142721** (see Scheme 25).

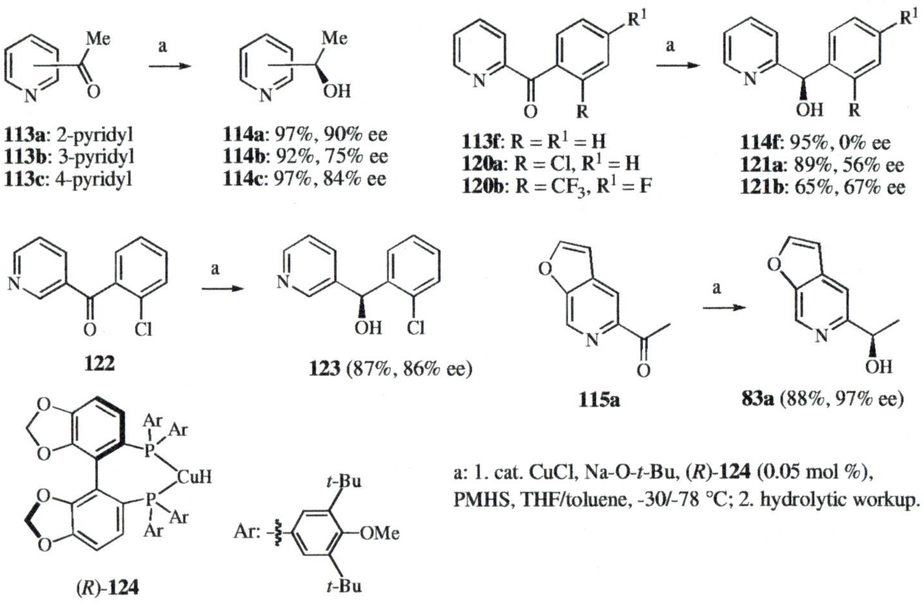

Scheme 36

2.5.2.3. Hydrogenation

Ruthenium complexes combined with *t*-BuOH in 2-propanol act as very effective catalytic systems for the hydrogenation of a diverse range of ketones.[62] A variety of ruthenium complexes used in the enantioselective reduction of pyrydyl ketones are shown in Figure 6.

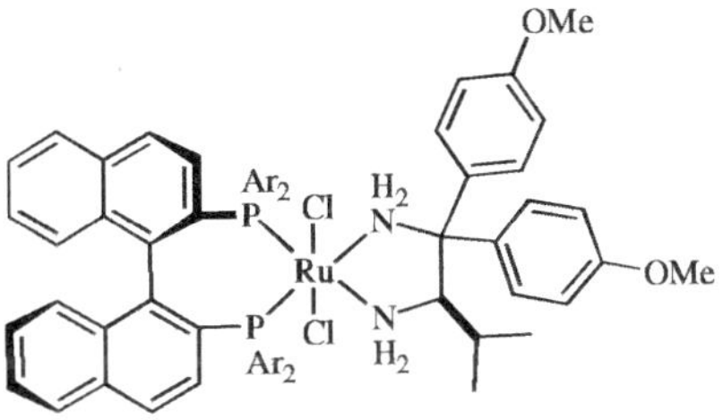

(*R,R*)-**125a** (Ar = 3,5-Me$_2$C$_6$H$_3$)
(*R,R*)-**125b** (Ar = 2-MeC$_6$H$_4$)
(*R,R*)-**125b** (Ar = C$_6$H$_5$)

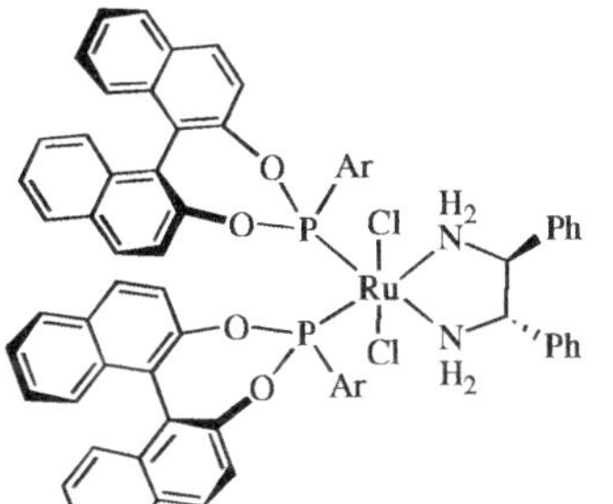

(*S,S,S,S*)-**126** (Ar = 2-BrC$_6$H$_4$)

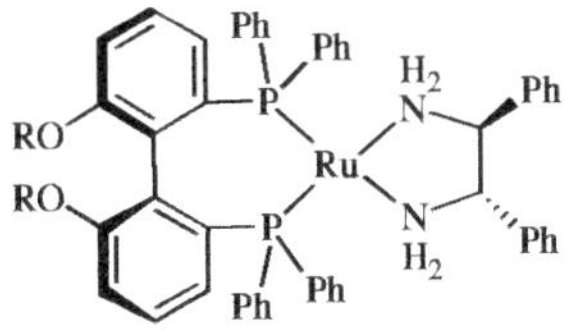

(*S,S,S*)-**127** (R = -PEG-OMe)

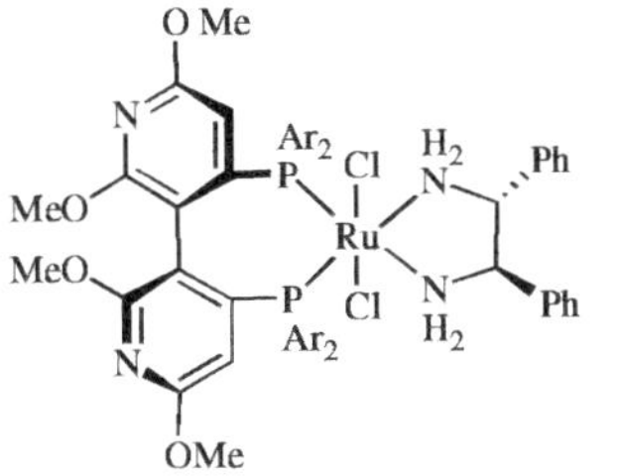

(*R,R,R*)-**128** (Ar = 3,5-Me$_2$C$_6$H$_3$)

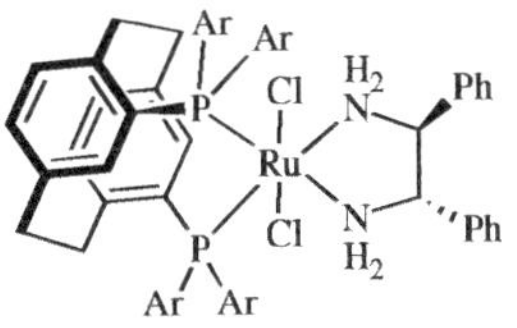

(*S*)-**129** (Ar = 3,5-Me$_2$C$_6$H$_3$)

Figure 6

Table 4. Enantioselective hydrogenation of pyridyl ketones (Figure 5).

ketone	catalyst	S/C	H$_2$ (atm)	time (h)	alcohol	ee (%)	reference
113a	(*R,R*)-**125a**	2000	8	3	**114a**	96	63
113b	(*R,R*)-**125a**	5000	8	12	**114b**	99.6	63
113c	(*R,R*)-**125a**	5000	8	12	**114c**	99.8	63
113d	(*R,R*)-**125a**	2000	8	3	**114d**	94	63
77a	(*R,R*)-**125a**	10000	8	17	**79a**	100	63
113b	(*R,R,R*)-**128**	4000	23.8	10	**114b**	97	69
113c	(*R,R,R*)-**128**	4000	23.8	24	**114c**	97.9	69
113b	(*S,S,S,S*)-**126**	2000	49.4	20	**114b**	70	68
113c	(*S,S,S,S*)-**126**	2000	54.3	8	**114c**	93	68
113a	(*S*)-**129**	1500	5.5–8	18	**114a**	78	65
113b	(*S*)-**129**	1500	5.5–8	18	**114b**	99	65
113c	(*S*)-**129**	1000	5.5–8	0.5–2.5	**114c**	96	65
113b	(*S,S,S*)-**127**	1000	20	12	**114b**	55.4	66

Noyori *et al.* reported that the complex *trans*-RuCl$_2$[(*R*)-xylbinap][(*R*)-daipen] (*R,R*)-**125a** (Figure 6) acts as an efficient catalyst for the asymmetric hydrogenation of hetero-aromatic ketones such as pyrydin-2-yl ketones.[63] Hydrogenation of the 2-acetylpyridine in 2-propanol containing (*R,R*)-**125a** (substrate/catalyst of 2000), *t*-BuOK and B[OCH(CH$_3$)$_2$]$_3$ at 8 atm [ketone:(*R,R*)-**125a**:borate=2000:1:20] afforded (*S*)-1-(2-pyridyl)ethanol in 96% ee and 100% yield. The higher analogue 2-methyl-1-(pyridin-2-yl)propan-1-one was hydrogenated smoothly without addition of the borate. The 3- and 4-acetylpyridine were hydrogenated equally with an S/C of 5000 to give the related (*S*)-alcohols with >99.5% ee and in high yield. Double hydrogenation of 2,6-diacetylpyridine **77a** (S/C=10000) gave only (*S,S*)-1,1'-(pyridine-2,6-diyl)diethanol **79a** among the three possible stereoisomers.

The catalyst (*R,R*)-**125a** (1.0 mol%) was applied to the asymmetric hydrogenation of aryl heteroaryl pyridyl ketones to afford the related secondary alcohol in almost quantitative yields (Scheme 37).[64] Decent enantioselectivity despite the isosteric nature of the aromatic group and pyridyl ring were obtained. Thus, 4-benzoylpyridine and 3-benzoylpyridine afforded the corresponding (*R*)-alcohols **130** and **37** in 57 and 73% ee, respectively (Scheme 37). Interestingly, the sense for the asymmetric induction of 2-benzoylpyridines was opposite to that of 3- and 4-benzoylpyridines, as the (*S*)-alcohols were obtained, suggesting that the asymmetric induction may be controlled by the coordination of the pyridine nitrogen to the catalyst. In the reductions of 2-benzoylpyridines the enantioselectivity for the hydrogenation was greatly influenced by the *para*-substitution on the phenyl ring. Thus, the alcohol **101** (R^1=Cl) was obtained in lower ee [60.6 *vs* 74.8% for **113f** (R^1=H)], whereas **104** (R^1=OMe) was obtained in higher ee (89.5 *vs* 74.8% for **113f**). On the other hand, the introduction of the methyl group *ortho* to the carbonyl moiety improved the enantioselectivity dramatically, as reduction of (pyridin-2-yl)(*o*-tolyl)methanone gave alcohol **131** in 98.7% ee. A similar result was obtained for isoquinolin-1-yl(phenyl)methanone, which gave alcohol **132** in 91.4% ee. The methyl group and the benzo fused ring in the parent ketones of alcohols **131** and **132** apparently generates sufficient stereobias between the aromatic groups to effect good enantiodifferentiation for the reduction. This procedure was also effective for obtaining the pyridyl thiazolyl alcohol (*R*)-**133** quantitatively and in 94.4% ee. This procedure was also applied to the synthesis of the chiral alcohol **135**, which serves as the penultimate intermediate to Roche's antimalarian drug (*R,S*)-mefloquine (Scheme 37).

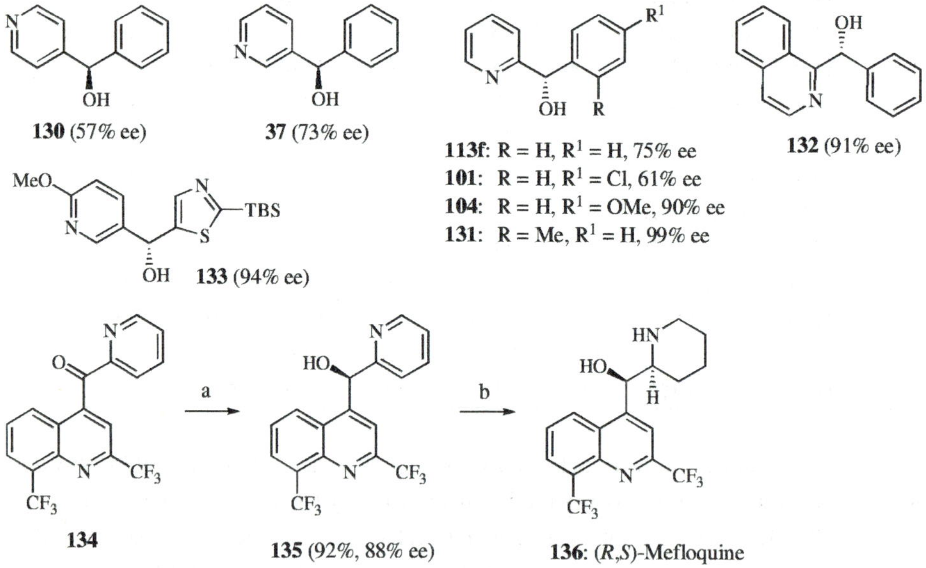

130 (57% ee)

37 (73% ee)

113f: R = H, R^1 = H, 75% ee
101: R = H, R^1 = Cl, 61% ee
104: R = H, R^1 = OMe, 90% ee
131: R = Me, R^1 = H, 99% ee

132 (91% ee)

133 (94% ee)

134

a

b

135 (92%, 88% ee)

136: (*R,S*)-Mefloquine

a: (*S,S*)-**125c** (1.0 mol%), K$_2$CO$_3$ (25 mol%), *i*-Pr$_2$NH/THF (4:1), 25 °C, H$_2$ (1.7 atm); b: Pt, H$_2$, HCl.

Scheme 37

Asymmetric hydrogenation of ketone **134** under the usual conditions proceeded uneventfully to give (*R*)-**135** in 88% ee and 92% isolated yield. PhanePhos-ruthenium-diamine complex (*S*)-**129** (Figure 6) catalyzes the asymmetric hydrogenation (*i*-PrOH, substrate/catalyst =1500–1000/1, *t*-BuOK (base/Ru=50/1), 18–20 °C, 5.5–8.0 atm initial H_2 pressure, for 0.5–2.5 hours) of a wide range of aromatic, heteroaromatic and α,β-unsaturated ketones with high activity and excellent enantioselectivity.[65] Among heteroaromatic ketones, the three isomeric acetylpyridines were examined. Pyridyl ethanols were obtained in good to very high enantiomeric excesses (78–99%) (Table 4).

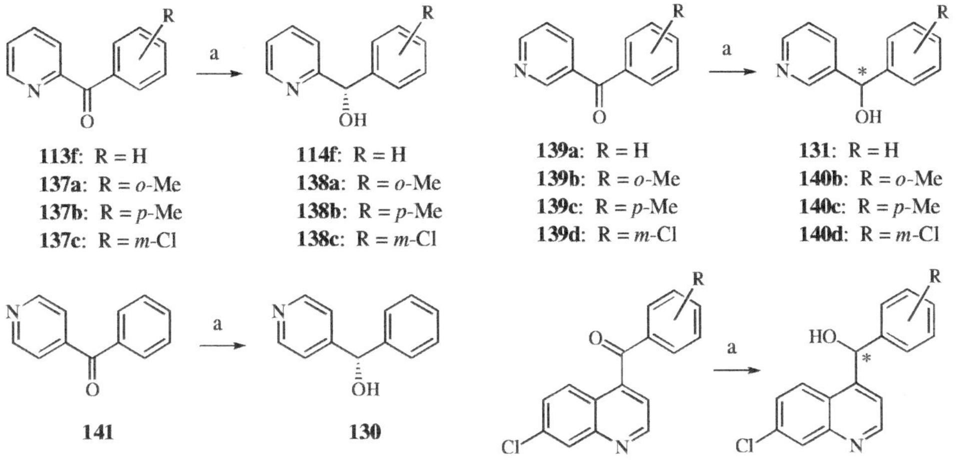

113f: R = H
137a: R = *o*-Me
137b: R = *p*-Me
137c: R = *m*-Cl

114f: R = H
138a: R = *o*-Me
138b: R = *p*-Me
138c: R = *m*-Cl

139a: R = H
139b: R = *o*-Me
139c: R = *p*-Me
139d: R = *m*-Cl

131: R = H
140b: R = *o*-Me
140c: R = *p*-Me
140d: R = *m*-Cl

141

130

142a: R = H
142b: R = *o*-Me
142c: R = *p*-Me

143a: R = H
143b: R = *o*-Me
143c: R = *p*-Me

a: (*R,R*)-**125a**, (*R,R*)-**125b** and (*R,R*)-**125c**, ketone/KOH/[Ru = 1/41000/5/1, *i*-PrOH, 30 °C, H_2 (19.7 atm).

		catalysts		
ketone	alcohol	**125a** ee (%)	**125b** ee (%)	**125c** ee (%)
113f	**114f**	70 (*S*)	33 (*S*)	14 (*S*)
137a	**138a**	99 (*S*)	34 (*S*)	20 (*S*)
137b	**138b**	60 (*S*)	40 (*S*)	8 (*S*)
137c	**138c**	50 (*S*)	9 (*S*)	15 (*S*)
139a	**131**	69 (*S*)	50 (*S*)	41 (*S*)
139b	**140b**	68 (*R*)	79 (*S*)	93 (*S*)
139c	**140c**	78 (*S*)	50 (*S*)	42 (*S*)
139d	**140d**	55 (*S*)	24 (*S*)	19 (*S*)
141	**130**	54 (*S*)	32 (*S*)	17 (*S*)
142a	**143a**	90 (*S*)	3 (*S*)	31 (*S*)
142b	**143b**	13 (*R*)	25 (*S*)	20 (*S*)
142c	**143c**	80 (*R*)	25 (*R*)	12 (*S*)

Scheme 38

A soluble polymer (MeO-PEG) supported biphenylbisphosphine (BIPHEP)-Ru/chiral diamine (1,2-diphenylethylenediamine) complex (*S,S,S*)-**127** (Figure 6), in which the polymer is attached to the two phenyl rings of BIPHEP ligand, has been prepared and shown to be highly active with good enantioselectivity for the catalyzed asymmetric hydrogenation of aryl ketones.[66] In this contest, the reduction of 3-acetylpyridine [2-propanol, ketone:ligand:Ru:diamine:*t*-BuOK = 1000:1.1:1:1:20 (mol ratio), H_2 (20 atm), 25 °C, 12 hours] was found to proceed with a high conversion (>99%), but providing the optically active (*R*)-1-(pyridin-2-yl)-ethanol with moderate enantioselectivity (55% ee).

Mortreux and co-workers have recently reported on the enantioselective hydrogenation of a series of pyridyl aryl ketones performed in the presence of Noyori type catalysts (*R,R*)-**125a**, (*R,R*)-**125b** and (*R,R*)-**125c** (Figure 6), which differ from the substituents on the phenyl residues on the phosphorous atoms.[67]

The hydrogenation proceeds under mild conditions providing chiral pyridyl aryl methanol derivatives with consistently high yields (80–95%) and moderate to excellent enantioselectivities (up to 99% ee) according to the structure of the chiral diphosphine. The results are summarized in Scheme 38. From this study emerged that the BINAP based catalyst (*R,R*)-**125c** affords commonly lower selectivities (8–42% ees) than those bearing XylBINAP (*R,R*)-**125a** or *p*-TolBINAP (*R,R*)-**125b**. Still, one substrate, **139b**, was hydrogenated with 93% ee in the presence of BINAP. In addition, it was roughly estimate that aryl ketones bearing 3-pyridyl moieties **139a–d** were hydrogenated with higher selectivities than their homologues with 2-pyridyl **113f**, **137a–c** and 4-pyridyl **141** or chloroquinolinyl **142a–c** residues.

A series of BINOL-derived ligands have been prepared and incorporated into ruthenium(II) complexes containing the diamine ligand (*S,S*)-1,2-diphenylethane-1,2-diamine (DPEN).[68] Among them, the complex (*S,S,S,S*)-BrXuPHOS-Ru-DACH (*S,S,S,S*)-**126** (Figure 6) proved to be an excellent catalyst for the asymmetric hydrogenation of ketones, giving reduction products with enantiomeric excesses up to 99%. In this contest, 3-and 4-acetylpyridines were reduced [2-propanol, *t*-BuOK (0.5 mol%), S/C=2000, H_2 (50–55 bar), 0 °C, 8–20 hours] to give the expected alcohols in 70 and 93% ees, respectively (Table 4).

The dipyridylphosphane/diamine-Ru complex *trans*-RuCl$_2$[(*R*)-Xyl-P-Phos][(*R,R*)-Dpen] (*R,R,R*)-**128** (Figure 6) combined with *t*-BuOK in 2-propanol acts as a very effective catalytic system for the enantioselective hydrogenation of a diverse range of simple ketones including heteroaromatic ketones such as 3-and 4-acetylpyridines.[69] The rate of the hydrogenation of 3-acetylpyridine was faster than that of 4-acetylpyridine, thus the former was hydrogenated with an S/C of 4000 under 350 psi H_2 to provide the (*S*)-1-(pyridin-3-yl)ethanol in 99.2% conversion and with 97.0% ee after 10 hours, while the latter required 24 hours for total conversion into (*S*)-1-(pyridin-4-yl)ethanol with similar enantioselectivity (97.9% ee) (Table 4).

2.5.3. Stoichiometric diastereoselective reduction

Chiral pyridyl alcohols and the related phosphinites and phosphites (Schemes 39 and 40) have been prepared from Moberg group and used as catalysts for the addition of diethylzinc to benzaldehyde and in the palladium-catalysed allylic substitutions.[3,70] Diastereoisomeric alcohols **145a** and **145b** were obtained by reaction of neomenthyl-1-nitrile **150a** with 2-pyridyllithium, followed by reduction of the ketone **144a** and chromatographic separation of the products (Scheme 39). The derivatives **146a** and **146b** were prepared by reaction of (6-bromopyridin-2-yl)lithium **4** with **150a** to give the ketone **144b**, followed by reduction with NaBH$_4$ to give a mixture of the (*S*) and (*R*) isomers in a ratio of 4:1. When **4** was treated with a 1:1 mixture of epimeric nitriles **150a** and **150b**, only the isomer **148** with (*R*) configuration at the carbinol atom was obtained in 24% yield. 6-Aryl substituted pyridyl alcohols **147a–c** were obtained in high yields (64–93%) by Suzuki cross-coupling of the bromo derivative **146a** with arylboronic acids. Pyridyl alcohols **152** and

154a–c were obtained analogously using methyl mandelate **161** as the chiral precursor (Scheme 40). Nickel-mediated dimerization of the TBDMS-protected **155** and the analogous lactic acid derivative afforded, after deprotection, bipyridine compounds **156** and **157**, respectively. Phosphinites **158, 159** and **160** were obtained in the usual way from **152** or **154a**, respectively. Moreover, other phosphinites were prepared from **152**.

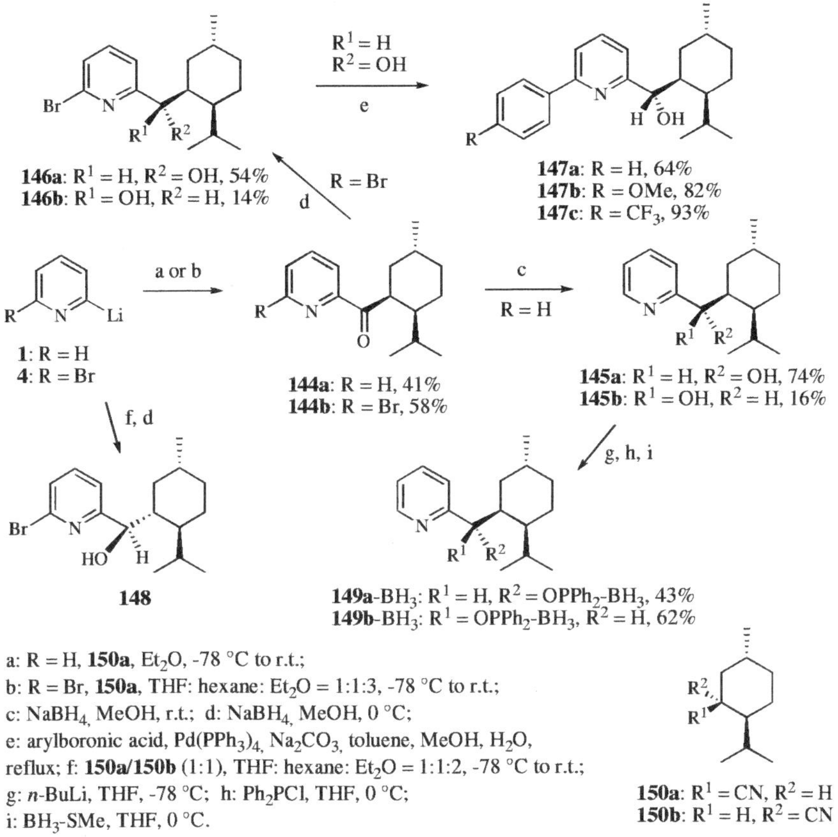

Scheme 39

Oxidation of the mixture of **44a–d** (for their synthesis see Scheme 12) with sodium dichromate afforded a mixture of the equatorial ketone **45a** and the axial ketone **45b** in 76% yield (Scheme 41).[18] Base equilibration of these ketones with *t*-BuOK in DMSO led to almost complete formation (>20:1) of the termodinamically more stable isomer **45a**. Reduction of **45a** in THF with NaBH$_4$ or LiAlH$_4$ proceeded with low diastereoselectivity, but the more hindered reagent lithium tri-*tert*-butoxyalluminium hydride gave a 14:1 mixture of **44a** and **44b** in 93% yield.

2.5.4. Catalytic diasteroselective reduction

A very practical synthesis of ephedrine analogues in high yields and enantiopurities was realized by a highly diastereoselective Meerwein-Ponndorf-Verley reduction of protected α-amino aromatic ketones using catalytic aluminum isopropoxide.[71] Among the examined substrates, an example of pyridyl ketone was reported. Thus, pyridine **162** was reduced to the alcohol **163** in good yield (82%) and with a distereomeric ratio > 99:1.

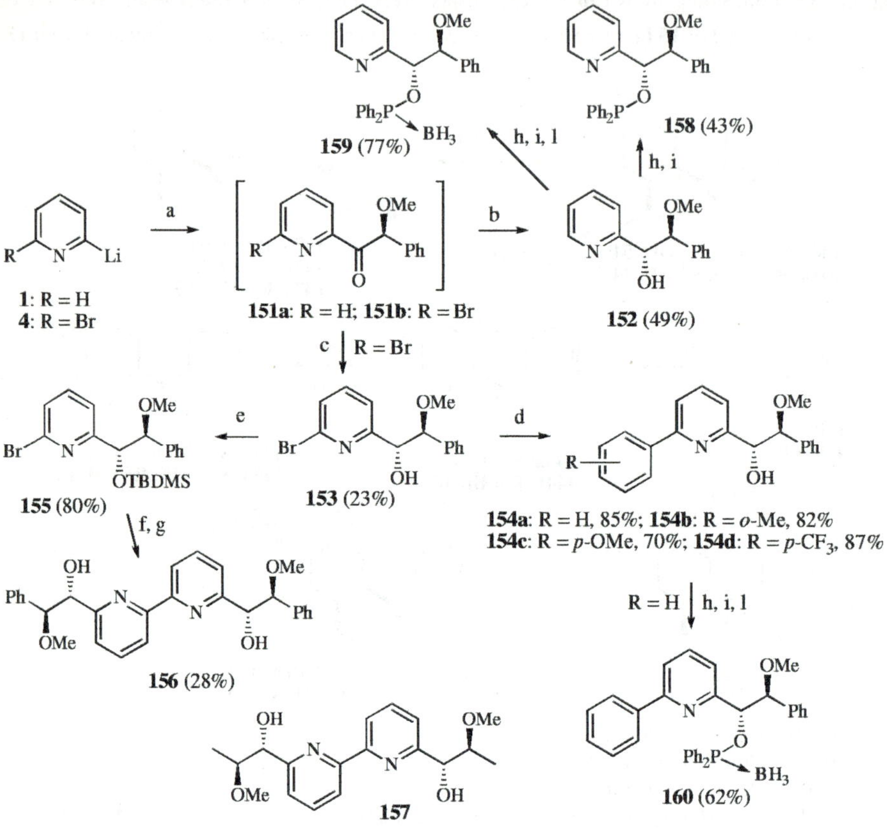

a: **161**, Et_2O, -78 °C; b: $NaBH_4$, MeOH, -78 °C ; c: $NaBH_4$, MeOH, 0 °C;
d: arylboronic acid, $Pd(PPh_3)_4$, Na_2CO_3, toluene, MeOH, H_2O, reflux;
e: TBDMSCl, imidazole, DMF, r.t.;
f: PPh_3, $NiCl_2$, Zn, DMF, 70 °C; g: TBAF, THF;
h: *n*-BuLi, THF, -78 °C; i: Ph_2PCl, THF, 0 °C;
l: BH_3-SMe, THF, 0 °C.

161 =

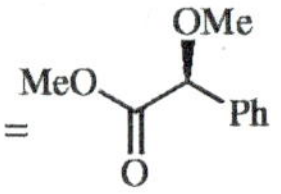

Scheme 40

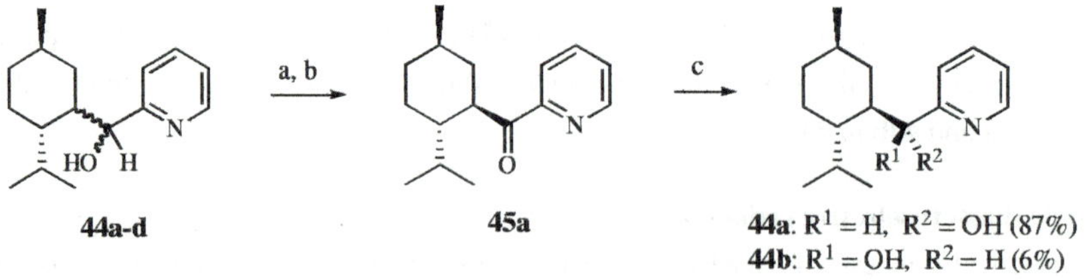

a: Et_2O, $Na_2Cr_2O_7$, H_2O, H_2SO_4, r.t., 12 h, 76%; b: *t*-BuOK, DMSO, r.t., 12 h, 93%;
c: THF, Li(*t*-BuO)$_3$AlH, r.t., 16 h, 93%.

Scheme 41

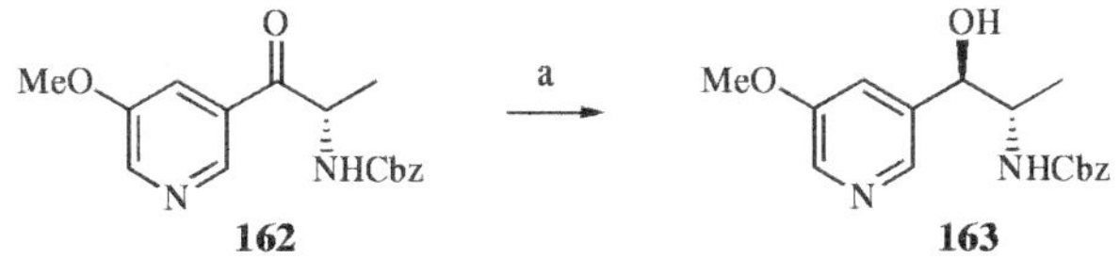

a: Al(O-i-Pr)$_3$ (0.6 equiv), i-PrOH (11 equiv), MePh, 50 °C, 48 h, 82%, >98% de.

Scheme 42

2.6. Cyclotrimerization of chiral 2-hydroxynitriles

2-Pyridyl alkanols have been prepared by the Chelucci group[72] starting from natural occurring chiral compounds and exploiting the cobalt(I)-catalyzed cyclotrimerization of chiral O-protected α-hydroxynitriles with acetylene in the key step.[73,74] The synthesis of the pyridyl alcohol **113a**[72a] starts from the aldehyde **165**, which was prepared in 82% overall yield from the commercial (S)-2-hydroxypropanoic acid methyl ester **164** by reaction with *tert*-butylchlorodimethylsilane in THF containing NEt$_3$ and 4-(dimethylamino)pyridine, followed by reduction at −78 °C of the protected alcohol using diisobutylaluminum hydride (Scheme 43). Compound **165** was converted into the nitrile **167** *via* the formation of the oxime followed by dehydration with N,N'-carbonyldiimidazole. In the next step, the nitrile **167** was cyclized with acetylene by using (π-cyclopentadienyl)cobalt 1,5-cyclooctadiene [CpCo(COD)] as the catalytic precursor and toluene as the solvent to give the crude pyridine **168**. The latter was treated with 10% hydrochloric acid to give the enantiomerically pure hydroxy pyridine **113a** in 60% overall yield based on **165**.

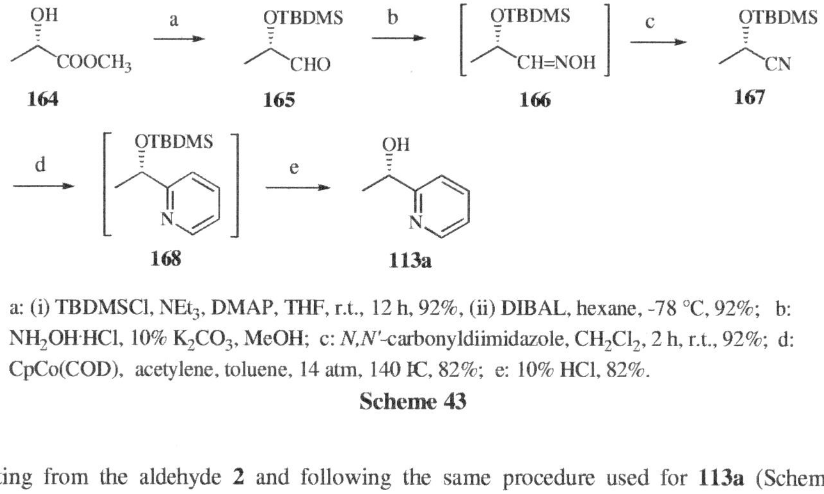

a: (i) TBDMSCl, NEt$_3$, DMAP, THF, r.t., 12 h, 92%, (ii) DIBAL, hexane, -78 °C, 92%; b: NH$_2$OH·HCl, 10% K$_2$CO$_3$, MeOH; c: N,N'-carbonyldiimidazole, CH$_2$Cl$_2$, 2 h, r.t., 92%; d: CpCo(COD), acetylene, toluene, 14 atm, 140 °C, 82%; e: 10% HCl, 82%.

Scheme 43

Starting from the aldehyde **2** and following the same procedure used for **113a** (Scheme 43), the pyridine **171** was obtained in 85% yield (Scheme 44).[72a] Deprotection of the acetonide group afforded 2-(1,2-dihydroxyethyl)pyridine **172** in 81% overall yield based on **2** and 98% enantiomeric excess.

Another example of this protocol is represented by ($1R,2R$)-1,2-bis(pyridin-2-yl)-1,2-ethanediol **177**, whose synthesis is reported in Scheme 45.[72b,c] Dimethyl 2,2-dimethyl[1,3]dioxolane-4,5-dicarboxylate **173** was converted into the corresponding dinitrile **175** by treatment with gaseous NH$_3$ in ethanol followed by dehydration of diamide **174** with p-toluenesulphonylchloride in pyridine. Cyclotrimerization of **175** with acetylene, followed by hydrolysis, completes the synthesis.

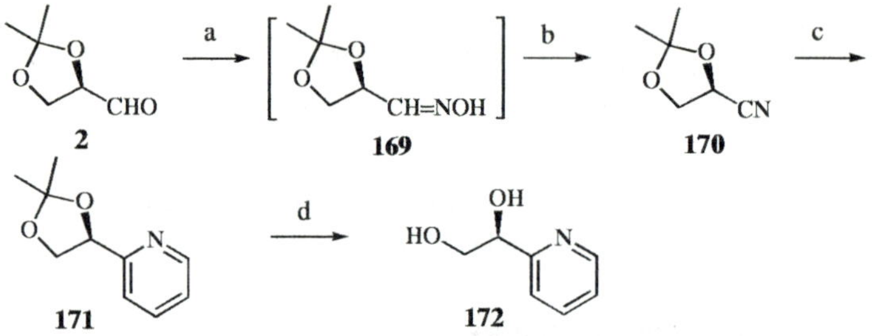

a: NH$_2$OH.HCl, 10% K$_2$CO$_3$, MeOH; b: *N,N'*-carbonyldiimidazole, CH$_2$Cl$_2$, 2 h, r.t., 87%;
d: CpCo(COD), acetylene, toluene, 14 atm., 140 °C, 85%; e: 6% HCl, 95%.

Scheme 44

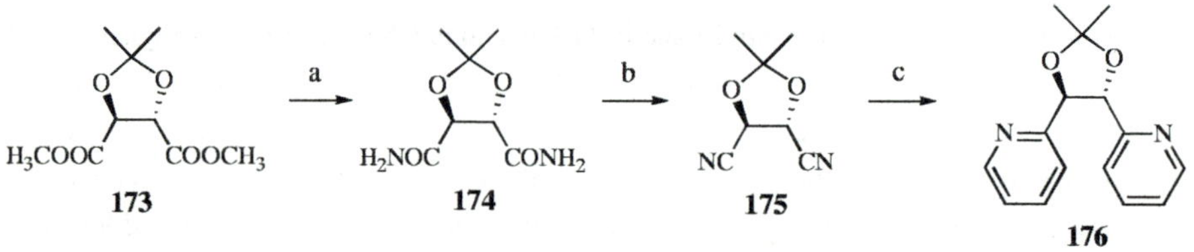

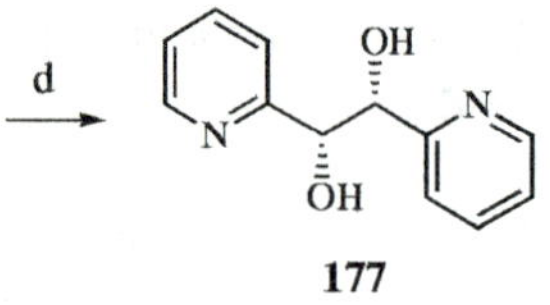

a: NH$_3$, EtOH, r.t., 10 d, 98%; b: *p*-CH$_3$C$_6$H$_4$SO$_2$Cl, pyridine, 100 °C, 5 h, 81%; c: CpCo(COD), acetylene, 8 atm., 140 °C, 72%; d: 0.1 M H$_2$SO$_4$, reflux, 4 h, 85%.

Scheme 45

Recently, Heller *et al.* have developed a two-step catalytic approach for the synthesis of chiral pyridyl alcohols,[75] in which the assembly of the pyridine moiety is carried out under mild conditions by a racemization-free photochemical cyclotrimerization process (Scheme 46).[76]

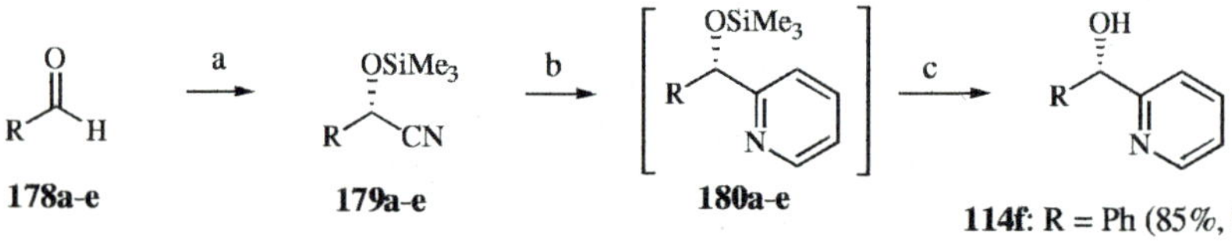

R = **a**: Ph, **b**: 4-MeOC$_6$H$_4$, **c**: 3-MeOC$_6$H$_4$, **d**: 4-ClC$_6$H$_4$, **e**: *t*-Bu

179 = **a**: 78%, 91% ee; **b**: 71%, 91% ee
    c: 92%, 93% ee; **d**: 91%, 88% ee
    e: 94%, 81% ee.

a: **182**, Me$_3$SiCN, CH$_2$Cl$_2$, r.t., 24 h;
b: CpCo(COD), acetylene, irradiation
($\lambda \sim$ 420 nm), r.t., 12 h;
c: MeOH, Bu$_4$NF (cat), r.t., 24 h

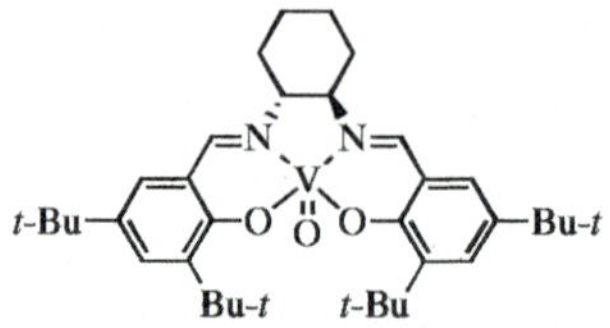

114f: R = Ph (85%, 91% ee)
181b: R = 4-MeOC$_6$H$_4$ (72%, 91% ee)
181c: R = 3-MeOC$_6$H$_4$ (84%, 92% ee)
101: R = 4-ClC$_6$H$_4$ (81%, 88% ee)
71:   R = *t*-Bu (91%, 82% ee)

Scheme 46

334

In the first step, the synthesis of cyanohydrins by cyanation of aldehydes in the presence of chiral transition-metal-salen complexes was investigated using either hydrogen cyanide or trimethylsilyl cyanide as cyanide sources. In both cases, the acetyl- or trimethylsilyl-protected cyanohydrins were obtained in high yields and good enantiomeric excesses. The behaviour of *O*-acetyl cyanohydrins in the light-promoted cobalt(I)-catalyzed cyclotrimerization with acetylene has been next studied, but the results were disappointing and yields did not exceed 25%. In contrast, under the same cyclotrimerization conditions, *O*-trimethylsilylprotected cyanohydrins **179** gave the corresponding pyridines in excellent yields and with the same enantiomeric excess of the starting material The *in situ* deprotection of silylated pyridyl alcohols **180** afforded in high yields the corresonding pyridyl alcohols, which were isolated in enantiomerically pure form by recrystallization.

Recently, the cyclotrimerization of nitriles with various silicon-tethered diynes was used by Schreiber *et al.* to generate a library of potential inhibitors of neuregulin-induced neurite outgrowth.[77] The reaction of silyloxaheptadiynes or silyoxaoctadiynes with nitriles catalyzed by CpCo(CO)$_2$ (25–30 mol%) at 140 °C in THF provided good yields of the desire bicyclic silylpyridines (Scheme 47).

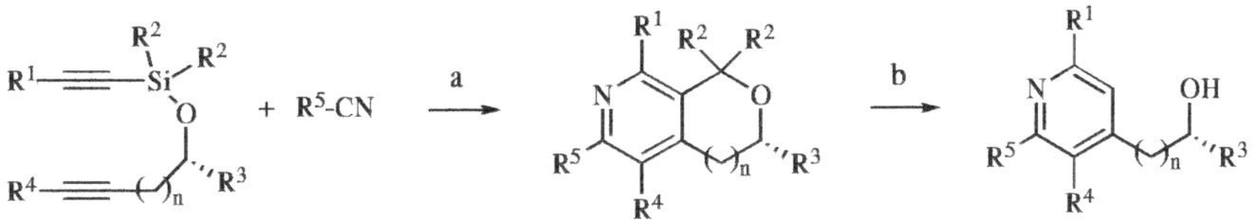

a: CpCo(CO)$_2$ (25-100 mol%), THF, 140 °C; b: TBAF, THF, 23 °C, 18 h.

Scheme 47

In some cases, better results were obtained by using a stoichiometric amount of the Co-complex. The bicyclic silylpyridines and monocyclic pyridines, prepared by treating the former with TBAF, were tested for the aforementioned biological activity (some representative examples are reported in Figure 6). Using this assay, 1-[2-phenyl-6-(3,4,5-trimethoxybenzyl)pyridin-4-yl]propan-1-ol (Figure 7) was discovered to be a potent inhibitor of the neuregulin/ErbB4 pathway, with an approximate EC$_{50}$ of 0.30 M, while its bicyclic precursor was inactive at 20 μM.

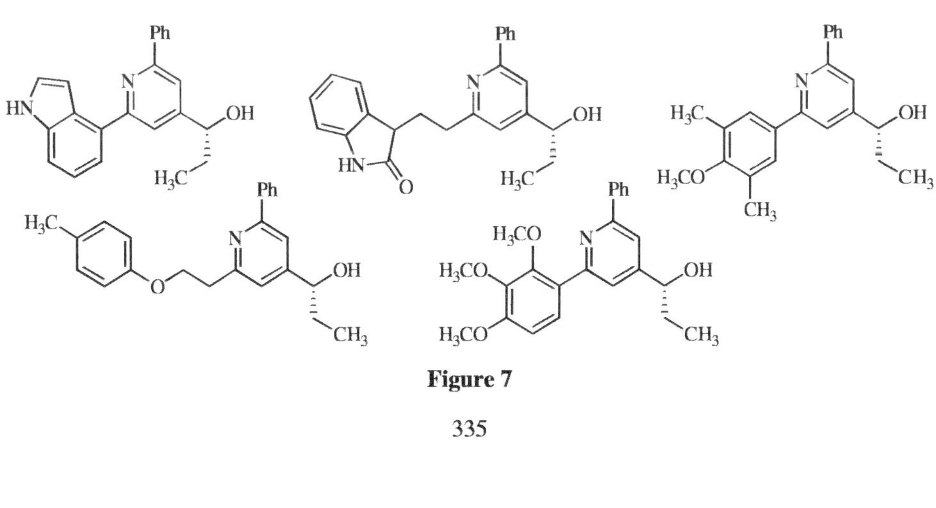

Figure 7

335

2.7. Asymmetric dihydroxylation

Wilson and Pfaltz exploited the asymmetric dihydroxylation[79] of alkenylpyridine as a key step to prepare bipyridine **185** (Scheme 48)[78] and quinolinyl phosphane **188** (Scheme 49).[33]

Asymmetric dihydroxylation of pyridine **183** with AD-mix-β in 12 hours afforded diol **184** in 57% yield (after transformation into the corresponding ketal) and 90% ee (Scheme 48).[78] The yield and rate of the AD reaction was improved later by performing the reaction with 1.0 mol% of potassium osmate dihydrate and 5.0 mol% of the ligand $(DHQD)_2PHAL$. This procedure afforded the corresponding ketal in 81% yield and 90% ee after 2 hours.

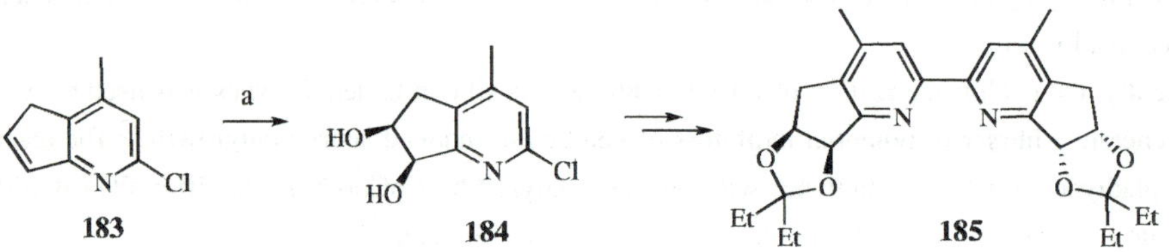

183 **184** **185**

a: AD-mix-β, *t*-BuOH/H_2O (1:1), 0 °C, 12 h or 1.0 mol% $K_2OsO_4{\cdot}2H_2O$, $(DHQD)_2PHAL$ (5.0 mol%), $K_3[F_2(CN)]_6$, K_2CO_3, *t*-BuOH/H_2O (1:1), 0 °C, 2 h.

Scheme 48

Using this, protocol Pfaltz *et al.* obtained the diol **187** by dihydoxylation of 2-vinylquinoline **188** in 46% yield and 94% ee (Scheme 49).[33]

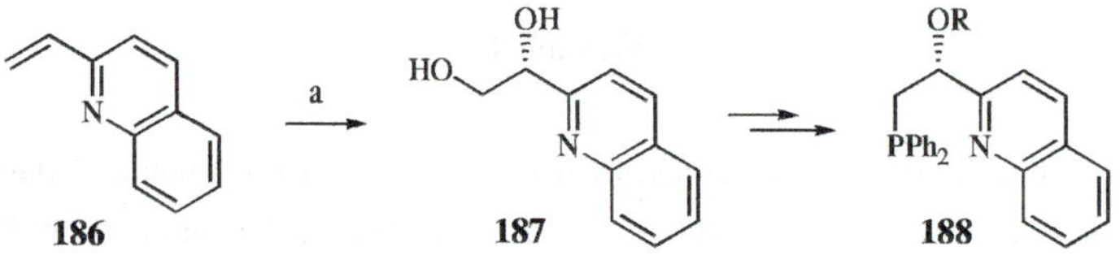

186 **187** **188**

a: $K_2OsO_4{\cdot}2H_2O$ (1 mol%), $(DHQD)_2PHAL$, $K_3[F_2(CN)]_6$ (3 equiv), K_2CO_3 (3 equiv), *t*-BuOH/H_2O (1:1), 0 °C to r.t.

Scheme 49

Contrary to what observed with 2-vinylquinoline, the AD-mix β-catalyzed direct dihydroxylation of 2-, 3- and 4-vinylpyridine gave moderate enantiomeric excesses (63–79%). However, the very low yields of the resulting diols precludes any further use of them (Scheme 50).[80]

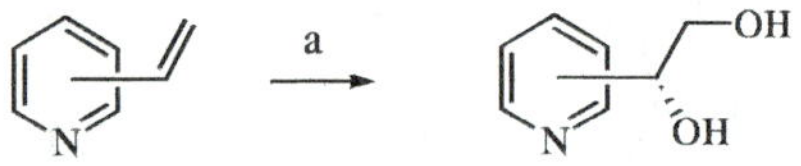

189a-c **190a-c**

a: 2-pyridyl, **b**: 3-pyridyl
c: 4-pyridyl

190a: 8%, 79% ee; **190b**: nd, 72% ee
190c: 10%, 63% ee

a: AD-mix-β (12 mg)/olefin, *t*-BuOH/H_2O (1/1), 0 °C, 23 h.

Scheme 50

Very recently, Chelucci and Sanfilippo have investigated the asymmetric synthesis of a rare example of axially locked 2,2'-bipyridine, using the Sharpless asymmetric dihydroxylation and Ullmann atropodiastereospecific coupling as the key steps.[81] Thus, (E)-1,2-di(2-bromopyridin-3-yl)ethene (E)-193, obtained in 63% yield by Wittig reaction between 2-bromopyridine-3-carboxaldehyde 191 and phosphonate 192, was asymmetrically dihydroxylated according to the Sharpless procedure using AD-mix-α in a two-phase system to give diol (S,S)-194 in 74% yield and >99% ee after recrystallization (Scheme 51). The Ullmann intramolecular coupling[82] of 194 (Cu powder, DMF, reflux, 1 hour) gave a clear reaction since the disappearance from the reaction mixture of the starting material was observed, but it was not possible to isolate the expected (5S,6S)-5,6-dihydroxy-5,6-dihydro-1,10-phenanthroline 195. However, when the dioxocine 196, formed from the diol 194 by treatment with NaH and then with 1,2-bis(bromomethyl)-benzene, was subjected to the usual Ullmann reaction, the conformationally fixed 2,2'-bipyridine 197 was successful obtained in 38% yield (Scheme 51).

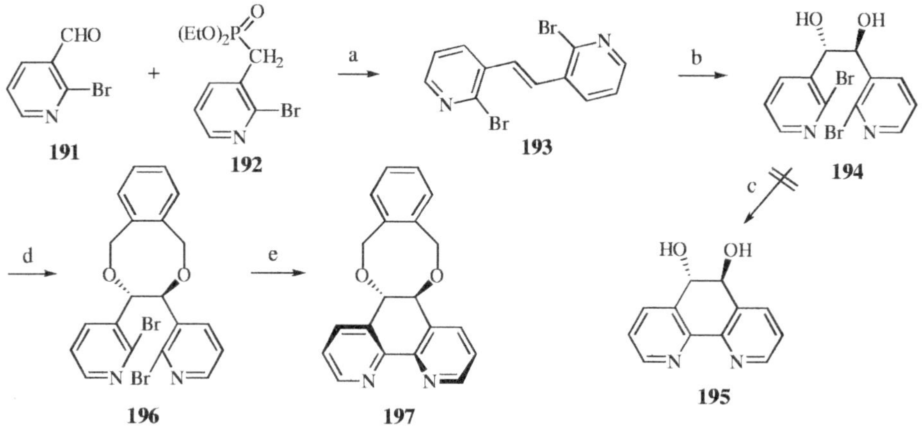

(a) NaH, 1,4-dioxane, 90 °C, 1 h, 63% yield; (b) AD-mix-α, t-BuOH/H₂O, 48 h, r.t., 74%; (c) Cu, DMF, reflux, 1 h; (d) 1. NaH, THF, r.t., 1 h; 2. 1,2-bis(bromomethyl)benzene, r.t., 72 h, 28% yield; (e) Cu, DMF, reflux, 3 h, 38%.

Scheme 51

2.8. Chiral aziridine and epoxide ring opening

Savoia *et al.* have recently reported the synthesis of the 2-pyridyl-substituted aziridine 199 by addition of chloromethyllithium to the pyridineimine 198 derived from (S)-valinol[83a] and the subsequent stereoselective opening of the aziridine ring by attack of nitrogen, sulfur and oxygen nucleophiles.[84b] In particular, heating a mixture of aziridine 199 and *p*-toluenesulfonic acid (20 mol%) in 9:1 acetonitrile/water at the reflux temperature for 6 hours gave a 82:18 mixture of the regioisomeric ring-opening products 200 (77% yield) and 201 (14% yield), which were separated by column chromatography (Scheme 52).[84b] Slightly better results were obtained using cerium trichloride heptahydrate (30 mol%); with the other experimental conditions remaining constant an improved regioselectivity (86:14) was in this way obtained.

Nicolau *et al.* reported the total synthesis of the antibiotic thiostrepton by the union of a number of suitable fragments (Figure 8). Among them, the synthesis of a quinaldic acid macrocycle with a 5,6,7,8-tetrahydroquinoline portion bearing two stereocentres with C-OH groups was described.[24] Scheme 52 expounds part of the synthesis of this unit where the two C-OH stereocentres are generated by borane-promoted asymmetric reduction and epoxide ring opening.

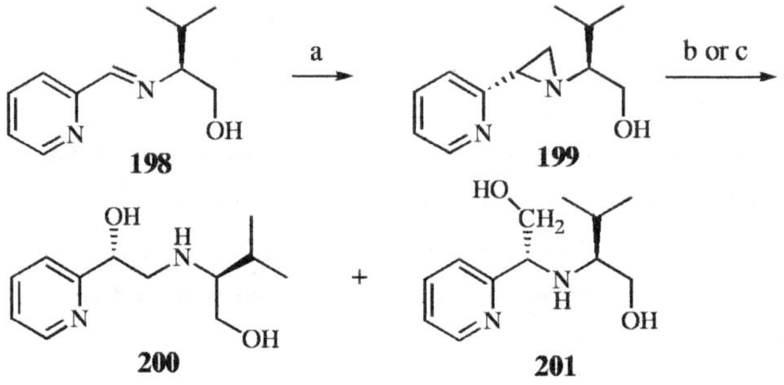

a: (i) CH_2ICl, LiBr, THF, -78 °C, (ii) MeLi, -78 to 20 °C, (iii) NH_4F, $MeOH/H_2O$,
b: p-$MeC_6H_4SO_2OH$ (20 mol%), $MeCN/H_2O$ (9:1), reflux, 6 h, **200** (77%), **201** (14%), **200/201**= 82/18; c: $CeCl_3 \cdot 7H_2O$ (30 mol%), $MeCN/H_2O$ (9:1), reflux, 9 h, **326** (71%), **201** (8%), **200/201** = 86/14.

Scheme 52

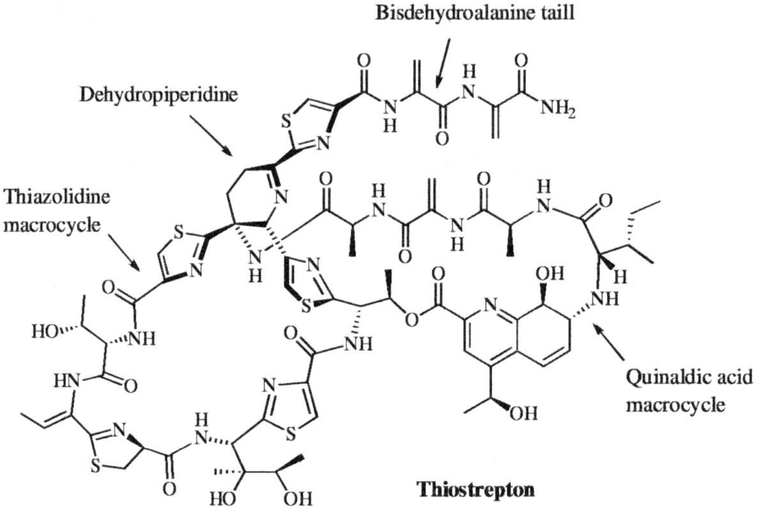

Figure 8

Nucleophilic addition of an acetyl radical species *para* to the protonated pyridine nitrogen atom of methyl 5,6,7,8-tetrahydroquinoline-2-carboxylate **202** proceeded in nearly quantitative yield to generate the prochiral acetyl derivative **203**. A modified CBS procedure, requiring the *in situ* preparation of *B*-methoxy-oxazaborolidine catalyst, was employed in the asymmetric reduction of the diacethylated pyridine **203**.[44] This methyl ketone was reduced by using chiral ligand **94** (4.0 mol%) and a stoichiometric amount of BH_3-SMe_2 to provide alcohol **204** (95% yield, 90% ee), which was submitted to TBS protection (TBSOTf, 2,6-lutidine, 90%). The protected alcohol **205** was then oxygenated to afford **206** in 65% overall yield

338

through a Boekelheide-type sequence involving a) MCPBA-mediated *N*-oxide formation; b) TFAA-induced acylation of the generated *N*-oxide; c) NaHCO₃-facilitated rearrangement–hydrolysis of the resulting trifluoroacetate.[21] Alcohol **206** was then dehydrated by treatment with Burgess reagent to afford olefin **207** in 68% yield (Scheme 53).

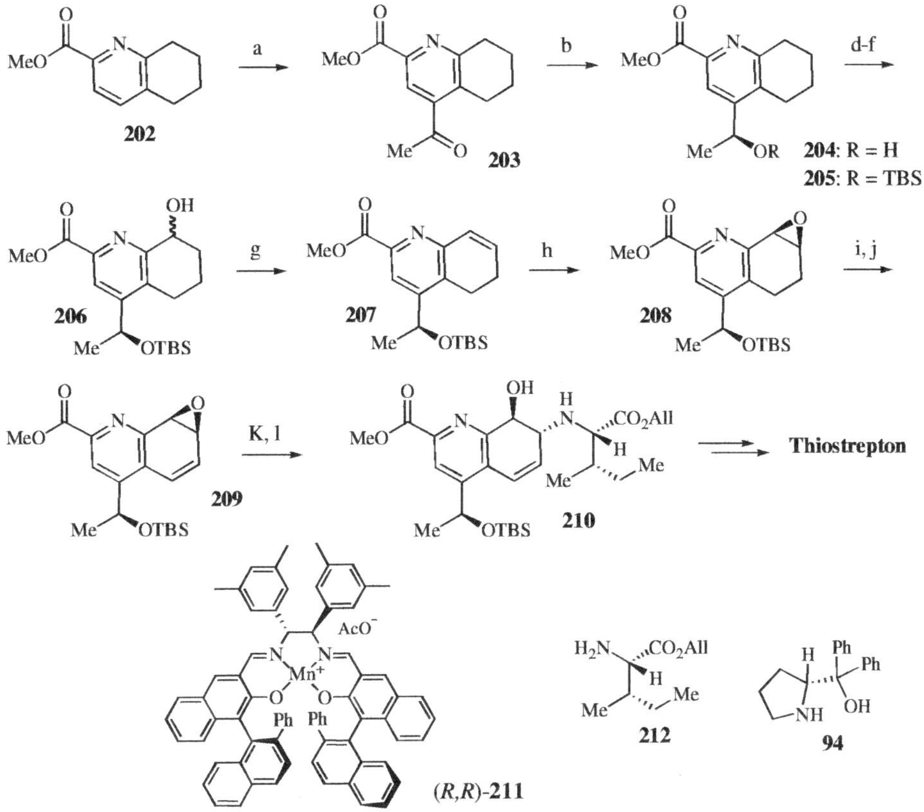

a) MeCHO, H₂O₂ (2.0 equiv), FeSO₄ (0.1 equiv), CF₃CO₂H (1.0 equiv), 25 °C, 2 h, 99%; b) **94** (4 mol%), B(OMe)₃ (0.12 equiv), BH₃-SMe₂ (1.0 equiv), THF, 25 C, 4 h, 95%, 90% ee; c) TBSOTf (2.0 equiv), 2,6-lutidine (2.0 equiv), CH₂Cl₂, 0 °C, 4 h, 90%; d) MCPBA (1.0 equiv), CH₂Cl₂, 25 °C, 12 h; e) TFAA (3.0 equiv), CH₂Cl₂, 25 °C, 12 h; f) NaHCO₃ (2 M), CH₂Cl₂, 8 h, 65% (three steps); g) Burgess reagent (1.2 equiv), THF/benzene (1:1), reflux, 3.5 h, 68%; h) (R,R)-**211** (0.01 equiv), 4-Ph-py-*N*-oxide (0.1 equiv), NaOCl (0.79 M), phosphate buffer adjusted to pH 11.5 with NaOH (2.0 M), CH₂Cl₂, 25 °C, 1 h, 82%, 87:13 ratio of products; i) NBS (1.1 equiv), AIBN (0.1 equiv), CCl₄, 80 °C, 40 min, 40% (and 26 % recovered starting material); j) DBU (1.1 equiv), THF, 25 °C, 2 h, 96%, l) **212** (3.0 equiv), LiClO₄ (5.0 equiv), MeCN, 60 °C, 22 h, 69%.

Scheme 53

The product **207** served well as the precursor to the desired compound **208** in a diastereoselective epoxidation reaction brought about by NaOCl in the presence of the Katsuki manganese-salen catalyst (R,R)-**211** and 4-phenylpyridine *N*-oxide.[84] Under these conditions, epoxide **208** was obtained in 82% yield as a 87:13 diastereomeric mixture with the desired diatereomer predominating. Chromatographic separation of **208** followed by radical bromination led to a mixture of diastereomeric bromides (40% yield plus 26% recovered starting material) that were eliminated by exposure to DBU to afford epoxide moiety of **209**. This compound was then regio- and stereoselectively opened by the free amine of L-isoleucine allyl ester **212** in the presence of lithium perchlorate to afford aminoalcohol **210** in 69% yield (Scheme 53). Presumably, the observed regioselectivity in this reaction is due to coordination of the lithium ion with the quinaldic acid nitrogen atom that simultaneously activates the epoxide moiety and deactivates the benzylic site through destabilization of the incipient carbocation at that position. Compound **210** was further elaborated and jointed to other fragments to afford thiostrepton as the final target compound.

3. Conclusion

This account is the first part of a series of reviews dedicated to chemical and bioorganic syntheses and application in asymmetric catalysis of chiral non-racemic pyridyl alcohols. In this occasion, we have outlined the stereoselective syntheses of a wide range of optically active pyridyl *sec*-alcohols through the description of a variety of methods allowing to generate a *sec*-carbinol chiral centre bound to a pyridine ring. The production of optically active pyridyl alcohols from racemic precursors by optical activation methods is described in the second part.[85] Notwithstanding many routes toward the asymmetric syntheses of these compounds with comparably simple structures have been developed, new opportunities are yet to be explored. It is hoped that this review will stimulate further research on this subject so that new chiral pyridyl alcohols can be prepared and applied in many areas of the organic and organometallic chemistry.

4. Note added in proof

Soai and co-workers reported an efficient one-pot system of asymmetric catalysis in which the chiral catalyst **214a** with only 1% ee self-improves its enantiomeric excess by asymmetric autocatalysis and then acts as a highly enantioselective chiral catalyst for the asymmetric addition of diisopropylzinc to 3-pyridyl carboxaldehyde derivatives, to provide the corresponding *sec*-alcohols with very high enantiomeric excesses (up to 99%) (Scheme 54).[86]

29b: R = R^1 = H ⟶ **30h**: R = R^1 = H, 84%, 96% ee

33b: R = CO-N(i-Pr)$_2$, R^1 = H ⟶ **34b**: R = CO-N(i-Pr)$_2$, R^1 = H, 82%, 86% ee

215: R - R^1 = CH=CH-CH=CH ⟶ **218**: R - R^1 = CH=CH-CH=CH, 87%, 99% ee

216: R = Br, R^1 = H ⟶ **219**: R = Br, R^1 = H, 71%, 90% ee

217: R = Me, R^1 = H ⟶ **220**: R = Me, R^1 = H, 73%, 94% ee

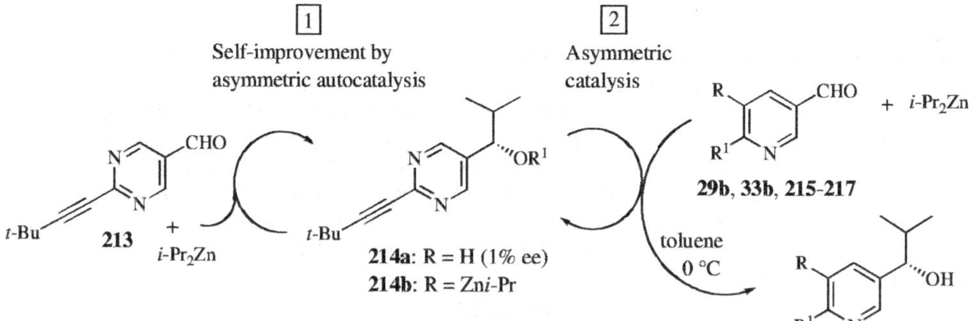

Scheme 54

Ishihara *et al.* developed a highly efficient enantioselective dialkylzinc addition to aldehydes by using a conjugate Lewis acid-Lewis base BINOL-Zn(II) catalyst bearing phosphine oxides, phosphonates or phosphoramides at the 3,3'-positions.[87] Among the examined aldehydes, pyridine-3-carboxaldehyde **29b** and pyridine-4-carboxaldehyde **29c** were ethylated in the presence of the ligand (*R*)-**221** to give (*R*)-1-(pyridin-3-yl)propan-1-ol **30b** and (*R*)-1-(pyridin-4-yl)propan-1-ol **30c**, respectively, in low yields (30–41%), but with good enantioselectivities (79–81% ees) (Scheme 55).

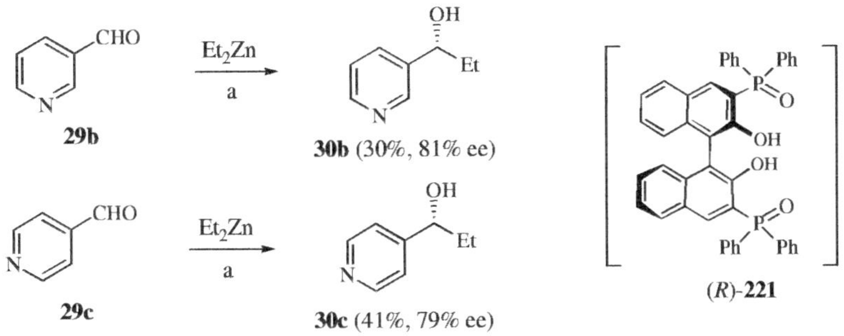

a: Et$_2$Zn (3.0 equiv), (*R*)-**7** (10 mol%), THF-toluene, r.t., 0.5 h

Scheme 55

Knochel and co-workers showed that borane protected pyridyl and (iso)quinolyl aldehydes underwent the asymmetric addition of a number of organozinc reagents in the presence of Ti(Oi-Pr)$_4$ and the chiral catalyst (1*R*,2*R*)-bis(trifluoromethanesulfonamido)cyclohexane **226**, to give the corresponding alcohols with good to excellent enantioselectivities (Schemes 56 and 57).[88]

The use of borane protected pyridine derivatives allows the asymmetric addition of diorganozincs without the interference of the basic nitrogen, whose strong coordinating ability to the titanium metal centre causes deactivation of the chiral catalyst and favours non-asymmetric addition pathways. Thus, 3-pyridyl alcohols **30b**, **30g** and **223a–d** were obtained with 54–96% yields and 69–93% ees from pyridine-3-carboxaldehyde-BEt$_3$ complex **222** (Scheme 56), whereas 4-pyridyl alcohols **30c** and **225a–d** were obtained with lower yields (35–88%) and slightly reduced enantioselectivities (66–91% ees) (Scheme 56). Interestingly, functionalized diorganozincs like [(CH$_2$)$_4$OPiv]$_2$Zn and [(CH$_2$)$_5$OPiv]$_2$Zn gave the corresponding alcohols in useful yields and with excellent enantioselectivities (85–93% ees) with both starting points. In the case of the pyridine-2-carboxaldehyde-BEt$_3$ complex, no addition reaction was observed due to the steric hindrance resulting of the complexation with BEt$_3$. The method was also extended to quinoline- and isoquinoline-4-carboxaldehyde **227a** and **229a**, as well as to the corresponding borane complexes **227b** and **229b**, respectively (Scheme 57). The unprotected quinoline **227a** added Et$_2$Zn and Pent$_2$Zn with good yields and 72–81% ees, whereas the use of the corresponding BEt$_3$ complex **227b** led to disappointing yields and enantioselectivities.

These results were explained by the lower stability of the BEt$_3$ adduct due to the steric repulsion between BEt$_3$ and the hydrogen in position 8. This instability causes the formation of free BEt$_3$ that reacts with the organozinc reagent to form mixed trialkylboranes and dialkylzincs. On the contrary, Et$_2$Zn or Pent$_2$Zn reacted with isoquinoline **229a** to produce the corresponding addition products with 44–71% yields in racemic form, whereas the corresponding BEt$_3$ complex **229b** gave satisfactory yields and enantioselectivities (92–93% ees).

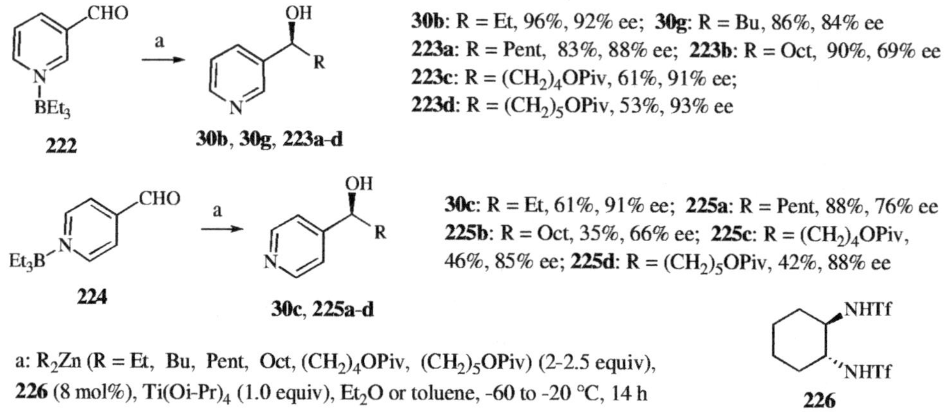

30b: R = Et, 96%, 92% ee; **30g**: R = Bu, 86%, 84% ee
223a: R = Pent, 83%, 88% ee; **223b**: R = Oct, 90%, 69% ee
223c: R = (CH$_2$)$_4$OPiv, 61%, 91% ee;
223d: R = (CH$_2$)$_5$OPiv, 53%, 93% ee

30c: R = Et, 61%, 91% ee; **225a**: R = Pent, 88%, 76% ee
225b: R = Oct, 35%, 66% ee; **225c**: R = (CH$_2$)$_4$OPiv,
46%, 85% ee; **225d**: R = (CH$_2$)$_5$OPiv, 42%, 88% ee

a: R$_2$Zn (R = Et, Bu, Pent, Oct, (CH$_2$)$_4$OPiv, (CH$_2$)$_5$OPiv) (2-2.5 equiv),
226 (8 mol%), Ti(Oi-Pr)$_4$ (1.0 equiv), Et$_2$O or toluene, -60 to -20 °C, 14 h

Scheme 56

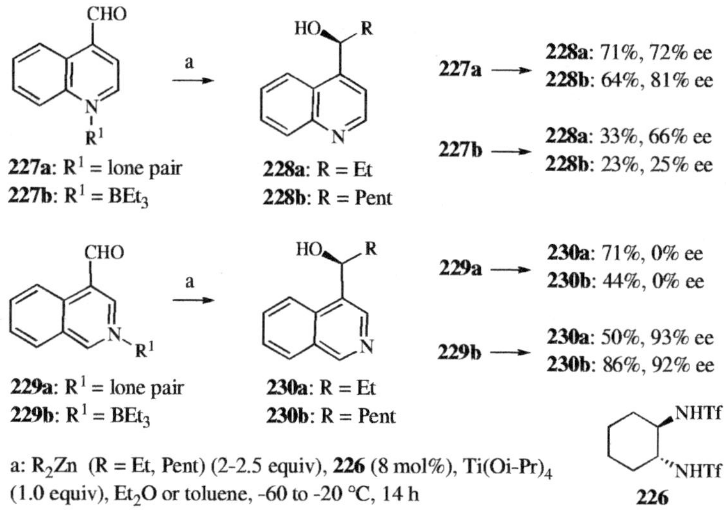

227a: R^1 = lone pair **228a**: R = Et
227b: R^1 = BEt$_3$ **228b**: R = Pent

227a → **228a**: 71%, 72% ee
228b: 64%, 81% ee

227b → **228a**: 33%, 66% ee
228b: 23%, 25% ee

229a: R^1 = lone pair **230a**: R = Et
229b: R^1 = BEt$_3$ **230b**: R = Pent

229a → **230a**: 71%, 0% ee
230b: 44%, 0% ee

229b → **230a**: 50%, 93% ee
230b: 86%, 92% ee

a: R$_2$Zn (R = Et, Pent) (2-2.5 equiv), **226** (8 mol%), Ti(Oi-Pr)$_4$
(1.0 equiv), Et$_2$O or toluene, -60 to -20 °C, 14 h

Scheme 57

Schneider and co-workers described the asymmetric synthesis of the C$_2$-symmetric 1,10-phenanthroline **236** (Scheme 58).[89] The synthesis starts from the TBS-protected 2-bromo pyridyl alcohol **64a** (99% ee), readily prepared through enantioselective Noyori hydrogenation of the corresponding ketone in good yield and with excellent enantioselectivity (see Scheme 35)[54] and subsequent silylation. The introduction of the aldehyde moiety was achieved by regioselective metalation of the pyridine ring in **64a** with LDA in THF at −78 °C and subsequent treatment with DMF, to afford the 3-pyridyl carboxaldehyde **231** in 50% yield.[90] Aldehyde **231** was then converted into sulfonyl hydrazone **232** by treatment with *p*-toluenesulfonyl hydrazide. Sulfonyl hydrazone **232** was in turn converted under phase-transfer conditions into the corresponding diazoalkane, which was subsequently dimerized through the action of catalytic amounts of Rh$_2$(OAc)$_4$ to afford a 4:1 mixture of Z- and E-alkenes **233** in 94% yield.[91] These isomers were separated by chromatography after removal of the silyl ethers with TBAF in THF, to afford the pure Z-stilbene **234** in 67% yield. Finally, intramolecular coupling of **234** was carried out in the presence of PdCl$_2$(PhCN)$_2$ (20 mol%) and stoichiometric amounts of tetrakis(dimethylamino)ethylene, used simultaneously as ligand and reductant, to afford the chiral phenanthroline **236** in about 70% yield.

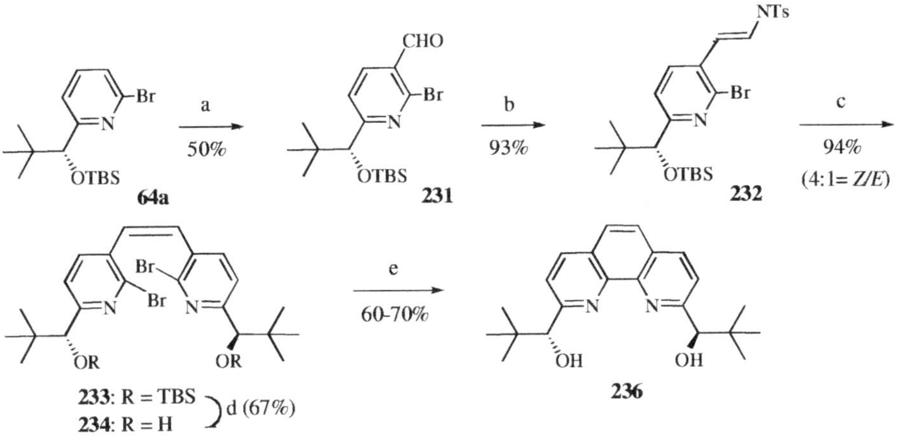

a: 1. LDA, THF, -78 °C, 2 h; 2. DMF, -78 °C, 4 h;   b: p-TsNHNH$_2$, CH$_2$Cl$_2$, 2 d;   c: 1. (C$_6$H$_5$CH$_2$)Et$_3$N$^+$Cl$^-$ NaOH (aq., 14%), toluene, 65 °C, 2 h; 2. Rh$_2$OAc)$_4$, -20 to 0 °C, 16 h; d. TBAF·3H$_2$O, THF, 6 h (pure Z); e: PdCl$_2$(PhCN)$_2$ (20 mol%), TDAE (tetrakis(dimethylamino)ethylene), DMF, 75 °C, 48 h .

Scheme 58

Acknowledgments

Financial support from MIUR (PRIN 2003033857-Chiral ligands with nitrogen donors in asymmetric catalysis by transition metal complexes. Novel tools for the synthesis of fine chemicals) and from the University of Sassari is gratefully acknowledged.

References

1. For some examples of the use of pyridyl alcohols in asymmetric catalysis, see: epoxidation of allylic alcohols: (a) Hawkins, J. M.; Sharpless, K. B. *Tetrahedron Lett.* **1987**, *28*, 2825. Conjugate addition of dialkylzinc reagents to α,β-unsaturated ketones: (b) Bolm, C.; Ewald, M. *Tetrahedron Lett.* **1990**, *31*, 5011. (c) Bolm, C. *Tetrahedron: Asymmetry* **1991**, *2*, 701. (d) Bolm, C.; Ewald, M.; Felder, M. *Chem. Ber.* **1992**, *125*, 1205. Nucleophilic addition of dialkylzinc reagents to aldehydes: (e) Ishizaki, M.; Fujita, K.; Shimamoto, M.; Hoshino, O. *Tetrahedron: Asymmetry* **1994**, *5*, 411. (f) Ishizaki, M.; Hoshino, O. *Tetrahedron: Asymmetry* **1994**, *5*, 1901. (g) Ishizaki, M.; Hoshino, O. *Chem. Lett.* **1994**, 1337. (h) Bolm, C.; Schlingo, H.; Harms, K. *Chem. Ber.* **1992**, *125*, 1191. (i) Chelucci, G.; Soccolini, F. *Tetrahedron: Asymmetry* **1992**, *3*, 1235. Strecher-type reaction: (j) Chen, Y.-J.; Chen, C. *Tetrahedron: Asymmetry* **2008**, *19*, 2201. Ring-opening reaction: (k) Tschöp, A.; Nandakumar, M. V.; Pavlyuk, O.; Schneider, C. *Tetrahedron Lett.* **2008**, *49*, 1030. (l) Chen, Y.-J.; Chen, C. *Tetrahedron: Asymmetry* **2007**, *18*, 2201. (m) Boudou, M.; Ogawa, C.; Kokayashi, S. *Adv. Synth. Cat.* **2006**, *348*, 2585. Asymmetric hydrogenation: (n) Verendel, J. J.; Andersson, P. J. *Dalton Trans* **2007**, 5603. (o) Liu, Q.-B.; Yu, C.-B.; Zhou, Y.-G. *Tetrahedron Lett.* **2006**, *47*, 4733. (p) Bell, S.; Wustenberg, B.;

343

Kaiser, S.; Menges, F.; Netscher, T.; Pfaltz, A. *Science* **2006**, *311*, 642. (q) Kaiser, S.; Smidt, S. P.; Pfaltz, A. *Angew. Chem. Int. Ed.* **2006**, *45*, 5194.

2. Biologically relevant pyridyl alcohols. Ceramide: (a) Karlsson, K.-A. *Chem. Phys. Lipids* **1970**, *5*, 6. (b) Hannun, Y. A. *Science* **1996**, *274*, 1855. (c) Ariga, T.; Jarvis, W. D.; Yu, R. K. *J. Lipid Res.* **1998**, *39*, 1. (d) Mathias, S.; Peña, L. A.; Kolesnick, R. N. *Biochem. J.* **1998**, *335*, 465. (e) Perry, D. K.; Hannun, Y. A. *Biochim. Biophys. Acta* **1998**, *1436*, 233. (*S*)-Carbinoxamine: (f) Roszkowski, A. P.; Govier, W. M. *Pharmacologist* **1959**, 1, 60. (g) Barouh, V.; Dall, H.; Patel, D.; Hite, G. *J. Med. Chem.* **1971**, *14*, 834. *allo*-Heteroyohimbine: (h) Uskokovic, M. R.; Lewis, R. L.; Partridge, J. J.; Despreaux, C. W.; Pruess, D. L. *J. Am. Chem. Soc.* **1979**, *101*, 6742. (*S*)-Naproxen: (i) Franck, A.; Rüchardt, C. *Chem. Lett.* **1984**, 1431. (j) Salz, U.; Rüchardt, C. *Chem. Ber.* **1984**, 117, 3457. (*R,S*)-mefloquine: (k) Ohnmacht, C. J.; Patel, A. R.; Lutz, R. E. *J. Med. Chem.* **1971**, *14*, 926. Herbicides and fungicides: (l) Rempfler, H.; Cederbaum, F.; Spindler, F.; Lottenbach, W. Patent WO 96/16941, 1996; CA 125, 114501. (m) Coqueron, P.; Desbordes, P.; Mansfield, D. J.; Rieck, H.; Grosjean, M. C.; Villier, A.; Genix, P. Eur. Patent 1 548 007 A1, 2005; CA 143, 97399.

3. Rahm, F.; Stranne, U.; Brenberg, U.; Nordström, K.; Cernerud, M.; Macedo, E.; Moberg, C. *J. Chem. Soc., Perkin Trans. 1* **2000**, 1983.

4. Chung, J.; He, L.; Byun, H.-S.; Bittman, R. *J. Org. Chem.* **2000**, *12*, 7634.

5. Son, S. U.; Jang, Y.; Han, J. W.; Lee, I. S.; Chung, Y. K. *Tetrahedron: Asymmetry* **1999**, *10*, 347.

6. Han, J. W.; Jang, H.-Y. J.; Chung, Y. K. *Tetrahedron: Asymmetry* **1999**, *10*, 2853.

7. Son, S. U.; Jang, U.-Y.; Han, J. W.; Lee, I. S.; Chung, Y. K. *Tetrahedron: Asymmetry* **1999**, *10*, 347.

8. Fort, Y.; Gros, F.; Rodriguez, A. L. *Tetrahedron: Asymmetry* **2001**, *12*, 2631.

9. Choppin, S.; Gros, P.; Fort, Y. *Org. Lett.* **2000**, 2, 803.

10. (a) Pu, L. *Tetrahedron* **2003**, *59*, 9872. (b) Pu, L.; Yu, H.-B. *Chem. Rev.* **2001**, *101*, 757.

11. Braga, A. L.; Paixão, M. W.; Lüdtke, D. S.; Silveira, C. C.; Rodrigues, O. E. D. *Org. Lett.* **2003**, 5, 2635.

12. (a) Soai, K., Shibata, T.; Sato, I. *Acc. Chem. Res.* **2000**, *33*, 382, and references therein. (b) Soai, K.; Niwa, S.; Hori, H. *J. Chem. Soc., Chem. Commun.* **1990**, 982. (c) Shibata, T.; Morioka, H.; Tanji, S.; Hayase, T.; Kodaka, Y.; Soai, K. *Tetrahedron Lett.* **1996**, *37*, 8783. (d) Sato, I.; Nakao, T.; Sugie, R.; Kawasaki, T.; Soai, K. *Synthesis* **2004**, *9*, 1419.

13. Pagenkopf, B. L.; Carreira, E. M. *Tetrahedron Lett.* **1998**, *39*, 9593.

14. Huang, W.-S.; Pu, L. *Tetrahedron Lett.* **2000**, *41*, 145.

15. Bolm, C.; Muñiz, K. *Chem. Commun.* **1999**, 1295.

16. Racherla, U. S.; Liao, Y.; Brown, H. C. *J. Org. Chem.* **1992**, *57*, 6614.

17. (a) Ramachandran, P. V.; Chen, G.-M.; Brown, H. C. *Tetrahedron Lett.* **1997**, *38*, 2417. (b) Brown, H. C.; Chen, G.-M.; Ramachandran, P. V. *Chirality* **1997**, *9*, 506.

18. Cunningham, D.; Gallagher, E. T.; Grayson, D. H.; McArdle, P. J.; Storey, C. B.; Wilcock, D. J. *J. Chem. Soc., Perkin Trans. 1* **2002**, 2692.

19. Cheemala, M. N.; Knochel, P. *Org. Lett.* **2007**, *9*, 3089.

20. Ricci, G.; Ruzziconi, R. *Tetrahedron: Asymmetry* **2005**, *16*, 1817.

21. Fontenas, C.; Bejan, E.; Haddou, H. A.; Balavoine, G. G. A. *Synth. Commun.* **1995**, *25*, 629.

22. Collomb, B.; von Zelewsky, A. *Tetrahedron: Asymmetry* **1995**, *6*, 2903.

23. Kröhnke, F. *Synthesis* **1976**, 1.

24. (a) Nicolau, K. C.; Zak, M.; Safina, B. S.; Lee, S. H.; Estrada, A. A. *Angew. Chem. Int. Ed.* **2004**, *43*, 5092. (b) Nicolau, K. C.; Safina, B. S.; Zak, M.; Estrada, A. A.; Lee, S. H. *Angew. Chem. Int. Ed.* **2004**, *43*, 5087. (c) Nicolau, K.; Safina, B. S.; Funke, K. C.; Zak, M.; Zécri, F. J. *Angew. Chem. Int. Ed.* **2002**, *41*, 1937.

25. Midland, M. M.; McLoughlin, J. I. *J. Org. Chem.* **1984**, *49*, 1316.

26. (a) Brown, H. J.; Chandrasekharan, J.; Ramachandran, P. V. *J. Am. Chem. Soc.* **1988**, *110*, 1539. (b) Brown, H. J.; Ramachandran, P. V.; Weissman, S. A.; Swaminathan, S. *J. Org. Chem.* **1990**, *55*, 6328.

27. Vedejs, E.; Chen, X. *J. Am. Chem. Soc.* **1996**, *118*, 1809.

28. Bolm, C.; Edwald, M.; Felder, M.; Schlingloff, G. *Chem. Ber.* **1992**, *125*, 1169.

29. Tschöp, A.; Marx, A.; Sreekanth, A. R.; Schneider, C. *Eur. J. Org. Chem.* **2007**, 2318.

30. Hawkins, J. M.; Dewan, J. C.; Sharpless, K. B. *Inorg. Chem.* **1986**, *25*, 1501.

31. Ramachandran, P. V.; Chen, G.-M.; Lu, Z.-H.; Brown, H. C. *Tetrahedron Lett.* **1996**, *37*, 3795.
32. Jiang, Q.; Van Plew, D. Murtuza, S.; Zhang, X. *Tetrahedron Lett.* **1996**, *37*, 797.
33. Drury, W. J. III; Zimmermann, N.; Keenan, M.; Hayashi, M.; Kaiser, S.; Goddard, R.; Pfaltz, A. *Angew. Chem. Int. Ed.* **2004**, *43*, 70.
34. Wishka, D. G.; Graber, D. R.; Seest, E. P.; Dolak, L. A.; Han, F.; Watt, W.; Morris, J. *J. Org. Chem.* **1998**, *63*, 7851.
35. Ohno, A.; Nakai, J.; Nakamura, K. *Bull. Chem. Soc. Jpn.* **1981**, *54*, 3482.
36. Seki, M.; Baba, N.; Oda, J.; Inouye, Y. *J. Am. Chem. Soc.* **1981**, *103*, 4613.
37. Seki, M.; Baba, N.; Oda, J.; Inouye, Y. *J. Org. Chem.* **1983**, *48*, 4613.
38. Vasse, J. L.; Charpentier, P.; Levacher, V.; Dupas, G.; Quéguiner, G.; Bourguignon, J. *Synlett* **1998**, 1144.
39. Soai, K.; Niwa, S.; Kobayashi, T. *J. Chem. Soc., Chem. Commun.* **1987**, 801.
40. Samaddar, A. K.; Konar, S. K.; Nasipuri, D. *J. Chem. Soc., Perkin Trans 1* **1983**, 1449.
41. Schmitzer, A.; Perez, E.; Rico-Lattes, I.; Lattes, A. *Tetrahedron: Asymmetry* **2003**, *14*, 3719.
42. Schmidt, M.; Amstutz, R.; Crass, G.; Seebach, D. *Chem. Ber.* **1980**, *113*, 1691.
43. Kopilov, J.; Kariv, E.; Miller, L. L. *J. Am. Chem. Soc.* **1977**, *99*, 345.
44. For reviews, see: (a) Cho, B. T. *Tetrahedron* **2006**, *62*, 7621. (b) Glushkov, V. A.; Tolstikov, A. G. *Russ. Chem. Rev.* **2004**, *73*, 581. (c) Corey, E. J.; Helal, C. J. *Angew. Chem. Int. Ed.* **1998**, *37*, 1986.
45. Quallich, G. J.; Woodall, T. M. *Tetrahedron Lett.* **1993**, *34*, 785.
46. Masui, M.; Shiori, T. *Synlett* **1997**, 273.
47. Xie, Y.; Huang, H.; Moa, W.; Fan, X.; Shen, Z.; Shen, Z.; Sun, N.; Hua, B.; Hua, X. *Tetrahedron: Asymmetry* **2009**, *20*, 1425.
48. Corey, E. J.; Helal, C. J. *Tetrahedron Lett.* **1996**, *37*, 5675.
49. Hayakawa, Y.; Kato, H.; Uchiyama, M.; Kajino, H.; Noyori, R. *J. Org. Chem.* **1986**, *51*, 2400.
50. Berenguer, R.; Garcia, J.; Vilarrasa, J. *Tetrahedron: Asymmetry* **1994**, *5*, 165.
51. Stepanenko, V.; De Jesús, M.; Correa, W.; Guzmán, I.; Vázquez, C.; Ortiz, L.; Ortiz-Marciales, M. *Tetrahedron: Asymmetry* **2007**, *18*, 2738.
52. Gladiali, S.; Alberico, E. *Chem. Soc. Rev.* **2006**, *35*, 226.
53. Okano, K.; Murata, K.; Ikariya, T. *Tetrahedron Lett.* **2000**, *41*, 9277.
54. Ishikawa, S.; Hamada, T.; Manabe, K.; Kobayashi, S. *Synthesis* **2005**, 2176.
55. Baratta, W.; Herdtweck, E.; Siega, K.; Toniutti, M.; Rigo, P. *Organometallics* **2005**, *24*, 1660.
56. Nordin, S. J. M.; Roth, P.; Tarnai, T.; Alonso, D. A.; Brandt, P.; Andersson, P. G. *Chem. Eur. J.* **2001**, *7*, 1431.
57. Everaere, K.; Carpentier, J.-F.; Mortreux, A.; Bulliard, M. *Tetrahedron: Asymmetry* **1999**, *10*, 4083.
58. Matharu, D. S.; Morris, D. J.; Clarkson, G. J.; Wills, M. *Chem. Commun.* **2006**, 3232.
59. Lipshutz, B. H.; Lower, A.; Noson, K. *Org. Lett.* **2002**, *4*, 4045.
60. Lee, C.-T; Lipshutz, B. H. *Org. Lett.* **2008**, *8*, 4187.
61. Lipshutz, B. H.; Lower, A.; Kucejko, R. J.; Noson, K. *Org. Lett.* **2006**, *8*, 2969.
62. (a) Johnson, N. B.; Lennon, I. C.; Moran, P. H.; Ramsden, J. A. *Acc. Chem. Res.* **2007**, *40*, 1291. (b) Noyori, R.; Ohkuma, T. *Angew. Chem. Int. Ed.* **2001**, *40*, 40.
63. Ohkumu, T.; Koizumi, M.; Yoshida, M.; Noyori, R. *Org. Lett.* **2000**, *2*, 1749.
64. Chen, C.; Reamer, R. A.; Chilenski, J. R.; McWilliams, C. *Org. Lett.* **2003**, *5*, 5039.
65. Burk, M. J.; Herms, W.; Herzberg, D.; Malan, C.; Zanotti-Gerosa, A. *Org. Lett.* **2000**, *2*, 4173.
66. Chai, L.-T; Wang, W.-W; Wang, Q.-R; Tao, F.-G. *J. Mol. Catal. A* **2007**, *270*, 83.
67. Maerten, E.; Agbossou-Niedercorn, F.; Castanet, Y.; Mortreux, A. *Tetrahedron* **2008**, *64*, 8700.
68. Xu, Y.; Clarkson, G. C.; Docherty, G.; North, C. L.; Woodward, G.; Wills, M. *J. Org. Chem.* **2005**, *70*, 8079.
69. Wu, J.; Ji, J.-X.; Guo, R.; Yeung, C.-H.; Chan, A. S. C. *Chem. Eur. J.* **2003**, *9*, 2963.
70. (a) Rahm, F.; Fischer, A.; Moberg, C. *Eur. J. Org. Chem.* **2003**, 4205. (b) Adolfsson, H.; Nordström, K.; Wärnmark, K.; Moberg, C. *Tetrahedron: Asymmetry* **1996**, *7*, 1967.
71. Yin, J.; Huffman, M. A.; Conrad, K. M.; Armstrong, J. D. *J. Org. Chem.* **2006**, *71*, 840.

72. (a) Chelucci, G.; Cabras, M. A.; Saba, A. *Tetrahedron: Asymmetry* **1994**, *5*, 1973. (b) Cabras, M.A.; Chelucci, G.; Giacomelli, G.; Soccolini, F. *Gazz. Chim. Ital.* **1994**, 124, 23. (c) Chelucci, G.; Falorni, M.; Giacomelli, G. *Gazz. Chim. Ital.* **1990**, 120, 731.

73. (a) Heller, B.; Hapke, M. *Chem. Soc. Rev.* **2007**, *36*, 1085, and references cited therein. (b) Chopade, P. R.; Louie, J. *Adv. Synth. Catal.* **2006**, *348*, 2307. (c) Kotha, S.; Brahmachary, E.; Lahiri, K. *Eur. J. Org. Chem.* **2005**, 4741. (d) Varela, J. A.; Saá, C. *Chem. Rev.* **2003**, *103*, 3787. (e) Dzhemilev, U. M.; Selimov, F. A.; Tolstikov, G. A. *Arkivoc* **2001**, *ix*, 85.

74. For a review on the synthesis of chiral pyridines by this method, see: Chelucci, G. *Tetrahedron: Asymmetry* **1995**, *6*, 811.

75. Heller, B.; Redkin, D.; Gutnov, A.; Fisher, C.; Bonrath, W.; Karge, R.; Hapke, M. *Synthesis* **2008**, 69.

76. (a) Heller, B.; Sundermann, B.; Fischer, C.; You, J.; Chen, W.; Drexler, H.-J.; Knochel, P.; Bonrath, W.; Gutnov, A. *J. Org. Chem.* **2003**, *68*, 9221. (b) Heller, B.; Sundermann, B.; Buschmann, H.; Drexler, H.-J.; You, J.; Holzgrabe, U.; Heller, D.; Oehme, G. *J. Org. Chem.* **2002**, *67*, 4414. (c) Heller, B.; Heller, D.; Oehme, G. *J. Mol. Catal. A* **1996**, *110*, 211. (d) Heller, B.; Oehme, G. *J. Chem. Soc., Chem. Commun.* **1995**, 179.

77. Gray, B. L.; Wang, X.; Brown, W. C.; Kuai, L.; Schreiber, S. L. *Org. Lett.* **2008**, *10*, 2621.

78. Lyle, M. P. A.; Wilson, P. D. *Org. Lett.* **2004**, *6*, 855.

79. Kolb, H. C.; VanNieuwenhze, M. S.; Sharpless, K. B. *Chem. Rev.* **1994**, *94*, 2483.

80. Genzel, Y.; Archelas, A.; Broxterman, Q. B.; Schulze, B.; Furstoss, R. *J. Org. Chem.* **2001**, *66*, 538.

81. Chelucci, G.; Sanfilippo, C., in preparation.

82. For a leading reference on the coupling of pyridine by the Ullmann reaction, see: Newkome, G. R.; Patri, A. K.; Holder, E.; Schubert, U. S. *Eur. J. Org. Chem.* **2004**, 235.

83. (a) Savoia, D.; Alvaro, G.; Di Fabio, R.; Gualandi, A.; Fiorelli, C. *J. Org. Chem.* **2006**, *71*, 9763. (b) Savoia, D.; Alvaro, G.; Di Fabio, R.; Gualandi, A. *J. Org. Chem.* **2007**, *72*, 3859.

84. (a) Ito, K.; Yoshitake, M.; Katsuki, T. *Tetrahedron* **1996**, *52*, 3905. (b) Sasaki, H.; Irie, R.; Hamada, T.; Suzuki, K.; Katsuki, T. *Tetrahedron* **1994**, *50*, 11827. (c) For a pertinent review in this area, see: Katsuki, T. *Curr. Org. Chem.* **2001**, *5*, 663.

85. Part II: Chelucci, G. In *Targets in Heterocyclic Systems*; Attanasi, O.; Spinelli, D., Eds.; Società Chimica Italiana: Rome; submitted.

86. Sato, I.; Urabe, H.; Ishii, S.; Tanji, S.; Soai, K. *Org. Lett.* **2001**, *3*, 3851.

87. Hatano, M.; Miyamoto, T.; Ishihara, K. *J. Org. Chem.* **2006**, *71*, 6474.

88. Lutz, C.; Graf, C.-D.; Knochel, P. *Tetrahedron* **1998**, *54*, 10317.

89. Nandakumar, M. V.; Ghosh, S.; Schneider, C. *Eur. J. Org. Chem.* **2009**, 6393.

90. Melnyk, P.; Gasche, J.; Thal, C. *Synth. Commun.* **1993**, *23*, 2727.

91. Ravi Shankar, B. K.; Shechter, H. *Tetrahedron Lett.* **1982**, *23*, 2277.